# Sustainable
# Livestock and Poultry Breeding

## The Authors

Dr. Arun Kumar Tomar is Ph.D. (Animal Breeding) from CCS, HAU, Hisar. He did B.Sc. (Ag.) and M.Sc. (Ag.) in A.H. & Dairying from J.V.College, Baraut, Bagpat (U.P.). He was selected as Scientist in the descipline of Animal Genetics & Breeding of A.R.S. (ICAR) through All India Competitive Examination conducted by A.S.R.B. He was posted as Scientist (Animal Genetics & Breeding) at C.S.W.R.I. Avikanagar (Rajasthan), and subsequently promoted to the post of Senior Scientist and Principal Scientist (A.G.B) at Project Directorate on Cattle, Meerut (U.P.). He was selected as Head, Division of Animal Genetics & Breeding, Central Sheep & Wool Research Institute, Avikanagar, via Jaipur, Rajasthan where he is working at present. He has published some books & booklets and many research papers and has experience of editorial work in the capacity of Assistant Editor of Research Journals.

Professor Sukhvir Singh Tomar is Ph.D. (A.G.B.) from N.D.R.I., Karnal. He did B.Sc. (Ag.) and M.Sc. (Ag.) in A.H. & Dairying from J.V.College, Baraut, Bagpat (U.P.). He started his career as Lecturer, J.V. College, Baraut (Meerut) and subsequently at P.A.U, Hisar campus. Later on, he joined ARS of I.C.A.R. as Scientist (AGB) at C.S.W.R.I. and worked as Senior Scientist and Principal Scientist (A.G.B.) at N.D.R.I., Karnal from where he was retired having 40 years of research experience making significant contribution in the area of Animal Breeding Research and teaching experience to B.Sc., M.Sc. and Ph.D. Classes. He has published about 300 research articles and 3 text books. He has made notable contribution in research in the area of Animal Breeding, particularly pertaining to threshold characters, replacement rate, selective value, and other aspects of sire evaluation. He has long experience of about 28 years as Editor of Research Journal and guided a number of M.Sc. and Ph.D. students. He is still actively engaged in research and publication work.

The authors have published already two books. The first book "Animal Genetics and Breeding" was written as per syllabus of B.V.Sc. & A.H. of Indian Agricultural Universities. The second book "Genetics of Livestock Population" covered both qualitative and quantitative genetics in population as per syllabus prescribed for post-graduate classes in the discipline of Animal Breeding. Both these books covered both the theory as well as practical problems in very simple lucid and illustrative language.

# Sustainable Livestock and Poultry Breeding

**Dr. Arun Kumar Tomar**
M.Sc. (Ag.), Ph.D., A.R.S.
Head of Division (Animal Genetics & Breeding)
ICAR Central Sheep & Wool Research Institute,
Avikanagar, via Jaipur
Rajasthan - 304501
And
**Prof. Sukhvir Singh Tomar**
M.Sc. (Ag.), Ph.D., A.R.S.
Former Principal Scientist (AGB),
ICAR National Dairy Research Institute, Karnal

**2016**
**Daya Publishing House®**
*A Division of*
**Astral International Pvt. Ltd.**
**New Delhi - 110 002**

**Cataloging in Publication Data--DK**
Courtesy: D.K. Agencies (P) Ltd. <docinfo@dkagencies.com>

**Tomar, Arun Kumar, author.**
Sustainable livestock and poultry breeding / Dr. Arun Kumar Tomar (M.Sc. (Ag.), Ph.D., A.R.S.) and Prof. Sukhvir Singh Tomar (M.Sc. (Ag.), Ph.D., A.R.S.).
pages cm

ISBN 9789351309253 (International Edition)

1. Livestock--India. 2. Poultry--Breeding--India. 3. Animal breeding--India. 4. Sustainable agriculture--India. 5. Agricultural development projects--India. I. Tomar, Sukhvir Singh, author. II. Title.

SF55.I5T66 2016 DDC 636.00954 23

*Published by* : **Daya Publishing House®**
*A Division of*
**Astral International Pvt. Ltd.**
– ISO 9001:2008 Certified Company –
4760-61/23, Ansari Road, Darya Ganj
New Delhi-110 002
Ph. 011-43549197, 23278134
E-mail: info@astralint.com
Website: www.astralint.com

*Laser Typesetting:* SSMG Computer Graphics, Delhi - 110 084

*Printed at* : Replika Press Pvt. Ltd.

# Preface

The domestication of farm animals has changed the genetic structure of animal population and the environmental conditions, and consequently it has brought the changes in morphological, production and reproduction traits as well as behavior of domestic animals. Man had been making continuous efforts since long to exploit the genetic variability by use of selective breeding and developed new breeds of livestock species and poultry with high performance to meet human needs.

The livestock improvement involves the genetic improvement of population and development of new strains/breeds of livestock species including poultry based on the application of the principles of population genetics by exploiting the genetic variability using the biometrical techniques and optimum breeding plans (selection and mating systems). This is the subject matter of animal breeding. The task of animal breeding stems on and accomplishes in two phases starting from identification of superior germ plasm which is a difficult task and ending to its multiplication.

The traditional animal breeding rested on "breeding best with the best" based on phenotypic superiority which could be genetic and environmental in nature making the identification of superior germplasm a very difficult task. Lare on, with the advancement of the knowledge of the principles of population genetics the identification of superior germplasm rested on genetic superiority by estimating the breeding worth (breeding value) of the animals. However, this also could not be found successful to identify the genetically superior germplasm. In view of the shortcomings in conventional methods of genetic evaluation, recent biotechniques based on molecular genetics are in progress to identify the superior germplasm based on molecular markers for estimating exactly the real B.V. The superior germ plasm once identified is multiplied by mating of the selected superior animals following some mating system (Inbreeding or outbreeding) depending on the breeding strategy being followed to achieve an objective.

It was felt the need of the day to compile the entire subject matter related to sustainable animal breeding and production in a single volume to meet the requirement as per syllabus of Post Graduate programme of animal breeding taught in State Agricultural Universities in India. The related matter has been collected from different published sources covering the theoretical and practical aspects and presented in a simple and illustrative language. The entire subject matter of animal breeding in this book has been covered in 3 parts comprising 27 chapters. The first part of 8 chapters has been devoted to the domestication of animals, animal husbandry development and education in India, breed improvement programmes for different species in India, and contribution of livestock. The second part comprises 11 chapters covering the conventional animal breeding techniques (selection and mating systems) based on the principles of population genetics. The last third part comprising 8 chapters of the book deals with the essentials of sustainable animal breeding and production covering the documentation and conservation of AnGR, strategies for breeding of different species of livestock and poultry, biotechniques in animal breeding, reproduction, nutrition, management and health care.

Hope this book will be of immense use to the post graduate students, teachers and those appearing in different All India Competitive Examinations like NET, SRF, and ARS in Genetics and Animal Breeding conducted by ASRB (ICAR) as well as by UGC and UPSC. The subject matter has been presented in single volume in a more meaningful and desired manner and simple language with numerical examples. We do not claim any originality as the subject matter has been collected and compiled from various published sources. The authors are highly thankful to these sources.

The authors are highly thankful to Dr. D. Sunderasan, Dr. R.M. Acharya, Dr. A.L. Chaudhary, Dr. R.N. Singh, and Dr. S.B. Basu for their valuable guidance, kind cooperation and help. Thanks are also due to Smt. Shakuntla Tomar, Dr. Ajay Kumar Tomar, Dr. Amit Kumar Tomar and other family members of the authors for their consistent encouragement and kind cooperation from all corners at every step.

Further, we welcome the readers to point out the mistakes that are likely to be and give suggestions for further improvement of the book.

**Dr. Arun Kumar Tomar**
**Prof. Sukhvir Singh Tomar**

# Contents

*Preface* v

## Part I. Animal Breeding Programmes in India

**1. Domestication of Farm Animals** **1-22**

1.1 Plant – Animal – Human Relationship 1
1.2 Domestication and Civilization 2
1.3 Definition of Domestication 3
1.4 Ancient Literature 3
1.5 Origin and Domestication of Different Species 4
1.6 Methods of Domestication 8
1.7 Evolutionary Sequences Under Domestication 9
1.8 Changes Brought by Domestication in Evolutionary Process 10
1.9 Changes in Animal Characteristics Under Domestication 12
1.10 Basic Characteristics of Domestic Animals 16
1.11 Classification of Organisms 16
1.12 Place of Domestic Animals in Animal Kingdom 18
1.13 Breed Formation 21
1.14 Theories About Origin of Present Day Species 22

**2 Animal Husbandry Development & Education in India** **23-44**

2.1 Milestones of Indian History 23
2.2 Ancient Period Animal Husbandry 24
2.3 Medieval India (500 – 1500 Ce) Animal Husbandry 33
2.4 Modern India (1500 – 1947) Animal Husbandry 34
2.5 National Five Years Plans and Animal Husbandry - At a Glance 40
2.6 Animal Science Research Institutes (ICAR) 44

**3. Breed Improvement Programmes I. Bovines** **45-73**

3.1 Place of Bovines in Animal Kingdom 45
3.2 Chromosome Profiles of Bovines 47
3.3 **Cattle** Breeds 48
3.4 Panchgavya (Cow Therapy) 49
3.5 **Buffalo** Breeds 50
3.6 Importance of Buffalo 50
3.7 Cattle and Buffalo Improvement Progrmmes 51
3.8 ICAR-Research Institute on Cattle and Buffaloes 67
3.9 **Mithun** 69
3.10 Utility of Mithun 70

3.11 Improvement Programme on Mithun 71
3.12 ICAR-National Research Centre on Mithun 71
3.13 **Yak**: Ship of High Hills or Snow 71
3.14 Utility of Yak 72
3.15 ICAR-National Research Centre on Yak 73

**4. Breed Improvement Programmes II. Ovines and Rabbits 74-87**

4.1 Differences Between Sheep and Goats 74
4.2 **Sheep** 74
4.3 Utility of Sheep 76
4.4 Sheep Improvement Programmes 77
4.5 ICAR-Central Sheep and Wool Research Institute 79
4.6 Important Milestones of C.S.W.R.I. 80
4.7 **Goats** 81
4.8 Utility of Goats 82
4.9 Goat Improvement Programmes 83
4.10 ICAR-Central Institute for Research on Goats 84
4.11 **Rabbits** 84
4.12 Productive Characters of Rabbit 86
4.13 Utility of Rabbit 86
4.14 Development Programme of Rabbit 87

**5. Breed Improvement Programmes III. Camel, Equine and Pig 88-99**

5.1 **Equines** 88
5.2 ICAR-National Research Centre on Equines 90
5.3 **Camel**: The Ship of Desert 91
5.4 Utility of Camel 92
5.5 Special Adaptive Features of Camel 92
5.6 ICAR-National Research Centre on Camel 93
5.7 **Pigs** 94
5.8 Utility of Pigs 95
5.9 Pig Production Systems 96
5.10 Piggery Development Programmes 97
5.11 ICAR-National Research Centre on Pig 99

**6. Breed Improvement Programmes IV. Poultry 100-107**

6.1 Chicken Breeds 100
6.2 Other Poultry Species 101
6.3 Chromosome Numbers 102
6.4 Poultry Utility 103
6.5 Poultry Development Programmes 103
6.6 ICAR-Research Institutes on Poultry 105

**7. Constraints of Development Programme** **108-114**

7.1 Development 108
7.2 Livestock Development Programmes 108
7.3 Limitations of Improvement Programmes 109
7.4 Backward State of Livestock In India 110
7.5 Suggested Measures for Livestock Improvement 111
7.6 Scope of Animal Husbandry 111
7.7 Limitations of Livestock Rearing 111
7.8 Extension Education Programmes 112

**8. Livestock Contribution** **115-137**

8.1 Agriculture Vs Livestock 116
8.2 Changes in Food Habbit in India 119
8.3 Livestock and Poultry Population 120
8.4 Livestock Contribution 122

**Part II. Conventional Animal Breeding Techniques**

**9. Response to Selection** **138-162**

9.1 Selection Objective and Selection Unit 138
9.2 Selection and Culling Criteria 138
9.3 Importance of Selection from Relatives Performance 140
9.4 Genetic Effect of Selection 140
9.5 Sequential Schemes of Selection 141
9.6 Selection Differential (Phenotypic Superiority) 142
9.7 Response to Selection 146
9.8 Accuracy of Selection 151
9.9 Generation Interval 154
9.10 Factors Hampering the Genetic Progress 156
9.11 Constraints in Genetic Progress 158
*Numerical Solved Examples* 159

**10. Measurement of Response** **163-178**

10.1 Purpose and Assumptions in Measurement of Response 163
10.2 Problems in Measurement of Response 163
10.3 Methods of Measurement of Response 167
10.4 Genetic Change by Direct Selection 173
10.5 Selection Between Populations 174
10.6 Random Sample Test 175
*Numerical Examples* 176

**11. Bases of Selection** **179-207**

11.1 Single Trait Selection 179
11.2 Aids to Individual Selection 181
11.3 Methods of Selection (Multi-Trait Selection) 183
11.4 Tandem Selection 183
11.5 Independent Culling Levels (ICL) Method 184
11.6 Total Score or Selection Index 186
11.7 Comparative Efficiency of Different Methods of Selection 192
11.8 Types of Selection Indices: 195
11.9 Restricted Selection Index 195
11.10 Multitrait – Multisource Selection Index 199
*Numerical Examples* 201

**12. Pedigree Selection** **208-212**

12.1 Basis of Pedigree Selection 208
12.2 Guides to Pedigree Selection 208
12.3 Practical Difficulties to use Pedigree Selection 209
12.4 Merits of Pedigree Selection 209
12.5 Demerits of Pedigree Selection 209
12.6 Breeding Value Based on Pedigree Records 209
12.7 Accuracy of Pedigree Selection 211

**13. Family Selection** **213-221**

13.1 Sib Selection 213
13.2 Family Selection 214
13.3 Factors Affecting Accuracy of Sib and Family Selection 215
13.4 Advantages of Family Selection 217
13.5 Disadvantages or Limitation of Family Selection 217
13.6 Within Family Selection 217
13.7 Combined Selection 218
*Numerical Problems* 219

**14. Progeny Selection** **222-235**

14.1 Genetic Principles of P.T. 222
14.2 Superiority of Progeny Selection Over Other Criteria 222
14.3 Advantages of P.T. 223
14.4 Limitations of P.T. 223
14.5 Sources of Error in P.T. 223
14.6 Precautions in Progeny Testing 224
14.7 Requirements for P.T. 224
14.8 Constraints in P.T. and their Solutions 225
14.9 Methodology P.T. in Organized Herd 226

14.10 Accuracy of P.T. 228
14.11 Field Progeny Testing 228
14.12 Sire Evaluation 230
*Numerical Problems - Progeny Testing* 233

**15. Selection for Different Kinds of Gene Action** **236-257**

15.1 Selection for Additive Gene Action 236
15.2 Selection for Non-Additive Gene Action 237
15.3 Concept of GCA and SCA 237
15.4 Genetic Principle to Exploit Non Additive Gene Action (Heterosis) 239
15.5 Selection Methods for Combining Ability 240
15.6 Effectiveness of RRS 245
15.7 Comparison of RRS and WLS for Crossbred Response 250
15.8 Selection and Breeding for Threshold Characters 251
*Numerical Examples: Reciprocal Recurrent Selection (RRS)* 254

**16. Selection Experiments** **258-275**

16.1 Choice of Model Animal for Selection Experiment 258
16.2 Relevance of Lab Experiments to Large Livestock 259
16.3 Consequences of Selection Experiments 261
16.4 Total Response and its Determinants 267
16.5 Nature (Expectation) of Selection Limit 269
16.6 Factors Affecting Selection Limits 271
16.7 Methods to Overcome the Limits to Selection 273
16.8 Relevance of Limit to Selection 275

**17. Mating Systems I. Inbreeding** **276-315**

17.1 Genetic Relationship 276
17.2 Concept of Gene's Identity 277
17.3 Measurement of Relationship 278
17.4 Methods to Compute Rxy 280
17.5 Factors Affecting the Degree of Relationship 285
17.6 Use of Degree of Relationship 293
17.7 Inbreeding 294
17.8 Types of Inbreeding 294
17.9 Disadvantages of Inbreeding (General) 299
17.10 Applications of Inbreeding 299
17.11 Ways to Minimize Inbreeding 300
17.12 Genetic Effect of Inbreeding 300
17.13 Phenotypic Effect of Inbreeding 300
17.14 Measure of Degree of Inbreeding 301
17.15 Methods to Compute Inbreeding Coefficient ($F_x$) 303

*Solved Example* 304
*Numerical Solved Examples* 308
*Exercises* 315

**18. Mating System II. Outbreeding** **316-333**
18.1 Definition of Outbreeding 316
18.2 Forms of Out Breeding 316
18.3 Effects of Outbreeding 321
18.4 Results of Crossbreeding 323
*Numerical Examples* 333

**19. Poultry Breeding** **334-361**
19.1 Place of Poultry in Animal Kingdom 334
19.2 Chromosome Numbers 335
19.3 Measurement of Quantitative Characters 336
19.4 Methods of Mating in Poultry 342
19.5 Pedigree Record Keeping 343
19.6 Selection Based on Phenotypic Characters 344
19.7 Culling Criteria 345
19.8 Hatching 346
19.9 Basis of Selection - Genetic Selection 347
19.10 Mating Systems 351
*Numerical Examples* 352

**Part III. Sustainable Animal Breeding & Production**

**20. Sustainable Animal Production** **362-370**
20.1 Milk Production in India 362
20.2 Constraints of Low Productivity 363
20.3 Animal Production Systems 364
20.4 Sustainable System 367
20.5 Determinants of Sustanable Animal Production 368
20.6 Ways of Achieving Sustainable Animal Production 369

**21. Documentation and Conservation of Germplasm** **371-391**
21.1 Biodiversity 371
21.2 Animal Genetic Resources (AnGR) 372
21.3 Special Features of Angr of India 372
21.4 Main Features of Cattle And Buffalo Population in India 373
21.5 Unique Features in Indigenous Breeds of Livestock 375
21.6 Documentation (Management) of AnGR 376
21.7 Animal Data Bank 379

21.8 Conservation Vs Preservation 380
21.9 Reasons of Threat to Genetic Diversity 380
21.10 Endangered Level of A Breed 382
21.11 Needs of Conservation 383
21.12 Principles of Conservation 384
21.13 Methods of Conservation 384
21.14 Utilization of Indigenous AnGR 388
21.15 Strategies for Conservation 388
21.16 Agencies for Improvement and Conservation Programmes 389
21.17 ICAR-National Bureau of Animal Genetic Resources, Karnal, Haryana, 1984 391

**22. Animal Breeding Strategies** **392-411**

22.1 Breed Improvement 393
22.2 Breeding Strategies for Optimum Production 402

**23. Biotechniques for Animal Breeding** **412-421**

23.1 Biotechnology 413
23.2 QTL Mapping Designs 414
23.3 Applications of Molecular Markers 416
23.4 Molecular Markers in Breeding Programmes 416
23.5 Application of Molecular Genetics 420

**24. Biotechniques in Animal Reproduction** **422-431**

24.1 Reproductive Characters of Different Species 422
24.2 Conventional Practices of Animal Reproduction 425
24.3 Reproductive Biotechnologies 427
24.4 E.T. Work in India 431

**25. Biotechniques in Animal Nutrition** **432-449**

25.1 Major and Micro Nutrients 432
25.2 Feed Resources 434
25.3 Energy Requirement 436
25.4 Feeding Requirements of Dairy Animals 437
25.5 Feed and Fodder Requirement 438
25.6 Increase in Forage Production 439
25.7 Quality of Feeds and Fodders 441
25.8 Feed Processing Technology: Increase of Feed Quality 441
25.9. ICAR-National Institute of Animal Nutrition and Physiology, Bangalore, 1995. 449

**26. Sustainable Animal Management** **450-458**

26.1 Components of Livestock Management 450
26.2 Care And Management of Animals of Different Categories 451
26.3 Animal Housing 453
26.4 Adverse Effect of Heat Stress and its Amelioration 456

**27. Animal Health Care** **459-476**

27.1 Health 459
27.2 Signs of Good and Ill-Health of Animals 460
27.3 Benefits of Control Measures 462
27.4 Disease Resistance 462
27.5 Symbiosis Among Organisms 462
27.6 Disease Causing Agents 463
27.7 Control of External Parasites 467
27.8 Control of Internal Parasites 468
27.9 Control of Contagious Diseases 470
27.10 Mastitis in Dairy Animals 473
27.11 ICAR-Research Institutes on Animal Health 474

# 1 Domestication of Farm Animals

There are ample evidences to suggest that animals were domesticated before the beginning of the written history. The excavation findings, old scriptures, and edicts (Orders of authority) are used to trace out the history. There are ample information on animal care, health management, and treatment of diseases in ancient Indian literature (Indus Valley Civilisation, Holy Vedas, Puranas, epics, etc). Since ancient India the livestock keeping had been an inviolable (unbreakable) tradition rather than an economic enterprise. The most important cultural and religious traditions of the Hindus are linked to domesticated animals and particularly the cow occupied an important place for cultural ethos rather than for the nutritive lacto-products they offer.

There are about 100000 species of animals listed in literature but man has domesticated a mere handful of only 40 species of domestic animals including poultry to serve various purposes. The domestication of animals of nearly all the species started and the agriculture was also started in a crude way when man used to live in huts or wooden houses. The domestication of animals by man has been certain long before the time of recording historical data, though its exact time is not known.

The first animals to be domesticated probably were the dogs and goats about 8500-9000 years ago from today whereas the cattle and sheep about 6000-6500 years ago from today. However, the domestication of sheep and goats took place in the pre-agriculture period.

## 1.1 PLANT – ANIMAL – HUMAN RELATIONSHIP

Man is a member of the Mammalia class, Primate order, and Hominoidea superfamily of animal kingdom and hence considered as an animal, but with a difference of wisdom and culture. There had been a relationship of man with animals. The art of the cave dwellers had depicted different stages of man-animal relationship during the Hunting age, Pre-historic time and Early Historic time. During hunting age, the animals were valued for their protein-rich meat, skin, bone, horn, and antlers (branched horns) but during the Pre-historic time the animals were reared for their products, transportation, and farming whereas during the Early Historic time man enslaved animals and misused them even in the battlefield. In modern days the animal rearing has become a business in all parts of the world. There is a close link among plants, animals and human beings.

The cultivation of plants is closely linked with the domestication of animals in a way that the development of animal husbandry is not possible without crop production. The animals not only provide animal protein and fat in the form of milk and meat to us but also provide draught animal power for agricultural operations and transport, and organic manure required for crop production. Thus, the animals are the integral part of agriculture whereas the crops in the green stage, in the form of seeds and the crop residues provide feed and fodders to the animals. The large ruminants in particular require and depend on such feeds and fodders from crops in addition to natural vegetation. Thus, the agriculture and animal husbandry together form the integral part of our living and economy, particularly for the rural community.

All the living beings on earth depend on the solar energy (radiant energy of sun) which is the primary energy. The plants are capable to capture solar energy and to make its use in production of vegetable matter and seeds (grains) which are the food materials for animals and human. The stored energy of food materials (plants and the seeds) is consumed by animals through eating the plants and seeds. Thus, the plants are the primary food source of animals. The animals by eating the plants in turn convert the energy they derive from plants in to animal protein and fat. The plants thus conserve and build up nutrients (constituents of food) and energy in the form of food whereas the animals degrade and utilize the nutrients and energy. Thus, by way of domestication of plants and animals, man had empowered the control over the primary source of food energy as a nutritional security for him. This has lead to an enormous increase in human population over the times.

## 1.2 DOMESTICATION AND CIVILIZATION

The domestication of plants and animals had played prime role in starting and developing the human society and had laid the foundation of human civilization and modem agriculture on which had rested the social and scientific progress of today. This is also well evidenced from the relative use of domestic animals by man in different parts of the world which has affected the degree of development of ancient civilization. The cattle, buffaloes, sheep, goat, camel, horses, and asses had their significant contribution in social and economic development of human society/ civilization in early days of human settlement. The domesticated animals made their contribution in crop production, transport and supply of human food in the form of milk and meat. In the beginning of domestication, the draft animal power was though more important for carrying out the agricultural operations and a means of transport compared to the animal products like milk, meat, skins and hides which were started using by man subsequently.

The original citizens (aborigines) of Australia and the Red Indians in North America who had only the dogs with them in ancient time, could not adopt the agriculture but remained hunters whereas a high level of civilization, the Maya culture was developed by the Indians of Mexico, Central America, Peru and Bolivia who domesticated Llama, alpaca and vicuna, and though they had no domestic animals like bullocks, horses, camels, asses and elephants that are more efficient than Llama and alpaca but started cultivation of crops with use of digging sticks and spades. Therefore, Egyptian, Mesopotamian and Indian civilization were more advanced than the Meso- American civilization.

## 1.3 DEFINITION OF DOMESTICATION

The domestication is the man's activity of taming of wild animals which had started by capturing of infant wild animals by man and raising them as pet animals. Therefore, the domestication is defined as the taming of wild animals and to control their breeding, feeding and care (management and health control). The domestication however, involved some biological changes in morphological, physiological or behavioural attitudes of the animals. Thus, domestication differs from taming. The taming is defined as the elimination of tendencies of animals to flee from the man. The wild animals (species occupying natural habitats) may be tamed, whereas individuals of most domestic species require taming only if their association with man is limited during early age. The untamed domestic animals should be differentiated from feral animals or domestic animals which have reverted to the wild state.

According to Lush (1945) the domestication of animals includes several aspects viz., tameness, raising them at farm which implies to keep the domestic animals in or near man's own dwellings (residential places), to bring their growth and reproduction under man's control, to make use of animal's products or services to man's own advantages and dependency of animals on man's care for their existence.

## 1.4 ANCIENT LITERATURE

The cow "***Kam Dhenu***" has been mentioned in ancient writings. There is mention of the use of cow ghee in religious ceremonies in Vedas (Rig Veda), and sculptures of "***Nandi Bull***" in Shiva Temples are found.

The cow has been said to be the mother of ***Rudras***, the daughter of ***Vasus*** and the sister of Aditayas in Vedas.

In Hindu mythology, the male buffalo *(Bhainsa)* has been referred to as the vehicle of the God of death, "***Yama***" and also regarded as the physical form of the demon Mahishasur, a symbol of inertia - ***"Tamas".***

It is generally believed that domestication must have started at the end of the Old Stone Age, ***Paleolic period*** (1000-8000 BC) or in the beginning of New Stone Age. Man used to hunt and eat the animals in Old Stone Age. This period was marked by the use of stone and bone implements without agriculture.

The new Stone Age, ***Neolithic period*** (8000-6000 BC) was marked by the use of ground and polished bone and stone weapons and tools. The period of New Stone Age extended up to about 3000 BC.

## 1.5 ORIGIN AND DOMESTICATION OF DIFFERENT SPECIES

The available information on the origin of different species of domestic animals in terms of their wild ancestors (progenitors) from which the present day species have descended, the probable time and place of their domestication are as under -

### Order of Domestication

(*i*) *Pre-agricultural period:* Mammals were domesticated like dog, reindeer, goat and sheep,

(*ii*) *Early agricultural period:* The mammals of different categories were domesticated for different purposes as –

(*a*) Mammals which were crop destroyer and used for human food like cattle, buffalo, gaur, banteng, yak and pig.

(*b*) Mammals were domesticated for transport and labour like elephant, horses, camels and ass.

The Llama, alpaca, guinea pig and turkey are native to the America. The domestication of other species took place in the Old World, largely in central or Western Asia with some domestication in Egypt and Europe. The chickens and elephants were domesticated in India and one center of domestication of swine was in China. It is also likely that domestication may have been in several regions for some species like sheep and swine. Upper Egypt or Ethiopia was one of the centers of domestication for the ass.

### 1.5.1 Cattle

The cattle were domesticated during New Stone Age (Neolithic period, 8000-6000 BC) both in Europe and Asia. The domestic cattle are of two types: *Bos taurus* (humpless cattle of temperate zones) and *Bos indicus* (humped cattle of tropical countries). The cattle of United States are purely of European origin.

According to Herre (1958), the wild cattle, *aurochs* or urus (B. primigenius) which are generally considered the progenitors of all domestic cattle including the ***Zebus, originated in India*** and then spread over most of Asia and Europe and across northern Africa. Mason (1985) and others had mentioned that *Bos primigenius* was a very large animal of the elephant size but of a bull shape and was a wild ox of Europe.

Another progenitor (ancestor) of present day breeds of European origin has been reported by Rice (1962) and Lush (1945) as the *Bos longifrons* (brachyceros) which was in Europe as a domesticated animal early in Neolithic period. This was small type. The present day European or American breeds of cattle may have been developed as a result of various degrees of crossing between the ancestors.

The cattle of India and Africa (*B. indicus*) were domesticated as early as 2100 BC and thought to be descended from the wild Malayan Banteng (Rice, 1962). The humped cattle or zebu were produced by manipulating the morphology of domestic cattle for adaption to a hot, dry environment with two new features - a large hump and a long hanging dewlap.

Lush (1945) have further mentioned some more names of the ancestors of domestic cattle. This was primarily the *B. namadicus* which was contemporary with man in India in the early Pleistocene (about 30,000 years ago).

*Domestication of wild cattle (Aurochs) probably first occurred in India* (Baluchistan, Sindh and Afghanistan), the Near East (Iran, Iraq, Jordan, Syria, Turkey, and USSR) and Egypt from 6000 to 4000 BC. The seals of Sivikotaka in Gujarat and Kalibhanga in Rajasthan bear images of animal favourite of Harappans (2200 - 1600 BC). In Siwaliks occur fossils of *Bos aceutifrons* regarded as the possible ancestor of the Pleistocene wild cattle that is *Bos namadicus* of India and Europe. It is assumed that domestic cattle were brought by immigrants for Southern Europe from the New East during Neolithic time (8000-6000 BC) and these were probably descended from Asiatic wild cattle or from interbreeding between European and Asiatic races.

A schematic development of modern breeds of cattle from wild cattle by domestication is given here:

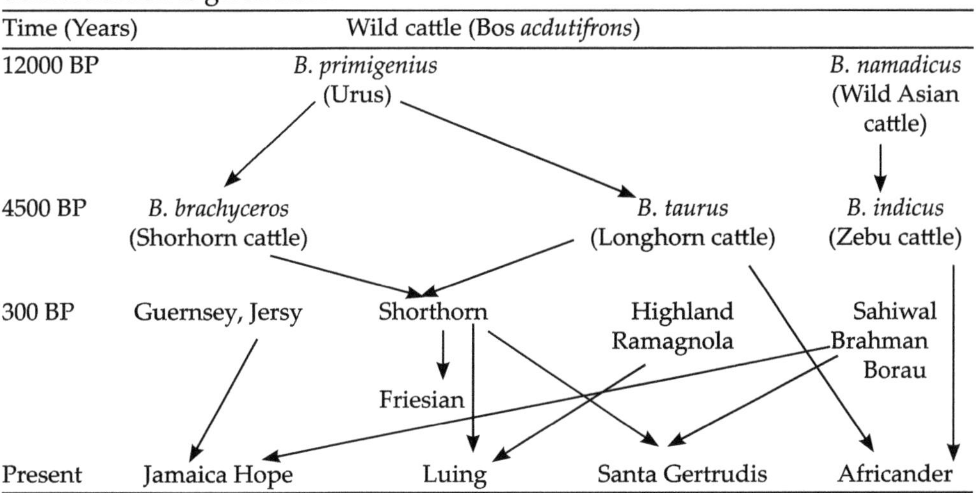


B.P. = Before Present

### 1.5.2 Tribes of cattle

Gaur, Gayal-Mithun, Yak, Bantang, and Kouprey belong to the sub genera **Bibos or Bibovine**. The Gaur, Gayal and Banteng are all restricted to South-Eastern Asia and nearby islands.

**Gayal (Mithun)** is quite similar to **gaur** in appearance, colour, etc. The mithun had been domesticated about 8000 years ago. The Gayal may be only the domesticated form of gaur. It is also said that the gaur of Assam, Bangladesh and Burma has been domesticated as Mithun or gayal exclusively for sacrificial purposes. There are conflicting views about the origin of Mithun. Some consider it as the hybrid of gaur and cattle while others regarded it as the domestic form of gaur.

The **Yak** is long haired and busy tailed cattle, domesticated from its wild progenitor, *B.mutus*. It was originated in China which still has 90 % of the total world yak population of 14 million.

The **Banteng** is the common work ox of Java, Bali and Borneo.

American bison, European bison, and yak belong to sub genera **Bison or bisontine**. The **European Bison** (wisent) and the **American bison** have never been really domesticated.

### 1.5.3 Buffalo

The original domesticated buffaloes of India, Pakistan and elsewhere (Asian buffaloes) are believed to have been derived from the Indian wild buffalo (*Bubalus arnee)*. The **arni** is thus the ancestor of domestic buffalo. The domestic Indian buffalo is called *Bubalus bubalis* which was earlier called as Bubalus *arnee.* It was domesticated in India during pre-historic times between 2500 – 3000.

Asian buffaloes (Bubalus bubalis) or water buffaloes have two sub species which are river and swamp types. *Riverine and Swamp buffaloes were domesticated almost together (around 2000 BC), former in the Indus Valley Civilization* and the latter in the Yangtze Valley. Domestic buffalo reached South-East Europe by 12$^{th}$ century; later taken to the America and Australia, and Africa in the 20$^{th}$ century. Riverine breed development has largely centered in Indo-Pak subcontinent.

Cockrill (1982) had mentioned that all buffaloes of river group, including Egyptian and Mediterranean breeds have their origin in Indian subcontinents 5000 years ago or even earlier whereas the swamp buffaloes were probably first domesticated in China some 5000 years ago.

### 1.5.4 Sheep and goats

The two genera (ovis and capra) of sub family bovinae of family Bovidae are closely related.

**(i) Goat:** The goat was the earliest ruminant to be domesticated in pre-agricultural period as early as 6700 BC in Baluchistan, though Iran is an alternate place of its domestication as these goats existed in Iran 6000 BC.

The domestic goat have descended from wild goat or bezoar (*Capra hircus aegagrus*) of south west Asia. The wild ancestors of goat were found in barren hills of Baluchistan and western Sind during Pre-Harappan period and it is the ancestor of various breeds of goats. **Gaddi goats of Himalayan region resemble to the ancestral wild goats and are still used for carrying goods in the higher Himalayan region of India.** The wild goat is divided into 3 groups:

(*a*) Bezoar of Pasang (C. *aegagrus),*

(*b*) *Ibex* (C. *ibex)* and

(*c*) Markhor *(C. falconeri).*

The bezoar goat (*Capra hircus aegagrus)* is most important and spread from Sind to Iran to Russian Turkistan and Caucasus, west of Asia Minor to Cyclades in the west.

The Markhor goats *(C. falconeri)* found in Turkistan, Afghanistan, Baluchisthan and Kashmir, has replaced the wild goat in Pakistan, and also has descended the caucassian goat.

Lush (1945) mentioned that domestic goats descended from C. *aegagrus* or E. *prisca* which is known from Pleistocene fossils from Greece.

(*ii*) **Sheep:** The original center of domestication of sheep was the Aralo-Caspian steppe and Turkestan. Sheep keeping spread early into Iran and later into Mesopotamia and Baluchistan. The sheep kept in India, Tibet, and other countries of East and South Asia were of western derivation and basically of Urial stock. In India, sheep keeping was practiced evidently from Pre-Harappan period. The dominant form of sheep rearing still remained of nomadic nature. Domestication of sheep, besides ensuring a permanent meat supply, also improved the supply of skin, hair (wool), fat, and bones. Although these animal by-products are available from other sources, production of wool, however, remained a monopoly of the sheep. Sheep rearing is an exclusive occupation of a class of herders traditionally marked out as a pastoral caste.

The domestic sheep (*ovis aries*) have descended from 3 species of *Ovis viz.* Mouflon (*ovis musimon*) and urial (ovis vigna) found in Europe and Asia, and *ovis orientalis* which has many wild species was domesticated in Indian sub-continent and argalis group (ovis poli) of central Asia, etc. The domesticated wild sheep was Arkal of the species *Urial*. The sheep of India, Tibet, east and south Asia are basically of *Urial* stock. The descendents of *musimon* are found in northern Iran. The *Ovis anon,* the argali is found in central Asia. The vital role in domestication of ruminants and particularly the sheep has been played by the dogs which are still used to control large flock of sheep.

Rice (1962) mentioned that sheep probably originated in Europe and cooler regions of Asia in Pleistocene era.

### 1.5.5 Camel

The single hump camel (Camelus *dromedrius*) was domesticated probably in Arabia, Indus valley and north eastern Africa during Egyptian civilization whereas the two hump Bacterian camel (C. *bactriansus*) was domesticated in central Asia somewhere between Iran and the Gobi desert.

Skeletal remains found in excavation at Kalibanga (Rajasthan) suggest that camel was known during 2300 – 1800 B C. The Archiology Survey of India reported the age of paintings of camels in Diamabad (Ahmedabad, MS) to be as old as 1300-1200 B C.

### 1.5.6 Horse

The origin (ancestry) and date of domestication of horse is not clear. However, it has been mentioned that the domestic horse is the result of probably two or three wild types *viz.*, steppe horse, desert horse and forest horse. The domesticated horses of Europe and Western Asia have probably descended from the Tarpan or Mongolian horse. It has been also mentioned in literature that horses were present in Europe in Paleolithic times about 25,000 years ago. The Greece and Egyptian records of the domestication of the horse date from about 2000 BC and 1600 BC, respectively. Probably the last animal to be domesticated by man was the horse whereas the dog was the first animal domesticated.

### 1.5.7 Pigs

The present day breeds of domestic pig (*sus domesticus*) have originated from two wild ancestors:

(*i*) *Sus scrofa* (European wild boar), the progenitor of the Northern European breeds. Its crossbred progeny with domestic swine are fertile. It was domesticated in Neolithic times around Baltic Sea.

(*ii*) *Sus vitatus* (collared Malayan pig) which is native to Middle Eastern Asia, domesticated in China and the progenitor of the breeds of Southern Europe, Asia and Africa.

The Indian wild pig is found in Siwaliks and tarai area of northern India. *The pigs were domesticated in India* and east Turkistan and spread to Europe. The European pigs are wild.

### 1.5.8 Poultry

The red jungle fowl of India (Gallus gallus) is considered as the progenitor (ancestor) of domestic breeds of fowl (*Gallus murghi).* The domestic fowl were domesticated in India (Indus valley) about *2000 B.C.* from where they spread to the east and west. The red jungle fowl is still found in many parts of India.

## 1.6 METHODS OF DOMESTICATION

The starting of domestication had certainly been need based. Man thought the need of certain animals and attempted to domesticate them. The needs of man could have been for religious rites (as sacrifice to the God) and for meeting his requirements like meat for food and skin for clothing. Thus animals of economic value to man were started to be domesticated viz., sheep, goat, cattle, pig and fowl. These could have been another reason of domestication of animals by man like attraction of some animals (dogs) by waste of food available around man's dwellings. Such animals acted as scavengers of man's surroundings and they slowly became pet.

There can be two methods of domestication. The first is the imprinting method and second is hunger sympathy method.

1. ***Imprinting method***: This involves to bring home a few young as pets or to capture the cripples (those which are unable to run away). Such animals are well fed and cared. The rearing of pets was started for the tendency of women and children to keep pets. This suggests the most probable stimuli to domestication. The wild animals are easily caught, tamed in young age and well cared and fed. The hand rearing become attached to human being and respond to them as they might do to their own species and prefer their human foster parents. This forms the social bond. This is called as imprinting. The evidence to support this view is that primitive

(ancient) people in all parts of the world used to tame and rear the wild animals. However, there was a problem to provide milk to rear the infant mammals. But there are reports indicating that there are still women of tribes in tropical America and South West Asia who nurse pups, kids and piglets. The human nurses were gradually replaced by lactating animals of one domesticated mammalian species. This helped to domesticate another species. Once the goat and sheep were domesticated the milk became available for orphaned calves and colts. This helped to domestication of large species.

2. ***Hunger sympathy method***: In this method a pit is dig out in the jungle and then covered with grasses. The wild animals are either carefully derived in a planned way to that pit or they themselves came across that pit and fall down in the pit while crossing it. After few days a man visits that site and provides some feed and fodder to that animal in the pit. This process goes on for few days till the animal starts recognizing and loving that man. The man then brought home the animal probably in a cage and takes care. The wild elephants are captured by this planned drives method. This is the effective method of taming the most unruly among the wild ones. The tame elephant is then used to bring more number of wild ones and to teach them to work. However, one point in domestication play very important role in taming the wild animals and the same is the temperamental aptitude of wild animals. This makes the success or failure of attempts of domestication.

The domestication had been compleeted in multi stages viz., loose contact of animals with man in the beginning, confinement of the animals within human environment where they bred at random in captivity, selective breeding by man, and finally the development of breeds with desired characteristics. The development of breeds started as early as BC 3000.

## 1.7 EVOLUTIONARY SEQUENCES UNDER DOMESTICATION

The evolutionary changes have occurred in different species of animals due to the major forces (mutation, migration, genetic drift and selection) responsible to alter the gene frequencies. The basic requisite of these forces is the mutation.

The animals of different species have migrated in small numbers' from their centers of domestications to other countries having different cultures and environmental conditions. Thus, migration and genetic drift (small population size) had played their role.

The geographical isolation by distance determined by dense forests, mountains, hilly rivers, deserts etc., the physical environment, ecological constraints, social preferences, disease prevalence and the natural selection have resulted in the origin and development of different regional breeds/types because these had been inter mating within the geographical region. The artificial selection had played its role in changing the gene frequencies, in order to meet the needs of different geographical areas.

The evolutionary progress is favoured by the availability of unoccupied environmental niche. A range of physical environment in which a species is found or multidimensional space within which the environment permits an individual or species to survive is called the ecological or *environmental niche*. In ancient time there was human culture of food collection by hunting the animals and this culture was followed by the culture of plant cultivation. This has created an unoccupied niche which favoured rapid domestication of a number of animal species. There had been then the selection between species to serve human culture and needs viz. milk producers, draught animals, scavengers, pack and pat animals etc.

***Divergent selection***: The domesticated species have diverged for different functions. The dogs have diverged into types selected for herding, bird hunting, trail hunting, guarding and draught purposes; horses selected for speed and riding; chickens selected for fighting, meat and egg production. The divergent selection divides a well established breed into different types. Short horn cattle had diverged into beef and milk production to the extent that they converged with other breed selected for the same function. The dairy Short horn now resembles the Friesian breed more than it resembles the beef Short horn which now resembles the Angus breed. Improvement was started for a specialized function *viz.*, increased milk/egg/ wool production, speed etc. This developed the breeds and varieties by morphological diversification and imposed reproductive isolation.

## 1.8 CHANGES BROUGHT BY DOMESTICATION IN EVOLUTIONARY PROCESS

The domestication is an evolutionary process. The domestication has shifted the emphasis from natural selection to artificial selection, from random mating to the mating systems, migration of animals in small numbers and hence genetic drift. The domestication under these evolutionary processes had changed the genetic structure of populations by differential reproduction of genotypes. Moreover, the domestication has also made the changes in the environment. It is known that neither selection nor the change in environment from wild to domesticated conditions create any new gene, but the change in environment allows the genetic differences to express themselves accordingly. The changes in environment exploit the hereditary differences and this makes the selection more easy and accurate.

### 1.8.1 Change in population size, selection and mating systems

The domestication had intensified forces of evolution which bring change in genetic structure of population. Before domestication, the wild animals used to move widely in groups and bred at random. After domestication, animals were confined in captivity within human environment in *small numbers* and the man chose the breeding pattern as to which animal should be saved for breeding. Thus, this had resulted in a change in *population size* (which became small) and in the *mating system* together with *artificial selection*.

Man had the idea of resemblance among relatives since very beginning. Keeping this idea in mind, man started breeding the most desirable animals and castrating others. The practice of castration is an ancient practice. No doubt that man had practiced the artificial selection for thousands of years. The process of artificial selection by man had resulted in the genes of favoured animals to become more abundant in the population compared to the genes of less favoured animals. Thus scientifically speaking, the artificial selection has changed the gene frequencies and hence the genetic structure of population. Now, the selection is known to be the main force of changing the genetic structure of population. Under domestication, the animals were more narrowly restricted by herding or fencing to grow, remain and reproduce in the same area where they were born, this together with small population size resulted in increase in inbreeding and relatedness among domestic animals. The inbreeding resulted from restricted movement and from limiting the number of breeding males had been an important force to produce diverse races of domestic animals. The concentration on a few good animals for breeding inevitably led to some inbreeding and thus a particular type, colour and conformation tended to become fixed. This had thus resulted into differed breeds of today.

*Migration:* Man had transported the breeding animals far beyond the area in which they were born and could have wondered before their domestication. A number of examples on this aspect may be cited to support that how this process went on during historic times. The cattle were brought from Holland to England across water, Merino sheep' from Spain to many lands, Angora goat from Turkey to USA and South Africa, Zebu cattle from India to Brazil and USA, Short horn cattle from England to Argentina and Australia etc. The invading armies and the migrating people brought with them the animals from their native lands. Alexander, the Great, brought large number of livestock with his armies form Greece to. India. Genghis Khan during I300$^{th}$ century with his armies traveled with enormous reserves of cavalry horses all the way from China to Central Europe. Such cases of migration of animals are known since the beginnings of domestication. Thus domestication had resulted in an increased amount of out breeding by transporting of animals by men with him. Therefore this process had changed the genetic structure of domestic animals.

### 1.8.2 Change in environment

The environment plays its role in changing the expression of a character. The change in environment after domestication allowed the differences in heredity to express more clearly than could have been in the environment of wild animals. The genetic differences are not clearly expressed in poor environment as expressed in better environment. The example of feeding regime on fattening ability is well fitted in this case. The genetic differences in the feed conversion efficiency are more clearly expressed in a well managed feed lot experiment than under the condition of malnutrition/under feeding/feed scarcity. After domestication, the environment has improved in terms of nutritional changes (quantity and quality of feed, feeding times etc.), way of life, protection from extreme climate, better management practices etc. This permitted the hereditary differences to express themselves and this has been exploited by man to develop better animals.

A new phenotypic expression of the genotype can be developed under drastically different than normal environment. There may occur morphological changes in muscle and fat, skin and bone proportions during growth under nutritional changes and the changes related to way of life like more sedentary life (habit of too much sitting) *e.g.* body conformation of local *lalas* (shop keepers) is different than a farmer or labourer. The body conformation of modern pigs fed on a low plane of nutrition is similar to the wild boars.

The high quality feed and better management practices like protection from adverse climate influence the growth rate and sexual behaviour. The growth rate is high and the sexual maturity is earlier and continue longer (prolonged breeding season).

In mammals, more thorough milking may prolongs the lactation. The maternal care may also be prolonged if provided to new young at appropriate interval. There is interference in milk let-down in case of disturbance in the activation of the secretion of adrenaline due to mishandling of animal.

The removal of eggs from the nest immediately to laying prolongs the egg laying in many species of birds. Under modern management conditions, the unselected jungle fowl hatched and reared more eggs than in natural habitat.

## 1.9 CHANGES IN ANIMAL CHARACTERISTICS UNDER DOMESTICATION

The domestication has brought changes in economically useful animals with respect to morphological characters, production and reproduction traits, and psychology/behavioural attributes. The changes resulted from the domestication can be observed by comparing the wild ancestors with domestic animals. The degree and trend of change have been under control of man to a great extent. The traits of interest to man have been subject to greater change due to directional selection and breeding. Secondly, the trend of changes had been depended on the objectives set by man viz. the change in size of domestic animals of different species. The size of cattle has decreased but of the horses has increased. Man has influenced the evolutionary process of domestic animals (population size, selection, migration and mating systems). These evolutionary changes have caused differentiation in anatomy, morphology, physiology and behaviour of animals which are inherited with slow change from generation to generation.

### 1.9.1 Morphological Characters

The changes have occurred under domestication in size and shape of body, colour and type of colour etc. The size of domesticated cattle has decreased whereas the size of horses has increased as compared to wild ancestors as a result of selection and breeding of cattle for smaller body size as the milk production is not related to body size but the draught power is related to body weight. The birds have also increased in size.

The proportions of different ***body parts*** have also been subjected to change along with changes in body size. The changes have occurred in skeleton and limbs depending upon the purpose for which the animals had been domesticated. The draught animals tended to have larger limbs and compact skeleton whereas the animals developed for meat had deep body and shorter legs. The parts in which the most valuable meat is to be found (loins and buttocks) have become more developed in domestic swine. There has been shortening of facial part in skull under domestication.

There had been changes in ***body colour*** from wild colour which was typically one which blended well with the environmental background so as to act as protective colouring favourable to survival from predators. The deviations from protective colouring are unfavourable to the survival of animals under wild conditions whereas domestic animals need no protective colouring. As a result various coat colours were developed.

The ***hair coat*** of ancestral forms has also changed form mixed hair type with medullary and fluffy (soft) hairs to increased amount of fluffy hairs e.g., in case of sheep fine wool breeds have been developed.

The ***bones*** of animals after domestication have been changed in their shape and compositions. The bones of domestic animals have become smoothed from nodosity and sharp edged crest for muscle attachment in wild ancestors. The proportion of hard matrix has been reduced with increasing spongy matrix. The internal organs have also been modified in domestic animals. The intestines of domestic rabbit have become shorter while in case of domestic pigs are longer than wild boar. The weight of heart in domestic sheep has reduced to half of their wild ancestors.

### 1.9.2 Reproductive Characters

The reproductive processes of domestic animals have undergone significant changes. The wild animals are seasonal breeders so as their young ones are born in a favourable season of the year whereas most of the domestic animals do not have the seasonality. The puberty in domestic animals is earlier. The litter size has also increased in litter bearing domestic animals e.g., the domestic swine has twice as many young per litter as the wild boar and has two breeding seasons. Thus the breeding season has prolonged under domestication.

### 1.9.3 Production Characters

The ***egg laying capacity*** of domestic chicken has increased significantly. A domestic hen may lay an egg daily against laying and brooding of about 10 eggs by a wild hen.

After domestication, man has developed many breeds of livestock through selection and breeding for their ***high productivity***. The nature has provided the udder and teats to the mammals for nursing their young ones. Thus a wild cow (Urus) produced as much milk as required to her calf in first or two months of life and can not produce more than 500-700 kg milk whereas the milk production of a cow at present has increased to 10,000 kg during particular lactation.

The ***wool yield and quality*** in case of sheep has increased tremendously.

### 1.9.4 Behaviour Characteristics

***Relative stability of behavioural components***: There are specific aspects of behaviour viz., patterns of behaviour, effective stimuli which bring (elicit) behaviour, and the level of response indicated by the intensity or frequency at which appropriate stimuli release specific response pattern. The behaviour patterns are less subject to change and hence conservative than the stimuli which release them whereas the levels of response are least conservative and most readily modified. There is no evidence that the behaviour patterns of domestic ungulates, carnivores and gallinaceous birds differs from those of their wild progenitors. The wolves do bark but less frequently than some breeds of dogs. The domestic rabbits under semi-natural conditions having no prior experience of digging, also burrow and construct nest like that of wild forms. Likewise, there is no basic difference in social organization of red jungle fowl and domestic hens.

***Behavioural changes:*** The domestication may bring change in behaviour through learning or physiological adjustments. The animals get signals for some events through ***hearing under confinement***. *e.g.,* animals learn to expect feeding time by moving to the area where fed, preparation of milking like rattling of milk pails and concentrate feeding may elicit the release of milk before the udder is actually manipulated. Likewise, the stimulus for milk let-down has shifted form the calf to man'. The animals rapidly learn to locate feed, water and their mates, under confinement.

The ***physiological adjustments*** under domestication may also bring change in behaviour. For example, the physiological needs for energy and water are changed under changed environmental conditions which follow the changes in feeding behaviour. The level of feed intake and selection of specific nutrients are changed for increase in growth rate and egg production. There is an increase in water and feed consumption during milk production.

There are certain unique stimulus conditions which may produce changes under domestication e.g., suitable conditions for optimal nesting behaviour are absent under confinement and man provides artificial nests for birds or nesting materials for pigs. Even some birds do not use these nests and require man's efforts to train them to use nests. Placing of eggs in nests attract the hens to the nest.

Some ***husbandry practices*** are also important to bring change in behaviour e.g. separation of two sexes, castration etc. A number of males with marginal levels of sexual response, if kept separate, behaved like those which have higher sexual potential. The castration of non-breeders reduces aggressiveness and can be kept in large groups.

The selective breeding has played its role in shaping behaviour. The selective pressure under domestication has resulted in presumed ***reduction of flight, swimming and tree resting*** (taking rest on tree) in Muscovy ducks. The territorial patterns of behaviour tend to break down under domestication and the selection pressure favours the animals with more flexible territorial and social behaviour.

The feral (wild, brutal) horses started living in groups with one stallion and his harem. Likewise, fighting cocks in pen areas show some territorial behaviour and each territory has a single cock and a harem of females. Another examples is of wild mallards (ducks) which are monogamous in the natural habitat but become polygamous failing to develop pair-bonds with specific females under confinement.

Lorenz (1940, 43) studied the consequences of domestication among ducks and found that in contrast to wild mallard (duck), the male domestic ducks forced copulation with females which try to escape.

The selection changes the behavioural thresholds. The examples of successful selection for increased sexual and fighting behaviour as well as the decreased broody, sexual, fighting and emotional behaviour can be cited. The non broody strain of turkeys under selective breeding became broody and stopped laying eggs if eggs were left on the nest continuously.

Selection of bulls for refined characteristics has often nearly eliminated masculine secondary sexual characteristics, having feminine appearance and they are very docile and often show reduced sexual behaviour.

It is well known that castration eliminates or at least greatly decreases sexual activity for the reason of reduced supply of androgenic hormones. The castration also reduces the aggressiveness and makes possible to live in large groups.

Some domestic animals contain many females who fail to provide adequate maternal care to their offspring, due to the unnatural living condition which do not provide the stimuli necessary for the expression of full-fledged parental behaviour. The domestication had a deleterious effect upon the biological requirements for parental behaviour, for the reason of selection for economic characters. For example, the selection in domestic fowl has been deliberately done against broodiness (a form of parental or care taking behaviour).

The domestic breeds are more tractable (docile) than a related wild species. There are evidences for inheritance of differences in behaviour viz. variation in sexual behaviour, parental behaviour, agonistic behaviour, learning ability etc. Any quantitative behavioural character can be changed by changing the genotype and similarly that any change in the genotype probably will have some effect on behaviour.

Environmental forces can drastically modify the future behavioural and physiological capabilities of the organism. The ability to learn, to copulate or to survive under severe stress depends on the characteristics of the environment prevailed during early life. The environment actively interacts with the psychobiological processes of the being organism and this interaction changes these processes. The environment forces have broader and longer lasting effects when acts early in life.

## 1.10 BASIC CHARACTERISTICS OF DOMESTIC ANIMALS

The domestic animals have certain basic characteristics related to their adaptation, behaviour and reproduction favouring domestication. These are:

1. *Social group*: The domestic animals and birds are highly social and their wild relatives form elaborately (with great care or labour) organized groups under natural conditions.
2. *Adaptability*: They have the ability to adapt to a wide range of environmental conditions other than those in which they normally used to live. Thus they do not have any specialized habitat requirement.
3. *Sexual behaviour*: They live in large numbers together in the form of herds or flocks including both the sexes. This helps the animals to have little social stress under confinement as they have minimum fighting attitude and they can mate. They do not have any pair-bonds between two sexes but any female can mate to any male. The males dominate over the females. This reduces the conflict between animals and facilitates the mating. They provide sexual signals by movements or posture rather than colour patterns or morphological structures which commonly change under domestication and would result in failure to breed in the absence of essential stimuli.
4. They adapt the sexual pattern in a limited space and have the ability to reproduce in confinement.
5. *Food:* They can live on the by-products of agriculture (herbivorous) and hence do not have specialized dietary habit.
6. The parent young interaction is such that the female accept the new other young soon after parturition or hatching. Removal of young one from their parents during early life and cared for by man, establishes a bond to man. The rearing of young ones in groups away from their parents under intensive human care make them to bond to their own species and adapted to successful reproduction under confinement.
7. They are little disturbed by man or sudden change in environment. The wild animals have the tendency to flee by seeing man from a specific distance which may differ for different species. This flight distance can be reduced, by careful handling and approach, to zero by taming. Thus they respond to man.
8. They provide service or useful products with economic value.
9. They are tame and have evolved by selection from wild type.

## 1.11 CLASSIFICATION OF ORGANISMS

All living organisms vary considerably in their form (size and shape), structure and mode of life. It is not possible to arrange and study the heterogenous collection of all the organisms without proper classifications and without keeping them in systematic order. It was felt necessary to arrange and classify the animals and plants into certain groups and subgroups to have a broad idea and minute details of life in the world.

Man may have started giving names to animals and plants as soon as the power of speech was acquired. Upnishads and Vedas (1500-600 BC) have references of classification of organisms. However, Aristotle (384-322 BC) started the work of biological taxonomy writing 9 volumes discussing history of plants and laid foundation for the study of floral morphology.

The arrangement or classification of different kinds of organisms and their, relationship amongst them is known as systematic, which is as old as human civilization, and covers taxonomy (law of arrangement). The taxonomy is a Greek word formed from taxis means an arrangements order, and namos means a law. Taxonomy is the study of classification and identification. The classification is a system of arrangement of organisms into groups showing relationship while the identification is to know the correct place (position) of an organism in the plan of classification. After an organism is identified (find a place in classification) it is then given a name by which it is referred to. This is called as nomenclature which is a process of giving scientific names to animals and plants. The scientific/ botanical/zoological name of an organism consists of two words *viz.*, generic name (first word) and the species name (second word). This is the binomial system of nomenclature proposed by Linnaeus (1958), a Swedish Botanist who is often referred to as the father of taxonomy. The importance of binomial system of nomenclature laid in the fact that same animal or plant is known by different local names in different countries, states and localities. This makes the same and only one name of a particular plant or animal scientifically all over the world. For example, the scientific name of zebu cattle is *Bos indicus* which means that the animal (cattle) belongs to the genus Bos and to the species indicus.

**Hierarchy of taxonomy:** All the plants and animals are ranked/arranged in a hierarchy of categories (groups, sub groups) such as kingdom, phylum (for animals) and division (for plants), class, order, family, genus and species. The hierarchy means a series or succession of different ranks. The hierarchy of classification is a system of arrangement or framework for classification in which the different categories are placed in order to logical sequence.

The **kingdom** is the highest taxonomic category. All animals are included in **animal kingdom** while all the plants in the plant kingdom. The lowest base category in both the kingdom is the ***specie***s.

The **species** is a basic unit in taxonomy as well as in evolution. The species is the lowest member of a classificatory system. The term strain refers merely to source. A species is a group of individuals (animals or plants) which are similar in morphological characteristics, reproductively isolated and genetically distinct from other groups of individuals (species).

A **genus** comprised the related species having more features in common than in other closely related groups of species. All the species of a genus have been evolved from a common ancestor.

A **family** includes related genera and in separated from other related families by important characteristic differences.

The **order** is the next higher taxonomic category which includes related families. The organisms of related orders are kept under the next higher category named as **class.** The classes of animals having some features in common are included in a **sub-phylum** and **phylum** which constitutes the kingdom.

## 1.12 PLACE OF DOMESTIC ANIMALS IN ANIMAL KINGDOM

The animal kingdom has 10 phyla. The domestic animals are kept under the phylum **chordate** (which has 4 sub-phylum) and in its sub-phylum **vertebrate or craniata** (with back bone). The vertebrate have the characteristic features of body with bony or cartilaginous endoskeleton, presence of vertebral column, cranium or brain box distinct, head well developed provided with sense organ, paired lateral appendages being usually present.

The sub-phylum vertebrate has two super classes, **Agnata** having animals with no jaws and no lateral appendages (includes 50 species) and second is **Gnathostomata** having animals in which jaws and lateral appendages are present. The super class gnathostomata has 6 classes of animals viz.

Mammalia (3700 species),

Aves (8600 species),

Amphibian (2500 species),

Reptila (6300 species),

Osteichthyes (30000 species- marine or fresh water), and

Chondrichthyes or Pisces (550 species – all are marine).

### 1.12.1 Classification of Class Mammalian

| | | | |
|---|---|---|---|
| Class: | Mammalia | | |
| Sub-class: | 1. Eutheria | | |
| Order: | 1. Artiodactyles (Even toed hoofed) | | |
| Sub-order: | 1. Suina: Families - | 1. Hippopotamidae | |
| | | 2. Suidae: Genus: Sus (non ruminat) | |
| | | | S. scrofa (wild boar) |
| | | | S. domestica (domestic swine) |
| | 2. Tylopoda: Family – | Camelidae (ruminant) | |
| | | Genus: Camelus | |
| | | Species: C. bactrianus (Asian-two humped) | |
| | | C. dramedrius (Arabian-one humped) | |
| | 3. Pecora: Families | 1. Cervidae (Deer) | |
| | | 2. Giraffidae (giraffes) | |

3. **Bovidae** (hollow horned)

Sub family: Bovinae (family bovidae)

Genus: (1) Bos (Bovis)

Species: (*i*) Taurinal (Bos): *B. taurus* (European cattle); *B. indicus* (Zebu cattle)

(*ii*) Bibovinal (Bibos) *B. gaurus* (gaur); *B. frontalis* (gayal-Mithun) *B. sondaicus* (Banteng)

(*iii*) Bisontinal (Bison) *B. grannies* (yak) *B. bonasus* (European Bison) *B. bison* (American Bison)

(iv) Bubalinal or Bovini (Bubalus)

*B. bubalis* (Indian Water buffaloes)

*B. depressicornis* (Celebes of Indonesia)

*B. mindorensis* (Tamaraw or Mindora)

(v) Syncerinal or Bovini (Bubalus)

S. caffer caffer

(African cape and Congo buffaloes)

Genus: (2) Ovis (Sheep)

Species: Domestic sheep *O. aries*

*O. Canadensis*

*O. vignei* etc.

Genus: (3) Capra (goat)

Species: *C. hircus* (bezoar), *C. falconeri*, *C. caucasia*, *C. pyrenaica* and *C. ibex*

Order 2: Perissodactyles (Even toed)

Families: I. Rhinocerotidae

2. Equidae (non ruminant)

Genus: Equus

Species: *E. caballus* (Horse)

*E. asinus* (Ass)

*E. kiang* (half asses)

*E. zebra* (Zebra)

Order 3: Rodenta Rabbits; Mice

Order 4: Carnivora

Suborder: Fissipedes

Families: Felidae - Cat

Ursidae - Bears

Camidae - Dogs, Wolves, Jackals, Foxes

Order 5: Primates

Suborder I: Anthropoidea: Monkeys; Man

Sub class 2: Metatheria

Order 6: Marsupialia - Kangaroo

There are 5 orders more in class mammalian.

***Ruminant and non-ruminant mammals:*** The domestic animals of the class mammalian are either ruminants (four stomach) or non-ruminants (single stomach or mono gastric).

(*i*) **Ruminants:** Cattle, buffalo, bison, neel gai, sheep, goat, camel, Llama, alpaca, giraffe, and deer,

(*ii*) **Non ruminants**: Horse, ass, elephant, rhinoceros, hippopotamus, pig, dog, monkey.

### 1.12.2 Features of mammals

All the farm animals except poultry are kept under the class mammalian. The mammals have the following features –

(*i*) Body is covered with hair

(*ii*) Presence of mammary glands, sweat and subaceous glands

(*iii*) Presence of external ears in pair

(*iv*) Digits usually evading in claws, nails and hoofs

(*v*) Foetus is nourished by the mother through placenta

(*vi*) Heart is four chambered

(*vii*) Presence of teeth in jaw bones

(*viii*) Animals are warm blooded

(*ix*) Higher animals are viviparous

### 1.12.3 Species and breeds

The morphological and physiological characters formed the basis of dividing the animal kingdom into species and breeds within a species. The first division is into species. The most important feature of division is the reproductive discontinuity which means that two species do not interbred or the progeny of two species do not produce fertile progeny when they are further mated together. For example, horses and asses are two species and on mating produce viable but sterile progeny, except few rare cases. The interspecies crosses have variable fertility e.g., the female progeny produced by mating yak and cattle are fertile but their male ones are sterile. However, fertile male progeny of yak and cattle can be produced by two back crossing to cattle. Therefore, the reproductive discontinuity as a dividing criterion between species in some cases is liquidated.

A species is further divided into breeds or sub species. Different breeds have both qualitative and quantitative characters which differ for one breed to the other. These qualitative characters are morphological such as colour and type of hairs, shape and size of horns, presence or absence of some body parts like zebu breeds of cattle are humped whereas European breeds are hump less for which they are regarded as subspecies of the same genus, Bos. The quantitative characters can be considered as differences in type and size of the body, milk production and fat per cent in milk etc. However, there is no clear dividing line between breeds for quantitative characters for the reason that these traits show continuous variation under the control of polygenes.

Each species of farm animals contains a number of breeds. A group of animals of the same species having same origin (related by descent) and having common characteristics like general appearance, body colour, feature, size, configuration etc. is said to be a breed. The mating of animals within a breed maintains the purity of breed. A breed is a genetic entity developed over a long period of time as a result of planned mating and selection. Therefore a breed has certain well defined physical conformation different from other breeds in the vicinity and has distinct local names.

The breeds are not included in the Linnaeus Taxonomic classification arid hence the binomial nomenclature of different breeds of a species is same.

There is further division of a breed into strains based on isolation from each other due to geographical conditions or due to different aims of breeding the animals. This is called as ***strain breeding***. The individuals of a breed or strain are more closely related to each other than others due to interbreeding. This group is called a line. A line is called an inbred line when the inbreeding coefficient is reached at least 0.375 as a result of two generations of full sib mating. The full sib or half sib group of animals in cattle is the sire dam or sire family and all the individuals of a family are equally closely related to each other.

## 1.13 BREED FORMATION

The different breeds and strains of domestic animals and birds have developed over thousands of years during the process of evolution in wild and domesticated stages as well as through intensive systems of production to meet the marketing demands. They must have gone through stress conditions for developing different genes and gene combinations in them regarding their adaptability to different geo-climatic conditions, resistance to diseases and parasites, heat tolerance and utilization of locally available roughages. The natural selection and genetic drift caused the breed differences. Secondly, man's attempt to influence the evolutionary process of domestic animals by selecting the animals with desirable characteristics, migrating and crossing to produce desired animals over a period of time also caused breed differences. All these caused the evolution of new types of animals with distinct appearance within a species. The natural selection increases an individual's fitness over others and this develops a stable population in a given environment. Thus the geographically isolated groups eventually form the breeds. The migration plays its role to develop new type of animals with differential survival and reproduction. Highly specialized breeds and strains have been produced and developed in developed world by recent advances in animal breeding techniques of genetic manipulation (A.I. with frozen semen and E.T.T.). These techniques have made easy, fast and intensive to introduce new genetic resources all over the globe.

The local agro-climatic conditions and socio-cultural preferences have played significant role in evolving the Indian breeds. Indian people prefer milk with high fat content and this has caused the buffalo breeding for milk. The type of cattle breeds evolved in the country had been the utility based viz., draught, milk and dual purpose breeds have been evolved. The major factors in development of breeds of different livestock species have been the farmer's preference, geographic isolation, natural selection and the intended utilities.

## 1.14 THEORIES ABOUT ORIGIN OF PRESENT DAY SPECIES

There are two schools of thought about the origin of the present day domesticated animal viz. monophyletic and polyphyletic origin. The monophyletic theory states that the animals have descended from a single wild species whereas the polyphyletic theory states that they have been descended from two or more wild species. However, there is dispute in this regard for the reason of scantiness of the evidence. It is not certain to say that which contemporary wild species was the ancestor for the reason that the original wild ancestor has become either extinct or the domesticated animals have been changed too much.

The polyphyletic origin stands on the fact of tremendous diversity of domesticated animals *e.g.* there are too many contrasts between different breeds of dogs or species. On the contrary, it is said that the variations among races of domesticated animals are largely man-made as enough of variability in wild populations is hereditary. The selection having diverse goals and other breeding practices like inbreeding can produce distinctly contrasting races from a single wild population after several generations. Therefore, the observed diversity does not seem to be an explanation in favour of the polyphyletic origin of species.

❑❑❑

# 2 Animal Husbandry Development & Education in India

The progress in future is difficult even to imagine without reviewing the development of the past. The review is an essential aspect of critically evaluating the past and formulation of future plans based on past experiences. Therefore, ***the future progress lies in the past***. Therefore, it is essential to review the animal husbandry development in India.

## 2.1 MILESTONES OF INDIAN HISTORY

- **Ancient India**: The history of India begins with evidence of human activity of Homo sapiens, as long as 75,000 years ago, or with earlier hominids including Homo erectus from about 500,000 years ago. Isolated remains of Homo erectus in Hathnora in the Narmada Valley in central India indicate that India might have been inhabited since at least the Middle Pleistocene era, somewhere between 500,000 and 200,000 years ago.

  The ancient India covered **prehistoric era** of Stone Age between 7000 - 3000 BC and Bronze Age from 3000 – 1300 BC with early Indus Valley Civilisation (Harappan period extending from 2600 – 1900 BCE) around Indus river and extended to Ghaggar-Hakra river valley, Ganges-Yamuna Doab, Gujarat, Punjab, Rajasthan, Pakistan (Sindh, Punjan and Balochistan) and Southeast Afghanistan; **Vedic period** between 1750 BCE to 500 BCE having Indo-Aryan culture with texts of Vedas and also includes the two epics, the principal texts of Hinduism i.e., Ramayana (c. 2000 BC) and Mahabharata (c. 1400 BC) for which this is known as **Epic period** which is the later Vedic period (2000–1000 BC) known as the age of Iron and painted Grey Ware); **Mauryan period** 322 BC – 232 BC, and Budhism; ***Gupta Empire*** (c. 320–550 CE); ***Second Urbanization period*** also known as Golden Age between 200 BC to 700 CE;

- **Medieval India**: This covered late Pauranic period from 500 – 1500 CE including Mugal Empire and post Mugal period of Maratha and Sikh Empires, and

- **Modern India**: It is also known as Colonial period (1500 – 1947 CE) and independent India. It is divided into East India British Company Rule (1657 – 1857), British India Empire (1858 – 1947), and post independence India (1947 onwards).

## 2.2 ANCIENT PERIOD ANIMAL HUSBANDRY

The animal husbandry was started in India during ancient period. This is well evidenced from the *domestication of some species of animals in India* like dogs, cattle, buffalo, camel, pigs and fowls between 6000 and 4500 BC. The breed improvement and animal health practices in general during ancient period had been as under –

(*a*) ***Breed Improvement:*** There had been a number of production systems both for crops and livestock in India at different places during different times. As a result some breeds of livestock particularly the cattle, buffaloes and horses have been evolved in India. These are unique resources of livestock which are being used by farmers for multiple purposes. They have their special characteristics of disease resistance,etc.

The history of the breeds has indicated the pratice to infuse the good qualities of other breeds in ancient days. The pure breeding was practised in different pockets of the breed even centuries ago and hence each breed has specific characteristics and qualities. The traditional breeding techniques were specific to the region where the pressure on land was within its carrying capacity, animal numbers were only to be increased and there was no time limit for breed improvement programmes like today scenario of time scale bound and saturated carrying capacity of the land. Indian breeds like Gir and Sahiwal are amongst the hardiest of high-yielders. The cows of Hallikar breed are poor milk producers but its bullocks are excellent for draught purpose. The traditional approach of breed improvement to meet the requirements of local people using local resources has made the indigenous breeds to be highly adaptable. The formation of breeds for milk, draught and dual purpose was based on utility of breed.

The traditional technique used for breed improvement in cattle and buffaloes had been the practice of castration of all males before maturity. This was because the bull was considered more important for breed improvement. A vast majority of the ancient pictures and clay models of bulls rather than of cows also show the importance of the bull. The purity of the breed was maintained by not allowing the stray and unknown bulls to mix with the herd. The best breed males were donated to temple trusts for their further use. The temple trusts also had Gaushalas to preserve the best germplasms.

The breed improvement was always linked to housing, nutrition and hygiene improvement. The quality animals were used to display on the occasions of Yagnas and religions festivals. The concept of today's livestock shows seems to have come from this practice. There had been the practice of livestock census through the village administration to keep account of livestock number.

The prescribed feeding schedules were followed for feeding of animals according to their productivity. Underfeeding, poor management and cruelty to cattle were regarded as penal offences. Grazing lands were regarded as common property resources for the purpose. The temple trusts had grazing lands for the community. The animal houses were constructed following the prescribed guidelines.

The modern principles of animal breeding, feeding and management are generally in conformity to the traditional practices of ancient period.

(*b*) ***Animal Health Practices:*** During the post-Vedic era medicine occupied an important position. Charaka and Shusruta Samhitas were followed from about seventh century B.C. Nakula Samhita is considered the first treatise having information on treatment of animals with herbal preparations during the Mahabharata period. The first hospital for treatment of animals anywhere in the world was established in India during Ashoka period and later in Rome, Italy. The animals were treated during the early medieval period with drugs of vegetable and animal origin, minerals and rasa sastra. The modern system of allopathy was introduced in India by the British. In the latter part of the 20th century, rejuvenation of the science began in India.

## PERIOD-WISE ANIMAL HUSBANDRY DEVELOPMENT

The ancient period is divided in different stages like Stone Age, Bronze Age, Vedic Age, Epic Period, etc as discussed below:

(*i*) **Stone Age (7000 – 3000 BC) animal husbandry:** The first animals to be domesticated probably were the dogs and goats about 8500-9000 years ago from today whereas the cattle and sheep about 6000-6500 years ago from today. However, the domestication of sheep and goats took place in the pre-agriculture period.

Palaeolithic man hunted animals for food, which had evidence in Belan Valley of Mirzapur in the provincial state of Uttar Pradesh in North India. Goats and sheep were first domesticated near Iraq and United Arab Emirates 8700 years ago, much earlier than the advent of agriculture.

The Neolithic man of new Stone Age tamed and confined the animals around 2500 BC. People who belong to Chalcolithic age reared goats and other animals. Goats were domesticated earlier for their milk and other products (meat and fiber) and therefore, appropriately called poor man's cow.

(*ii*) **Bronze Age (3000 – 1300 BC) animal husbandry during Harappan culture (2600 – 1900 BC):** Bronze Age begun around 3300 BC with early Indus valley Civilization. This civilisation flourished from 3300 to 1300 BCE (Bronze Age) and was the first major civilization in South Asia of 5000 to 6000 years old. It was the world's earliest civilization along with Mesopotamia and ancient Egypt. It was a sophisticated and technologically advanced urban culture. There are archaeological evidences (excavations) of advanced urban culture in Mohanjo-daro in Larkana district (Sindh) and Harappa in Montgomeri district (Pb) (now in Pakistan) and certain other places in northern India with agricultural and animal husbandry practices in Indus valley during Harappan period of 2600 to 1600 BC. Inhabitants of the ancient Indus river valley, the Harappans, developed new techniques in metallurgy and handicraft (carneol products, seal carving), and produced copper, bronze, lead, and tin.

The famous Indus Valley civilization was better known for highly developed culture and organized society. The people of the Indus Valley civilizations had humped and humpless cows, buffaloes, elephants, horses, goats, sheep, fowl, dogs, etc. as evidenced from seals recovered from Indus Valley. The food grains, fibres, remnants of animals (bones), and seals and clay toys of Mohenjodaro-Harappan depicting the animals (humped bull with an enormous dewlap having folds and pendulous sheath) recovered in the excavations indicated the cultivation of land using plough and bullocks. The people were though fond of mutton, beef, chicken, and meat of tortoise, but fish was their main animal food.

People of the Indus Valley civilization were quite familiar with domesticated fowl. At the same place, small pieces of earthen hen toys were recovered. Domestic fowl was also found in Harrappa. Indus Valley people kept birds for games and breeding for meat. When Aryans invaded India around 2500 BC, they appreciated cocks.

Early Aryans (2400 – 1500 BC) of pre-vedic period in India were a race of agriculturists with their dependency solely on agriculture and livestock for their livelihood. The animal husbandry in Indus Valley civilization is supported by the seal resembling goats during Harappan culture. They had knowledge of a number of wild-game and animal products, such as milk, curd, ghee, and meat. The existence of some breeds of cattle (***Kankrej*** and ***Tharparkar***) in Gujarat, Sind and Rajasthan that are still the prominent breeds of the area probably developed for more than 4000 years back during Harappan period indicates the Indus civilization.

(*iii*) **Vedic period (1750 – 500 BCE) animal husbandry:** This period had Indo-Aryan culture with texts of Vedas. Among the four Vedas, the ***Rigveda*** is an oldest and first Vedic text and holy book of Aryans of the Indian subcontinent. During the Rigveda (1500–1200 BCE), Aryan society consisted of largely tribal and pastoral groups which were distinct from the Harappan urbanization. This society became increasingly agricultural and was socially organised around the hierarchy of the four varnas, or social classes. The Kuru kingdom was the first state-level society of the Vedic period around 1000 BCE. On the downfall of Kuru kingdom, the center of Vedic culture shifted to their eastern neighbours, the Panchala kingdom. During the Late Vedic Period, the kingdom of Videha emerged as a new center of Vedic culture, situated even farther to the East (Nepal and Bihar state in India).

(*a*) ***Vedas:*** There are 4 ***Vedas*** and all of them have information about animal wealth and their keeping during Vedic Age. The 4 Vedas and the information they contain are as under -

***Rigveda*** mentions that cattle husbandry was well developed during this period. The cow (Kamdhenu) was a lovely, adored (worship), and the 'best wealth' of human as the source of their good fortune, happiness, and good health. This was because the cow milk provided special energy, strength, and intelligence while cow dung and urine nourished agriculture farming whereas bullock power used in agriculture operatioms like carrying draft, transportation, and cottage industry. The skin from dead animals was the support to leather industry and handicrafts.

***Ayurveda*** is the science of knowledge of medicine and life. It mentions that Vedic Saints had knowledge of medicines and they were the first animal healers or veterinarians. Surgical treatment of animal diseases was routinely done during Vedic period. They knew the medicinal values of the herbs and used their medical knowledge to keep the cattle free from ailments.

***Atharvaveda*** (1500 – 500 BC) was composed around 1000 BCE and hence came to existence during the early part of Epic period but during the Vedic period. This contains the literature on medicine and some information on veterinary science (traditional medicine including information about ailments of animals, their treatment, herbal medicines, the value of medicines in curing the diseases, and also describe divine healing powers).

***Yajurveda*** cites importance of growth and development of medicinal plants. The ancient Indians had such a good knowledge of herbs that even Alexander acquired some of the skills from them, particularly for treatment of snakebite.

(*b*) ***Samudramanthan:*** Amongst 18 gems recovered in Samudramanthan during Pauranic (ancient) age by Gods and Demons, included *Airavat elephant, Ucchasrava horse* and Kamdhenu cow. All these three were acquired by Gods from Demons.

***Cows*** were regarded as mother (*Gau-mata*), wealth and the backbone of the economy of Aryans in ancient India. The literature on cows (*Gau Ayurveda*) is also available. Cows were high milk-yielders and were milked thrice a day by women (*Duhitras*) who also used to feed the cows dry hay and green fodder. Cow prosperity started declining with increasing human population and socio-economic conflict. Wars were fought for acquiring cows.

Cattle were maintained by Aryans on pastures near to their dwellings and by cuttng the jungles to graze cattle there. Aryans preferred cows compared to the buffaloes. Castration of males was practiced by crushing the testicles between two stones and oxen were used for farm transport, ploughing and irrigation. There was great emphasis on protection of cows by Aryans. Although the cow slaughter was banned and deadly sin but horseflesh was eaten occasionally at the time of religious sacrifice called *Ashvamegha yagna*. Slaughter of animals was more or less a sacrificial act.

The "cow science" is a unique gift of India to the whole world. The whole cow family (*gau vansh*) was considered as essential for existence of humanity, its protection, nourishment, development, and culture. Milk, fruits, vegetables, and grains formed bulk of the food of Vedic Indians though they were meat eaters.

The animal houses of good quality and their management has been described through Mantras in Vedas (Atharvaveda) and that Vrihaspati Deva knew the animal behavior and management well.

Treatment of weak, infertile, and unproductive cows for making them productive was well described.

***Horses*** were introduced by Aryans for rapid transportation. The pack, riding, chariot, war, race, and even ploughing horses were frequently mentioned in the Vedic age (1500–1000 BC). The Aryans took advantage of the trained horses to march into the fertile land of Iran and Mesopotamia. The Aryan chariot (ratha) is depicted at Sanchi.

***Buffaloes*** are referred in the form of a furious demon, Mahishasur and docile beast, the ride of Death God Yamraj in the prehistoric ancient Indian literature. Taming and domestication of buffalo has been mentioned during the Epic era of Ramayana and Mahabharata and true domestication during the Indus Valley civilization. The Indian subcontinent is the richest habitat of riverine buffaloes (dairy) whereas East and Southeast Asian countries are dominated by draft type swamp buffaloes. In South India buffaloes were used for ploughing lands after which they wallow in the pond to reduce tiredness. Buffalo keeping was a symbol of prosperity in Southern India. It is the utility of the buffalo that has increased its prospects.

***Sheep*** were reared mostly for wool and goats for milk and their meat was consumed by men and offered to their Gods.

***Cock*** is mentioned in Atharvaveda and Yajurveda, but not in Rigveda. During 1000 BC, eating hen meat was prohibited, possibly for religious reasons. The study of Northwest Indian coins indicates that cocks were favored.

Vedic Aryans did not prefer fish like the Indus Valley people. They used to keep dogs for guarding houses and for hunting of boars.

(*c*) ***Epic period*** (2000 – 1000 BC) includes the two epics i.e., Ramayana (2000 BC) and Mahabharata (1000 BC) which are the principal texts of Hinduism, depicting the religious, social, political, and economic life of the Aryans. This was the age of Iron and painted Grey Ware. This is also known as Early Pauranic period or early classical age or Golden Age.

*Ramayana* is the oldest literature of Sanskrit. The treatment of various ailments using medicinal herbs and surgical procedures and various uses of oil as preservative and treatment are mentioned. Surgical procedures like caesarean section, hysterectomy, etc. were known to be performed by trained *vaidhyas* or physicians. Fruit juices, flower extracts, and wines made from fruits were said to have great medicinal properties. Medicinal herbs like *Arjuna* (*Terminalia arjuna*), *kutaja* (*Holarrhena antidysenterica*), *kadamba* (*Anthocephalus cadamba*), *sarja* (*Vateria indica*), neem (*Azadirachta indica*), *ashoka* (*Saraca asoca*), *asana* (*Pterocarpus marsupium*), etc. were used widely to cure ailments of men and animals. There is description of diseases like leprosy, tuberculosis, mental disorders, etc. along with their treatment.

*Mahabharata* has references of many domestic animals including cattle, sheep, goats, dogs, elephants, and horses and their uses. Cow dung was used as manure. Lord Krishna (900–1000 BC) was an expert of cow husbandry. During Krishna's era, a cow named Surabhi was famous. . A lot of information about cow husbandry is available in Shrimad Bhagwat, poetry of Suradas and paintings of that era. Gokul and Mathura were famous for excellent breeds of cows, high milk production, quality curd, butter and other products. The milk and milk products in lieu of tax were provided by the people of Brij region to their king Kansa. Lord Krishna started worship of cows on the day after *Deepawali* festival in place of Lord Indra. Since then this day is celebrated as *Gopastami* and cows are worshiped.

*(d)* ***Ethno-veterinary medicine***: The term ethno-medicine is used for human treatment based on the similar healing art of medicine throughout the world whereas the term ethno-veterinary medicine for animal treatment. In India, ethno-veterinary practices were in vogue (fashion) since time immemorial. The treatment of animal diseases using Ayurvedic medicine has been mentioned in Agni Purana, Matsya Purana, Skanda Purana, Devi Purana, Gurud, Vishnu and Linga Purans. The Indian medical literatures like Atri-Samhita, Charaka Samhita, Susruta Samhita, and Harita Samhita contain references about care of disease animals and treatment for a variety of ailments with medicinal plants.

Ancient medical veterinary literature has information dealing with all aspects of surgery like preliminary surgical methods, dressing and bandaging of wound, surgical grafting, and treatment of fractures, dislocations, and fistula, methods of suturing and plastering. However, there appears to be no mention of anaesthetic techniques.

*(e)* ***Prominent authors on veterinary science in ancient India:*** Salihotra, Palakapya (or Dhanvantri), Susruta, Rajputra, Mrgasarma, Vyasa, Garga, Nakula, Sahadeva were the first and prominent authors on veterinary science of Epic period.

*Salihotra* (2350 BC) is said to live in a country Salutar near Kandhar. He is said as the son of the Sun God, learnt the science of horses from Brahama or also said the son of one Hayaghosa. Shalihotra, the father of Indian veterinary sciences (first known veterinary doctor), the first and foremost known author and greatest physician of the world on veterinary science, was an expert in horse husbandry and medicine and composed a text Haya Ayurveda. He was teacher of Susruta.

*Palakapya* (1000 *BC* or 2000 – 3000 *BC*) was a sage and an expert dealing with elephants and composed a text Gaja or Hastya Ayurveda (Elephantology) which is older than Ramayana. Hastya Ayurveda mentions about anatomy of elephant, treatment of different kinds of diseases (medicine and surgery), training of elephant, and also classification of elephants on the basis of a number of characteristics. Some eminent scholars think that *Palakapya* and *Dhanvantri* was one and the same person belonging to remote past. Various techniques of surgical operations along with instruments have been dealt in detail in the work of Shalihotra and *Palakapya* who were both contemporaries.

*Susruta* was the son of Salihotra or his disciple or son of Visvamitra as recorded in different literature. He was the father of Hindu surgery and disciple of Dhanvantari who was the Vedic father of medicine. Susruta belonged to 600 BC. He made great improvement in the general techniques of surgery and performed many new and major operations. "Susruta Samhita" is the earliest known work dealing with surgery.

*Nakula* and *Sahadeva*, the two Pandava brothers of Mahabharata period were experts of horse and cattle husbandry, respectively.

Ancient India had the distinction of having the advanced center of toxicological and herbal research in Taxila. The Hindus were pioneers to use the snake venom in medicine. During the ancient period, it was popularly known that snake venom in a very small dose given orally was one of the most powerful stimulants. There were prepared few of the most important medicines during the period from young and healthy black cobra for the healing of sick humans and animals.

(*iv*) Second Urbanisation period (800 – 200 BC) animal husbanrsy: Between 800 and 200 BCE was the period of ***second urbanization*** in India after Indus Valley Civilisation. ***Jainism*** and ***Buddhism*** (Mahavira 549–477 BCE, proponent of Jainism, and Buddha 563-483, founder of Buddhism) in Magadh were originated and ***Upanishads*** known as Vedanta (conclusion of the Vedas) were written during 700 - 600 BCE. During 600 – 300 BCE was the *Iron Age Vedic Civilization* with the rise of 16 ***Mahajanpaths*** spread across the Indo-Gangetic Plain from modern-day Afghanistan to Bengal and Maharastra. ***Nalanda***, one of the first great universities in recorded history, was the centre of Buddhist learning and research in the world from 450 to 1190 CE. The earliest Buddhist text *"Suttanipata"* describes cattle as a giver of food, beauty, and happiness *(annada, vannada,* and *sukhada)* and therefore deserves to be protected. Buddha was a great protector of all kinds of animals and birds (including game) in ancient India as he preached lessons of non-violence to masses. The pictures of humped bulls (rarely cows) inscribes on coins by rulers from 600 BC show their importance and utility.

***Persian & Greek conquests***: *Alexander the Grea*t (Greek king) had conquered Asia Minor and defeated King *Porus* (Persian king) in the Jhelum (Pakistan) in 326 BCE but he returned back after learning about the might of Nanda Empire of Magadh. Graeco-Romans imported livestock from India after invasion by Alexander. During 310 BC, a Prince of Punjab presented a few silver coins to Alexander, which had legends of cocks along with spur. These descriptions are available in *Indika*, a book authored by Megasthenes, the ambassador of Seleucus Nikator, king of Mecedonia in the court of Chandragupta Maurya.

***Maurya Empire*** (322–185 BCE) was established in Magadha (Bihar) by Chandragupta Maurya when he overthrew the Nanda Dynasty. Maurya Empire flourished under the reign of Chandragupta's grandson, Ashoka the Great who ruled the Maurya Empire for 37 years from 268 BCE until he died in 232 BCE. The Lion Capital of Asoka at Sarnath, is the national emblem of India. During this period the high quality steel called Wootz steel was developed in south India and was later exported to China and Arabia. There had been advancements in ancient India's science, mathematics, astronomy, religion, and philosophy during Maurya Empire (322 – 185 BCE) and Gupta Empire (320 – 550 CE).

There had been various middle Kingdoms for the next 1500 years. It was a Hindu religion period and was considered the Indian ***"Golden Age"*** (200 BC – 700

CE) also known as ***classical Age***. Chandragupta's minister (guide and political advisor) Chanakya (Kautilya) had authorship of the *Arthashastra*, a treatise on economics, politics, foreign affairs, administration, military arts, war, and religion.

***Gupta Empire*** (c. 320–550 CE) had scholars like Kalidas, Aryabhata, Vishnu Sharma, and Vatsyayana who made great advancements in many academic fields (science, technology, engineering, dialectic, logic, mathematics, astronomy, religion, art, and philosophy constituting Hindu culture). A positional numeral system (Hindu-Arabic numerals) with only 9 symbols existed upto 600 – 800 CE when ***the symbol for zero was developed in India*** and was later transmitted to the west through the Arabs. Chandragupta I (c. 319–335), Samudragupta (c. 335–376), and Chandragupta II (c. 376–415) successfully resisted the northwestern kingdoms until the arrival of the Hunas, who established themselves in Afghanistan by the first half of the 5th century, with their capital at Bamiyan.

*Arthashastra* (science of economics) mentions a lot of information about different animal (elephant, horse, and cow) departments, grazing lands, rules of meat science, livestock products like skin and fur, and veterinary jurisprudence. This knowledge flourished during the great Hindu kings of the Gupta period up to 800 AD before Islamic followers invaded India.

***Cow*** according to Kautilya's Arthashastra, was a worshiped animal during Mouryan period. The killing of cow was a deadly sin. There was capital punishment for stealing or hurting a cow. Even the bull fighting with another bull was not allowed and a person found guilty of bull fighting was fined. Injury to a bull by any person was heavily fined. Veterinary services were essential services during the Mauryan period. Arthashastra mentions to provide 4 bulls for every 10 cows/ buffaloes in a breeding herd. The King used to identify and provide enough land for pastures near each village and the *Gopa* (village accountant) was supposed to keep the details of the pasturelands as feeding of animals on pasture was the main practice. The animals were let loose for stable grazing after crop harvesting and grazing in common land preferably by a hired man of village people for the purpose of proper grazing and watering, and guarding the animals by wall enclosure at night. In Arthashastra, there is separate mention about feeding of green *(yavasa)* and dry fodder *(trina)* and oil cakes to the animals. There is description of separate rations for cow, buffalo, mule, camel, etc. Milking was done either once in the morning during summer and spring or twice (morning and evening) during rainy season, autumn, and first part of winter and cattle provide draft power for agricultural operations.

***Buffalo*** also became a recognized dairy animal by this period particularly for higher butter yield. Difference in the fat content in buffalo and cow milk was well known. Milk fat was usually used as *ghee* or butter oil.

***Goat*** has been described in Arthashastra as an important milch animal like cows and buffaloes. ***Sheep*** were raised for wool. In this period, ***asses*** were used to carry loads. ***Horses*** were used to yoke different kinds of chariots like festival chariots, battle chariots, and traveling chariots. In the stables, different kinds of horses were kept separately.

***Horses*** were used in the later Vedic period (1000–600 BC), Buddhist period (600 BC), and Mauryan period (400 BC. Equine husbandry made tremendous progress in the Mauryan age (322–232 BC) and these were used for riding and for war. The royal horses were under the charge of a superintendent of horses *(Asva adhyakacha)*, who used to register the breed, age, color, and place of origin. Detailed accounts of housing and feeding of horses were mentioned in Arthashastra. Veterinary doctors and horse trainers were assigned free endowment. In the Gupta dynasty (300–550 AD), horses were given more importance than elephants in Samudragupta's army because of their speed and easy maneuvrability. He also performed *Ashvamedha yagna* to proclaim his imperial power and issued a gold coin depicting a horse. King Skandagupta (455–467 AD) was shattered by Huns, who were expert horse riders. The Kannauj Empire (606–647 AD) has also been mentioned to use saddled horses in warfare. *Horses* were regularly trained for warfare. There were horses of many breeds. Arthashastra has graded them as best, middle, and ordinary quality. Thoroughbred horses were recommended parched rice, drippings, minced meat, red rice-powder, and grasses. ***Mules*** have also been mentioned in Arthashastra in the Mauryan period.

***Elephants*** were very important animals in the Mauryan period. They were used in warfare, as they were very useful for storming fortresses; breaking upon massive doors and to move even in dense forests and marshy lands. There were about 6000 elephants with Nandas and 9000 with Chandragupta Maurya. Elephants for war and riding were housed inside the fort. Person found guilty of elephant killing was sentenced to death. Tusks of an elephant were considered precious.

**Ashoka period** (268 - 232 BCE): The present-day Veterinary Council of India adopted its insignia, the sculpture of a bull and a part of the text of the stone edict from the period of Emperor Ashoka, which projected the veterinary profession as its "best heritage". Ashoka was the grandson of Chandragupta. He turned to Buddhism after Kalinga battle and gave veterinary science a new turn in India. It is described that the first veterinary hospital existed in Ashoka's regime in the world. The 'Baniyan Hospital' of Suratis is believed to be one of them, which consisted of a large piece of land enclosed by high walls. Provision for keeping indoor patients was made inside to accommodate animals. The great king Ashoka arranged cultivation of herbal medicines for men and animals in his empire and adjoining kingdoms.

The meat science had a sound basis in ancient India. During Ashoka period, meat eating was not banned but non-violence *(ahimsa)* was a state policy. Pregnant or milking goat, sheep, pig, and piglets up to 6 months of age were banned from slaughter. Slaughterhouse was regulated by a superintendent. Butchers selling meat from sick or dead animals and spoiled meat were severely punished.

Chanakya (324-187 BC), Manu (200 BC-200 AD), Kalidas (5$^{th}$ century of Gupta period), and Sukra (12 or 14 AD) had their work or book on veterinary science.

Satyamitra (100–200 AD) engraved fowl on coins with palm leaves. During this period, India had trade with Western Asia, Arabia, and Egypt through sea and land route, which was instrumental in the dissemination of red jungle fowl throughout the world.

## 2.3 MEDIEVAL INDIA (500 – 1500 CE) ANIMAL HUSBANDRY

(*i*) ***Late Pauranic Period*** or ***Late classical age*** is the medieval period. The livestock improvement work was given some attention during 535 to 1300 AD in south India.

In 12 th and 13 th centuries, Turks and Afgans invaded parts of northern India and Muslim rule started in some parts of North India with the establishment of Delhi Sultanate in 1206 CE which declined in late 14 th century.

In 1398, ***Timur*** attacked and defeated Sultan of Tughlaq Dynasty (Mehmud Tughlak) in Delhi. He sacked, destroyed, ruined Delhi for 3 days and night, and killed 100,000 war prisoners.

A draught breed of cattle, ***Hallikar*** dominated in South India. This breed was used for agricultural operations and cart transport. ***Ongole*** breed of cattle found in Nellore and Guntoor districts was also used for draft power. A milch strain of cattle ***Amritmahal*** was developed by Mysore Emperor from Hallikar cattle as a result of selection for milk by establishing a farm of Hallikar cattle.

The ***horses*** were introduced in India by Arabs during 800 AD to 1300 AD and used for riding, sports as well as for warfare whereas the ***asses*** were used to carry loads and primarily used by washer man.

The ***Nellore breed of sheep*** was the tallest used for mutton. The lower group of people also used to raise ***poultry for meat***.

(*ii*) MUGHAL PERIOD ANIMAL HUSBANDRY (16 – 17$^{th}$ century):

***Babar***, a descendant of Timur, came from central Asia and established Mughal Empire in 1526. The Mughal Empire covered Afghanistan, Pakistan, India and Bangladesh. But, Himayun, son of Babar was defeated by Afgan warrior Sher Shah Suri in 1540. Later on, Akbar's forces defeated and killed Hindu King Hemu in the Second Battle of Panipat on 6 November 1556. The Mughal dynasty ruled most of the Indian subcontinent by 1600 but it went into a slow decline after 1707. The Mughals suffered several blows due to invasions from Marathas, Sikh and Mysore Kingdoms, and Afghans, causing the Mughals as puppet rulers by 1757 and the Mughal dynasty were finally defeated during the Indian Rebellion of 1857.

The ***horses*** being used in warfare were imported from Iraq, Iran, Turkistan, Turkey, Tibet and Central Asia. More than 10,000 horses were with King Akbar. The horse breeding was started in Agra and Mewat region as well as in Punjab. The Mules were also produced particularly in Kashmir.

The ***camels*** were used for riding, racing and carrying load and also a part of Mughal army used in warfare.

The ***cows*** were given greater attention as a source of food, draft animal power and to provide manure. The cattle breed Kankrej found in Gujarat was excellent in speed.

The ***buffaloes*** in addition to cows were also used for agricultural operations and transport. The milk production of buffaloes of Punjab was higher than cows.

The ***goats*** were used to be kept by nomadic people of especially arid and semi arid areas for the sale of milk and surplus animals for their livelihood. Gaddi nomaidic people in Himalayan region used to maintain the Gaddi goat with long hair. Small flocks of Barbari and Surti goat for milk were also maintained. King Jahangir had nterest in animal husbandry as he preferred goat meat over wild animals.

The ***sheep*** particularly of Lohi, Bikaneri and Marwari breeds were maintained in large large flocks by the people of Rajasthan, Gujarat and Punjab because of their main occupation of wool production during the Mughal empires as most of the part of the country was pastoral. The shawl and hand knitted carpet industry was encouraged by Mughal.

***Informal education:*** There was informal nature of veterinary education in India during Pre-British period in the form of treating mostly the systematic diseases using natural herbs to treat various gastro-intestinal, respiratory and cutaneous diseases. The traditional Veterinarians engaged in treating the animal health problems were known as *Salhotries* after the name of first veterinary doctor, *Salhotra.* There were no nutritional deficiency diseases and toxic problems, and the animals were of little bigger size with more milk production than the present day animals. In those days, the animals were used to graze on communal lands of a village, under the control of a herdsman hired by village people with supplementation of animals by home grown feeds.

## 2.4 MODERN INDIA (1500 – 1947) ANIMAL HUSBANDRY

This is also known as ***colonial period.*** The Mughal Emperor, ***Jahangir,*** allowed British East India Company to trade in India in 1617. After three wars from 1746 to 1756 among French, Britain and Indian rulers and battle of Plassey in 1757, British acquired India and appointed Robert Clive as its first Governor of Bengal in 1757. East India Company controlled most of Indian subcontinent by 1850 with a policy of Divide and Rule. There was Indian Rebellion in 1857, due to dissatisfection with company rule, after which the British Provinces of India were directly administered by the British Crown till India became independent in 1947.

This period of modern India can thus be divided into Animal Husbandary during East India Company Rule from 1657 to 1857, British Indian Empire from 1858 – 1947, and Post Independence India after 1947.

(*i*) **Animal husbandry during East India Company (1657 – 1857)**

The livestock breeding was started during 18 $^{th}$ century by the work of Robert Bakewell (1725 – 1795) who laid the foundation for the Shire horses, Longhorn cattle, and Leicester sheep and appreciated the inbreeding as most effective tool to produce refinement and fixing type. The herd book containing the records of pedigrees of winning race horses was started in

1791. Later on, the herd books of cattle (Shorthorn in 1822, Hereford in 1846, Aberdeen Angus in 1862), and swine in 1876 were published. The first *cow testing association* in the world was established in 1872 in Denmark.

In defense sector of East India Company, the horses were given major importance. The overall breeding and feeding work of cavalry horses was taken care by the Army Officers. This was because the veterinarians were not available as the veterinary science was not developed in England. The Veterinary College, London was started in 1791.

*Veterinary Medicine* begun dates back to the last half of 18 th century with the spread of equine diseases and cattle plague in the early period of East India Company. Lt. W. Frazer established a stud farm of the army in 1774 -76 near Pusa (Bihar). Later on, this place was not found suitable to breed the horses due to development of deficient bones in horses. The Bengal stud was established to breed cavalry horses when the British could not get better horses. They also rejected the indigenous knowledge for treating the horses and started allopathic treatment.

The first training course in Veterinary Science was organized in Calcutta in 1821 by Veterinary Surgeon (Dr. Hodgson, J. T.) of the Governor-General's Bodyguard. The objective of the course was the recruitment of sub-assistant Veterinary Surgeons in British Indian Cavalry.

The Army Veterinary Department in India was established in 1827 to meet the shortage of qualified veterinarian.

(*ii*) **Animal Husbandry during British Indian Empire (1858 – 1947)**

The different animal husbandry programmes started during British Indian Empire are given here as under -

(*a*) **Premier *Veterinary Education:*** First Army Veterinary School was established *in 1862* at College of Science, *Pune* for veterinary education which offered one year course in Urdu for training for remount depots and military farms.

The first Veterinary Journal (Quarterly) in the world was published in 1882 from Army Veterinary School, Poona (India).

First civil veterinary school was established in 1877 at *Babugarh (Hapur), U.P.* to impart training to agriculturists in Urdu medium of instruction. This was the outcome of the recommendation made in 1871 to establish a civil veterinary school by the Indian Cattle Plague Commission which was constituted in 1869 under the chairmanship of Col. Hallen J.H.B. on the outbreak of ***cattle plague (rinderpest)***.

Subsequently, the veterinary schools were established in *Ajmer (1881), Lahore (1882) and Shimla (1888)* to control plague.

Punjab Veterinary School, Lahore was started in 1882 as the first non-military veterinary school by combining veterinary school of Ajmer and given the status of Punjab Veterinary College, Lahore. This College was started in Urdu medium for providing formal veterinary education and based on 3 years course awarded Licentiate in Veterinary Practice (LVP) in 1882.

The first Indian Veterinary College in English medium for three years training was established at Parel, Bombay in 1886 with its first Principal Prof. Steel, J.H. when the local Society for Prevention of Cruelty to Animals (SPCA) offered its building. The veterinarians after training were awarded the diploma in Veterinary Science (MRCVS) and they were allowed to do veterinary practices.

The ***disease investigation work*** on outbreak of ***Surra in horses and camel*** was started in India at College of Science, Pune as Imperial Bacteriological Laboratory on December 9, 1989. However, due to unsuitable climate of Poona for vaccine and sera production, it was transferred to Mukteswar (Kumaun hiils of U.P.) after establishing a bacteriological laboratory in 1893, to start manufacturing anti-sera against Rinderpest. This lab at Mukteswar got fire accident in 1899 and hence a new laboratory was established at Izatnagar, Bareilly (U.P.) in 1913, and the former IBL was re-designated as Imperial Veterinary Institute, and late on as I.V.R.I., Izatnagar (Bareilly).

The hand cream Separator invented by de Laval was introduced in India in 1889 by Dairy Supply Company, London.

The ***veterinary courses*** started by other colleges were the Graduate Courses like Graduate of Bengal Veterinary College (GBVC) *Calcutta* in 1893-94, Graduate of *Madras* Veterinary College (GMVC) in 1903, Graduate of Bihar Veterinary College (GBVC) *Patna* in 1927-1930. These were the premier institutes of veterinary education before independence of the country and subsequently two more Veterinary College at Hyderabad (1946) and Mathura (1947) were started. There were 9 Veterinary colleges in India before independence.

The teaching of veterinary science was started in Agriculture College, Saidapet (Madras) and at College of Science, Pune. The students after High School Certificate were eligible to do 3 years training leading to GPVC/GBVC. Later on, a three years diploma, Licenciate in Veterinary Practice (LVP) was started.

Civil Veterinary Department came into existence in 1902.

After independence and partition of India, Veterinary College at Hissar was started in 1948 as bifurcation of Veterinary College Lahore and also two Veterinary colleges at Jabalpore and Assam.

Madras Veterinary College was the first for its affiliation to a University in 1935 with the start of Master's programme in 1955, though Master's and Doctoral programmes through research work were started at IVRI in 1936 with the awarding of degrees by various universities in India and U.K.

***(b) Crossbreeding Programmes:*** The crossbreeding of indigenous species with exotic breeds was started in the end of 19 $^{th}$ century.

***Crossbreeding Zebu cattle:*** The crossbreeding of Indian cattle with introduction of exotic inheritance from European breeds of cattle was started at the end of 19$^{th}$

century and ***Taylor*** breed of cattle was evolved as early as in 1875 at Patna. The crossbreeding of zebu cattle with bulls of exotic breeds was started at M.D.F. in 1889; at N.D.R.I. Bangalore in 1910; at G.L.F., Hosur in 1919; and at Allahaba Agricultural Institute, Allahabad in 1923 (see details in chapter 18).

***Crossbreeding sheep:*** In 19th century, sheep development programme was started by the East India Company by importing exotic breeds (Cape Merino) of sheep for crossbreeding of indigenous sheep breeds around Poona.

In the beginning of 20th century, the crossbreeding of sheep with Romney Marsh was undertaken in plains and hilly areas. Crossbreeding with South Down and Cape Merino was undertaken in Punjab, Bengal, Madras and Mysore. Bikaneri ewes were crossed with Merino rams at GLF, Hisar and a fine wool breed with ¾ exotic inheritance was evolved by the name *'Hisardale'*.

***Crossbreeding poultry:*** The poultry improvement was first initiated by Christian Missionaries by establishing first poultry farm at Etah (UP) in 1912 and Katpadi (TN) during 1920's by importing exotic breeds from UK, America and Australia.

Selective breeding of indigenous fowl was started in 1940 at IVRI Izatnagar (UP). The crossbreeding was also initiated at the same time using exotic breeds.

(*c*) **Royal Commission on Agriculture:** The Government of British-India in 1925 when Lord Reading was the Viceroy realized the need to give new impulse to develop agriculture in India in relation to agriculture education and co-operation through local self government in the provinces and constituted Royal Commission on Agriculture in 1926. The decision to constitute R.C.A. was undertaken in view of the facts that India resides in villages to which due attention has not been given to the rural poor and farmers and that the working committee of Indian National Congress in 1920 passed a resolution on the removal of untouchability, promotion of communal unity and use of Khadi, etc. The R.C.A. was entrusted the job to examine and report on the existing conditions of agriculture and rural economy of British India to promote welfare and prosperity of rural India.

The R. C. A. analyzed that about ¾ of the farming community has small and marginal land holdings and takes only one crop in a year with his family labour and a pair of bullocks to meet the needs of the family. Therefore, it was needed to make provision of scientific development of agriculture through research, education and training. The R.C.A. gave its recommendations in detail about the agricultural research and education related to crop production, animal husbandry, fisheries, forestry, cooperatives, and marketing of agricultural products along with agricultural finance (credit societies, land mortgage banks), consolidation of land holdings, communication, public health and village development. This also included the improvement of crop varieties and their seed production, plant protection, use of organic manures, checking of soil erosion, improvement of livestock through selection, fodder production and better health care by upgradation of veterinary education.

On the recommendation of Royal Commission of Agriculture (1928), 4 years Veterinary programme instead of 3 years Diploma was developed. Madras Veterinary College for the first time in Asia in 1936 initiated a 4 years degree programme in Veterinary Science (B.V.Sc.). Later on, the degree was renamed as Bachelor of Veterinary Science and Animal Husbandry (B.V.Sc. & A.H.).

The RCA also emphasized on strong agricultural research organization. This lead to the establishment of *Indian Council of Agricultural Research* (I.C.A.R.) to promote, guide and coordinate agricultural research throughout India by granting a non lapsing *ad hoc* fund to be used for granting the research scholarships and to support agricultural research schemes on agriculture and animal husbandry. The I.C.A.R. in 1930 started its two Research Journals viz. Indian Journal of Agricultural Sciences and Indian Journal of Veterinary Science and Animal Husbandry (renamed as Indian Journal of Animal Sciences).

(*d*) ***Herd Registration Scheme***: An international convention of herd registration held in Rome in 1935, recommended that one herd book for each of the breed in a country should be kept to avoid conflicting methods and standards of recording. Accordingly, a central scheme under Ministry of Agric, G.O.I. was initiated in India in 1941 for breed registration. However, the scheme of herd registration as a central scheme was implemented in 1949 after independence.

(iii) **Post Independence Animal Husbandry**

After independence there had been appreciable improvement in agriculture education and research including veterinary science and animal husbandry. Salient achievements in this regard are given here as under –

**SAUs and *Veterinary colleges after independence:*** In the beginning of $20^{th}$ century it was realized the importance of the Science and Technology for agricultural development. Rural Universities for the development of agriculture and rural life in India were established on the recommendation of Radhakrishnan Commission (1949).

The I.C.A.R. in 1949 emphasized on the followings :

- Preservation of general utility animals combining draft and milk quality,
- Improvement of well defined breeds by selective breeding,
- Improvement of milk production from recognized milch breeds and propagation of milch breeds in under developed area of the country.

First Joint Indo-American Team constituted by GOI in 1955 realized the need of agricultural education in India which actually laid the foundation of the development in agriculture including veterinary education and a post-graduate college of animal science was started in IVRI in 1958 for leading to the award of Master's and Doctorate degrees under the Veterinary Faculty of Agra University.

Second Joint Indo-American Team in 1960 under the chairmanship of Dr. M.S. Randhawa, the then Vice-President of ICAR, made the specific recommendation to establish Agricultural Universities with the objective of composite education in

Agriculture, Veterinary/ Animal Husbandry and Home Science including three components of teaching, research and extension. This leads to the establishment of first Agricultural University, Pantnagar (Nainital), U.P. on the US model of *Land Grant System* of education with a start of a College of Veterinary Science as one of the University College.

The Education Wing established in 1952 under ICAR developed a uniform pattern of agricultural education including Animal Science education. It was recommended in 1960 to improve the infrastructure of veterinary colleges so as to meet manpower requirement.

Second National Educational Commission (1964-66), headed by Dr. D.S. Kothari, Chairman of UGC stressed on the recommendation to establish at leat one Agricultural University in each state of the country. This leads to the establishment of State Agricultural Universities (SAU's) for agriculture and allied sciences including Veterinary College as one of the constituent college of SAU's.

The ICAR extended its role on the agricultural education in 1966 like UGC for general universities and consequently a division of Agriculture Education was established in ICAR under DDG (Education).

The ICAR constituted first and second Dean's committee (1966, and 1981) to examine various aspects of degree programmes of B.V.Sc. & A.H. And then a third Dean's Committee in 1995 to examine curriculum of various degree programmes in agriculture, veterinary and A.H. and Dairy Technology. The veterinary education became the integral component of Land Grant system of education through ICAR-SAU network.

***Common Academic Regulations (2009):*** The ICAR has approved the Common Academic Regulations (2009), course curriculum and Syllabi for PG (Master's and Doctoral) Education in Agriculture and Allied Sciences to ensure uniform system of education, duration of PG degree programme, residential requirements, eligible criteria and mode of admission, credit load requirement (distribution between coarse and research work), system of examination, etc. under semester system of education.

***Veterinary Colleges:*** There are now 54 Veterinary Colleges in India. Of these 24 are the constituent colleges of 18 SAU's, 26 are of the constituents of Veterinary Universities, 2 are affiliated to traditional universities of Pondicherry (Pondicherry University) and Agartala (Tripura University), and 1 under Central Agricultural University, Aizwal with its jurisdiction extending to all the N.E. hilly States of Sikkim, Arunachal Pradesh, Tripura, Meghalaya, Mizoram, and Manipur. RAGACOVS is autonomous veterinary college in India under the administrative control of Board of Governors and is registered under the Societies Registration Act. Some more veterinary colleges under private sector are affiliated to veterinary universities recognized by V.C.I. among which Rajasthan is the first state to start private veterinary colleges in India.

Now, there is at least one Veterinary College in each state. These colleges are awarding 4 years degree in Veterinary Science (B.V.Sc &A.H.). Most of these colleges are also awarding 2 years M.V.Sc. degrees, and 3-5 years Ph.D. degrees in various disciplines of Veterinary Sciences, and Animal Sciences.

***Veterinary Council of India:*** Indian Veterinary Council Bill – 1981 was passed by GOI on the recommendation of the National Commission of Agriculture and the same became Act in 1984 which leads to the establishment of Veterinary Council of India (V.C.I.). The V.C.I. has prepared under the provisions of Indian Veterinary Council Act 1984, Minimum Standard of Veterinary Education (MSVE) Regulations in 1993 and implemented in various colleges under ICAR-SAU system which are mandatory on the part of the State Govt/ Universities.

**Veterinary *Course curriculum:*** The curriculum for B.V.Sc. & A.H. was restructured and revised during 1994-95. Under these regulations (VCI-MSVE Regulations 1993 norms), 17 departments need to be created in veterinary college. These regulations were revised again in 2008.

| | Discipline Groups | Disciplines |
|---|---|---|
| 1. | Basic Veterinary Science: | Vet. Anatomy & Histology; Veterinary Physiology Vet. Bichemistry; Vet. & A.H. Extension. |
| 2. | Vet. Paraclinical Science: | Vet. Microbio.; Vet. Parasitology; Vet. Pathology; Vet. Pharmacology & Toxico.; Vet. Public Health. |
| 3. | Vet. Clinical Sciences: | Vet. Clinical Med.; Vet. Epidermiology & Preventive Medicine; Animal Reproduction; Gynae & Obst.; Vet. Surgery & Radiology; Ethics & Jurisprudence. |
| 4. | Livestock Production: | Animal Breeding, Nutrition, LPM, LPT, Poultry Science. |

***Veterinary Universities:*** There are now 12 *Veterinary Universities* also in India. The first one was the veterinary and animal science university, Chennai (T.N., 1989), followed by at Kolkota (W.B., 1995), Nagpur (M.B., 1999), and Mathura (U.P., 2000). During this century 8 new Veterinary Universities have been established at Tirupati (A.P., 2004), Bidar (Karnataka, 2005), Ludhiana (Punjab, 2005), Jabalpore (M.P., 2009), Bikaner (Rajasthan, 2010), Mannuty (Kerala, 2010), Hissar (Haryana, 2010), and Raipur (Chhattisgarh, 2012).

## 2.5 NATIONAL FIVE YEARS PLANS AND ANIMAL HUSBANDRY - AT A GLANCE

The development of animal husbandry including veterinary science and dairying was given proper attention under *National Five Years Plans.* The plan number, plan period, budget allocation and different development programmes started are given below without their details but only at a glance-

### First Five Year Plan (1951 – 56)

- Rs. 77.8 millions were allotted towards dairy development,
- Central Council of Gosamvardhan set up a central coordinating agency on cattle development (1952),

- The ***Key Village Scheme*** (KVS) was launched in 1952. The other schemes were Gosadan Scheme and Goshala development scheme at the same time,
- ***NDRI*** was shifted to Karnal from Bangalore in 1955.

## Second Five Year Plan (1956 – 61)

- Rs. 120.5 millions were allotted for livestock sector,
- Development of dual purpose breeds was undertaken,
- ***Crossbreeding*** programme in hilly and high rainfall areas using exotic bulls of Jersey breed at 12 locations of 10 states was started in 1955. A Jersey cattle breeding farm was set up in H.P. to meet the requirement of Jersey bulls for crossbreeding,
- Formation of **breed societies** for registration of cattle, milk recording and supply of breeding stock of Gir, Hariana, Ongole, and Murrah was undertaken,
- ***Delhi Milk Scheme*** was started in 1959,
- ***Establishment of SAUs*** on the recommendations of the Joint Indo-US consultative committee on agriculture on the pattern of Land Grant Universities in USA, involving faculties of Agriculture, Livestock, Home Science, and Agricultural Engineering. The separate Directorate of Research and Directorate of Extension were also established.

## Third Five Year Plan (1961 – 66)

- Rs. 336 millions were allocated for dairying,
- Scheme of ***progeny testing*** *(P.T.)* of bulls of Hariana cattle and Murrah at Hissar was started. Coordinated P.T. Programme on cattle was started at 9 State Govt. Farms viz. Chakganjaria (UP) for Sahiwal, Bharatpur (Raj.) for Hariana cattle, Hesserghata (Karnataka) for Red Dane cattle, Gauriakarma (Bihar) for Jersey cattle, Durg (MP) for Murrah buffaloes,
- ***Crossbreeding*** of Hariana cows with exotic bulls (Jersey) at Harringhata (W.B.) was started in 1963 by ICAR. This crossbreeding programme at Harringhata was merged in 1968 with UNDP/FAO/ICAR supported programme by including HF and BS bulls in the programme
- ***Intensive Cattle Development Project*** (ICDP) or Area Development Programme was started in 1965 at 15 centres of 10 states with crossbreeding policy in milk shed areas of large dairy plants,
- ***Central Frozen Semen Bank*** at Hesserghata (Karnataka) was set up in 1963 and 4 more frozen semen stations were set up at Amritsar, Gurgaon, Indore and Bangalore by the assistance of Danish Govt.,
- **Central Sheep and Wool Research Institute**, Avikanagar (Rajasthan) 1962

## Fourth Five Year Plan (1969 – 74)

- Rs. 787.5 millions were allocated to A.H. and Dairying,
- ***AICRP*** on cattle was launched (1969),
- Launching of ***Operation Flood I*** (July, 1970), P.T. Programme sponsored by Agric. Ministry, GOI,
- Decision to set up 6 large ***central cattle breeding farms*** (CCBF) for some selected breeds of cattle and buffaloes at Suratgarh (Raj.) for Tharparkar, Chiplima (Orissa) for Red Sindhi, Koraput (Orissa) for Jersey cows, Dhamrod (Gujarat) for Surti buffaloes, Alamadhi (T.N.) for Murrah buffaloes. Sixth CCB Farm at Andeshnagar (U.P.) was though started in 1976,
- Central Sheep Breeding Farm, Hisar, was established to supply rams and bucks (Beetal goats were introduced in 2003-04),
- ***National Commission on Agriculture*** was set up in 1970 by GOI under the chairmanship of Late Sh. Nathu Ram Mirdha.

## Fifth Five Year Plan (1974 –79)

- Rs. 944.5 millions were allocated for A.H. and Dairying,
- Operation Flood II (1979) was launched,
- ICDP centres were equipped with ***semen freezing facilities***, Govts of Switzerland and Australia, and World Bank gave assistance, more number of frozen semen banks were established in different states by their own resources and also under cooperative sector under Operation Flood,
- **Central Institute for Research on Goats**, Makhdoom, Farah (Mathura, U.P.) 1979.
- **Central Avian Research Institute**, Izatnagar, U.P. 1979

## Sixth Five Year Plan (1980 – 85)

- Rs. 4362.9 millions were allocated for A. H. and Dairying,
- Two cattle strains Karan Swiss and Karan Fries were developed in 1980,
- ICAR started giving technical/scientific guidance and financial support to Military Dairy Farms in 1985 ,
- A centrally sponsored project on ***Extension of Frozen Semen Technology*** was initiated by providing assistance to State Govts. for infrastructure,
- Centrally assisted programme of ***Progeny Testing of crossbred bulls*** was initiated in Assam, Orissa, MS, AP and Kerala,
- Centrally sponsored scheme for ***development of indigenous breeds of cattle and buffaloes*** was initiated in 1984-85 to provide funds to state Govt farms to increase herd size for production of superior males,
- Centrally sponsored scheme for giving 50 % ***assistance to Gaushalas*** for production of high quality heifers. The scheme was discontinued in seventh plan,

- **National Bureau of Animal Genetic Resources**, Karnal, Haryana. 1984
- **National Research centre on Camel**, Bikaner (Raj). 1984.
- **National Research centre on Equines**, Hisar, Haryana. 1985
- **Central Institute for Research on Buffaloes**, Hissar, Haryana. 1985

## Seventh Five Year Plan (1985 – 90)

- Rs. 5887.9 millions were allocated for A. H. and Dairying,
- ***National level ETT project*** of Deptt. of Biotechnology was started in 1988 at NDDB, IVRI, NDRI and NII,
- **Project Directorate on Cattle**, Meerut (U.P.). 1987
- **Project Directorate on Poultry**, Hyderabad (A.P.) 1988
- **National Research centre on Mithun**, Jharnapani, Nagaland. 1988
- **National Research centre on Yak**, Dirang, Arunachal Pradesh. 1989

## Eighth Five Year Plan (1992 – 97)

- Rs. 13075.1 millions were allocated for A. H. and Dairying,
- Deptt. of A.H. and Dairying came in to existence (Feb., 1991),
- ***Centre of Advance Studies*** in different disciplines at different SAUs and ICAR Institutes were started by ICAR,
- **Integrated Dairy Development Project (IDDP)** in non-operation Flood, hilly and backward areas was launched ,
- ***National Bull Production Programme*** was started in 1991 by merging the development programme of indigenous breeds of cattle and buffaloes initiated in sixth plan,
- **National Institute of Animal Nutrition and Physiology**, Bangalore, 1995

## Ninth Five Year Plan (1997 – 2002)

- Rs. 1965 crores were allocated for A. H. and Dairying,
- India became the number one milk producing country by producing 74.7 MMT during 1999,
- Conservation of registered breeds
- National Project for Cattle & Buffalo Breeding ***(NPCBB)*** initiated from Oct. 2000,
- ***Central Feed Development Organization*** was set up with 7 centres,
- **National Research centre on Meat,** Hyderabad (A.P.) 1999,
- **Project Directorate on Animal Disease Monitoring and Surveillance,** Bangalore. 2000,
- **Project Directorate on FMD, Mukteswar,** Nainital (U.K.). 2000
- **National Research Centre on Pig**, Rani, Guwahati, Assam. 2002

### Tenth Five Year Plan (2003 – 08)

- Rs. 2500 crores were allocated for A.H. and Dairying
- Scheme for "Strengthening infrastructure for ***quality and clean milk production***"
- "Assistance to Cooperatives" has approved 24 rehabilitation proposals of milk unions of 11 states,
- National project for improvement of poultry and small animals.
- Centrally sponsored fodder development scheme – for grassland development, fodder seed production, central fodder development organization to establish 7 regional stations and central fodder seed farm at Hessarghata,
- Conservation of threatened livestock breeds,
- Livestock insurance
- ***Veterinary Type Culture Collection Centre (ICAR),*** NRCE Campus, Hissar (Haryana) 2004

### Eleventh Five Year Plan (2009 –13)

- Rs. 8174 crores were allocated for A. H. and Dairying,
- In 2012, the first phase of the ***National Dairy Plan*** began.

## 2.6 ANIMAL SCIENCE RESEARCH INSTITUTES (ICAR)

The I.C.A.R. has established 19 animal science research institutes, out of which 4 are of national level, 9 are on animal species specific, 2 are on poultry sciences, and rest 4 are of specific purpose based on veterinary sciences.

The two national institutes on veterinary science (IVRI) and dairy science (NDRI) have got the status of Deemed Universities, engaged in research, education and development work and awarding M. V. Sc. and Ph.D. degrees in different disciplines of Veterinary Sciences, Animal Sciences and Dairy Sciences. The NDRI is also awarding 4 years degree in B.Tech. (Dairy Technology).

The names, locations, years of establishment, growth and salient achievements of these institutes in developing the transferrable technologies in different discipline of animal sciences and veterinary sciences have been given in brief in this volume at appropriate places.

❑❑❑

# 3 Breed Improvement Programmes I. Bovines

The cattle and buffaloes having close zoological relationship belong to the sub family Bovini of the family Bovidae, sub-order artiodactyles, order ungulata (hoofed mammals), sub-class utheria, and class mammalian.

## 3.1 PLACE OF BOVINES IN ANIMAL KINGDOM

The cattle are much broader term than thought in general. It is rather a sub family known as bovinae or a genus (Bos or Bovis) of the family bovidae (hollow horned animals). The genus Bos (Bovis) includes the following 5 sub genera or sub groups having 9 species (Zebu cattle, humpless cattle, mithun, yak, gaur, bantang, kouprey, bison, and buffaloes):

| Genera | Species |
|---|---|
| (1) Bos or Taurine | Humpless and humped (zebu) cattle |
| (2) Bibos or Bibovine | Gayal-mithun, Gaur, Bantang, Kouprey |
| (3) Bison or bisontine | Yak, American bison, European bison |
| (4) Bubalina (Bubalus) | Asian buffaloes. |
| (5) Syncernia | African Cape and Congo buffalo |

The above classification of genus Bovis or Bos can be simplified as:

- **True cattle** which are further divided in following 6 species - Aurochs (humpless and humped cattle), Gayal or mithun, Yak, Gaur, Banteng, and Kouprey.

  All these 6 species are inter-breedable with male offspring as infertile.

- **Buffaloes** divided in two sub-genera viz. Asian buffaloes and African buffaloes.

  The Asian buffaloes have the following 3 species -

  (*i*) Bos (bubalus) Bubaline – the Arnee (Water buffaloes): Two sub-species – Riverine and Swamp buffaloes

  (*ii*) Bos (*bubalus*) mindorensis - the Tamaraw

  (*iii*) Bos (*Bubalus*) depressicornis - the Anoa or Celebes buffalo

The African buffaloes have the following 2 species -

(*i*) *African buffaloes* (*Syncerus cafer cafer*), and

(*ii*) *Congo buffaloes* (*Syncerus cafer namus*)

- **Bison** which are of the following two species –
  (*i*) *European bison* (*Bison bonasus*), and
  (*ii*) *American bison* (*Bison bison*).

### 3.1.1 Cattle

The ***domestic cattle*** *(aurochs)* are of the following two types:

(*i*) *Bos taurus* (humpless cattle of temperate zones). The cattle of United States are humpless and these are purely of European origin.

(*ii*) *Bos indicus* (humped cattle of tropical countries, zebu cattle). The zebu cattle have dropping ears, upright horn, long face and slender (long and thins) legs, have variation in body colour from white to grey and black. The jaws are large and hind limbs are sloping. They have comparatively low basal metabolic rate, better capacity for heat dissipation through cutaneous evaporation and resistance to some diseases like tick born diseases. The zebu cattle's voice is more of a grunt than a low. The humped cattle or zebu have two new features - a large hump and a long hanging dewlap.

### 3.1.2 Buffaloes

The ***buffaloes*** belong to the following two sub genera/groups *viz.* Bubalina or Bovina or Bubalus (*Asian buffaloes*) which are mostly domesticated and Syncerina or Syncerus (*African buffaloes*) which are wild. No interbreeding appears to occur among Asian buffalo (Bovina) and African buffalo (Syncerina).

1. **Subgenus bubaline or bovina** or **bubalus** (Asian buffaloes) has the following 3 species -

   (*i*) **Bos (bubalus) Bubaline** – the **Arnee:** The **arni** is the ancestor of domestic buffalo of India, Pakistan and elsewhere which are known as Asian buffaloes (Bubalus bubalis) or water buffaloes. These buffaloes have two sub species which are **river** and **swamp types**. India is home to river as well as swamp type buffaloes.

   (*ii*) ***Bos (bubalus) mindorensis*** - the **Tamaraw**: It is native to the island of mindori in the Philippine and is a dwarf buffalo (3-3 ½ feet height). It is nocturnal, wild and fierce animal. It lives in mountain forests.

   (*iii*) ***Bos (Bubalus) depressicornis*** - the **Anoa** or Celebes buffalo: It is the smallest member of the bubaline group in the world. There is no evidence of its existence elsewhere other than the island of Celebes, Indonesia. The size of anoa varies from 2 feet to 3 feet depending on its three races. It is widely hunted by the local people.

2. **Subgenus syncerus** has only one species which is African buffaloes, known as *Bos (syncerus) caffer* or Bos (Bubalus) caffer. It is also called as the *African* cape and Congo buffalo for its natural habitat extending from the cape to the south bank of the Congo River. The cape buffaloes are still wild and

have not been domesticated. They live in bushy country and graze during the first half of night and hide in shady places during the day.

***Differences Between Riverine and Swamp Buffaloes***: The two sub-species (riverine and swamp) of Asian buffaloes differ in chromosome numbers, morphology, habitat and their purpose they serve.

(*a*) The riverine buffalo has diploid number of 50 chromosomes while swamp buffalo has only 48 chromosomes. The amount of chromosomal material is similar in two types because swamp buffalo lacks one of the small chromosome pair but had longer first chromosome pair probably arisen due to translocation between chromosome 4 and 9 of riverine buffaloes. However, the two subspecies are inter-fertile and produce progeny with 49 chromosomes. The crosses between two types are fertile.

Of river buffaloes, 5 pairs of autosomes are metacentric/submetacentric while rest 19 pairs of autosomes and both sex chromosomes are acrocentric. Regarding the chromosome profile of swamp buffaloes 5 pairs of autosomes are metacentric/ submetacentric while 18 pairs of autosomes and both sex chromosomes are acrocentric.

Among the lesser known buffalo breeds, Parlakhemundi, Manipuri and Mizorami buffaloes have 48 chromosomes. Recently cytogenetic profiling and mitochondrial DNA analysis revealed Manipuri buffaloes as pure swamp type. The Manipuri buffaloes in the hilly region are mainly used for draught and meat purposes. Paralakhemundi buffaloes are medium sized animals and are reported to be swamp type as they were found to possess 48 chromosomes.

Several breeds of buffaloes like Toda, Chilika, and Assamese etc were earlier presumed to be swamp type based on their morphological appearance, but cytogenetic studies on these buffaloes proved them to possess 50 chromosomes typical to that of riverine type.

(*b*) The morphology of two types of buffaloes (river and swamp) differs considerably. Swamp buffaloes are small in size and have body weight of 300 to 450 kg while the riverine buffaloes are generally larger in size (400 to 800 kg).

(*c*) The river buffaloes prefer to wallow in clean water and rivers, and hence found through out India where clean water of rivers, irrigation canals and ponds is available to wallow. Riverine buffaloes prefer to enter in clean water. It is very much docile. The river buffaloes have been developed by selection in India and Pakistan into many dairy breeds. The Swamp buffaloes prefer marshland to wallow in mud. It is found in Assam in India.

(*d*) **The river buffaloes** are primarily used for milk production whereas the swamp buffaloes are primarily used for draught and meat production.

## 3.2 CHROMOSOME PROFILES OF BOVINES

***Cattle*** (both species, *Taurus* and *Indicus*) and ***Bos bantang*** (*Balinese cattle*) have 30 pairs of chromosomes (2n = 60). All the 29 pairs of autosomes are acrocentric and the X - chromosome is submetacentric in both the species. The Y chromosome in zebu cattle is acrocentric but submetacentric in exotic cattle.

***Yak*** has 30 pairs of chromosomes (2n = 60). All the 29 pairs of autosomes are acrocentric or telocentric like cattle and goats. Both members of sex chromosome are submetacentric.

***Mithun*** has 29 pairs of chromosomes (2n = 58). This makes Mithun genetically different from cattle and yak which have 60 chromosomes. Out of 28 pair of autosomes, one pair is metacentric/sub-metacentric and rest 27 pairs are acrocentric/ telocentric while both members of sex chromosome are submetacentric.

***Asian buffaloes*** differ in chromosome numbers as the *riverine buffaloes* have diploid number of 50 chromosomes while *swamp buffaloes* have only 48 chromosomes.

***African buffaloes*** (*Syncerus cafer cafer*) have 26 pairs of chromosomes (2n = 52); 4 pairs of autosomes are submetacentric/ metacentric, while 21 pairs of autosomes and both sex chromosomes are acrocentric.

***Congo buffaloes*** (*Syncerus cafer namus*) have 27 pairs of chromosomes (2n = 54); 3 pairs of autosomes are submetacentric/ metacentric while 23 pairs of autosomes and both sex chromosomes are acrocentric.

***European bison*** (*Bison bonasus*), and *American bison* (*Bison bison*) have 30 pairs of chromosomes (2n = 60). All the 29 pairs of autosomes are acrocentric and X chromosome is submetacentric in both the species. The Y chromosome of European bison is metacentric while in case of American bison it is acrocentric.

## 3.3 CATTLE BREEDS

***Cattle breeds of India*** (Bos indicus - Zebu cattle): The zebu breeds are classified on the basis of utility as follows –

*(i)* ***Milch Breeds:*** Sahiwal, and Red Sindhi (Organized farms), Gir (Gujarat), Tharparkar, and Rathi (Rajasthan),

*(ii)* ***Dual purpose***: Hariana (Haryana, UP, Raj), Mewati (UP, Raj), Kankrej (Guj), Malvi (MP), Deoni, and Krishna-valley (AP, Karnatak, MS), Ongole (AP), Kangayam (TN),

*(iii)* ***Draft breeds***: Kherigarh, and Ponwar (UP), Kenkatha (UP, MP), Nagori (Raj), Bachur (Bihar), Nimari (MP), Dangi (Guj, MS), Gaolao (MP, MS, AP), Khillari, and Red–Kandhari (MS), Amritmahal, and Hallikar (Karnatak), Bargur, and Umblechery (TN), Punganur (AP), Vachur (Kerala), Siri (Sikkim).

North East states (Manipur, Meghalaya, Mizorum, Nagaland, Tripura, Assam in the centre, Arunachal Pradesh in the North, and Sikkim little away in the North West bordering China and Bhutan) have some non-descript cattle. The NE region shares boarder with Bangladesh, Myanmar, Bhutan, Tibet and China. The cattle of NE region include: *Assam local cattle* also known as *Lohitya* (of Brahamputra valley) and *Tho-Tho* cattle of Nagaland. These are reared under poor management by rural households.

Some non-descript cattle are also found in Southern states of India like Alamadi, Pullikulam, and Malai madu (TN), Malnad Gidda (Karnatak), Kasarcode, and Dwarf cattle (Kerala).

***Exotic Cattle Breeds*** *(Bos taurus)*:

1. ***Holstein-Friesian*** (two province of Netherland viz. North Holland and West Fries land – imported from Europe and USA by Dutch in 1621);
2. ***Brown Swiss*** (Native of Switzerland - introduced into United States in 1869);
3. ***Jersey*** (Originated on the Island of Jersey – imported in England in 1811 and US in 1850);
4. ***Ayrshire*** (developed in Ayrshire country of Scotland during 18th century by crossing many strains);
5. ***Guernsey*** (Developed in Guernsey from Brown & White cattle of Brittany and large Brindle cattle of Normandy).

## 3.4 PANCHGAVYA (COW THERAPY)

Cow is popularly known for *Panchgavya* which means five main products received from cows. The five products from cow are: ***milk, curd, ghee, urine, and dung***. It has been mentioned in the holy Vedas that *Panchgavya* or cow therapy is a holistic approach of treatment. The five products are known for remedial values by their consumption or application externally or spraying in the environment. These elements when consumed individually or collectively enhance the immune responses of the body as per their scientific evaluation.

The ***cow milk, curd,*** and ***ghee*** are known for high nutritive value and effective in a number of human ailments. Curd and buttermilk are good appetizers and keep the digestive system normal through sustainable maintenance of pro-biotic bacteria. Cow ghee has been reported to improve memory and reduce mental tension.

In Ayurveda, the ***cow urine*** is also termed as ***'Sanjivani'***. The cow urine prepared as *Kamdhenu Ark* is effective in treatment of kidney disorders and diabetes mellitus. It has also been found helpful in prevention and control of bacterial infections by increasing phagocytosis by macrophages. The cow urine has antioxidant property and protects DNA damage due to mitomycin-C induced chromosomal aberrations.

Similarly, the ***cow dung*** keeps the environment free from pollution and does not allow any radiation effect. Most of the skin diseases can be cured by its application.

Inspite of all the above, *Panchgavya* therapy is useful only when the elements of *Panchgavya* are collected from a pastured Indian native or desi cow (Zebu cattle).

## 3.5 BUFFALO BREEDS

Indian riverine buffaloes are grouped as recognized breeds and lesser known buffalo populations:

(*i*) *Well known breeds*: Murrah, Nili Ravi, Bhadawari, Mehsana, Surti, Jaffarabadi, Nagpuri, Pandharpuri, Marathwada, Toda, Banni, Chilika and Kalahandi.

(*ii*) *Unrecognised breeds*: Tarai, Godawari, Jerangi, Sambhalpuri, Manda, Parlakhemundi, South Kanara, Gojri, Sikkimese, Kuttanad, Assamese, Swamp, Manipuri and Kujang.

The buffalo breeds are also classified according to body size and geographical distribution as follows –

*Classification based on body size -*

(*i*) Large size: Murrah, Nili-Ravi, Jaffarabadi, and Banni

(*ii*) Medium Size: Mehsana, Nagpuri, Bhadawari, Sambalpuri and Tarai

(*iii*) Small size: Surti, Manda, Kalahandi, Jerangi, Assamese and Mainpuri

*Geographical Classification –*

(*i*) North-western India: Murrah (Pb, Haryana), Nili-Ravi (Pb), Mehsana, Surti, Jaffarabadi, and Banni (Guj),

(*ii*) Central India: Bhadawari (MP, UP), Tarai (UP), Nagpuri/Berari/Ellichpuri (MS, MP, AP), Padharpuri/Marathwadi (MS),

(*iii*) East India: Manda/ Ganjam, Jerangi, and Kalahandi/ Paddakiwadi, (Orissa, AP), Sambalpuri (Orissa & MP), Chilika, and Paralakhemundi (Orissa),

(*iv*) South India: Toda (TN), South Kenara, and Dharwari (Karnatak), Kuttanad (kerala), Godawari (AP),

(*v*) North East India: Buffaloes of this area are of small size (< 400kg). These are: - *Assamese buffaloes* (Swamp type, the buffaloes of Brahamputra valley having 48, 49, and 50 chromosomes. Thus these buffaloes are distinct from riverine breeds and also from swamp buffaloes);

- *Manipuri buffaloes* (swamp buffaloes with 48 numbers of chromosomes which are of two types viz. hill type (*Tamgi-eroi)* and valley type (Chingi-*eroi*);
- *Mizoram buffaloes* are similar to Manipuri buffaloes; and
- *Sikimese buffaloes.*

## 3.6 IMPORTANCE OF BUFFALO

(*i*) The buffalo is the backbone of dairy industry in India and considered as India's milking machine contributing about 52 % of total milk produced in the country despite the fact that buffalo number is less than half of the country's total cattle population.

(*ii*) Buffalo produces high quality milk rich in fat (7 %) and various minerals like calcium, phosphorus and iron but low cholesterol value compared to cow milk.

(*iii*) Buffalo in India contribute about 30 % of total meat production of the country which is equal to the total meat produced from cattle.

(*iv*) The buffalo is a unique species for its feed conversion efficiency using low grade fibrous feeds.

(*v*) Buffalo sustains under harsh climatic conditions, and has resistance to diseases.

## 3.7 CATTLE AND BUFFALO IMPROVEMENT PROGRMMES

The Royal Commission on Agriculture realized that about ¾ of the farming community have small and marginal land holdings with their poor living in India and hence gave its recommendations for agricultural research and education including animal husbandy in India which lead to the establishment of ICAR to promote, guide and coordinate agricultural research throughout India.

The following programmes for improvement of cattle and buffalo in India were started from time to time for their implementation through Department of Agriculture, I.C.A.R., Various State Govt., Military Dairy Farms, Dairy Cooperatives and NDDB, Central Goshala Development Board, Central Council of Gosamvardhan.

### 1. Military Dairy Farms

The first Military Dairy Farm (MDF) in India was started in 1889 at Allahabad (UP) to supply milk to the British troops and their families. Later on, some other MDF's were setup. In 1975, there were 24 MDF with 9835 animals comprising 7231 cows and 2604 buffaloes.

The crossbreeding of Zebu cattle with European breeds was started on large scale first time in the country at MDF in 1907 with first importation of Ayrshire stock for the reason of low productivity and breeding efficiency of Indigenous cattle. Later on, other breeds were imported.

The crossbreeding at MDF's was discontinued in 1952 for the reason of dependence on foreign countries for importing exotic bulls. The crossbred females were mated with bulls of indigenous breeds. The cows of indigenous breeds (Sahiwal, Red Sindhi, Tharparkar, Gir and Hariana) were introduced at MDF's.

But again the ban on crossbreeding at MDF's was lifted in 1958.and the crossbreeding was undertaken with Friesian bulls. The breeding policy of MDF's was:

(*i*) Purebreeding of Sahiwal and Red Sindhi cows yielding 2000 kg or more milk in a lactation and crossbreeding to low yielders of these breeds and that of Tharparkar, Gir and Hariana cows with Friesian breed.

(*ii*) Forward crossing of crossbred cows with 62.5 % or less exotic inheritance with superior breed (Friesian) and backcrossing of crossbred cows with over 62.5 % exotic inheritance with inferior breed (Sahiwal).

(*iii*) To develop a herd of purebred Friesian cows at MDF Meerut by use of Friesian bulls of high genetic merit.

The main objective was to evolve a strain of Holstein-Sahiwal cattle. At present, the Directorate of MDF has collaboration with Project Directorate on Cattle, Meerut to fulfill the objective of evolving a new strain named as *"**Frieswal**"* since 1987 which could yield 400 kg of milk per lactation. Cows with 5/8 Friesian inheritance have been designated as Frieswal.

The MDF's has also a programme of selective breeding of buffaloes. As a result of culling level of replacement stock which was set 1350 kg for the first lactation and subsequently revised to 1600 kg in 1966, 1800 kg in 1968 and 2400 kg in 1978, the lactation milk yield of buffaloes has increased from 1443 to 2080 kg. Buffalo improvement had been carried out at military dairy farms following the policy of culling and selection since beginning and still going on.

## 2. Central Herd Registration Scheme (CHRS)

An international convention on registration of cattle was held in Oct. 1936 in Rome, with its main recommendation that there should be only one herd book for a single breed in each country so as to avoid the conflicting standards and methods of performance recording. Accordingly, a central scheme under Ministry of Agric, GOI was started in India in 1941 for the registration of animals of Sahiwal and Red Sindhi breeds.

The herd book contains a list of animals qualifying the breed characteristics and production standard and thus purebred animals of high merit are enlisted in the book. This helps the owners in sale and purchase of the breeding stock.
Herd book is a list of cows/buffaloes which qualify the conditions of production and breed characteristics as laid down under "Herd Book Rules and Regulations". Through this process of certification meritorious pure bred animals are registered and selective breeding is followed for multiplication of superior germplasm.

In India the herd books were opened by 1949 for 8 more breeds of cattle and 2 breeds of buffaloes *viz.* Gir, Tharparkar, Ongole, Hariana, Kankrej and Kangayam breeds of zebu cattle; HF and Jersey breeds of exotic cattle; Murrah and Surti breeds of buffaloes. The herd registration work in the field was started during 1963 – 64 as a central scheme. The CHRS Units established at different centres were -

- In 1963 at Rohtak (Haryana) for Hariana and Murrah covering Haryana, U.P. and Delhi;
- In 1969 at Ahmedabad (Gujarat) for Gir, Kankrej and Surti for Gujarat and Maharastra;
- In 1978 at Ongole (Andhra Pradesh) for Ongole breed covering A.P. and T.N.; and
- In 1979 at Ajmer (Rajasthan) for Hariana, Tharparkar, Gir and Kankrej and Murrah covering Rajasthan.

*Milk production standards:* The animals to be registered should meet its breed characteristics like type, colour etc. in addition to producing minimum amount of milk in a lactation of 300days, depending on the breed of the animals. The standards set for milk production in 1949 were: Sahiwal (1900 kg), Red Sindhi and Tharparkar (1600 kg), Gir and Hariana (1400 kg), Kankrej and Ongole (1000 kg), Kangayam (700 kg) and Murrah (1600 kg).

However, these criteria had been subjected to change over time depending on the utility of the breed and the improvement in production levels. The standards were revised in 1963. At present the standards of milk production for different breeds are: 1900 kg for Sahiwal and Gir; 1600 kg for Sindhi, Tharparkar and Hariana; 1300 kg for Kankrej and 1200 kg for Ongole breed of cattle; 2500 kg for HF and 200 kg for Jersey cattle; 200 kg for Murrah, and 1600 kg for Surti and Jaffarabadi.

The main objectives of the Central Herd Registration Scheme are:

(*i*) To locate superior germplasm of indigenous breeds in breeding tracts.

(*ii*) To introduce milk recording for further propagation.

(*iii*) To regulate sale and purchase within the country and abroad.

(*iv*) To propagate and awaken consciousness among the breeders for scientific breeding, development and preservation of cattle, which would improve their (breeders') socio-economic conditions.

During third five year plan (1961-66) the breeding tracts of important breeds were covered to organize breeder's societies for taking up this work.

The CHRS was reviewed in 1998 and made the following recommendations:

(*i*) To register only elite animals which should be purchased by national semen grid or its state components for breed improvement. The criteria for declaring the elite animals were revised for Sahiwal, Red Sindhi and Gir (3500 kg & above), Hariana (2700 kg & above), Kankrej (3000 kg & above), Ongole (2500 kg & above), Murrah (3800 kg & above), Jaffarabadi (4000 kg & above), Mehsana (3300 kg & above), and Surti (2900 kg & above).

(*ii*) To give incentives to the farmers for maintaining elite calves and dams as : Rs. 1000 per year for the dam for 4 years, Rs. 2000 for male calf for first year.

(*iii*) To register the bulls used for natural service in the area.

The four centres mentioned above were established and a total of 92 Milk Recording Centres in U.P., Delhi, Haryana, Rajasthan, Gujarat, MS and A.P. are functioning for recording milk yield of Gir, Kankrej, Hariana and Ongole cattle, Murrah, jaffrabadi, Mehsana and Surti buffaloes.

### 3. Key Village Scheme

This scheme was launched in 1952 (first five year plan, 1951-56) for the development of cattle and buffaloes for improvement in productivity of milch animals. It was based on all India scale taking a Key Village Block in the breeding tract of bovines. The prime objective of this KVS was the multiplication of superior germ plasm from established State Livestock Farms and to use the superior germ plasm (bulls) in breeding tract of important breeds of cattle and buffaloes together with increased production of feeds and fodder and scientific management practices, to provide prophylactic measures against contagious diseases, and marketing facilities for selling of produce, and to contact farmers for advising them. Inferior males were castrated or removed, superior quality bulls were supplied in the villages and records of breeding, feeding, milk production and diseases were maintained by the staff appointed by the government. The breeding policy was improvement through selection in indigenous breeds, grading up of non-descript population with superior bulls of indigenous breeds in areas with inadequate feed and fodder resources, and the crossbreeding of non-descript population with exotic breeds in areas of better feeds and fodder resources.

A Key Village Block had one main centre comprising 4 key village units to cover 10,000 breedable cattle/buffaloes with A.I. station to supply semen to the key village units. Later on, the centres were also set up in areas of non-descript cattle and buffaloes for their grading up. A total of 146 KVB were set up during I plan period and 197 new KVB and 64 urban A.I. centres were started during II plan period, totaling 583 KVB by the end of IV plan.

Some of the KVBs were transferred to ICDP during III plan period (1961 – 66) and some with Operation Flood project of State Dairy Federation.

The existing position of rest of the KVBs is that about 400 are functioning at present, out of which maximum are in *Orissa* (238 centres, each covering 1000 cows of Hariana and Red Sindhi) followed by 48 KVBs in *MS* which were established from I to VI plan periods covering 1.93 lakh cows and 2.17 lakh buffaloes with breeding policy of crossbreeding of non-descript with exotic breeds and grading of buffaloes with Murrah as an improver breed.

There are functioning 39 KVBs in *Gujarat* which were established from 1953 in Junagarh, in 1954 in Malwa, in 1960 in Bhavnagar and Godhra Districts, to 1965 in Amreli District and covered a population of 43900 cows and 37300 buffaloes with breeding policy of pure breeding in descript breeds of cattle and buffaloes, and crossbreeding of non-descript cows.

There are 31 KVBs with 400 sub-centres in *Karnataka* from II plan period with frozen semen facilities since 1972. Hallikar and Amritmahal breeds are under selective breeding whereas the non-descript population is being under crossbreeding programme with HF and Jersey bulls. There are only 5 KVBs in Pondichery with 59 sub-centres involved in crossbreeding with exotic cattle.

In eastern part of the country, there are 30 KVBs in *Arunachal Pradesh* covering a population of 25000 cattle with no definite breeding policy whereas in *Mizoram* are 50 KV Centres covering 12000 cattle.

The KV centres in *Haryana* are functioning in Rotak, Sonepat, Mahindergarh and Ambala districts to improve Hariana and Murrah buffaloes through selective breeding and non-descript population by crossbreeding with HF and Jersey bulls whereas Murrah is used as an improver breed of buffalo.

## 4. Crossbreeding Scheme

The details of improvement programmes through crossbreeding of zebu cattle with exotic breeds of European origin before and after independence of India (1947) have been given in chapter 17.

## 5. Intensive Cattle Development Project (ICDP)

Another large scale programme, intensive cattle development programme (ICDP) was an extension of the Key Village Scheme. The ICDP was started on the lines of Intensive Agr. Distt. Programme in 1963 of III plan period (1961 – 66) by the Govt. of India at 15 centres of 10 states for increasing milk production as an area developed approach and taking care of every aspect of cattle improvement viz. breeding, feeding, health care, extension activities and marketing facilities. The ICDP was located in the breeding tract of indigenous breeds of cattle and buffaloes. Each ICDP was linked with a dairy project for processing and marketing of milk. Each project covered up to one lakh breedable cows and buffaloes. At present about 130 ICDP's exist.

The areas selected to implement the ICDPs were of good potential for cattle development and milk production with the aim to cover 70 % population and an increase of milk production by 30 % in 5 years period. Each unit of ICDP had 100 sub-centres assisted by stockman covering 1000 breedable cows and buffaloes and thus each unit was responsible to cover 100,000 breedable cattle and buffaloes.

This project had a 100 % central assistance in the beginning (1965) but it was transferred to states after 2 years of its initiation and since then the project is being run by the States through their animal husbandry departments. The IIM, Hyderabad in 1967, found this scheme to be beneficial to continue. A study on the ICDPs was also carried out by the Programme Evaluation Organization of the Planning Commission in 1970 – 71 and observed some drawbacks in providing the inputs as per plan of the project and a setback due to its transfer to states.

The present position of the project is that most of the ICDPs have the necessary facilities of semen freezing with liquid nitrogen plants. The liquid nitrogen is also purchased from fertilizer plants. The ICDPs provide breeding services through insemination facilities with frozen semen of exotic breeds and also cover other aspects of feed and fodders to supply fodder seeds and compound feeds, health, and marketing of milk. The ICDPs have become major project for cattle and buffalo improvement. In addition to state A.H. Deparment, the dairy cooperatives and voluntary organizations *e.g.* BIAF have also made their contribution in this project. The breeding programme having A.I. with frozen semen by dairy cooperatives under Operation Flood programme in some states (Gujarat, MS, Karnataka and Kerala) has been successful.

The State A.H. Department of *Gujarat* has 9 ICDPs in its 9 districts (Ahmedabad, Bhavnagar, Rajkot, Junagarh, Mehsana, Surat, Valsad, Vadodhra and Himmatnagar) having 649 sub-centres covering a breedable population of 0.45 million. The Dairy Cooperative at Anand and Palanpur is also running this project. The genetic improvement of Gir and Kankrej breeds is through selection in their breeding tract and in rest of the area the crossbreeding is used with HF/ Jersey. The selection is the breeding tool for improvement of buffalo breeds (Mehsana, Jaffarabadi and Surti) in their breeding tract and upgrading the non-descript population with these improver breeds

The ICDPs in *Maharastra* are located at 8 places (Pune, Dhulia, Nagpur, Amravati, Jalana, Miraj in Sangli, Bhandra and Chiplun in Ratnagiri) covering a breedable population of 0.56 million cows and 0.38 million buffaloes. In MS, the exotic breeds used for crossbreeding are HF and Jersey.

There are only 2 ICDPs in *Karnataka* since 1964 located at Bangalore with its 100 subcentres covering 100,000 breedable populations and at Dharwad with 50 substations covering 50,000 breedable cow populations.

The ICDPs in *Orissa* were also established in 1964 at 4 places and cover a population of 0.70 million. The local cows are being improved by crossing with HF / Jersey bulls whereas the local buffaloes are being upgraded with Murrah.

There is only one ICDP in *Nagaland* located at Dimapur since 1975 which cover about 50,000 breedable cow populations being bred with HF/Jersey.

## 6. Goshalas

The gaushalas are maintained in different parts of the country mostly in existence in Northern states based on religio-economic considerations for the last two decades. There is great scope of improving the livestock of Gaushalas if proper assistance and guidance is provided. These can be used as a source of *in situ* conservation of indigenous breeds and can also be associated in progeny testing programme. In fact, some of the goshalas are already participating in Associated Herd Progeny Testing Scheme. Moreover, Goshalas maintaining non-descript animals can be engaged in crossbreeding programme so as to enhance the milk production and genetic improvement. There are about 1020 organized Gaushalas in 21 states maintaining about 1,30,000 cattle and 1400 breeding bulls giving employment to about 11000 persons. Some of the Gaushalas have very good resources and facilities viz. a Gashala on an average, possessing about 60 ha of grazing area and 25 ha of cultivable land.

The Govt. of India in 1949 set up a Central Gaushala Development Board for development of Gaushalas as centers for cattle breeding. Later on (1952), the Govt. of India established the *Central Council of Gosamvardhana* (CCG) to act as the coordinating and advisory body on cattle development. The CCG, after conducting a survey on the resources of Gaushalas, sponsored an *ad-hoc* scheme for their development and proper functioning. The technical assistance and guidance was extended to the Gaushala managements by appointing Gaushala Development Officers by some of the states. The assistance was given to 242 Gaushalas against the target for the development of 346 Gaushalas during second plan (1956-61).

This scheme continued under Third plan to develop more Gaushalas. But the scheme was transferred to state sector from Fourth plan but allocation of funds received low priority. Due to paucity of funds, these gaushalas could make little impact on the improvement in milk production except very few Gaushalas located at Nasik, Urli Kanchan, Indore, Ahmednagar and Amritsar.

During the sixth plan, the assistance was given to gaushalas under the centrally sponsored scheme for the production of high quality indigenous and crossbred heifers. This assistance was given for growing of fodders after purchasing the land, improving irrigation facilities and purchase of agricultural equipments with 50 % cost to be borne by the state governments. However, this scheme was discontinued in 7 $^{th}$ plan for unsatisfactory progress.

## 7. Milk Producers Cooperatives

The first milk cooperative society was formed in Bengal in 1917 to supply milk to the society of Calcutta. This was followed by Gujarat and Madras. The subsidized milk distribution system was initiated by the Govt. of Bombay in 1943 but it was closed in 1947.

The milk producer's cooperative societies in Gujarat were initially formed for the purpose to get the reasonable price of their surplus milk but subsequently extended their activities for enhancing the milk production.

*1946*: **AMUL:** Milk production in India is mostly in the hands of small farmers and landless people who keep small number of milch animals. There was no cooperation among these milk producers and the marketing of milk was dominated by middlemen who used to get the major share of the benefit. Sardar Vallabh Bhai Patel, the iron man realized this and he encouraged the small dairy farmers of Kaira District of Gujarat to form the milk cooperative in 1946. The milk producer's cooperative societies formulated at village level and district level founded the **AMUL (Anand Milk Producer's Union Ltd.)** with the objective to extend the help and guidance in milk production and marketing to the rural dairy household so as the milk is reached directly to the urban market without middlemen. The headquater of AMUL was located at Anand, a distict town of Kaira district (Gujarat). The district union (AMUL) represented all the village societies.

*1956*: ***Milk Producer's Cooperatives:*** The milk marketing and dairy development system in India was founded by Milk producer's cooperatives in 1956. Under this system the dairy cooperatives were formed as village level societies and district level unions. The unions at district level were federated into State level cooperatives milk marketing federation. A milk collection centre in each village or for some villages was set up where the milk was being collected, measured and tested for fat percentage. This system has better impact on milk production as well as on crop production as a consequence of the income from sale of milk. Not only this but this system of forming the dairy societies at village and district level promoted the social leadership quality, brotherhood feelings based on equity principle and cooperation among the members of these societies.

*1964 – 65: **National Dairy Development Board*** (NDDB): The cooperative dairy system impressed the then Hon'able PM Sh. Lal Bahadur Shastri on his visit to Anand in Oct. 1964 and on his suggestion to form a registered non-profitable cheritable trust leads to the foundation of ***National Dairy Development Board*** (NDDB) in 1965 at Anand (Gujarat) as an autonomous body to extend the milk producers cooperatives in major milk sheds of the country, under the chairmanship of Dr. V. Kurian with a Board of Directors of dairy experts.

*1969 -70: **Operation Flood:*** The NDDB in 1969 formulated the programme of the development of dairy industry in India and launched a Dairy Development

Programme in 1970 in the name of *Operation Flood*. The *objective of the operation Flood* was to supply the milk to urban consumers at stable and reasonable price.

The NDDB received an aid of huge surplus of dairy products, milk powder and butter oil from European Economy Community (EEC), which could help the dairy development programme from sale of these products. This founded the ***Indian Dairy Corporation*** (IDC) authorized by the GOI as an agency to implement the project of receiving the dairy products from EEC and to sell these products because the NDDB being a charitable trust could not do this job. Dr. Kurein was also the chairman of IDC. The IDC played its role in financing and promoting the dairy development in the country, but subsequently this job was given to NDDB.

In 1968, the union had only 924 members of 13 village milk societies whereas in 1994-95 the union had 532670 members of 954 cooperatives with a collection of 229210 tones of milk. The Kaira union in 1964 set up balance cattle feed plant and started to provide the facilities of animal health care and breeding by running a semen production centre to make the A.I. facilities to the milk producers.

**Operation Flood**: The Anand pattern dairy cooperatives had been successful in organizing the milk production, procurement, processing and marketing. The Operation Flood was a large dairy development project to organize milk producers through out the country based on AMUL Model of milk producers cooperative societies. The operation Flood was implemented in three phases as:

***Operation Flood - I (1970-80):*** It was launched in July 1970 following an agreement with the World Food Programme which provided skin milk powder (124000 tones) and butter oil (40000 tones) as an aid to finance. The emphasis was to set up the Anand pattern rural milk producer's cooperatives for procuring(collection, cold storage and onward transportation), processing and marketing of milk together with providing technical inputs and services for milk production. These included fodder seeds, compound feed, A.I. and health coverage of the animals of the members of the societies. However, the major emphasis was given on milk collection and marketing so that the marketing facilities and remunerative prices paid will encourage the producer members to increase the milk production. About 13300 Dairy Cooperative Societies in 27 milk sheds with membership of 1.8 million farmers were organized.

***Operation Flood – II (1980-85):*** This was for the sixth plan period (1980-85) and launched on October 1979 with an outlay of Rs.2730 million. It was designed to build on the foundation of Operation Flood – I involving most of the states and UT with a larger area in each state with the following objectives-

To build a viable self sustaining dairy industry involving about 10 million rural milk producer's families.

To develop a National Milch Herd of about 14 million crossbred cows and upgraded buffaloes.

To develop link of rural milk sheds with major demand centers of urban population of about 150 million by erecting a National Milk Grid.

During this OF-II, the number of village cooperative societies increased to 34500 enrolling 3.6 million farmer members in 136 milk sheds.

***Operation Flood – III (1985-96)***: It was launched in 1985 with major emphasis to consolidate the achievements gained during the earlier phases by improving the productivity and efficiency of cooperative dairy sector. It also provided the animal health and breeding facilities for improving milk production. During OF-III, 72744 Dairy cooperative societies in 170 milk sheds of the country with 94 million members were organized. The numbers of Village Milk Producer's Cooperative Societies established by the end of 2008 were 117575 federated into 170 milk unions and 15 state milk federations procuring on an average 21.5 million liters of milk daily. There are 12.4 million dairy farmers as members of the milk cooperative societies.

The OF had contributed to sustained increase in milk production because of assured market of milk, remunerative price for raw milk to producers, technical inputs and services including A.I., balanced cattle feed and emergency veterinary health services to the dairy farmers.

## 8. State Cattle Breeding Farms

The government took the decision of establishing Cattle Breeding Farms of indigenous breeds for their genetic improvement through selection and to produce the genetically superior bulls in sufficient numbers, for a number of reasons viz. production of breeding bulls on large scale was not taken care in private sector due to small herd size, selection of males was not possible because the performance recording was not available on parent generation, and cost of raising the bulls up to breeding age. Moreover, the requirement of genetically superior bulls was increased with the start of KVS for cattle development in the first plan period.

The State Govt. Farms which started P.T. were:

- State Cattle Breeding Farm, Chakganjaria (Lucknow) for *Sahiwal;*
- SCBF, Hosur (TN) for *Red Sindhi;*
- SCBF, Junagarh(Gujarat) for *Gir;*
- SCBF, Bharatpur(Rajasthan) for *Hariana;*
- SCBF Durg (MP) and Banvasi (A.P.) for *Murrah;*
- Gauriakarma (Bihar) and Barapeta (Assam) for *Jersey*; and
- Hasserghata (Karnataka) for *Red Dane.*

However, these state farms were having small herd size because most of them were maintaining exotic, crossbreds and buffaloes also. There was thus no scope of selection for genetic improvement and also to produce the required number of males for breeding purpose. The position of the State Govt. Farms with respect to herd size is that there are 25 farms maintaining purebred indigenous animals of only 16 out of the recognized breeds of cattle and 13 farms are maintaining only buffaloes whereas rest of the farms are maintaining either crossbreds or crossbreds plus indigenous purebreds to produce crossbreds. Only 28 farms have breedable females above 150 and only 5 farms are maintaining the breedable population of

300 or more of a single breed in which the selection can be practiced to some extent and few breeding males can be produced. The successful P.T. programme requires maintaining 750 breedable females of a single pure breed so as at least 10 bulls can be tested in one breeding cycle with 25 recorded daughters.

As these state farms with small herd size were not found sufficient to serve the purpose of cattle development programme of fourth plan, GOI decided to set up 6 large central cattle breeding farms for genetic improvement through selection of bulls based on progeny test, and also to assist the state governments to develop facilities at state farms for P.T. programme.

## 9. Central Cattle Breeding Farms

Ministry of Agriculture, GOI took a decision to establish 6 large Central Cattle Breeding Farms for P.T. programme in the IV Plan period, for selection of bulls based on progeny test of different breeds of cattle and buffalo, with herd strength of 300 cows but these farms should have the capacity to maintain 750 animals with 1000 acres of land for fodder production. The aim was to progeny test 10 bulls at a time based on 10 daughters. The 7 Central Cattle Breeding Farms are:

- CCBF, Suratgarh for Tharparkar started in 1967;
- CCBF Chiplima (Orissa) for Red Sindhi started in 1968;
- CCBF Dhamrod (Gujarat) for Surti started in 1968; and
- CCBF Alamadhi (T.N.) for Murrah started in 1973.
- CCBF Koraput for Jersey started in 1972;
- CCBF Andeshnagar (UP) 1976;
- CCBF Hessarghata, Bangalore (Karnataka) 1962

The seventh farm at Hessarghata was established in 1962 as Jersey cattle breeding farm and later on transferred to Ministry of Agriculture with its changed name as CCBF in June 1975. The Jersey herd was transferred to Koraput (Orissa). During Feb. 1975, the HF animals (78 heifers and 20 bulls) were received from Denmark for maintaining at Hessarghata.

The primary objective of the farms was to produce at least 10 progeny-tested bulls in each farm by maintaining about 300 bredable females. This objective was never achieved and the programme to produce progeny tested bull was abandoned in 1988. The other objectives like genetic improvement of bull mothers of important cattle and buffalo breed and supply of high pedigree bulls also failed to achieve the target.

The Working Group on Animal Husbandry for the Tenth Plan has suggested that the goals of these farms should be changed and they should be used either for conservation of indigenous breeds or NDDB could take over these farms for implementing progeny testing programme using recent technologies like ETT /OPU-IVF.

Buffalo improvement programme through bull selection on the basis of progeny performance and distribution in the field has also taken up at the following state and central government farms -

- Alamadi Farm, Chennai; Livestock Farm Banawasi (A.P.); Govt. Breeding Farm Anjora;
- Govt. Livestock Farm Hisar for Murrah breed;
- Govt. Livestock Farm Dhamrod for Surti breed;
- Livestock Research Station (Sardarkrushinagar) for Mehsana, Junagadh for Jaffrabadi and Navsari (Gujarat) for Surti.

The CCBF Almadi started P.T. of Murrah buffaloes in 1977-78 but the programme was discontinued in 1989. This farm is now an associated herd of network project since 1993 for P.T. of bulls sponsored by ICAR with CIRB as nodal agency.

The P. T. of Murrah buffaloes at Banawasi (A.P.) was started in 1972-73. The breedable females were 160 mated with a batch of 5 bulls selected on the basis of dam's performance.

## 10. Progeny Testing Schemes

The P.T. programme was started during III *Plan* by Agriculture Ministry for the production and identification of superior breeding bulls. The bulls were tested on the basis of performance of their daughters rather than dam's milk yield. The tested bulls were used extensively through AI for achieving higher genetic gain. The progeny testing schemes were run in single herd in the beginning, then in associated herds and under field conditions.

*(i)* ***Single Herd P.T. Scheme***: The P.T. scheme was started in the beginning (III Plan) at GLF Hisar for Hariana and Murrah breeds.

During Fourth Five Year Plan, the scheme was started for other breeds viz.

- Sahiwal at SCBF, Chakgangaria (Lucknow) in 1971;
- Red Sindhi at SCBF, Hosur(TN) in 1972;
- Gir at SCBF, Junagarh(Gujarat) in 1972;
- Hariana at SCBF, Bharatpur(Rajasthan) in 1972;
- Murrah at SCBF Durg (MP) in 1971 and at Banvasi (A.P.) in 1972;
- Jersey at Gauriakarma (Bihar) and Barapeta (Assam) in 1972; and
- Red Dane at Hesserghata (Karnataka) in 1973.

*Small herd size:* The progeny testing scheme at none of these organized farms could yielded the desired results mainly for small herd size and hence a few sires could be tested. Therefore, under the situation of small herd size at these farms, there are two possibilities of genetic improvement:

(*a*) Associated herd P.T. programme by associating at least 750 female animals maintained at two or more organized farms or associating the breedable females from gaushalas and pinjrapoles or private agencies.

(*b*) Field P.T. programme in farmer's herds in the breeding tract of the particular breed. The village as a whole can be considered as a breeding unit in view of the similar feed resources/feeding and management practices in the village.

To overcome this problem of small herd size, it was thought testing the sires simultaneously based on the data off all herds. This can allow testing more number of sires and that too in different herds. This also increases the accuracy of selection. However, this scheme posses the problem of G-E interaction which can be taken care of.

(*ii*) ***Associated herd P.T. Scheme***: The associated herd progeny testing programme has been started for 3 breeds namely Sahiwal, Hariana and Ongole. The NDRI Karnal (Haryana) is the lead institute for running this programme for Sahiwal breed by associating 3 herds of NDRI Karnal, SCBF Durg and SCBF Chakganjaria (Lucknow). This programme is being run through frozen semen technology i.e. freezing of semen and distribution among different herds. The P.T. of 6 bulls in each set is the target and 2 best bulls to be selected for nominated mating to bring improvement.

The associated herd P.T. programme was started by Project Directorate on Cattle, Meerut, for Hariana and Ongole breeds through associated herds in their breeding tract. The associated herds for Hariana breed are located at GLF Hisar, Gaushala at Bhiwani, Jind and Kurukshetra whereas for Ongole breed the associated herds are located in Andra Pradesh at Lam (Gantur), Chintaladevi (Nellore), Ramtheertham (Parkasam) and Mahanandi (Kurnool). The breeding plan involve for each breed the mating of 75 breedable females with 4 to 5 superior proven bulls for production of future young bulls. About 8 to 10 bulls to be selected at maturity based on their expected breeding values and freeze about 3000 does of semen of each bull: About 1000 to be used for sire evaluation in associated herds and 1000 maintained for germplasm unit and PDC, each bull to be mated with 60 to 70 females in associated herds to get at least 20 daughters per bull. The herd strength of germplasm unit was 76 cows and 24 bulls for Hariana breed and 55 cows and 11 bulls for Ongole breed. The total breedable females in associated herds for Hariana and Ongole breeds were 1069 and 845 respectively.

The other important breeds of cattle are Tharparkar and Gir on which this programme will be started by the PDC Meerut. A similar programme for improvement of Rathi, Deoni and Kankrej breeds is under consideration under NATP at 3 centers viz. RAU Bikaner for Rathi, Marthwada Agri. University Parbani for Deoni and Dudhsagar Research and Development Association Mehsana for Kankrej.

***Buffaloes:*** The associated progeny testing through Network Project on Buffalo Improvement was initiated in 1993 with the objective to increase the intensity of selection of bulls from large population and increased number of progeny per bull for testing in Murrah buffaloes. 12-15 bulls are tested on about 800 bredable buffaloes located at six institutional herds participating in this project namely PAU Ludhiana, HAU Hisar, NDUAT Faizabad, CIRB Hisar, NDRI Karnal and IVRI Izatnagar.

Test mating from 13 sets of bulls (76 bulls) have been completed. Fourteenth set of 12 bulls started from January 2013 and shall continue up to July 2014. Progeny test evaluation of ninth set has been completed and total top ranking

24 bulls (25%) declared as progeny tested bulls of Murrah breed under Network Project. Till 31st March, 2013 about 74,909 frozen semen doses of progeny tested bulls are available at semen freezing laboratory at CIRB-Hisar, NDRI-Karnal and GADVASU-Ludhiana.

(*iii*) ***Field P.T. Scheme*: *Progeny Testing of Crossbred bulls:*** Centrally assisted scheme for P.T. of crossbred bulls was started in VI plan period (1989). This programme was started in A.P., Kerala, Maharastra, Orissa and Assam. Later on it was also started in T.N. but Assam has stopped the programme.

The P.T. of crossbred bulls in the ICDPs areas (fiel recording of milk production) had already been started in 1977 by Kerala Livestock Development Board. The objective was to make 1500 test inseminations from each bull to have at least 50 recorded daughters. The results have shown an improvement in lactation milk yield over the years which could be attributed due to selection and improved feeding and management. About 300 elite cows producing more than 2400 kg under field conditions are selected and used as bull mothers. This has lead to the development of a breed through *inter-se* mating, known as ***"Sunandini"***.

The PDC Meerut has also taken up the field progeny testing of ***crossbred bulls*** during VIII five year plan at 3 station viz. BAIF Urlikanchan (MS), PAU Ludhiana (Pb) and Kerala Agri. University, Mannuthy. The Holstein Friesian crossbred bulls having 50 to 75% exotic inheritance with a minimum of 4500 kg dam's milk yield are to be tested in batches for a period of 15 months and each batch will comprise 30 crossbred bulls. At least 40 daughters will be reached per bull spread over different units.

The Associated Progeny Testing on ***Buffalo*** Improvement under Network Project was reoriented from 2001-02 with the objective to strengthening the on going sire evaluation programme of associated herd progeny testing at institutional herds by involving performance recording on farmers animals using semen of bulls selected under Network Project.

Field unit of Murrah breed were initiated at GADVASU Ludhiana, NDRI karnal and CIRB Hisar. The field progeny testing was started in other buffalo breeds viz.: Jaffrabadi at JAU Junagadh, Surti at RAJVASU Vallabhnagar, Pandharpuri at MPKV Kolhapur and Godawari breed.

Mehsana District Cooperative Milk Producer Union initiated field progeny testing programme for genetic improvement of Mehsana buffalo during 1985 in association with NDDB, Anand.

## 11. Artificial Insemination Facilities

*Central Frozen Semen Production and Training Institute* at Hesserghata (Karnataka) was set up in 1963 and 4 more frozen semen stations were set up at Amritsar, Gurgaon, Indore and Bangalore by the assistance of Danish Government.

During V Plan period (1974-79), the ICDP centres were equipped with *semen freezing facilities*. Govts of Switzerland and Australia, and World Bank gave assistance. More number of frozen semen banks were established in different states by their own resources and also under cooperative sector under Operation Flood.

During VI Plan Period (1980-85), a centrally sponsored project on *Extension of Frozen Semen Technology* was initiated by providing assistance to State Governments for infrastructure.

The infrastructure facilities for A.I. including liquid nitrogen plants have been created. There are 55 semen production units out of which 42 are in Govt. sector and rest 13 are with NDDB, NGOs like BAIF, Pune and private agencies having 2472 bulls producing 39 million semen doses. These stations are maintaining about 40-50 bulls at each centre. There are about 65871 A.I. centres covering 34 million inseminations per annum.

## 12. All India Coordinated Research Project (AICRP)

The Indian Council of Agricultural Research (ICAR) had launched a coordinated research project in 1968 with the objective of evolving high yielding breeds of different species for different utility viz., cattle and buffalo for milk production, sheep for wool and mutton production, goat for milk, meat and mohair production, pig for meat production and poultry for egg and meat.

(*i*) **Cattle**: The AICRP on cattle involved the crossing of indigenous cattle breeds (Hariana, Ongole, Gir, Tharparkar, Sahiwal, Red Sindhi and Local) with superior exotic breeds (H.F., Jersey, Brown Swiss). The objective of AICRP on cattle was to know the optimum combination and level of exotic inheritance based on assessing the production and reproduction performance of the crossbreds in different agro-climatic regions. A control frozen semen bank was also established at Hessarghata (Karnataka) for collection, processing, storage and distribution of frozen semen. On the basis of the result obtained it was found that crossbreds with $^3/_4$ exotic inheritance did not do better than half bred due to disease problem, availability of required input and management problem etc. The better performance had been of those having 50 to 62.50% level of exotic inheritance than all other crossbreds.

(*ii*) **Buffalo**: The AICRP on buffalo was started in 1970-71 by ICAR for two buffalo breeds viz. Murrah and Surti at two centers for each breed. The centers for Murrah were NDRI Karnal and PAU Ludhiana whereas for Surti breed the centers were university of Udaipur at Vallabhagar (Raj) and UAS Dharwar. The objective was the progeny testing to identify the bulls of high genetic merit and to distribute the young bulls, produced from mating of elite buffaloes with proven sires, to various farms in the country. Later on during V plan the field units were attached to each of the centers for testing of sires under transfer to technology.

The breeding plan for all the 4 centers was to maintain 220 breedable females and a separate elite herd of 70 buffaloes. It was decided to evaluate 8 bulls at each centre and finally selecting only 2 bulls out of these 8 bulls for mating with elite buffaloes. But in 1980 the technical programme was changed to test 12 sires in testing herd every 2 years in the two herds of Murrah as well as Surti breed and finally selecting 2 bulls for mating in elite herds of both centers as one unit for both the breeds. Coverage of 3000-4000 breedable females in field units attached to each centre was also recommended. The semen of each bull had to be used in testing herds and 2 field units. Further, in was recommended to freeze a minimum of 2000 does of semen of each bull and to distribute 700 does from each bull to the 2 herds and other associated herd.

In 1985, the technical programme was changed:

- To keep Surti breed at Dharwad and Vallabnagar
- To keep Murrah at NDRI Karnal and to associate CIRB Hisar
- To change over to Nili Ravi buffaloes at PAU Ludhiana and associate herd of Nabha (Pb) centre of CIRB.

Under P.T. programme 68 bulls in 9 sets at CIRB Hisar, 53 bulls in 8 sets at PAU and 47 bulls in 8 sets at NDRI were evaluated till 1993. Three sets of Surti buffalo bulls at Vallabhnagar and 5 sets at Dharwad were progeny tested.

*Network project on Buffaloes*: During VIII plan period (1992-97), the efforts were made in collaboration with the various institutional and organized buffalo farms under the network project on buffaloes, w.e.f. 1993. The participating units taken for Murrah buffaloes were CIRB, Hissar (Haryana) with its substation Nabha (Pb); NDRI, Karnal; PAU, Ludhiana; HAU, Hissar; IVRI, Izatnagar and CBBF, Alamadi (TN) while those for Surti buffaloes the units were : RAU, Vallabnagar; GAU, Navsari and Anand; CCBF Dhamrod and UAS, Dharwad.

A total of 147 dams of Murrah and 56 of Surti breed from various units were selected as elite dams. All male calves born from elite mating were saved for breeding and kept at their respective units till 12 months of age. The bull rearing and semen processing units for Murrah were taken at CIRN, PAU and NDRI. In 1998, new centers added for Murrah buffaloes were GLF Hastinapur (UP), Banvasi and Mamnoor (AP).

During IX Plan (1992-2002), the network project included the improvement programme on Murrah buffaloes at institute farms of PAU, HAU, NDRI, IVRI, CIRB and CCBF Alamadi (TN) and Murrah buffaloes under field conditions around CIRB, PAU and NDRI.

The Institutional centres including field testing for different breeds were -

Jaffarabadi buffaloes at GAU Junagarh;

Nili-Ravi buffaloes at CIRB substation, Nabha;

Bhadawari buffaloes at IGFRI Jhani;

Murrah buffaloes at NDUAT Faizabad;

Surti buffaloes at LRS Vallabhnagar (Raj);

Godawari buffaloes BRS, ANGRAU;

Pandhanpuri buffaloes at MPKV Kohlapur/Rahuri; and

Swamp buffaloes at AAU Khanapara

## 13. Dairy Development Schemes

The Department of A.H. and Dairying for promotion of dairy activities in non-operation flood areas has implemented 3 schemes, besides NDDB continues its activities in operation flood areas.

*Intensive Dairy Development Project (IDDP)* in non-operation Flood, hilly and backward areas launched during 8th plan period (1993-94). This scheme covered 149 districts of 23 states and one U.T. benefited about 6.6 lakh farm families of 11000 villages.

The Department started the scheme "*Assistance to Cooperatives*" during 1999-2000, has approved 32 rehabilitation proposals of milk unions of 12 states, with 50:50 sharing basis between central Govt. and State Govt. concerned.

The Department had introduced a new Centrally Sponsored Scheme "*Strengthening infrastructure for quality and clean milk production*" during 10 th plan.

**Technology Mission on "Dairy Development**: The GOI launched Technology Mission on "Dairy Development" in 1988. This was assigned 29 research programmes at ICAR institutes, SAU's and NDDB. The emphasis was laid on optimization of resource use for cattle improvement in breeding farms, bull mother farms, semen freezing and A.I. facilities, veterinary health care facilities, etc. available with the state A.H. Deptt and under Operation Flood.

## 14. National Dairy Plan (NDP)

The National Dairy Development Board (NDDB) with Government of India and World Bank has developed a National Dairy Plan (NDP) to increase dairy productivity through access to high quality semen and creating linkage between rural milk producers and the commercial milk processing sector. The first phase of the plan to focus on 14 major milk producing states which are: Andhra Pradesh, Bihar, Gujarat, Haryana, Karnataka, Kerala, Madhya Pradesh, Maharashtra, Odissa, Punjab, Rajasthan, Tamil Nadu, Uttar Pradesh and West Bengal. These states account for over 90 percent of total milk production, 87 percent of the total buffalo population and 98 percent of total fodder production.

In addition to NDP, the Department of Animal Husbandry, Dairying and Fisheries, Ministry of Agriculture (MoA), approved the continuation of the following four schemes in financial year 2013-14 (April-March) during the 12th Five Year Plan:

- Intensive Dairy Development Program (IDDP)
- Strengthening infrastructure for quality and clean milk production
- Assistance to Cooperatives and
- Dairy Entrepreneurship Development Scheme

## 15. National Project on Cattle And Buffalo Breeding (NPCBB)

Government of India has initiated the programme in October 2000 for a period of 10 years by merging two Centrally-sponsored Plan Schemes: (a) Extension of Frozen Semen Technology and Progeny Testing Programme and (b) National Bull Production Programme. Presently, 28 states and one UT are participating in the project.

The salient features and objectives of the Project were:

(*i*) To arrange delivery of artificial insemination services at the farmers doorsteps;

(*ii*) To bring all breedable females under organized breeding through A I or natural service by high quality bulls within a period of ten years and

(*iii*) To undertake breed improvement programme for indigenous cattle and buffaloes so as to improve the genetic make-up as well as their availability.

The objective of above programmes was the overall improvement of buffaloes in all the states. The minimum standard for each breed have been laid down for selection of superior bulls and norms for germplasm production as under :

| **Breeds** | **Sires's dam lact.Yield** | | | **Daughter's average** |
|---|---|---|---|---|
| | **First** | **Best** | **Fat %** | **(I st lact.)** |
| Murrah, Mehsana, Nili-Ravi and Banni | 2400 | 3000 | 7 | 1800 |
| Jaffarabadi | 2800 | 3500 | 8 | 2000 |
| Surti | 1600 | 2000 | 7 | 1600 |
| Bhadawari, Pandarpuri | 1300 | 1800 | 7 | |

## 3.8 ICAR0RESEARCH INSTITUTE ON CATTLE AND BUFFALOES

There are three ICAR research institutes for conducting research on different aspects of animal science and dairy science. Some features of these institutes in brief have been given here as under -

### 1. ICAR National Dairy Research Institute,

Karnal, Haryana.1923

Regional stations: Bangalore (1923); Kalyani, W.B. (1964).

*Deemed University* (1989)

*Historical Background and Growth:* The institute has its origin as Imperial Institute of A. H. & Dairying, established at Bangalore in 1923 and it was renamed as Imperial Dairy Institute in the year 1936. The Headquarter of the institute was shifted from Bangalore to Karnal (Haryana) in 1955 and was renamed as National Dairy Research Institute. The institute has now 10 divisions. It has been granted the status of Deemed University since 1989 by U.G.C., G.O.I. The institute is awarding M.Sc. and Ph.D. Degrees in 12 disciplines. It is a premier institution of dairying in Asia.

The institute undertakes the research, teaching and extension activities for dairy development in the country, besides conducting basic and applied research to enhance animal productivity, and to develop new products and practices for dairy farmers and consumers.

The institute has developed many technologies like ***sire evaluation*** based on Dairy Search Index and based on mate's performance of sire, ***genetic analysis of threshold data*** without transformation, ***formulating the mating plan*** for proper/ better use and evaluation of more numbers of sires under test without wastage of data; ***adulterant detection in milk*** (Hansa serum test to detect adulteration of buffalo milk in cow milk, rapid method for detergent detection in milk); made use of ETT, *in-vitro* fertilization and production of buffalo and goat through ***improvement of ETT by embryo splitting***, production of world's first ***cloned buffalo calf*** (***GRIMA***) in 2009 through "hand guided cloning technique" which is an advanced modification of the cloning technique used originally to produce "DOLLY" sheep. The ***endocrine techniques*** for fertility enhancement, even in anoestrus/repeat breeding buffaloes; ***nutritional technologies*** (Urea molasses mineral block lick, urea molasses treatment of crop residues, manufacturing process of complete feed block containing roughages and concentrates, Degcure mixture, anionic mineral mixture to prevent milk fever, area specific mineral mixture) have been developed.

The institute has also developed ***dairy processing equipments*** (Continuous ghee making machine, paneer making machine, continuous khoa making machine); made various ***value added dairy products*** (Low cholesterol ghee, rasgolla mix powder, instant kulfi mix powder, ready to reconstitute kheer mix and rasmalai mix, Arjun herbal ghee, ready to reconstitute mix for fat oven baked gulabjamun, kund).

## 2. ICAR Central Institute for Research on Cattle,

Meerut (U.P.). 1987

The ICAR upgraded the *AICRP on cattle* and established the *"Project Directorate on Cattle"* at Meerut (UP), in 7 th Five Year Plan on 3rd November, 1987 to monitor, coordinate and support research and development activities for cattle improvement. The P.D.C. after about 27 years of its successful functioning has got the status of Central Institute for Research on Cattle (CIRC) by the I.C.A.R. since April, 2014.

The Project Directorate utilized the large population of HF x Sahiwal produced at MDF of the country with the purpose to develop a new breed of cattle *"Frieswal"* with 62.5 % HF inheritance with the aim of 4000 kg milk production of 4 % fat in a lactation period of 300 days.

In the year 1989, the Indigenous Breeds Project was initiated in collaboration with SAUs, State Government and NGO to conserve and improve the indigenous breeds for milk and draft through progeny testing in organized herds and farmer's animals (Field P.T.) as the indigenous breeds are known adaptability, disease resistance capability and to thrive on local feed resources. The genetic improvement work on Hariana, Ongole, Sahiwal, Gir and Kankrej was undertaken for progeny testing of bulls.

During the year 1991, the Field Progeny Testing (FPT) was started for genetic improvement in field cattle (HF crossbred cattle in farmer's herd) in collaboration with Guru Angad Dev Veterinary and Animal Sciences University, Ludhiana (Pb); Kerala Vet.& Animal Sciences University, Mannuthy; and Bhartiya Agro Industry Foundation (BAIF), Urlikanchan, Pune.

## 3. ICAR Central Institute for Research on Buffaloes

Hissar, Haryana. 1985

Regional station: Nabha, Punjab (1987).

The institute undertakes research work on all aspects of buffalo improvement including milk, meat and draft, both at institute herds and in farmer's herds and through Net Wok Project on buffalo improvement. The institute has developed a high pedigreed herd of Murrah buffaloes (500) at Hissar wherein an increase in milk production was observed from 1500 kg to 2300 kg in a lactation of 300 days, and a high pedigreed herd of Nili-Ravi buffaloes (400) has been developed at Nabha (Pb).

The institute has developed technologies efficient buffalo bull semen freezing protocol to increase post-thaw sperm motility, freezing of semen otherwise a problem, induction of lactation in infertile buffaloes, oestrus induction and ovulation synchronization in anoestrus buffaloes, area specific mineral mixture to improve reproductive and productive performance, packages of practices for raising surplus male calves for meat production, complete feed formulation using locally available agro-industrial byproducts, early pregnancy diagnosis using ultrasonography after 30 days of insemination and fetal sex determination after 55 days, super-ovulation of buffaloes and production of calves through ETT.

The institute has a network project on buffalo improvement involving elite herd of Murrah, Nili Ravi, Surti, Pandharpuri, Jaffarabadi, Bhadawari and swamp buffaloes at various centres for their genetic improvement through progeny testing involving about 4500 farmer's buffaloes in about 225 villages.

## 3.9 MITHUN

The Mithun or gayal (*Bos frontalis*) is a bovine species. It is also called as "Mountain cattle" or "Ceremonial ox" of north-eastern hill states.

### 3.9.1 Origin of Mithun

It resembles to domestic cattle in most of the physical, production and reproduction traits, but it has no hump, its dewlap is small, tail is short, horns are short and stumpy, and white stockings on all the four legs. It has an oily sweat gland which acts as insect repellent. The Mithun has massive body with prominent ridge on shoulder region.

There are different views about its origin. It is said as domesticated gaur, or a hybrid of crossing of gaur bull and zebu cow, or descendent form of wild Indian bovine. The mithun resemble with gaur in appearance, colour, body features and some habits and produce fertile crossbred progeny when crossed with gaur whereas it differs from zebu cattle in hemoglobin, transferrin and blood group genotypes. It is thus considered more to have originated from the gaur.

### 3.9.2 habitat and food habit of Mithun

Mithun adapt highly variable habitat with respect to topography. It can descend to 300 m above MSL and sometimes to still lower levels as well as it is found at and above 3000 m above MSL. Mithun predominates in 5 North East hilly states viz. Arunachal Pradesh, Mizoram, Manipur, Nagaland and Tripura. The mithun is also a domesticated food animal of Bhutan, Bangladesh (Chittagong hills) and Myanamar.

Mithun is adapted to humid climate and hilly terrain, preferring a cool atmosphere and shade. The mithun may not tolerate bright light and generally prefer to live in deep forests and dense grasses and appear on roadside during night.

They remain free grazing in the forest and on steep hilly slopes. Mithuns thrive on coarse fibrous feeds of tree leaves, shrub twigs and tropical grasses, for which these animals are of natural choice on 'low input system' and form an important component of hilly livestock production system. Mithuns graze in a herd and prefer to graze or brose on shrubs or small trees and has got the nibbing habit like goats. Mithun has the habit of roaming in search of fodder instead of staying at one place of grazing for long time.

### 3.9.3 Mithun Breeds

In India, Mithun are of two type *viz.* Arunachalees and Nagmees. The Arunachalees are found in Arunachal Pradesh, small in size than Nagmees, and developed for milk and meat. The Nagmees are found in Nagaland and also in Manipur, are heavier in size and weight (400 – 500 kg) with long and massive body, and produce meat.

### 3.9.4 Crossbreeding

Mithun is interbreedable with cattle and gaur. Mithun males are mated to cows and the hybrids are given the names according to the species involved and sex of ancestor. The $F_1$ males produced by crossing cows with mithun bull are called as ***Jatsa*** which are good for draught purpose but are sterile while the $F_1$ females are called ***Jatsamin*** which produce milk higher than mithun cows. The backcross is also made by crossing Jatsamin with mithun male and the two sexes produced are called as ***Nupsa*** (Male progeny) and ***Nupsamin*** (female progeny).

Siri cows were crossed with mithun bulls in Bhutan. The male and female $F_1$ are called ***Jechha*** and ***Jessam.*** The F1 females produced by crossing Jatsmin x Siri bull are called ***Yanka.***

### 3.9.5 Production characters of Mithun

The mithun weighs from 20 to 30 kg at birth attaining adult body weight of 450 – 500 kg with body height of 1.2 – 1.4 m.

Earlier it was not a practice to milk the mithun. Mithun cows produce 1.4 to 3.5 kg milk per day for a lactation period of about 6 months. Its milk contains 4 – 5 % fat, 3.5 – 4 % protein, 4 – 5 % lactose and 0.8 % ash.

## 3.10 UTILITY OF MITHUN

Mithun has a great potential as a *beef animal* for its massive and muscular body. Mithun is primarily reared for *meat* and its meat is preferred in NE region as a special delicious dish for the tribals. The meat of mithun has about 20 % protein, 0.6 to 0.9 % fat, and 0.5 to 0.9 % mineral matter. It is sacrificed only on religious and social functions i.e. marriage, feast of merit, celebration of victory over rival group.

The mithuns are used not only for meat but they are also used for field work and as a *draft animal* in some parts of India.

The mithuns have their *socio-economic importance.* Mithun is a *symbol of prosperity* and the number of mithuns is an indication of the status of the people of hill tribes. It is thus a *status symbol and measure of wealth,* called as currency being used in exchange of land, payment of ransom, tax, fine or bride price. Among some tribes the irrigated land can be purchased only in mithuns and likewise the house and orchards are purchased in mithuns. Mithuns provide socio-economic, religious and livelihood security to the hill people of NE region.

*Mithun horns* have decorative value to display on house walls and used for various craft purposes. They are also used as drinking cups, flask for keeping powder and valuables and making of some value added products. Mithun *skulls* are used to decorate the houses and their numbers in the house is considered prestigious. The mithun *bones* particularly the shoulder blades are used as implements like a hoe for tilling or shovels for scraping of garbage from house.

*Mithun hides* after removing hair is eaten, used as shields for chest covering as war dress, also used as sleeping mats and to make belts and purses as well as for making strings to tie the animals.

## 3.11 IMPROVEMENT PROGRAMME ON MITHUN

A farm of mithun has been set up by Govt. of Arunachal Pradesh at Kambi in west Siang district.

The ICAR has set up NRC on Mithun in 1988 in Phek district of Nagaland to conduct research on various aspects.

## 3.12 ICARONATIONAL RESEARCH CENTRE ON MITHUN,

Jharnapani, Nagaland. Established in 1988

The NRC on Mithun (Bos *frontalis,* a subspecies of cattle, also known as Gayal, the domesticated form of the gaur) has developed the improved germ plasm of all the 4 strains of Arunachali, Nagaland, Manipuri and Mizoram mithun; developed nutritional technologies to prepare low cost nutritious feed blocks, area specific mineral mixture, silage making using agro industrial byproduct (orange pulp after juice extraction) mixed with grasses and paddy straw; reproductive technologies for estrous synchronization, innovative method of semen collection by AV method after sprinkling urine from oestrus mithun cows over the perineal region of a cow not in heat otherwise mithun bull do not mount on cows which are not in heat, super ovulation by using FSH 400mg for 4 days on days 10 – 13 post estrus and embryo transfer after recovery from donor to recipient cows; value added products from mithun milk and meat, hides and leather product.

## 3.13 YAK: SHIP OF HIGH HILLS OR SNOW

The yak (B. *grannies* - another subspecies of cattle) belongs to the genus *Bos* of bovidae family of bovines. The yak is locally known as **banchor** for wild and **chor-gau** for domestic variety. Yak is a unique bovine species of Asian highlands, thought to be domesticated from wild yak (**Bos mutus**) about 4500 years ago in Tibetan plateau and adjoining mountains in Himalayas.

### 3.13.1 Habitat

The yak lives in high altitudes above 14,000 feet (1800 – 2100 feet) and can not live below 10,000 feet above mean sea level. Yak is found in extremely cold and high altitude, north east mountain of India. The yak can climb as high as 20,000 ft (about 6100 m). The largest population of yak is found in Tibet. In India Yak are found in two NE states of Arunachal Pradesh and Sikkim, and also in the states of J & K and Himachal Pradesh as well as in very limited numbers in high hills of U.P. It is adapted to extremely poor feeds.

### 3.13.2 Yak Breeds

No breed of yak has been described so far. The Indian yak has been described of 4 types viz. ***Ladakh yak, Himachal yak*** (Spity valley), ***Arunachal yak*** of Twang and Diarang region, and ***Sikkim yak*** which are of two types depending up on rearing community viz. ***Bho*** and ***Aho yaks.*** However, based on morphology, there are 5 different types of Yak viz. *comm. on type* which are small in size, mild and docile; *Bison type; Bare back type; hairy forehead type* and *pure white yak.* The yaks are distributed in Arunachal Pradesh and Sikkim of NE region, J & K and H.P. and also found in high hills of U.P. in very limited numbers

### 3.13.3 Crossbreeding

The interspecies hybridization of domestic yak with hill cattle (Galang) is common and the hybrids of yak and cattle are found throughout the entire yak breeding area, though the male hybrids are normally sterile. The yaks interbred with bison, banteng, gayal, zebu and European cattle. The hybrid males called as ***dzo*** are sterile but the hybrid females (**dzomo**) are fertile. The $F_1$ females are crossed either to yak or cattle. The yak sire is used in crossbreeding for better draught capacity and hardiness whereas the yak female is crossed to cattle sire for better milk and butter fat percentage. The crossbred females (from cattle male) have better milk production than local cattle. The $F_1$ hybrids generally grow faster and larger than pure yak and produce more milk and meat than pure yak as well to local cattle. The yak-cattle can survive at low hills because of their heat tolerance.

### 3.13.4 Morphology

The yak is a unique and sturdy bovine species with compact and long body (12 – 14 feet) covered with long hairs. The yak has short neck with no or rudimentary dewlap, narrow and long face, massive heavy head, bushy switch of tail, and short limbs (legs). The height of yak varies from 4 – 6 feet and average weight is about 250 kg. The birth weight of yak is about 14 kg. The hoofs of yak are large, hard and somewhat cup-shaped well suited for mountain climbing. Polled yak are also found. The horns are long generally curved outwards and upwards. There is a dense coat of thick black hairs on yak body and this protects from extremely cold climate and they can survive up to the temperature of – 40 $^0$C.

## 3.14 UTILITY OF YAK

Yak provides milk, meat, fibre, skin/hide and used as pack animals to carry woods and other supplies in the difficult, high altitudes of mountainous regions where horses and mules find difficulty. It can carry 55 – 70 kg load for 13 – 16 km daily. Yak has special ability to walk through soft snow. They are also used for riding and ploughing.

The *milk* production of yak is 1.1 to 1.6 kg daily having variable fat from 4.2 to 13.0 % with an average of 6.5 %, proteins 6.0 %, lactose 4.7 % and ash 0.8 %. The milk fat is golden yellow in colour. The milk is used for preparation of various milk products. The lactation period is of about 4 – 5 months.

The yak *meat* is tender and good for human consumption containing protein content equal to the meat of other species but highly rich in fat content. The average dressing percentage is about 50.

The yak *hairs* are plucked at the end of winter. The yak produces two types of *wool* viz. outer coarse wool and inner dense wool like pashmina which are about o.4 to 0.6 and 0.3 to 2.0 kg. The inner fine hairs are used to make fine garments. The outer rough hairs are used for making of ropes, tents, felt, caps, wall hangings and quality garments from hand spun yarn. The tail of yak is especially bushy with a bunch of long hairs called as *Chauri* used as *chawar* (holy duster) for religious purposes. The tail hairs are much soft and fine than that of horse. The white switch of tail is highly valuable.

The yak droppings (*dung*) are dry to the extent that these can be burnt without drying. The yak *skin* is used for preparing leather goods (shoes, bottles, thongs). The yak *horns* are used as decoration pieces and for making utensils.

## 3.15 ICARQNATIONAL RESEARCH CENTRE ON YAK

**Dirang, Arunachal Pradesh.** Established in 1989

This NRC on Yak was established for characterization and conducting research on various aspects on Yak.

The **NRC on Yak** has identified all the 5 different types of Yak (common type which are small in size, mild and docile; Bison type; Bare back type; hairy forehead type and pure white yak); developed area specific mineral formulation for yak feeding; reproductive management of semen collection, cryopreservation of semen and AI, standardization of ETT and estrus synchronization; health related technique by polyherbal wound healer and fly replant; value added yak milk (low fat paneer, yak milk whey beverages mix with kiwi fruit pulp) and meat products (sausage by incorporating pork and fat in yak meat); and techniques to prepare *Chawar* (Holy duster from bushy tail and white hair), caps, wall hangings and quality garments from hand spun yarn.

❑❑❑

# 4 Breed Improvement Programmes II. Ovines and Rabbits

Sheep and goats are called as small ruminants belonging to the family bovidae, sub-family bovinae. Sheep belong to the genus ovis while goats belong to genus capra.

Both ovine species (sheep and goat) are well adapted to hot and cold arid regions with relatively scant vegetation and drinking water. Goats prefer mountain areas and dry stepps.

## 4.1 DIFFERENCES BETWEEN SHEEP AND GOATS

A sheep is different from goat:

(*i*) By glands in both forefeet and hind feet

(*ii*) By the absence in sheep of strong goaty odors present in bucks, a gland beneath the tail known as subacious gland,

(*iii*) Absence of beard in sheep,

(*iv*) They also differ in horns which are generally spiral in opposite directions in two genera - the right horn of the sheep to the right like a corkscrew and the goats to the left,

(*v*) The tails of the two are different in the way that tails of goats are erect upwards but of sheep droop downwards,

(*vi*) Sheep have tear gland,

(*vii*) The two species have large differences in their ecological requirement. The sheep like surface grazing whereas the goats are browser, eating foliage from trees and shrubs than surface vegetation of grasses, legumes and forbs.

## 4.2 SHEEP

The sheep tolerate the diverse environmental conditions having desirable attributes of disease resistance and heat tolerance. They are more efficient in feed conversion, utilize uncultivable wastelands and weeds, and contribute food (meat) and material for human clothing besides its valuable dung used as manure.

Sheep (*Ovis aries*) has diploid (2n) chromosome as 54. Out of autosomes, 3 pairs are metacentric/submetacentric and rest 23 pairs are acrocentric. The X chromosome is acrocentric while Y chromosome is metacentric.

### 4.2.1 Sheep Breeds of India

The various breeds of sheep found in different ago-climatic zones are given as under-

| Zones | Breeds of sheep |
|---|---|
| (i) North Temperate | 1. Karnah (AW), 2. Kashmir Merino (AW), 3. Bhakarwal (CW), 4. Changthangi (CW), 5. Gurej (CW), 6. Poonchi (CW), 7. Gaddi (CW), 8. Rampu-Bushair (CW), |
| (ii) North-west Dry | 9. Munjal (M), 10. Hisardale (AW), 11. Nali (CW), 12. Chokla/ Sekhawati (CW), 13. Magra (Bikaneri, CW), 14. Sonadi (MCW), 15. Malpura (MCW), 16. Jaisalmeri (MCW), 17. Marwari (MCW), 18. Lohi, 19. Pugal (MCW), 20. Kheri (MCW), 21. Muzaffarnagri (MCW), 22. Jalauni (MCW), 23. Patanwari/ Kutchi/Kathiawari (CW), |
| (iii) Southern | 24. Nilgiri (AW), 25. Coimbatore (MCW), 26. Bilari (MCW), 27. Deccani (M), 28. Hassan (M), 29. Nellore (M), 30. Mecheri M), 31.Mandaya/Banur (M), 32. Madras red (M), 33. Kilakarsar (M), 34. Rmnad white (M), 35. Trichi black (M), 36. Vempur (M), 37. Kenguri (M), |
| (iv) Eastern | 38. Tibetan (CW), 39. Chota Nagpuri (MCW), 40.Ganjam (MCW), 41. Balangir (MCW), 42. Bonpala (MCW), 43. Garole (M), 44. Kendrapara (M), 45. Sahabadi (M). |
| | M = meat, CW = Carpet wool, |
| | AW = Apparel wool, MCW = Meat & Carpet wool |

### 4.2.2 Exotic breeds of sheep

The exotic sheep breeds are classed as –

(*i*) *Fine breeds sheep:*

1. ***Merino:*** Developed in Spain and imported in America in 1801.
2. ***Rambouillet:*** Originated at Rambouillet in France from *Spanish Merino* and imported in America in 1840.

(*ii*) *Medium Wool breeds:*

3. ***Corriedale***: Developed in New Zealand by crossing Merino ewes and Lincoln rams and also Licester rams subsequently. Imported in America in 1914.
4. ***Hampshire:*** Originated in England by crossing two strains and imported in USA in 1940.
5. ***Polworth:*** Native of Victoria and Tasmania (Australia). Developed by crossing Lincoln rams with Merino ewes.

6. ***Cheviot:*** Developed in Scotland from Licester, South Down and Merino.

(*iii*) *Long coarse wool breeds:*

7. ***Lincoln:*** Developed in Lincoln (England) in 1749 with Licester inheritance.
8. ***Licester:*** *Originated* in Licester (England) in 1760) by Robert Bakewell from local sheep by close inbreeding.
9. ***Romney Marsh:*** Originated in marshy district (England).

(*iv*) *Fur Breeds:*

10. ***Karakul:*** Native to Bokharo (Central Asia) near Karakul black lake. Originated by crossing two local breeds.

(*v*) *Mutton Breeds:*

11. ***Dorset:*** Developed through selection, native of England (Dorset region).
12. ***Suffolk***: Native of England (Suffolk region), developed from dark face South down rams used on Norfolk strain. Both ewes and rams are polled.
13. ***South Down:*** Originated in Southdown hills of England, from selection and interbreeding of local sheep.

## 4.3 UTILITY OF SHEEP

The sheep produce 46.5 million kg wool, 222 million kg mutton, 51 million skins, and 2977 million kg manure.

(*i*) ***Wool production***: The sheep is more important to produce ***wool,*** raw material for processing into cloth, knit wear, blankets and carpets. However, the development of synthetic fibres which are more versatile and cheaper has lowered down the economic importance of wool and other animal fiber.

(*ii*) ***Mutton production:*** The trend in sheep rearing has been changed from wool to ***mutton production*** as a result of synthetic fibres. The efforts are now being made to increase the fecundity, prolificacy, lamb survival, and growth rate to maximize mutton production. About 1/3 of the total sheep are slaughtered every year for meat purpose.

(*iii*) ***Pelt production***: The sheep also produce pelts. They are now preferred for ***pelt production*** particularly in arid regions with seasonal and low availability of feed. The most popular breed for producing lamb pelts is the Karakul in countries like USSR, Afghanistan, Iran and South Africa. Karakul is a coarse carpet wool breed with fat tailed thriving on poor vegetation and extreme climatic conditions. The process of pelt production by slaughtering the lams before or immediately after birth fetch more value in terms of the pelt and meat from such slaughtered lambs as well as the milk of their dams.

The lamb pelt production has certain advantages viz. high market value compared to other sheep products, killing of lambs for pelt production eliminates the risk of infant mortality, adult females can be rebred to produce lambs more often or they can be used as foster mothers, results in reduced pressure of inputs (feed/fodder) required for longer period for meat and wool production. Not only this benefit but it has an added advantage of removing the pressure on poor and seasonal grazing resources.

Skin from sheep is raw material for leather industry, being available about 50 million pieces of skins from sheep. Sheep is about 500% value addition in raw skin after processing and finishing for leather production.

## 4.4 SHEEP IMPROVEMENT PROGRAMMES

The productivity of indigenous sheep is low for want of proper selection of germplasm, limited feed supply and nutritional deficiency, heat stress, poor health monitoring and husbandry practices adopted.

In 19 $^{th}$ century, sheep development programme was started by the East India Company by importing exotic breeds (Cape Merino) of sheep for crossbreeding of indigenous sheep breeds around Poona.

In the beginning of 20 $^{th}$ century, the crossbreeding of sheep with Romney Marsh was undertaken in plains and hilly areas. Crossbreeding with South Down and Cape Merino was undertaken in Punjab, Bengal, Madras and Mysore. Bikaneri ewes were crossed with Merino rams at GLF, Hisar and a fine wool breed with ¾ exotic inheritance was evolved by the name *'Hisardale'*.

After independence, in 1950, the research work on sheep and wool development in India was taken up on regional basis by ICAR. The emphasis was on selective breeding within native stock and the crossbreeding with exotic fine wool breeds. The regional centers were set up in the hilly areas of UP, Rajasthan and Deccan plateau to conduct crossbreeding trials during first five year plan.

During 2 $^{nd}$ five year plan, three sheep breeding farms were established with the aim to prodce stud rams along with wool extension centers in some states so as to improve the field stocks by providing breeding rams of superior merit.

During 3 $^{rd}$ five year plan, the action taken were the strengthening of the existing sheep breeding farms, establishment of new sheep breeding farms (27), establishment of more number of sheep and wool extension centers (340), initiating a large scale sheep shearing and wool processing programmes in Rajasthan and other states of Northern and Western parts of the country, and establishment of Central Sheep and Wool Research Institute, Avikanagar (Malpura, Tonk Distt), Rajasthan under Central Govt. in 1962. The aims of establishing CSWRI were to undertake the fundamental and applied research in sheep production and wool utilization, and to provide post-graduate training in sheep and wool sciences. Later on, during 1965 the CSWRI was transferred under the administrative control of ICAR.

During 4th five year plan, 8 large sheep breeding farms were set up. There were 84 Sheep Breeding Farms, 52 sheep breeding centers and 602 sheep and wool extension centers in th country during fourth plan.

The GOI in 1970 constituted an *ad-hoc* committee on Sheep Breeding Policy. The committee suggested the importation of 20000 exotic sheep of fine wool breeds to produce breeding rams in order to meet the shortage of stud rams. Therefore, Indo-Australian Sheep Breeding Farm, Hisar was set up with 3000 ewes and 300 rams of Corriedale breed to meet the objective of producing and distribution of Corriedale stud rams to different states for crossbreeding work of improving wool and mutton production since Corriedale is a dual purpose breed. Later on, the Corriedales were replaced by Rambouillet.

Two Sheep Breeding Farms in private sector in Maharastra and Gujrat were setup to produce purebred and crossbred rams. Seven Central Sheep Breeding Farms in UP, MP, AP, Bihar, Karnataka and JK were established for same purpose of producing pure and crossbred rams.

During 5 th five year plan, some more breeding farms were established. Efforts were made for intensive sheep development through genetically upgrading the stock, providing health care, pasture improvement, machine shearing and wool processing.

***AICRP on Sheep Breeding***: The AICRP on sheep breeding were started in 1971 for fine wool and for mutton to achieve the objective by crossbreeding of native breeds with rams of exotic breeds. The centers for Fine Wool were located at CSWRI, Avikanagar (Raj.); GAU, Dantiwara; Sheep Breeding Farm, Tal and Sheep Breeding Research station, Sandynallah. The centers for mutton production were located at CSWRI, Avikanagar; APAU, Palampur (A.P.), CIRG, Makhdoom (U.P.); MPAU, Rahuri (M.S.) and TANVASU, Kuttupakkam (TN).

The above two coordinated research projects were merged into one project on Sheep Breeding in 1974 with the following objectives-

- *Fine wool component:* 2.5 kg GFW/year with average fiber diameter of 22 microns and modulation percentage below 5.
- *Carper wool component*: 2kg GFW/year with average fiber diameter of 30 microns and modulation percentage 25.
- *Mutton component*: 30 kg live weight at 6 months of age with 50 % dressing percentage and feeding efficiency of 15-20 percent.

***Net work project***: In 1990, all the ongoing units of AICRP on Sheep Breeding were converted into Network Project on Sheep Improvement (NWPSI). The mandate of NWPSI was survey, genetic evaluation and conservation of sheep genetic resources of the country. The breeding policy for NWPSI is the selective breeding in indigenous sheep. The NWPSI is functioning at 5 farm based cooperating units and 2 field based cooperating units. These are as under:-

Farm based units: CSWRI Avikanagar (Chokla for carpot wool)

CSWRI substation Bikaner (Marwari for carpet wool)

| | |
|---|---|
| | CIRG Makhdoom (Muzaffarnagri for wool and mutton) |
| | MPKV Rahuri (Deccani for wool and mutton) |
| | APAU palamner (Nellore for mutton) |
| Field based units: | RAU, Bikaner (Magra for carpet wool) |
| | TNVASU, Kattupakkam (Madras Red for mutton). |

In addition to NWPSI, the State Agric. Universities and state sheep breeding farms are involved in genetic improvement of sheep but with the major constraint of small population size. Some NGO's like Nimbkar Agr. Research Institute, Phalton (MS) are also involved in sheep improvement programme. This institute is working on Garole sheep.

The CSWRI have also the breeding programmes for indigenous sheep in addition to the breeds (Chokla and Magra) under NWPSI. These are as:

- Genetic improvement of Magra sheep under filed condition.
- Improving Malpura sheep for mutton production.
- Improvement of Awassi and Garole sheep for mutton, milk and prolificacy

In the state of Jammu and Kashmir, a new breed Kashmir Merino was evolved by crossbreeding of Gaddi and Bhakarwal breeds with rams of Delaine Merino, Soviet Merino and Rambouillet. This had 50-75% exotic inheritance.

## 4.5 ICAR CENTRAL SHEEP AND WOOL RESEARCH INSTITUTE,

Avikanagar, Malpura (Rajasthan), 1962

Regional Stations: Garsa, Kulu, H.P. (1963); Mannavanur, T.N. ((1965);

Bikaner, Rajasthan (1974).

The CSWRI was established in 1962 at Avikanagar near Malpura, Tonk District (Rajasthan). Subsequently, two regional stations one at Garsa, kulu District (HP) and the other at Mannavanur, Kodai Hills, Madurai District (TN) were established. One more cenre of CSWRI was started in 1975 at Bikaner, for Karakul pelt production.

This institute of ICAR at its headquarter had developed flocks of Malpura, Chokla, Avikalin, Patanwadi and its crosses, Garole and its crosses, and Kendrapada for their genetic improvement and other technology development. The institute at its regional station at Garsa, Kulu district (H.P.) has developed a unit of broiler rabbits of Soviet Chinchilla, White Giant, Gray Giant, Black Brown and Dutch genotype. This centre has 550 Angora rabbits (British Angora, Russian Angora and German Angora) which are producing annually 900 gm of fine special fibre. Another centre of the institute located at Mannavanur (T.N.) has the mandate to undertake research on fine wool sheep (Bharat Merino) and broiler rabbit. The institute's another station located at Bikaner has elite sheep flock of Magra and Marwari sheep (carpet wool quality breeds).

The institute has developed improved germplasm of Malpura breeds of sheep for mutton, Chokla which is best carpet wool breed, introgression of

Fec B gene of highly prolific Garole sheep into Malpura breed which has increased multiple births to 40 % compared to single birth otherwise; developed technique of indigenous intra-vaginal sponges for oestrus induction and synchronization; developed area specific mineral mixture for cattle, buffalo, sheep and goats of semi-arid region of Rajasthan, complete feed block of roughages and concentrate; developed technique for management of haemonchosis in sheep; developed a computer based programme to simulate life cycle of haemonchus conttortus in sheep which forecast the magnitude of parasitism 60 days in advance; technology for diagnosis of Johne's disease; developed carpet from indigenous wool and its blends, dying technology using natural dyes for wool; developed value added meat products *viz.* mutton nuggets, sausages, petties, pickles, soap enrobed eggs, mutton loaves and kofta.

The institute provides services through 3 transfer technology centres, better quality germplasm (rams), oestrus synchronization and Artifical Insemination, technologies related to fodder development and effective utilization of grazing resources, provides training for value addition of wool.

## 4.6 IMPORTANT MILESTONES OF C.S.W.R.I.

1963: Import of Romney Marsh, Southdown and Rambouillet sheep;

1969: Introduction of Corriedale sheep (Mannavanur) for crossbreeding with Coimbatore breed;

1971: All India Coordinated Research Project (AICRP) on sheep for fine wool and mutton; Canary colouration and manufacturing trails were started;

1971: Import of Soviet Merino sheep from Stavropol (erstwhile U.S.S.R.);

1974: Import of Dorset and Suffolk sheep;

1975: Import of Karakul from USSR;

1976: Avikalin and Avivastra strains were evolved;

1983: Avimans strain was evolved;

1986: Bharat Merino Merino strain was evolved;

1992: Conversion of AICRP in to Network Project on Sheep Improvement;

1994: Introduction of Awassi sheep in 1994 and Garole sheep in 1997 to cross with Malpura breed;

2008: Bharat Merino flock shifted to Mannavanur and Garsa;

2008: Introduced Patanwadi and produced three breed crosses (GM x Pattanwadi);

2009: Mega Sheep Seed Project was initiated; and

2010: Kendrapada sheep were introduced in prolific sheep project.

## 4.7 GOATS

The goat is known as poor man's cow. Mostly landless labourers in rural areas keep goats. The goat has been misunderstood as causing desertification. The goat eats the most inferior vegetation than other animals except camel. They thrive on herbs, weeds, wild plants, vegetable and fruits peelings and prickly plants and kitchen wastes. Goat rearing is advantageous. Goat keeping requires small financial investment, less consumption of feed, and goats are prolific. The goats are hardy and can withstand the rigororus weather better than other milch animals.

The goats (*Capra hircus* and *C. ibex*) have 2n (diploid) chromosome as 60. All the autosomes and X chromosome are acrocentric while Y chromosome is metacentric.

### 4.7.1 Goat breeds of India

The various breeds of goats found in India are classified according to agro-climatic zones as:

| Zones | States | Breeds of goats |
|---|---|---|
| (i) Temperate | J.K, H.P. Uttrakhand | 1. Gaddi, 2. Changthangi, 3. Chegu |
| (ii) North-West | Delhi, Haryana, Pb, UP, Rajasthan, Gujarat, MP, Chhatisgarh | 4. Beetal, 5. Barbari, 6. Jamnapari, 7. Sirohi, 8. Marwari, 9. Jhakrana, 10. Surti, 11. Kutchi, 12. Zalawadi, 13. Mehsana, 14. Gohilwadi |
| (iii) Southern- Pateau | MS, AP, TN | 15. Malabari, 16. Sangamneri, 17. Osmanabadi, 18. Kodaiadu, 19. Kannaiadu |
| (iv) Estern | Bihar, Orissa, Jharkhand, W.B. | 20. Ganjam, 21. Black Bengal |
| (v) NEH | Assam | 22. Assamese. There are two varieties of Assam goats viz. Assam local goat Asomi goats) and Assam Hill goats. |

### 4.7.2 Exotic breeds of goats

The important exotic breeds of goats are as under –

(*i*) ***Milch Breeds:***

1. ***Saanen:*** Originated in Saane and Simmental valleys of Switzerland.
2. ***Alpine:*** Originated in Nubia (Africa).
3. ***Anglo Nubian:*** Origin not known.
4. ***Toggenburg*:** Originated in Toggengurg valley of Switzerland.
5. ***Boer*:** It is from South Africa.

(*ii*) ***Mohair/Pashmina Breeds***:

6. ***Angora goat:*** Originated in Angora region of Asia Minor (Turkey).
7. ***Orenberg:*** Originated from local goats of Kazaakhistan.

## 4.8 UTILITY OF GOATS

The earliest discovery of man in prehistoric times was the ***goat,*** primarily as easy source of meat in the beginning but later on for its value for milk, hair and skin. The goats have survived to diversified agro-climatic conditions and sustained the poor people. The goat can sustain themselves on sparse (scattered) forage and extreme climatic conditions and especially useful to the people in semi arid and arid zones where droughts are common and other species of animals may perish (to an end). The goats produce a variety of products *viz.* milk, meat and fibres, though it is mainly meat animal in India.

(*i*) The ***goat milk*** production starts at early age. Goat milk is very similar to human milk in fat and protein percentage with less sugar. It has small fat globules and hence easily digestible. Goat milk is safe against TB. In comparison to cow and human milk, the goat milk has higher contents of calcium, phosphorus and chlorine but low in iron. The goat milk is white in colour because the goat converts the carotene into vitamin A. The goat milk has no goaty flavour if produced under sanitary conditions.

(*ii*) The ***goat meat*** is called ***Chevon***. Goat *flesh (Chevon)* is low in fat and preferred than other meat and hence fetch better price than mutton. The goat meat has its high acceptability over other type of meat and hence its demand is high. The goat meat is costliest of all meat. The goat is the most sacrified animal by Muslims as well as by Hindus. The dressing percentage varies from 45 – 55 per cent.

(*iii*) The goats also produce ***hairs***. Goat hairs fetch good price produced by Marwari breed. Chegu and Changthangi breeds produce elite quality fibres. Angora goats produce ***Mohair fibres*** which is long lustrous hair covering and fetch higher price than wool whereas ***Pashmina*** is produced by Kashmiri or pashmina goats. Pashmina is the undercoat of great fineness generally grows during winter months. Both the mohair and pashmina are used for the manufacture of superior cloths, fabrics and shawls.

(*iv*) ***Goat skin*** is also important for export earning foreign exchange. Indian breeds have superiority over other foreign breeds for skin quality. The processed goat skin is of light weight with fine texture. The goat skin after processing is used for gloves, jackets, book bindings, shoes and other items (various kinds of containers).

(*v*) Goat also provides ***manure***. There is a practice to leave the goats to graze in the field so that the field is enriched with goat manure and urine. The goat urine is rich in nitrogen and potash.

(*vi*) Goat also serves as a ***seed treatment machine*** for the hard coated seeds which do not germinate if shown directly. But when these hard coated seeds of xerophyete trees, after consuming by goats passes through their alimentary canal, the hard coat of the seeds is softened and gets pallated in the faecal matter and develop successfully into healthy sapling on their sowing in the soil.

The goats have shown an increase in population relative to sheep in spite of the fact that larger percentage of goats (70 %) slaughtered every year.

## 4.9 GOAT IMPROVEMENT PROGRAMMES

The Intensive Agr. Distt. Programme (Pb. Govt.) during 1964 imported 57 does and 12 bucks of Alpine (6 M + 56 F), Nubian (4 M + 14 F), and Toggenburg (2 M + 5 F) breeds from California. Saanen breed of goat was imported by NDRI, Karnal.

All India Coordinated Research Project (AICRP) on *goats* was initiated in IV Five year plan (1971) with its basic theme to improve the production performance of Indian breeds of goats through crossbreeding with exotic breeds of temperate countries. Two separate projects were incepted *viz.* AICRP on goats for milk and AICRP on goats for fiber (pashmina and Mohair) production.

Three centers for *milk production* were NDRI Karnal (Beetal goat) and KAU, Mannauty, Trichur (Malabari gaots) for crossing these breeds with bucks of Alpine and Saanen breeds; and MPKVV Rahuri for crossing of Black Bengal with Beetal and Jamnapari breeds. The project at NDRI, Karnal was discontinued during VII plan period while at KAU, Trichur was discontinued during VIII plan period.

Three centers for *fiber production* were: 2 for pashmina at IVRI, Mukteshwer (Chegu goats) and Upesi Farm, Leh (Changthangi goats) were started during IV plan and discontinued during VII plan period and one centre for Mohair at MPKVV Rahuri (local Deccani) to cross with Angora goat started during IV plan was discontinued during VIII plan period.

The third component of *meat production* was included in 1974 under AICRP on goats for meat. The centers for meat component were:

- CSWRI, Avikanagar (Sirohi x Beetal crossbreeding);
- Assam Agr. University, Khanapara, Gauhati (Assam local x Beetal crossbreeding);
- CIRG, Makhdoom (U.P.)

All the AICRP on goats for 3 components were merged into one project as AICRP on goats (milk, meat and fiber production) in 1974. The coordinating unit of AICRP on goats remained at CSWRI, Avikanagar up to Oct.1976 and later on shifted to National Goat Research Centre (IVRI), Makhdoom (UP). The number of units increased to 9 in VI plan period (1980-85) which declined to 6 in VIII plan period.

***Network Project on Goat***: The AICRP on goats was transformed in 1990 into network project in VIII plan with its mandate to define, characterize and improve the productivity of native goat breeds incorporating farmers flock in the home tract through selective breeding based on progeny performance. All the ongoing 6 centers of AICRP on goats were converted into the centers of Network Project on goat improvement. These centers were:

CSWRI, Avikanagar (Sirohi goats); RAU, Bikaner (Marwari goats);

KAU, Mannuthi (Malabari goats); MPKVV, Rahuri (Synthetic Angora goats);

Birsa Agr. University, Ranchi (Black Bengal goats ) and JNKVV, Mhow (Malwa goats).

In 1993, the network centers at BAU, Ranchi; JNKVV Mhow: MPKVV, Rahuri and KAU, Trichur were delinked. The existing centers at Bikaner for Marwari goats in their home tract and Avikanagar for Sirohi goats in farm flock were allowed to continue.

During VIII plan, three new centers were opened for Jamunapari and Barbari farm units at CIRG, Makhdoom, and Jamunapari field unit (Chakarnagar, Etawah, UP).

## 4.10 ICAROCENTRAL INSTITUTE FOR RESEARCH ON GOATS,

Makhdoom (Farah), Mathura, U.P., 1979.

This institute was established by ICAR to develop goat (a poor ma's cow) for livelihood security, poverty alleviation and employment generation. The institute established farms of Barbari breed (native breed of Agra area) and Jamnapari breed (native breed of Etawah District), in the year 1982, and Jakharana breed in the year 2004. The AICRP on goat improvement covers 10 breeds.

The institute has developed technologies for *frozen semen based A.I. techniques* which otherwise the A.I. is difficult in goats due to complex anatomy of reproductive tract; pelleted complete feed, area\specific mineral mixture, milk replacer for pre-weaning kids; developed value added products from goat milk (goat milk paneer, biscuits, shrikhand, ice cream, goat milk beauty shop, and meat products (emulsion based goat meat product such as sausage, patties, slices, kofta, shammi kabab and nuggets; non-emulsion based products such as meat pickle, cutlets and samosa; snack products like nimkee and murukku); Immune-diagnostics for timely detecting of brucellosis and Johne's disease. The institute provides service to stakeholders in the form of providing genetically selected and performance tested goats of 11 breeds at 14 locations across the country for improvement of farmer's flock under AICRP on goats improvement to state Govt, NGO's, progressive farmers.

## 4.11 RABBITS

The rabbits and hares belong to the order Rodentia. The rabbits belong to the genus *Oryctologus* whereas the hares belong to the genus *Lepus,* of the order Rodentia of sub-class Eutheria of the class Mammalia. *The rabbits belong to the species* Cuniculus. The zoological name of rabbit is *Oryctologus cuniculus.*

The rabbit is known to be domesticated in Africa, though about 3000 years ago the rabbits were reared for food in Asia. Based on production of good quality ear, wool and fur a number of breeds and strains have been developed through selection and breeding.

The origin of all domestic rabbits is the European wild rabbit. The Europe (USSR, France, Italy, and Spain) has about 85 % rabbit population of the world followed by China.

The rabbits are non-ruminant herbivorous animal with large gastro intestinal tract from mouth to anus. This constitute buccal cavity, oesophagus (simple tube for passing the food from mouth to stomach), stomach which has 4 types of glands, duodenum, caeca to digest fibrous matter and rectum (short and enlarge section of the intestine between caaaaeca and anus).

### 4.11.1 Difference between rabbit and hare

The rabbits are herbivorus, prolific and efficient convertor of fodder to food. The ***rabbits and hares*** are different from each other to the extent that:

(*i*) They belong to different genera of the family Leporidae of the order Legomorpha, super order Glires of the class Mammalia.

(*ii*) They (rabbits) are small running animals with relatively short ears and legs, giving birth to blind and naked young whereas the hares are larger hopping animals with longer ears and legs giving bith to open eyed furred young.

(*iii*) The rabbits are mostly nocturna, lives in holes dugged by them whereas the hares do not burrow but hidden in vegetation.

### 4.11.2 Rabbit Breeds

The rabbits have the following two types of breeds -

(*i*) ***Meat type rabbits:*** The meat type rabbits are New Zealand White rabbit, White Giant, Grey Giant and Soviet Chinchilla rabbit.

(*ii*) ***Fur type rabbit:*** Three strains of fur type rabbits are German Angora, Russian Angora and British Angora

### 4.11.3 Some Useful Terms

A Young rabbit whose eyes are not yet opened is called **Kit** and below 20 weeks of age is called **Bunny. A** newly weaned rabbit is called **weaner.** The culled rabbit is called **Roaster,** 10-12 weeks old rabbit ready for market is called ***Fryer*** and the rabbit grown for meat is called **broiler.** The mature males and females are called **Buck** and **doe,** respectively. The act of parturition is called ***Kindling.*** The wool is called ***Fur*** whereas the skin is called ***pelt.***

***Coprophagy or Caecotropy:*** The rabbits and rodents (rats, mice, squirrels and guinea pigs) which belong to the order Rodentia of super order Glires) have one similar habbit of passing food twice through G.I. tract. This is inherent phenomenon of digestion. This double passage of food is necessary for their life. This practice is known as ***Coprophagy or Caecotropy***. The rabbit excretes normal hard faeces (pellets) with about 50 % dry matter during day time whereas it excretes soft faeces with about 25 – 30 % dry mater mostly during night or dark environment of burrows in wild rabbits. These soft faeces contain large number of microbial cells and are suckled by the rabbit from their anus.

## 4.12 PRODUCTIVE CHARACTERS OF RABBIT

The ***rabbit meat*** is white meat and easily digestible. The average daily gain in body weight is 30-35 gm in broiler strains against 20-30 gm in fur strains. The active growth is between 3 – 10 weeks. The mature body weight varies between 3 0 to 4.0 kg. The rabbit is said to produce 10-15 times its body weight in a year.

The daily ***milk** production* is about 200 g.

The ***rabbit wool*** is called **Fur.** The annual wool yield varies from 230 to 1000 gm in different strains under different conditions. The females produce slightly more wool than males. The shearing is done normally 4-5 times at an interval of about 90 days. A number of factors like age, sex, shearing interval, feeding regime, management etc influence the yield and quality of wool.

The ***rabbit skin*** is known as **pelt.** The pelt of rabbit is highly valuable product, particularly in cold areas/countries, being used to manufacture woolen garments as well as fancy items for ladies. The pelt is graded depending on the size, colour and hair quality. The pelt with more than 18% wool of dry weight after removal of guard hair is considered good quality pelt rated as class I, more than 18% wool of dry weight of skin with few guard hairs is graded as grade II, 18% wool with intact guard hair as class III and less than 18% wool of dry weight of skin with high percent of intact guard hair as class IV.

The ***manure*** is also produced by rabbit. The rabbit droppings contain nitrogen (2.5 to 3.5 %), phosphorus (1.5 to 2.2 %), potassium (1.5 to 2.0 %) and calcium (2 - 3 %). This can be used in kitchen gardening for vegetable growing. The rabbit droppings after their processing can be used as animal feed.

## 4.13 UTILITY OF RABBIT

There are two types of rabbit viz. **Fur type** and **meat type.** The ***fur type*** is the Angora breed which has 4 strains viz. German Angora (GA), Russian Angora (RA), British Angora (BA), and French Angora. The Angora rabbits have white fur of about 15 cm length weighing 0.3 to 1.0 kg per annum in 4-5 clippings.

The ***broiler rabbit*** breeds are New Zealand White, Gray Giant, White Giant and Soviet Chinchilla. The first breed was originated in U.K. while the other 3 breeds are of Russian origin.

The ***rabbit fur*** is not considered to be a true fur and fetches much lower price than those of Mink and Foxes. The rabbit fur is a byproduct from animals reared mainly for meat production. However, rabbit fur is very important.

The ***rabbit fur skin*** (Pelt) constitutes more than 50 % of fur trade in international market. India imports about half a million rabbit fur skins annually from France and other countries to meet the demand of fur industry.

## 4.14 DEVELOPMENT PROGRAMME OF RABBIT

The *rabbit farming in India* is still in its infancy. The Kerala state had 77187 rabbits according to 1987 census. The numbers of Angora rabbits in H.P. during 1991 – 92 were 29982. The rabbits for meat are getting popular in south and eastern states.

***CSWRI regional station at Garsa, Kulu (H.P.)*** was the first in India to import 4 breeds of ***meat type rabbits*** from U.K. (New Zealand White rabbit) and USSR (White Giant, Grey Giant and Soviet Chinchilla rabbit) in 1979. The CSWRI supplied the breeding stock to different development agencies, universities and private entrepreneurs. There are about 30 – 36 rabbitries in Kulu district (H.P.) with an average population of 150 – 200. About 500 farmers of H.P., some from Punjab and UP are engaged in Angora wool production. The annual Angora wool production is about 15000 kg in HP.

A black brown strain of meat rabbit was developed at CSWRI, Garsa (H.P.) by crossing Soviet Chinchilla and Gray Giant. Another colour strain ***Dutch*** was also developed.

Three strains of ***fur type rabbit*** *viz.* GA, RA and BA were introduced at CSWRI, Garsa (Kulu) initially in 1980. All the three strains were crossed with German Angora. The fur quantity and quality attributes of purebreds and crossbreds were analysed. The German strain was found best fur producer for yield (610 g annual) and fineness (modulation 2.6 %). Russian Angora and British Angora produced on an average 310 g and 380 g fur annually under Indian conditions (Garsa, Hmachal Pradesh) whereas the modulation percentage averaged 2.9 and 3.1, respectively. The staple length averaged 5.5, 6.6 and 6.1 cm for GA, RA, and BA strains whereas the corresponding fiber diameter averaged 12.6, 12.7 and 13.2 micron, respectively. The litter size at birth averaged 5.5, 6.2 and 5.4 whereas litter weight at birth averaged 0.24, 0.30 and 0.27 kg, which increased to individual weight of 0.52, 0.56 and 0.60 kg at weaning for the three strains, respectively.

❑❑❑

# 5 Breed Improvement Programmes III. Camel, Equine and Pig

The details of the 3 species have been discussed separately regarding their breeds, utility, and breed improvement programmes.

## 5.1 EQUINES

The horse and ass belong to the genus Equus, family Equidae (non-ruminants), sub-order Perissodactyles (odd-toed), order Ungulata. The genus Equus has the following species of riding animals –

| Species | Chromosome number (2n) |
|---|---|
| *E. prezewalskii* (wild horse) | 66 |
| *E. caballus* (Horse) | 64 |
| *E. asinus* (Ass/ donkey) | 62 |
| *E. hemionus* onager (Persian wild ass) | 56 |
| *E. kiang* (half asses) | - |
| *E. zebra* (Zebra) | 44 |

The horses, donkeys (asses) and their inter-species hybrids, the mules are singletoed non-ruminant mammals, their toes being enclosed in hoofs. The horses are mainly of two types viz. slow moving called as ponnies and the fast moving, the riding horses used for racing.

The horse breeding in India was in the hands of Rathore's of Marwar and Maharawals of Kathiawad whereas the donkeys were kept by rural poor for their livelihood.

### 5.1.1 Horses

There are 6 important *breeds of horses*. These are Marwari, Katiawadi, Spiti, Zanskari, Manipuri and Bhutia. The two breeds have genes from Arabian horses whereas others have Mongolian inheritance. Due to mechanization, uncontrolled breeding, lack of interest and higher maintenance cost have caused most of the breeds to the point of extinction.

*Utility of horses: Hor*ses are used for draught/ cart, riding and pack animals. They are suitable for quick and easy transport, games/sports (polo clubs), defense and civilian use. They have important role to play in army, police and border security forces, ceremonial parades and cultural rites. In hilly and submountain areas, these are the only means of transport. They had important role to win battles in the war. The riding horses were being used to carry the urgent message before mechanized transport in old days. However, due to mechanization, small land holdings and abolition of *Jagirdaris,* the use of horses in transportation has decreased and their population too.

Milk P*roduction of horse:* The milk of mare contains 1.14 per cent fat, 2.0 per cent protein, 5.87 per cent sugar, 0.36 per cent minerals, 9.42 per cent total solids and 90.58 per cent water.

### 5.1.2 Pack Animals

The pack animals mainly include donkeys, mules and pony (Zanskari). These are reared by rural poor and also in hilly regions. They live and work in harsh conditions due to stress, overwork, overload capable of carrying 100 % of their body weight for short distances with poor nutrition. These animals are thus not properly cared and managed, provided no shelter, remained walking long distances, poorly handled in loading and unloading. Thus these animals suffer to improper handling, transport and husbandry practices.

1. ***Donkeys:*** The **ass** (*Equus asinus*) or **donkey** is available everywhere. It is most simple, hardy and unselfish animal.

   Three distinct types of Indian asses are Indian asses, Indian wild asses and Kiang. There are still **wild asses** (*Equus homionus)* found in India (Rann of Kutch, Gujarat and known as *Kulau)* and also found in Bikaner and Jaselmer (Rajasthan), Sind, Baluchistan, Iran and Iraq. *The Kiang is found in Sikkim and Ladakh.*

   The ass is low or zero input animal and hence economical to maintain as it needs no upkeep. It is let loose by the owner after taking days work for taking its food. It is capable to work on poor roughage, without rest and even without feed and water for longer hours. It is a pack animal and sure-footedness. It is used for various purposes like riding and agricultural operation in Arab countries, and particularly to carry load in India in narrow places where no other draught animal and even the vehicle can reach. It can travel long distances with load and used in defense forces particularly located in extremely remote area of high altitudes. It is thus called a beast of burden. The donkey is known for its stupidity and obstinacy, and bears ill treatment by man.

2. ***Mules:*** The reciprocal crosses between ass and horse have been made to produce the hybrid. The male ass (Jack) was mated with mare (female horse) and the resultant hybrid is called as **Mule.**

The mules can endure more neglect than horses and also endure hot weather better. They consume roughages that are refused by horses. The mule is intriguing (complicate) animal, said to be its own boss. They are shy with strangers and are free kickers. The mule is very much hardy and carry heavy load on narrow mountain paths.

The mules are used for draught, pack and transport particularly in hills. The mule has important role in defense services to carry loads of defense materials for forces including the food and fuel to high altitudes of difficult areas. It is not exaggerate to say that mules are the life lines of defense services because of difficulty to stay without mules in difficult areas of high altitudes.

Adult body weight of draught mule varies from 450 to 650 kg with 155 to 170 cm height. The size, speed, strength and sprit of mules are of horse type whereas sound footedness, lack of excitability, endurance (patience) and ability to survive on poor feed came from ass side. They can thrive on poor fodder, less susceptible to diseases and can endure thirst well. The life span is of about 20 years.

The hybrid produced from reciprocal cross of the female ass (Jennet) to stallion (male horse) was called **Hinny.** However, the Hinny was not found useful.

In Africa, the ass x zebra cross was made to produce **Asbra** which was also not found successful.

## 5.2 ICAR-NATIONAL RESEARCH CENTRE ON EQUINES,

Hisar, Haryana. 1985

Regional station: Bikaner, Rajasthan, 1989.

The NRC on Equines (ICAR) established the nucleus herds of the followings -

- Exotic donkeys of European breed (Poitu) by importing from France in 1990;
- Marwar horses at Bikaner;
- Zanskari ponnies under conservation programme to develop equine sanctuary at Bikaner; and
- Nucleus herd of indigenous donkeys established in 2010.

The following *categories of animals* are maintained *at NRC:*

(*i*) 140 animals of Marwari and Kathiawari horses, Zanskari and indigenous ponies, poitu and indigenous donkeys, mules at Bikaner and

(*ii*) 38 animals of Marwari horses and other ponies, poitu and indigenous donkeys, mules at Hissar.

The centre is working on different biochemical, physiological and hematological indices in equines to evaluate stress conditions of short term and long term feed and water derivation, heat stress, and work efficiency (load carrying capacity of donkeys), and indigenous breed characterization (phenotypic and genotypic) of all 6 breeds of horse and ponny breeds. The research work has been done on semen cryopreservation and A.I., early pregnancy diagnosis by ultrasonography, utilization of animal energy for agricultural operations, and use of donkey fibre to produce carpets by blending with sheep fibres. The NRC is working on nation-wide sero-surveillance and monitoring as well as disease investigation of infectious diseases for their control and eradication.

The NRC on equines is offering services and training to the farmers for A.I. and pregnancy diagnosis.

## 5.3 CAMEL: THE SHIP OF DESERT

The camel belongs to the genus camelus of family camelidae of ruminants, sub-order artiodactyle (even toed), order ungulata (hoofed mammals), sub-class eutheria of class mammalian. The alpaca and Llama belong to the family camelidae.

### 5.3.1 Species of camels

There are two species of camels:

(*i*) Singe humped or Arabian camel called Dromedary Camel (Camelus *dromedrius*)

(*ii*) Two humped or Asian camel called Bactrian Camel (C. *bacteriansus*).

*Chromosome profile:* Both the species of camel have same number of chromosomes (2n = 74) with same morphology with respect to the position of centromere. 5 pairs of autosomes are metacentric/ submetacentric and 31 pairs are acrocentric while both the sex chromosomes of both the species are metacentric.

(*i*) *Single hump camel*: The single hump camel in India is mainly important with adaptation in hot arid and semi-arid regions of Rajasthan, Gujarat and Haryana. The largest population (70 %) of camel is in Rajasthan. There are some camels rearing communities in India like *Raikas* and *Rabaries,* though the camel is reared by other farming communities.

The single hump camels are classified based on the topography *viz.* hill camel and plain camels. The plain camels are again of two type *viz.* riverine camels and desert camels.

The Indian camels can also be classified in another way based on their utility viz. baggage type and riding type.

*Baggage type*: The hill camel and riverine plain camels are the baggage type. They may carry 3 – 4 quintal load and are slow in movement. The hill type camels are more compact with short legs and more muscular development. The baggage camels have well developed hump, thick neck, big head, broad chest, strong legs and sound foot pads.

*Riding type*: The plain desert camels are riding type and hence used for riding. These are lighter animals, more active, have high legs with slender limb, small head, thin neck, small eyes, short ears, thin skin and better stamina capable to cover long distances with a speed of about 9 – 10 km per hour.

(*ii*) *Double hump camels*: Some double hump (Bactrian) camels are found in limited number in Leh (J & K). The double hump calmels have well developed muscular body.

### 5.3.2 Breeds of camel

There are 4 *well defined breeds* of camel viz. Bikaneri, Jaisalmeri, Mewari and Kutchi. The other breeds are the crosses of different breeds viz. Marwari, Mevati, Malvi, Sekhawati, and Sindh camel.

### 5.3.3 Body weight

The *birth weight* of Bikaneri camel is about 40 kg and of Jaiselmeri camel is about 35 kg. Two hump camels are lighter in weight at birth than single hump. The weaning is done at one year of age and *weaning weight* is about 210 kg.

## 5.4 UTILITY OF CAMEL

It is used for carrying load as baggage animal or carting for short distances of 30-40 km/day, agricultural operation including water lifting even without taking rest. The transport of fodder, water, fuel-wood and other material is being done by camels at the places in desert area where other means of transport is not available. The demand of camel as draught animal has presently been decreased due to development of roads and other advancements in agricultural operations, and also due to a change in nomadism into semi-sedentary system. While on work and in general also the eyes of camel are being covered with a leather cover so as it goes on working even without being driven by man while goes in circles for cane crushing and water lifting.

The camel is also being used by man for other purposes like racing, sport, music bands, ceremonial shows, performance in circus show, tourism to earn money, and as a zoo animal too.

Camel also provides milk, fibre and hides. The camel produces more than 1000 kg milk in a lactation. The camel milk has excellent nutritional and therapeutic value. The camel milk and its products do not differ too much organoleptically from cow and buffalo milk. The camel milk does not curdle without boiling up to 7 hours. The camel milk has lactoferrin to protect from infection like tuberculosis.

Camel produces coarser hair from outer coat and fine hair from under coat which can be blended with fine wool. The *wool yield* is about 4 - 5 kg from a camel. The fibre diameter is about 2 – 3 cm with about 20 $\mu$ fibre diameter. The camel hairs/wool is used in preparing blankets and war garments, cushion, bags, mats, cords and ropes from long hair of tail, etc. The camel hair is light, durable, has low conductivity of heat and very warm material. The camels are generally clipped in spring season.

The *camel skin* is about 8 per cent of live weight. The camel skin is of poor quality commercially and used for making bags, saddler, surrahi and containers for storing oil and animal fat, ropes and fancy items of handicraft industry like purses, gloves and lamp shades, etc.

## 5.5 SPECIAL ADAPTIVE FEATURES OF CAMEL

The camel is called as ship of the desert capable of travelling long distances on sand carrying men and material and it has *thick foot pads* for which it can walk and run on sand (hot and cold), gets protection from hot sand and also does not cause soil erosion due to disc like foot pads. The camel has *fine hair under coat* which acts as insulation to protect from high ambient temperature.

It has the *power to resist thirst and hunger* and hence it can live without feed and water for a long time without loosing appetite. It has adaptive mechanisms particularly the water retention in the body and ability to thrive on water deprivation. It has greater thickness of skin to reduce the water loss to the minimum. It has hump of fatty tissues which acts as an emergency food store.

The camel has *browsing habit* to eat top feeds (natural tree leafy vegetation) that are out of reach of other animals but it can access to branches of high trees due to *long neck*. It can also eat hard and thorny plants (bushes) that are not taken by other animals because it has *hard, mobile and bifid upper lips*. Thus, the camel has no competition for food with other animals and the camel can survive by taking tree tops when nothing is available to graze on ground. It has better feed conversion efficiency.

The camel has *long eye lashes and skin fold* over the eye and *long hairs in the ears* which protect it from blowing desert sand. Moreover, it may *close its nostrils* to prevent excessive moisture loss and entrance of sand during sandstorms.

The camel is a *mysterious animal*. The camel though has extraordinary patience if neglected and mishandled but to a certain limit after which it shows obstinacy and become revengeful.

There is a need of research on physiology for ecological adaptations and genetics of camel to use the camel genes for breeding of other animals through new genetic and reproductive techniques to make the other animals suited to the ecological and extreme environmental conditions under which the camel survive and work.

## 5.6 ICAR0NATIONAL RESEARCH CENTRE ON CAMEL,

Bikaner (Rajasthan), 1984.

The ICAR established National Research Centre on Camels at Bikaner. The NRC on camel came into existence by establishing Project Directorate on Camel at Bikaner on July 4, 1984 after transferring the erstwhile Camel Breeding Farm, comprising 149 camels of Bikaneri breed and 824 ha land and by upgrading this farm as NRC on Camel on September 20, 1995. The main focus of NRC on camel is on basic and applied aspects of one-humped camel for which an elite herd of 355 animals of Indian camel milch breeds (Bikaneri, Kachchhi and Marwari) and race breed (Jaisalmeri) was established at Bikaner.

The centre has developed complete feed blocks and pellets by mixing of fodders, agri- byproducts and concentrate, area specific mineral mixture (CAMMIN), evaluated milk production potential of Indian dromedary (one hump camel) to be 7 litres/ day to a maximum of 10.0 litres; with 2.8 % fat, 2.5 % protein, 4.25 % lactose and 9 % T.S.; and the average lactation length was found 16 months. The camel milk has higher minerals and protected proteins with low fat % having high quality higher chain fatty acids, higher percentage of calcium, vitamin C and micro minerals (iron, cu and Zn) for which the camel milk has beneficial value than cow and buffalo milk. The value added products from milk like cow/buffalo milk have been developed, skin products to manufacture belt, parts and container for storing oil, show pieces, bone products as a cheap substitute of ivory for production of jewellery and ornamental pieces and hence afforded employment opportunities for rural artisans.

The NRC is providing services to the stakeholders for quality germplasm, complete feed blocks and pellets, preparation of value added products and training in camel management.

## 5.7 PIGS

The pig belongs to genus *Sus* (non-ruminants), family suidae of sub-group suina of the sub-order artiodactyles, order ungulata, sub-class eutheria of class mammalian. There are two following species of pig -

(*i*) S. scrofa (wild boar) and

(*ii*) S. domestica (domestic swine).

The domestic pigs have 19 pairs of chromosomes (2n = 38). Of the autosomes, 12 pairs are submetacentric/ metacentric whiles rest 6 pairs are acrocentric. The X-chromosome is submetacentric while Y chromosome is metacentric. The European wild pigs have 18 pairs of chromosomes (2n = 36).

The pigs are now commonly found throughout the world except Muslim countries for the reason of religious taboo against pigs.

### 5.7.1 Pig Breeds

Domestic pigs found in India are believed to have a common origin with the wild pig (*Sus scrofacristatus*). Besides different exotic breeds of pig, *viz.* Large White Yorkshire (LWY), Hampshire, Duroc, Large Black and Landrace, there are several types of indigenous pigs available in India.

***(i) Native breeds:*** The available indigenous pigs are listed below.

| States | Breeds/local name | States | Breeds/ local name |
|---|---|---|---|
| Andman | Andman Wild | Nagaland | Suho (Nagaland local) |
| Nicobar | Nicobari Pigs | Mizoram | Zovwak |
| Tripura | *Doom, Mali* | Meghalaya | Niag Megha/ Khasi local |
| Assam | *Doom, Pigmy hog* | Bengal | Ghungroo (Regd. Breed) |
| *Sikkim* | *Lepchamoun* | *Orissa* | *Golla, Burudi* |
| *Manipur* | *Manipuri, Desi* (Burmese pigs) | Kerala, TN | Angamali |

The pigs found in UP, Bihar, MP, Punjab are nondescript (Desi).

*Population of pigs in India:* The livestock census (2007) revealed that the pig population is mostly concentrated in eastern and north eastern (NE) states (72.49 lakhs), followed by northern (15.53 lakhs), southern (11.12 lakhs), western (6.16 lakhs) and central Indian states (6.06 lakhs). The highest population is found in Assam (20 lakhs), followed by Uttar Pradesh (13.5 lakhs), West Bengal (8.15 lakhs), Jharkhand (7.3 lakh), Nagaland (6.98 lakhs), Bihar and Orissa.

(*ii*) **Exotic Breeds of Pigs**

| Name of Breed | Country developed | Migration/developed |
|---|---|---|
| 1. Yorkshire | Yorkshire (England) | USA (1893) |
| 2. Berkshire | Berkshire (England) | English hog with Chinese and Siamese sows |
| 3. Duroc | USA | Crossing pigs from Africa, Spain and Potugal |
| 4. Tamworth | Ireland | – |
| 5. Landrance | Denmark | Cross of English pigs on native pigs |
| 6. Polled China | USA | Cross of Berkshire, Hampshire, Tamworth and Russian pigs. |

### 5.7.2 Production characters in pigs

The *birth weight* is about 0.9 kg, *weaning weight* (2 months) is 4.15 kg, *litter size* at birth is 7.5 and *weaning percentage* is 54.0 percent. The indigenous pigs attain body weight about 30-45 kg at 18 months of age, though some local strains have higher body weight up to 70-80 kg.

The *dressing percentage* in swine is about 65-80 percent of their live weight which is higher than lambs/sheep (45-55 %). The edible meat in the carcass of hogs is more due to small production of bones.

## 5.8 UTILITY OF PIGS

Among the various livestock species, pig is the most potential source of meat production and more efficient food converter after the broiler chicken. The pigs make efficient use of farm and household by-products to convert into meat. Feeding of about 3.5 to 4.0 kg feed to pigs produces 1.0 kg pork. They have high fecundity, high feed conversion efficiency, short generation interval and relatively small space requirement. They are equally adapted to diversified and intensive agriculture.

The pork and lard are two products from pigs. The demand of lard has decreased after starting the use of the vegetable oils and fats in cooking and with the decrease of animal fat for soap making. Moreover, there has been a change in the liking of people preferring lean cuts and loins.

Pork production in India is largely supported by indigenous pigs (78.54%), followed by crossbred (20.14%) and exotic (1.32%) which include Yorkshire, Hampshire, Large Black, Landrace and Duroc. About 91% of indigenous pigs and 85% of crossbred pigs are reared in rural areas. While population of crossbred pigs registered an increase of 9.59% between the periods from 2003 to 2007, indigenous pig population witnessed a decline of 22.89% for the corresponding period.

The indigenous (desi) pigs which are smaller in size have been the basis of pig production over the years. There has hardly been any genetic improvement programmes for improving its economic traits, such as litter size, birth weight, weaning weight, average daily gain, feed conversion efficiency and carcass traits. Use of improved breeds for grading up as well as crossbreeding has gained momentum though lately so that crossbred pigs would form the basis for future pig production especially in rural areas to improve the productivity of small sized rural pigs.

## 5.9 PIG PRODUCTION SYSTEMS

The pig farming in India is practiced entirely by poor people belonging to lowest socio-economic strata with no or little resources in rural area who have practically no to undertake scientific pig farming with improved foundation stock, proper housing, feeding and management. Pig farming has been accepted socially and culturally by certain ethnic groups only, especially black ones, for festivals and ceremonial purposes. Interestingly, these ethnic groups are mainly concentrated in the North-Eastern Region where 40% of the country's total pig population exists. Piggery is unevenly distributed to almost all the eco-regions of the country and is an important occupation of the rural society especially the tribal masses.

The trend of pig farming is changing gradually due to urbanization. The farm size is usually small, but it is gradually coming up in the form of commercial farms. In comparison to other livestock, crop cycle of pig farming is shorter. Generally, two farrowings per sow per year can be obtained with harvest of 4-8 piglets in single farrowing. However, considerable piglet mortality during weaning period continues to remain as a challenge. Particularly in rural area, pig farming is based on agricultural by-products and kitchen wastes. Traditionally free-range feeding of pig is quite common among tribal people. Most household keeps 2–3 pigs and most of the pig farmer rear pigs in the scavenging (extensive) system.

Pig husbandry is practiced under a variety of production systems ranging from simple backyard piggery, pigs living on garbage belts, family operated farms to large scale integrated pig farming. According to holding size, 78.31% of pigs were with marginal and small farmers, 20.03% with semi-medium and medium farmers and only 1.66% with large farmers during 2006-07. The predominant is backyard or scavenging system with practically zero input. The system is characterized by low output resulting from low producer pigs. In the semi-intensive system, farmers mostly keep crossbred and indigenous pigs of better production potential, feeding them with locally available grains and feed ingredients mixed with oil cake and bran along with food waste, with provision of housing mostly as night shelter. Both backyard and semi-intensive system faces considerable losses because of piglet mortality. In the large scale production system, mostly purebred improved breeds are intensively reared with balanced diet and assured veterinary services.

The traditional small scale system is characterized by high mortality rate, low off take, absent or minimal health care, supplementary feeding and improper housing. Free range is practiced by small farmers mainly in rural areas where pigs are fed mostly on grass, brewery, cereal byproducts, waste products and food remnants. Free range and tethering systems have been on decline because of cropping intensity, diseases and worm infestations. Additionally, a commercial small-scale system is found and characterized by improved breeds fed concentrates and having relatively good performance. The large-scale system is characterized by improved breeds and they are confined mainly to government and institutional farms.

Mixed/integrated pig farming is also common, whereby local feeds such as cereal residues and potatoes are the main source of feeds to pigs. In turn they provide manure for land fertility. The integration of animals and intensive cropping in India offers good future prospects for economic and sustainable growth based on organic agriculture. The intensive pig production system is economically viable in areas with shortage of land to grow feeds and in large cities because of availability of industrial by-products.

## 5.10 PIGGERY DEVELOPMENT PROGRAMMES

There is acute absence of private organizations involved in pig breeding and pork production on commercial scale. The piggery development programmes in the country have been restricted to government and institutional farms. The pig farming received attention under the national Five-year Plans.

A coordinated piggery development programme was started under which 27 pig breeding farms, 105 piggery development blocks and 8 bacon factories were set up in different states.

During II and III Plan periods 8 bacon factories were established. These were at Aligarh (UP), Alwar (Rajasthan), Kharar (Pb), Ranchi (Bihar), Ernakulum (Kerala), Bombay, Hrishna distt (AP), and Mohanpur (W.B.).

Under III plan (1961 -66), 5 regional pig breeding cum bacon factories were set up whereas in the next IV plan (1969 – 74), 4 more were provided.

There are 123 ***pig breeding farms*** under veterinary/animal husbandry/animal resource development and Livestock Development Boards of different states catering to the needs of farmers. Majority of the farms (64 numbers) are located in north eastern states of India.

***AICRP on Pig:*** This was started in 1970 with the objectives of developing region specific technologies, at ten centres attached with different ICAR Institutes (IVRI) and SAU's (Assam Agr. Univ., Khanapara; APAU, Tirupathi; JNKVV, Jabalpur). These centres are spread over different agro-climatic regions of the country.

NRC on Pig is engaged in coordinating and collaborating the research and development of the AICRP centres both in terms of technical and financial aspects in consultation with ICAR. Major achievements of the centres are as follows:

(1) Successful raising and multiplication of exotic breeds of pig *viz.* Landrace, Large White Yorkshire and Hampshire under organized farm conditions,

(2) Genetic improvement of indigenous pigs through selective breeding especially for litter size and growth rate traits,

(3) Development of crossbreds of different exotic inheritance with consistent superior performance with respect to traits notably litter size and weight, growth rate and better feed conversion efficiency, and

(4) Successful incorporation of piggery in integrated farming system with considerable economic gain.

Under the AICRP on Pig, and National Research Centre on Pig, a number of crossbred groups (synthetic and experimental crosses) have been developed at constituent centres. Some of them are described below:

Birsa Agricultural University, Ranchi, has developed a new genetic group of pig named "T&D" by involving Tamworth breed (British) and desi pigs of Jharkhand. Animals of this genetic group are black in colour with lustrous skin which is preferred by the farmers of the state. It attains body weight of 90-100 kg at one year of age under village conditions as compared to 30-40 kg for Jharkhand desi pigs. Litter size at birth is 8-12 and farrowing interval is 180-200 days.

Kerala Veterinary and Animal Sciences University, Mannuthy, has developed 50 and 75 percent cross of Large White Yorkshire (LWY) with desi pigs of Kerala. Litter size at birth and weaning of 50 percent crossbred is 7.47 ± 0.50 and 7.20 ± 0.55. The crossbred with 75 percent LWY inheritance has litter size of 7.64 ± 0.38 and 7.40 ± 0.34 at birth and weaning, respectively.

Tamil Nadu Veterinary and Animal Sciences University, Kattupakkam, has produced and studied the performance of 50 and 75 percent crossbred of LWY with desi pigs of the state. Performance of 75 percent cross has been found to be consistently superior. Litter size at birth and weaning was 6.2 ± 0.45 and 5.84 ± 0.43 in 75 percent crossbred as compared to 5.94 ± 0.63 and 5.55 ± 0.63 in 50 percent crossbred. Similarly, litter weight at birth and weaning was 7.08 ± 0.52 and 50.59 ± 3.79 kg in 75 percent cross as compared to 6.18 ± 0.70 and 42.65± 4.79 kg in 50 percent crossbred. The centre has also developed a three-way synthetic pig of genetic composition Large White Yorkshire (25%), Landrace (25%) and Duroc (50%) for lean meat production. They have better feed conversion efficiency and low back fat thickness. Live weight at slaughter (8 months of age) is 90-100 kg.

Assam Agricultural University, Guwahati, has produced crossbreds of Hampshire and Assam local pigs with 50, 75 and 87.5 percent exotic inheritance. Performance of 75 percent crossbred has been found to be superior compared to 50 and 87.5 percent exotic inheritance.

***Megaseed Project on Pig:*** This project was started by ICAR in 2008. The main objective of this project is to produce and supply quality pig germplasm to the farmers. NRC on Pig is coordinating the project with four centres. Each centre has a production target of 900 piglets per year. The programme is progressing well since inception.

***Pig Development Scheme***: This is a central sector scheme which was implemented by the Department of Animal Husbandry, Dairying and Fisheries in 2010-11 through National Bank for Agriculture and Rural Development (NABARD). The Scheme envisages achieving the following objectives:

(*a*) Encourage commercial rearing of pigs by adopting scientific methods and creation of infrastructure,

(*b*) Production and supply of improved germplasm,

(*c*) Organizing stakeholders to popularize scientific practices, (*d*) Create supply chain for meat industry, and

(*e*) Encourage value addition for better income.

***Conservation of Threatened Breeds of Livestock***: This is a centrally sponsored scheme which also includes pig. This scheme is run by the department.

## 5.11 ICAR0NATIONAL RESEARCH CENTRE ON PIG,

Rani, Guwahati, Assam. 2002

There were 4 existing units of AICRP on Pigs (IVRI, Izatnagar; JNKVV, Jabalpur; AAU, Khanapara; and APAU, Triputi). These were to serve as centres of network and there was a proposal to include 4 more centres (ICAR complex, GOA; KAU, Mannuthy; BAU, Ranchi; and NRC, Shillong). It was decided that these centres will be coordinated by NRC on Pigs to be established at Shillong from January, 1993.

National Research Centre on Pig was established in 2002 in Guwahati, on the recommendations of review committee of AICRP on pigs (1995), considering the importance of pigs in the N-E Region. However, it started functioning in 2005 in Guwahati and later on shifted to its own building in 2009 in Rani (Guwahati). The objectives were to undertake studies on genetic resources, their conservation and improvement with other aspects of nutrition, reproduction, management, health, and meat technologies.

The NRC on pigs collected two strains of indigenous pigs (Ghungroo and Niang Megha) and characterized for their morphological, cytogenetic and DNA profile; crossed Ghungroo with Hampshire to produce crossbred variety; worked on boar semen preservation and AI in pigs for the first time in Assam; studied the unconventional feed resources and standardized the techniques for value added pork products (nuggets, patties, kebab, momos, cocktails, frankfurter, salami, ham slices). The NRC also provides services to stakeholders for improved germplasm (crossbreds), A.I. and value added products.

National Research Centre on Pig, Guwahati, has produced three experimental crosses, namely Hampshire x Ghungroo, Hampshire x Niang Megha, and Duroc x (Hampshire x Ghungroo).

The productive, reproductive, adaptive and carcass traits of developed crossbreds, namely Variety I (50%): Hampshire x Ghungroo, and Variety II (50%): Hampshire x Niang Megha has been studied in detail under both organized farm and field (extensive and semi-intensive) conditions. Performance of both the varieties has been found promising. However, variety I had been found to be superior performer. Average litter size at birth and weaning in variety I is 9.86 ± 0.25 and 8.81 ± 0.40 as compared to 6.94 ± 0.61 and 6.63 ± 0.61 in variety II. Weight at 8 months of age (slaughter age) in variety I is 71.56 ± 1.06 kg as against 57.49 ± 1.62 kg in variety II.

A three-way cross of 50 percent Hampshire x Ghungroo with Duroc as terminal sire is also being studied which has shown high potential for lean meat production with superior growth rate.

❑❑❑

6

# Breed Improvement Programmes IV. Poultry

The poultry belong to phylum chordate, sub-phylum vertebra, class aves, order galli form and ansari form (aquatic: ducks, geese, Muscovy) comprising many genera and species.

The term poultry includes many kinds of bird *viz.* chicken (domestic fowl) turkeys, guinea fowls, pea fowl, pigeons, pheasants, quail, swans *(hansa) or water fowl,* geese, ducks, ostriches, emu, and others. Though India is known as the home of domestic fowl, but the poultry keeping in India has been largely a rural cottage enterprise with scant care and attention to village hen.

The distinguish feature of poultry is the presence of comb. Fowls have relatively high breathing and pulse rate as well as high body temperature (105 – 109° F) than other domestic animal. The alternate term to the poultry science is *Ormithology* which is the study of birds other than poultry.

## 6.1 CHICKEN BREEDS

(*i*) Indian *breeds of domestic fowl* or native birds (chickens) of India found in different states are as under:

Kashmir (Faverola), Punjab and Haryana (Punjab Brown), U.P. (Brown Desi),

Gujarat (Ankaleshwar), Gujarat and MS (Bursa), M.P. (Khadknath or kalamasi),

AP (Assel, Denki, Kalasthi), AP and Karnataka (Ghangus), Kerala (Tellichery),

W. B. (Haringhata Black), Assam (Miri, Daothigir), N. E. region (Chitagong or Malay)

Hot and humid costal region, and N.E. region (Frizzle Fowl, Nacked Neck),

Nicobar Island (Nicobari), and all over country (Teni, Titri).

(*ii*) *Exotic Breeds of Chickens*: There are 5 classes of chickens based on their origin *viz.* Asiatic, American, English, Mediterian and Continental European. The important breeds of different classes are –

| Classes of chicken | Breeds |
|---|---|
| American | Rhode Island Red, Plymouth Rock, New Hampshire, Whyandotte |
| English | Australorp, Cornish, Orphington, Sussex |
| Mediterranean | Leghorn, Minorca, Andalusian, Spanish, Ancona |
| Asiatic | Brahama, Cochin, Langshan |
| Continental | Capine, Lakenvelder |

## 6.2 OTHER POULTRY SPECIES

The poultry species other than domestic fowl reared in India are:

1. ***Guinea Fowl*** (Numeda meleagris galeata -***Titri*** *or tatiri*): It is native of France. It is known by different names like Titri (Northern plains), Chittra (Western parts) and Chena murgi (South and Eastern India). Its main varieties are called as ***Pearl, Lavender and White.*** The pearl is most common in India.

2. ***Japanes Quail*** (Coturnix coturnix japonica -***Bater***)

   It is raised for both egg and meat. It is a small bird with rapid growth, early maturity, has high rate of lay (more than 250 eggs annually) and require less feed and space. The Indian native quail has 3 varieties viz. ***Rain quail, Grey quail and Button quail.*** Its adult body weight is only 40 gm. The Japanes quail were imported in India in 1974 from USA, Germany and Korea. The Japanes quail weigh between 100-300 gm.

3. ***Ducks*** (*Anas platyrhynchos* - ***Battakh***)

   The ducks are popular in NE states, WB, Orissa, AP, TN, Kerala and Tripura where it has shown increasing trend. The ducks are also found in planes of Assam and Tripura. Some breeds of ducks found in Assam are *Pati ducks* found near Brahamputra, *Nageswari* found in Barak valley, and *Chara Chemballi ducks.* These are reared for egg and meat. The age at first egg is 150 days.

   The important Indian breeds are Nageswari, Synthetemete, and Indian Runner producing about 200, 80-129 and 80-150 eggs per annum, respectively.

4. ***Turkeys*** (*Meleagris gallopavo*)

   The turkeys were first domesticated in West Central Mexico. These are exclusively raised for meat production. The males weigh between 5 – 8 kgs and the females between 3 – 4 kgs. They lay egg of about 60 gm and lay about 30 eggs before going broody. It takes about 12-14 weeks to produce turkey broilers.

   The important turkey breeds are: Bronze, White Holland Bourbon Red, Narragansett, Beltsville Small White, Broad Breasted Large White, Broad Breasted Bronze. The turkeys are reared at CARI Izatnagar; HAU Hissar; College of Agriculture Udaipur; UAS Bangalore; Central Poultry Breeding Farm Hessarghata.

5. ***Domestic Pigeon*** *(Columbia livia domestica -* ***Kabootar)***

   *It* is the first bird tamed by man 4500 B.C. The domestic pigeons have been descended from wild rock pigeons (Columbia livia) of Asia and Europe. The pigeons are most commonly present at the places of worship and memorial estates. The pigeon rearing is for recreation, pleasure and meat production. The Mughal kings introduced the pigeon farming in Indian sub continent.

   The pigeon breeds are of two types viz. fight and show purpose breeds and meat breeds. The fight & show purpose breeds are Kabli, Patiala, Lotan, Saharanpuri, Hyderabadi, Aseel, Lakka, Zeera, Maina Jog, Chandan Chuha, Nisvari, Kherni, Girbaz. The heavy weight or meat breeds are Naqab Poosh, Gola, Umer, Lahores.

   The *pigeon* rearing is very popular in Assam and many NE states for meat. The meat of pigeon squab is delicate and fetch good price in the market. It requires study of the domestic pigeon diversity of Assam.

6. ***Giant size birds*** (Ostrich and Emu)

   The adult body weight of *Ostrich (Struthio camelus) and Emu (Dramaius novehollandiae)* is about 11.5 kg and 55 kg with average height of 2.5 m and 1.6 m, respectively. The egg weight is about 1.6 and 0.6 kg with average number of eggs produced for breeding season is about 40 -100 and 45 – 50 for two species, respectively. They live longer for about 70 years and 50-60 years, respectively. The age at sexual maturity is 2-3 and 3-4 years for female and male Ostrich and the corresponding ages are 2 – 2.5 and 3 years for Emu female and males. Both these species are raised for feathers, nutritious lean meat, leather and oil.

   The Ostrich are kept at Maruti Ostrich Research & Training Farm, Bangalore and the Emu at Kainaka Retite Farm, Gorantla, Guntoor (A.P.).

## 6.3 CHROMOSOME NUMBERS

The genes which are the basic units of inheritance are located on chromosomes. All the species of poultry are diploids having chromosomes in pairs. The chromosome numbers of different species of poultry are given under section 19.2.

There are two types of chromosomes regarding their size in different avian species. These are of very large size known as macro-chromosomes which have definite recognizable features and second types are of small size known as micro-chromosomes. The cytologically distinct macro-chromosomes are 8 pairs including sex chromosome pair (Z and W) and the rest 30 pairs are of small size (micro—chromosomes) which are cytologically indistinguishable.

The sex chromosomes in poultry are designated as Z and W instead of X and Y of the mammals and Drosophila. The male sex in avian species is homo-gametic (ZZ) and the female sex is hetero-gametic (ZW), unlike mammals. Some genes are also present on sex chromosomes like mammalian sex chromosomes. These genes present on sex chromosomes are called sex-linked genes and the characters controlled by sex-linked genes are called as the sex-linked characters, like in case of mammals and Drosophila. The inheritance of the sex-linked characters is just opposite to those of sex-linked characters of mammals. Therefore, a hen transmits the sex-linked character to her sons but not to her daughters whereas the cock can transmit the sex-linked characters to his sons as well as to the daughters.

This information on transmission of sex-linked characters is very useful for auto-sexing in day old chicks. The rapid feathering chicks, produced from mating of homozygous males for rapid feathering to hens for slow feathering, will be all females but slow feathering will be males. Not only this information on inheritance pattern of sex-linked characters is used for qualitative traits but is also used for quantitative traits. This can make use by making the reciprocal crosses. In case of sex-linked inheritance, the mean of crossbred progeny produced from crossing of two strains or lines will be close to sire line than dam line.

## 6.4 POULTRY UTILITY

The various poultry species are reared for egg production and also for meat production. However, some breeds are reared for entertainment.

## 6.5 POULTRY DEVELOPMENT PROGRAMMES

The poultry improvement was first initiated by Christian Missionaries by establishing first poultry farm at Etah (UP) in 1912 and Katpadi (TN) during 1920's by importing exotic breeds from UK, America and Australia.

Selective breeding of indigenous fowl was started in 1940 at IVRI Izatnagar (UP). The crossbreeding was also initiated at the same time using exotic breeds.

A poultry development research scheme at Nagpur was undertaken by state Govt. in 1950's for evaluating the native chickens. The breeds evaluated were Aseel, Kadaknath, Black Bengal, Naked Neck and Haringhatta Black. The exotic breeds used for grading up/crossing were White Leghorn, Rhode Island Red, Black Australorp and New Hampshire.

The major work of poultry research was started during the last phase of first Five year plan. Scientific poultry rearing was started (during1957 of II Five Year Plan) by All India Poultry Development Project. About 30000 sexed day-old chicks from USA and 5000 hatching eggs of White Leghorns, Rhode Island Reds, New Hampshires, and White Cornish were imported and the foundation breeding stock was multiplied by setting 5 Regional Poultry Farms at Delhi, Simla, Bangalore, Bombay and Bhubaneshwar, and 24 State Government Farms under all India poultry development project.

In 1959, under USAID – India programme, 4500 White Rock chicks (meat strain) were imported from Israel. The White Rock Breeding Farm in Rome sent free of cost 2500 chicks to ICAR broiler scheme at Hyderabad.

In 1960's some private sectors viz. Arbor Acre, Hi-Breed, Rani Shaver and Unichix entered into foreign collaboration of Canada, Czechoslovakia and USA for commercial production of genetically superior chicks. The poultry business had also attracted some poultry farmer viz. Venketeshwara hatcheries, Kegg farms, CSR Ross Breeders, Poona Pearls, etc. The other leading international brand names available in India are Hy-line, Hypeco, Anak, Hubbard, Tegel, Dekalb, Vencobb, H & N, etc. The commercial egg type stocks are result of three/four ways crosses of White Leghorn whereas the meat stock involved the crosses of Cornish, Plymouth Rock and synthetic broilers strains.

The breeding stock was imported to develop poultry at Jorhat Agricultural College, Assam.

The ICAR had started improvement programme of native breeds like Aseel, Chittagong, Punjab Brown and Desi. The improvement programme also involved the upgrading of indigenous breeds with exotic breeds.

A number of projects in 1960's were started by ICAR on improvement of poultry for egg production by family selection, breeding and hybridization; improvement of broilers by selection and crossbreeding; and the improvement of native chickens. In late 1960's, the evaluation and improvement of indigenous fowl was started under PL-480 Project at two centers viz. PAU, Hisar and College of Agric. Udaipur.

In 1962 an importation of 2500 White Leghorns through Kansa State University-USAID-India programme was received from Rome at Osmania Veterinary College. 300 strain-cross White Leghorns day-old chicks were donated by West Germany to Osmania Veterinary College and also to College of agriculture, Ludhiana.

During mid-sixties, a breeding programme was started by importing 10, 000 chicks of 7 strains of high producing pureline stock from Australia under the Freedom From Hunger Compaign, to bred at Govt. of India Regional Farms. The Govt. of India had started Poultry Breeding on a systematic basis at four Central Poultry Breeding Farms (Hissarghata, Bombay, and Bhubaneshwar) to develop high egg producing hybrid strain and CPBF (Chandigarh) to develop fast growing broiler strains. The chicks over 2 millions produced upto 1971 comprising pure lines and crosses were given to State Farms and private breeders. The parent stock to hatcheries and hybrid chicks to farmers were being supplied by CPBF's.

The GOI decided to import a large number of purelines of layers and broilers and to set up large regional research units under AICRP on poultry for eggs and meat with coordinating unit at IVRI, Izatnagar.

***AICRP on Poultry Breeding*** for eggs and meat was started in 1971 at a number of research centres under SAU's and ICAR institutes. The programme for *egg production* was to test 13 layer strains at different centers with 3-4 strains at each centre, as purebreds and inter crosses together for their improvement through combined selection. Eight strains of *broilers* were tested, at different centers each with 3-4 strains, as purebreds as well as crosses and intra population selection based on 8 weeks weight. *Central Avian Research Institute*, Izatnagar (UP) was set up by ICAR in 1979 to upgrade the division of Avian Research of IVRI, to carry out research and training programmes on poultry and to coordinate two AICRPs on poultry. The institute included quail for egg and meat and subsequently other species of poultry (gunae fowl, ducks,, turkeys) were also included.

Regional Feed Analytical Laboratories were set up to provide facilities to farmers and feed manufacturers at Chandigarh, Bombay, Bhubaneshwar and Hissarghata.

The Random Sample Test Units were set up at Bombay, Bhubaneshwar and Hissarghata, to conduct egg laying test and broiler quality test for providing useful information to poultry breeders on the performance of various participating stock.

The Poultry Extension cum Development Centers each with a unit of 100 layers of improved breeds were set up at 269 places.

Central Training Institute for Poultry Production and Management was set up at Hessarghata to impart training to poultry farmers and Officers.

Central Duck Breeding Farm was also set up in Hissarghata for supply of day old ducklings of high egg producing breed of Duck (Khaki Campbell) to various states.

## 6.6 ICARORESEARCH INSTITUTES ON POULTRY

The followings are two research institutes on poultry working under the administrative and research control of ICAR.

### 1. ICAR Central Avian Research Institute,

Izatnagar, Bareilly, U.P. 1979

Regional station: Bhubaneswar (1992).

The ICAR established this institute at Izatnagar on 2nd November, 1979. It is a premier institute of ICAR on poultry to conduct research, education, extension and training for promoting productivity and profitability of Indian poultry industry.

**Chicken:** The institute has developed commercial layer chicken (*cari priya, cari sonali*), dual purpose chicken (*CARI Debendra*), commercial broiler chicken (*CARIBRO Vishal, CARIBRO Dhanraja, CARIBRO Mritunjay, CARIBRO Tropicana* and low input Desi chicken varieties (*CARI Nirbheek, CARI Shyama, UPCARI, Hitcari*).

**Japanese Quail***:* This institute has developed the farms of Japanese quail for egg and meat production throughout the country and developed *CARI Uttam, CARI Ujjawal, CARI Sheweta, Cari Brown, CARI Sunheri* for meat and *CARI Pearl* for egg.

***Guinea Fowl:*** It has improved 3 varieties of Guinea fowl (Kadambari, Chitambari and Swetambari) for meat and a variety (Guncari) for egg.

***Turkey:*** The Turkey, the choicest white meat, has been propagated in different parts of the country and developed *CARI Virat* strain.

***Ducks:*** Four breeds of Ducks (Moti, Desi, Khaki Campbell and White Pekin) are maintained at regional centre, Bhubaneswar.

The institute has developed some technologies for reduction of yolk cholesterol by supplementing Guggal alone or in combination with Arjun, Amla and Cinnamon in the diet of hen, supplementation of 125 mg Cu/kg diet as copper sulphate; developed diluents for chicken and quail semen; fertility prediction test for male Japanese quail; in-ovo feeding and vaccination technique for improved broiler production by administering the dietary constituents (amino acids, glucose and vitamins) and vaccines (Newcastle disease vaccine) into the amnion or yolk sac of embryo; technique of tenderization of tough hen meat by soaking carcass for 3 hours in 0.05 % papain or 0.075 % trypsin in combination with 1 0 % NaCl and 0.05 % sodium tripolyphosphate; oil spray coating for preservation of eggs with ground oil mixed with 200ppm BHT as an anti-oxidant or liquid paraffin, only 1.0 kg oil is enough for coating of 7500 eggs to preserve them safely for 30 days against 7 days at mean ambient temperature of 31° C and for 120 days self-life against 30 days at refrigeration temperature, respectively. The institute has developed processing technology for value added egg products and poultry meat products.

The institute is providing services for poultry production (layer and broiler, Quail, Turkey, backyard poultry), poultry nutrition and feeding, poultry processing and product technology, poultry housing and management, poultry disease control, poultry hatchery management, avian physiology, and marketing to public and private sector organizations, feed analysis on cost basis, orientation training programme of 6 days every year to start poultry farming as well as sponsored training courses on poultry production management on demand for the farmers nominated by the State A.H. Deptt, ATMA, NABARD, Bank officers, Veterinary field officers.

## 2. ICAR Project Directorate on Poultry,

Hyderabad (A.P.) 1988

*Historical Background*: The ICAR sanctioned two coordinated projects on poultry for eggs and meat which were merged to a single project known as AICRP on Poultry Breeding during V Plan at Izatnagar, Bareilly (U.P.) in 1970. It was elevated to the status of Project Directorate during VII Plan and shifted to Hyderabad at APAU Campus, Rajendernagar (A.P.) w.e.f. 1$^{ST}$ March, 1988, in 30 acres of land on 99 years of lease basis to ICAR.

A dual purpose poultry variety, ***Vanraja,*** was developed for rural and tribal areas fed on kitchen waste, natural vegetation and do not require any commercial feed. The birds attain adult body weight of about 2.0 kg at 12 weeks of age with annual 100 egg production.

Another variety, ***Gramapriya,*** layer type was developed for free range farming in rural and tribal areas, with an adult body weight of about 1.4 kg at 12 weeks age with annual production of 160-200 eggs annually.

Another multipurpose broiler, ***Krishibo,*** developed for small scale farming of rural areas. It has good quality meat. The chicks of all the three newly developed varieties have been distributed.

The Directorate also provides scientific and technical services on supply of chicks and fertile eggs to farmers, for specialized poultry farm operations to the public and private sectors, farmers, NGO's, extension agencies and other stake holders.

❑❑❑

# 7 Constraints of Development Programme

About 70 % of the milch animals are kept by landless and small/marginal farmers in India and they keep 1 – 2 milch animals under mix farming system. These animals are nondescript and low producers. However, the regular expenses of these farmers are met from sale of raw milk in the local market produced by cattle and buffaloes and this supports their livelihood during off season of earnings from cash crop. Not only this but animal rearing by tribal and nomads have been the only source of their earning and livelihood. Therefore, rural economy of India is mainly affected by livestock sector. Since the livestock sector has a potential and a great role in the upliftment of rural poor, the necessary efforts are needed to improve the productivity of rural cattle and buffaloes for the economic development of rural man.

## 7.1 DEVELOPMENT

The word development is generally used for human development and means to make provision for satisfaction of basic needs for human beings. Therefore, the developments means and comes from the satisfaction of basic needs of food, clothing, and shelter together with provision of drinking water, public transport, health and sanitation, energy (electricity), and education facilities.

The livestock and poultry have significant role in meeting two basic needs of food (milk, meat and eggs) and clothing (wool, pashmina) to human as well as shoes (skin and hides). Therefore, it is imperative to think of the development programmes of livestock and poultry. The development is made by use of natural, human and financial resources through formulation and execution of relevant programmes by the controlling authority (Government). Besides, some NGO's are also now a days helping in the execution of development programmes launched by the Government.

## 7.2 LIVESTOCK DEVELOPMENT PROGRAMMES

A number of extension education and development programmes for livestock and poultry improvement have been planned and executed, some by the British administration before independence and others by the Indian Government after independence. There has been tremendous growth of Animal Husbandry in

India after independence in animal science education and research in different disciplines (Animal Reproduction, Breeding, Nutrition, Management, and Health related aspects). A number of State Agricultural Universities, ICAR animal science research institutes, and other organizations (NGO's and or Govt. Departments) have been established in all the states, involving animal science and veterinary science education and research as one of the major component. The new technologies have been developed through research carried out by the animal scientists in different parts of the country for transfer to the livestock farmers. The livestock and poultry improvement programmes have been formulated by the planners and executed through various agencies (SAU's, ICAR, Department of A.H. & Dairying, NGO's and other agencies) for the benefits of the farmers. Thus, the farmers are ultimately the users of newer technologies and hence beneficiaries. As a result, the farmers have been benefited but still there had been some gaps, limitations and constraints in execution of development programmes.

## 7.3 LIMITATIONS OF IMPROVEMENT PROGRAMMES

*Farmer's status*: The farmers are not organized, keep few animals, have poor economic status (small land holding), and illiterate. As a result the farmers do not keep records and the breeding bulls. The farmers are thus forced to breed their female animals with the bulls available in or around the village irrespective of genetic merit (scrub and working buffalo bulls). The interest of the farmers is to get their female animals pregnant to initiate new lactation and to get the calf/lamb/ kid crop, rather than genetic improvement.

*Lack of knowledge*: The farmers are not aware about the techniques of enriching the low quality feeds and fodders. There is also taboo against sale of milk in some areas and the farmers are reluctant to accept new technology.

*Lack of facilities*: At farmer's herd, the major constraints have been unscientific breeding due to lack of infra-structure for A.I., and non-availability of the requisite numbers of semen doses of superior germplasm. Moreover, the conception rate of A.I. is low in field conditions. The facilities of AI and treatment of animals at farmer's door are not available.

*Lack of inputs:* There is lack of efficient system to get inputs like fodder seeds, cattle feed mineral mixture, and medicines. There is inadequate nutrition and poor management including health care. There is no supply of green fodder all round the year due to inadequate area under fodder crops. The health services are not available for immediate treatment. There is lack of disease diagnostic and surveillance facilities.

*Lack of market facilities:* The market facilities for disposal of milk are not available and hence there is monopoly of milk venders and middle man to pay low price of milk. The refrigeration facilities for milk are not available and there is difficulty in transporting milk from remote areas. There is problem of disposal of unproductive animals particularly the crossbred male calves.

*Risk:* There is high risk involved in keeping the dairy animals than growing crops and high cost is involved in cattle insurance.

*Lack of incentive:* There is no initiative and incentive to the farmers for milk recording and not to sell their cow and buffalo heifer of pedigreed sires before completing first lactation. The trained and well equipped man power (extension workers) is not available as and when required.

*Poor extension services*: The lack of awareness among farmers and communication gap cause delay in adopting improvement programmes. There is need to motivate the small farmers for participation in improvement programme including data recording of their animals and enhancing awareness among farmers especially farm women about improved package of practices for breeding, feeding, health care and management of cattle and buffaloes. It is necessary to develop awareness among farmers regarding participation in field progeny testing program / improvement program, use of quality germplasm, maintenance of elite animals, clean milk production, heat detection, correct time of A I, feeding strategies.

*Lack of coordination among Institutions*: Large number of institutions like Department of Animal Husbandry and Dairying, ICAR, CIRB, SAUs, State Livestock Development Boards, Cooperative Unions, NGO's etc. are involved in livestock improvement programmes. But there is lack of coordination between various institutions coupled with non-availability of semen for large number of defined population in the native tract.

## 7.4 BACKWARD STATE OF LIVESTOCK IN INDIA

The animal husbandry is in a backward state due to the followings –

(i) *Plenty and poverty*: The population of different species of livestock is more compared to land and other resources and the productivity is poor than their counterparts of American and European species. Moreover, the native animals are non-descript and among the recognized breeds the majority is of draught and dual purpose breeds of cattle.

(ii) *Poverty and illiteracy of farmers*: The livestock farmer in India is economically poor and illiterate. Thus, he can neither keep the high producing animals nor can provide better shelter to his animals to give them protection from environmental stress. He also can not appreciate the scientific method of livestock rearing with respect to bred his female animals with sire of high genetic merit, feeding of balanced ration, proper care and management as well as timely health care.

(iii) *Breeding Policy:* There is no breeding policy of animals at national level. There is lack of progeny tested bulls and A.I. facilities to cover the entire population. The animal owner remains interested to get his female animals conceived irrespective of breeding value of bulls locally available.

(iv) *Lack of feed and fodder:* The area of grazing and to grow fodder crops has decreased due to urbanization and growing demand of food grains for the growing human population. The more number of useless animals in the country has also resulted in scarcity of feed and fodder.

(v) *Poor Management:* The livestock rearing is a sde business and hence proper attention is not paid for their proper management. This results in weakness of animals and adverse effect on the efficiency.

(vi) *Health and disease problems:* The Indian livestock is weak due to malnutrition, dirty sheds with no provision of ventilation, clean drinking water and protection from cold and hot winds. This makes the animals susceptible to certain diseases. Ther veterinary facilities are also not available for proper and timely treatment of sick animals.

As a result of the above causes there is high mortality rates among young, late maturity, longer service period and dry period, and a number of reproductive problems as well as poor production efficiency.

## 7.5 SUGGESTED MEASURES FOR LIVESTOCK IMPROVEMENT

The quality of livestock can be improved by removing the causes responsible for their poor quality.

1. See section 22.1 (Proper utilization of AnGR).
2. To send the surplus, useless and old animals to Pinjarapoles and Gosadan.
3. To make provision of better feed and fodders:
   (*i*) To develop natural pasture land and grassland by reclaiming useless fallow lands,
   (*ii*) To grow good quality fodder of improved varieties,
   (*iii*) To establish feed and fodder storage for lean periods,
   (*iv*) To improve the quality of feeds and fodders by supplementation with leguminous crops, urea treated straw, urea molasses blocks, etc.
4. Control of diseases by timely vaccination and treatment of sick animals.
5. Improvement of animal housing systems.
6. To strengthen the extension education programmes.
7. To develop processing and marketing facilities.

## 7.6 SCOPE OF ANIMAL HUSBANDRY

There is a good scope of animal husbandry in India in view of the contribution of livestock in agricultural operations, source of food, wool, manure, etc., employment generation, transportation, source of income and entertainment.

## 7.7 LIMITATIONS OF LIVESTOCK REARING

The livestock rearing has certain limitations like that it is an expensive enterprise for poor people requiring large capital investment in purchase of animals and for making arrangement of animal house; requires hard labour and good knowledge of all aspects of breeding, feeding, health care and management practices; involve risk for repeat breeding and death of animals; requires green fodder throughout the year, etc.

## 7.8 EXTENSION EDUCATION PROGRAMMES

The extension education and training programmes help the farmers to empower them with appropriate technical knowledge, skill and facilities. It is therefore essential at this point to understand about preliminary aspects of extension education.

The education brings the desired changes in human behavior regarding knowledge, attitude, skills and action plan for betterment of the society. The extension is the communication of information and intervention used by an institution to help the people to induce desired change in behavior and to form sound opinion for taking good decisions. According to N.C.A., 1976, the extension is an out of school education and services for the farmers to enable them to adopt the improved practices in production, management, conservation and marketing. The sole objective of extension education is to teach the rural people in changing their behavior in desired way so as to raise their standard of living by their own efforts making use of their own resources with minimum help from Government. This can be achieved by realizing the rural people their needs and problems and mobilizing them about the use of recent technologies for raising their standard of living.

The extension education programmes are formulated keeping in view the ***national goals and objectives***. The national goals are the national food security, to reduce poverty and improving the rural livelihood, and sustainable use of natural resources. The objectives of the extension programmes in view of accomplishing the national goals either related to agricultural or livestock sector should include the technology transfer, increase of farm income, organization of farmers for specific activity, and to educate the farmers to use sustainably the natural resource management practices.

Now, the important point to consider is the methods to be used to educate the livestock farmers to bring the desired change in their behavior. In order to communicate the new ideas and techniques, there is immediate need to undertake the following *components of the extension education programmes* –

- Dissemination of the information related to scientific aspects of animal rearing through extension literature and media,
- Demonstration of new and innovative technologies for the farmers through farm visits,
- On-campus and field level training to the farmers on scientific livestock rearing.

(*i*) ***Dissemination of information*** is made through distribution of different literatures for communicating the information viz. pamphlets, leaflets, circulars, folders, bulletins, newspapers, magazines, newsletters, and journals. Attempts should be made that the information contained in these literatures is in a language which is simple and easily understandable to the rural people.

(*ii*) ***Demonstration*** programmes are carried out to demonstrate the new technology or practice developed to achieve a particular objective. The demonstrations create interest, improve the learning process, effective to practically show the mechanism/operation of new technology and give better results than traditional technique. The demonstration is learning by doing. The demonstration helps to convince the farmers to demonstrate the effect of the new technology, practice (like feed formulation, effect of drug/vaccine to cure/prevent a disease), breed, variety and to adopt the same.

(*iii*) ***Training*** improves the knowledge and skills. Thus, the training helps to work efficiently and play important role for human resource development (HRD) to face the challenges. The training programmes include identifying the training needs, development of the contents of the programme based on the skill, knowledge and attitude of trainees, arranging the topics of the complete training course in a proper sequence/schedule of time frame and duration. The training must address the training needs of the trainees.

## Frontline Extension Programmes

A concept of Frontline Demonstration (FLD) was evolved by ICAR during the inception of Technology Mission on Oilseed Crops during mid-eighties. The FLD are the field demonstrations conducted under the close supervision of scientists of NARS (National Agriculture Research System) to demonstrate the technology by the scientists themselves before taken up by the State Deptt. of Agriculture. The NARS included the KVK's, ICAR institutes, NRC, Project Directorates, and SAU's with their regional stations. These FLD are conducted before farmers and extension officers. These FLD generated the data on crop yields and constraints.

Several frontline extension programmes have been formulated and implemented with involvement of ICAR in agricultural extension. These include:

- *National Demonstration Projects* (1965) to demonstrate the production potentialities of technology package on major crops for training to farmers and extension workers;
- *Krishi Vigyan Kendra* (KVK, 1974) to provide vocational training to farmers and extention workers;
- *Operational Research Projects* (ORP, 1975) to identify technological and socio-economic constraints for formulation and implementation of technology modules;
- *Lab to Land Programme* (1979) of ICAR to assist selected farm families for generating employment and income;
- *Technology Assessment and Refinement* (TAR) – Institute-village Linkage Programme (IVLP) 1995 of ICAR to introduce technological interventions with integrated approach;

- *National Agricultural Technology Project* (NATP)/ National Agricultural Development Programme (NADP) or Rashtriya Krishi Vikask Yojna (RKVY) 2007 on the recommendation of National Development Council (NDC) to evolve a strategy to rejuvenate agriculture to achieve 4 % annual growth in agricultural sector during XI plan period;
- *Agricultural Technology Management Agency* (ATMA) - a registered society of key stakeholders in public sector extension to establish a district level coordinating agency in 24 districts with World Bank support. It involves block level technology team and farmer advisory committees, farmers group and self help groups and responsible for all the technology dissemination activities in the district with establishing linkage with all the line departments, research organizations, NGO's with a Project Director of ATMA as the Chairman;

❑❑❑

# 8

# Livestock Contribution

India is a rural, an agricultural, and a developing, diversified, an ancient with old civilization, largest democratic, and over populated country.

India lives in villages as about 75 % of its population resides in villages called "***Rural India***". Secondly, India is an ***agricultural country*** because the main occupation is agriculture for the livelihood of its people. According to an estimate ending 20 th century, about 60 % population of the country depends directly or indirectly on agricultural sector of crop and animal production; about 70 % of rural labour force are engaged in agriculture and depend directly or indirectly on agriculture sector, 5.5 % in animal husbandry, 0.31 % each in forestry and fishing, and rest 23.7 % in non-farm activities.

Thirdly, India is a ***developing country*** as it has not attained its full economic growth to the required level which is possible by sustaining agriculture growth. The growth in agriculture (including animal sector) is low than growth in other economic sectors. However, the growth in agriculture sector has proved much more effective to reduce poverty in rural India than growth attained in other economic sectors. Further, the growth in animal production sub-sector is faster than the crop sub-sector and also reduces poverty more than a similar increase in crop production. However, the crop and animal sectors are both the only source of food to human population.

Moreover, India is a ***diversified country*** than most of the countries in the world with respect to the followings:

- It has huge ***biodiversity*** of both plant and animal species. India grows varied *vegetation* of forest to the cultivation of various *field crops* (grains, pulses, oil seeds, fodder crops); *horticultural crops* (fruits including dry fruits, vegetables and flower production); *medicinal plants; spices; forest trees.*
- In India are present and reared all the species of *domestic animals* viz. *Bovines* comprising cows, buffaloes, mithun and yak; *Ovines* comprising sheep and goat; *rabbits; Equines* comprising horses & ponnies, asses/donkeys, and mules; *camels; pigs;* and *poultry species* comprising chickens, turkeys, ducks, fowl, guinea fowl-*titri*, Japanese quail-*bate*r, domestic pigeons-*kabutor* and big birds like emu and ostrich).

- India has the largest cattle and buffalo population and is the first in milk production in the world.
- India has different *agricultural and animal production systems:* dry land agriculture as well as irrigated crop cultivation, mixed cropping and mixed farming systems; extensive (landless or zero input), semi-intensive (medium input), and intensive (high input) animal production systems.
- India has also great diversity with respect to varied *topography* (hills, plains, rivers, lake, and sea).
- India has varied *climatic conditions* of temperate, sub-temperate, semi-arid and arid zone (hot dry climate).
- India has varios type of soils *viz.* sandy, loam, semi loam, rocky, black, red, and *various types of valuable stones, coal mines and various ores.*
- India has varied *seasons* having summer, rainy, autumn and winter.
- India varies in *economic status* of the people from farmers of different categories (landless and poor, medium and large farmers) involved in agriculture including animal husbandry and fisheries for their livelihood, working labourers, businessmen and those doing their jobs in Government and private sector.
- India also varied in various other aspects of different *languages, religions, casts and gotras within cast, culture and traditions,* and *social activities.*
- India has many *historical places* including *religious places, forts, etc.*

## 8.1 AGRICULTURE VS LIVESTOCK

The animal husbandry is an important sector and integral part of Indian agriculture. Agriculture depends on livestock and hence it comprises two main sectors viz. crop sector and animal sector. Crop cultivation and animal husbandry are dependent on each other. The crop sub-sector includes mainly the cultivation of field crops (grains, pulses, oilseeds, and fodder crops) and horticultural crops (fruits, vegetables, and flowers) whereas animal sub-sector includes livestock, poultry and fisheries.

In India the ***mixed farming system*** (crop-livestock farming) is practiced. The livestock are an integral part of this system as the crop cultivation and animal husbandry are dependent on each other. Mostly the farmers are small landholders and keep few numbers of useful animals for their livelihood and used in crop production. The crop production provides feed and fodder to the animals which in turn provide draught power and dung (manure) to maintain soil fertility for crop production. The importance of livestock has been recognized not only as providers of traction power, manure, fuel, food (milk, meat and eggs), and cloths (wool, mohair, fur and pelt) but also as producer of cash incomes to rural poor. In small and subsistence households the livestock are perennial source of wealth and an insurence during emergency and crisis period against crop failure. The integrated farming system of mixed crop animal farming is thus more useful, efficient and more sustainable for both of its components (crop and livestock production) in a synergetical manner. The livestock rearing has emerged as a fast growth sector of Indian agriculture and has become a business which was a way of life in the past.

The socio-economic and cultural ethos of the farming community is livestock associated. Further, the ***women participation in animal husbandry*** is dominating comprising about 70 – 90 % of the total workers in different parts of the country. On an average, about 44 % of farm households in our country are associated to dairying with increasing trend with increase in size of landholding (39 % marginal, 48 % small, 57 % medium and 68 % large farmers).

The contribution of animal husbandry sub sector is expected to be more potential and significant to reduce poverty among poor smallholders. This is because about two-thirds of farmers in India are associated in animal production and three-fourth of them as small landholders having land of ≤ 2 hectare. Small scale milk producers are more efficient in dairy production and they take up dairying on a commercial scale. There is dominance of small landholders (≤ 2 ha) in subsistence and commercial dairying. The small scale poultry producers are also as efficient as large scale producers. Production efficiency, marketable surplus of products, and approach to local and global market decide the benefit to small holders from growth in animal production. The regular expenses of landless labourers and small/ marginal farmers of India are met from sale of raw milk in the local market produced by cattle and buffaloes and this support their livelihood during off season of earnings from cash crop. They keep 1 – 2 milch animals under mix farming system and these constitute about 70 % of the milch animals. Not only this but animal rearing by tribals and nomads is the only source of their earning and livelihood. Therefore, in view of livestock contribution, this livestock sector has a potential and a great role in the upliftment of rural poor. The meat and eggs are mainly produced for the market but the marketed surplus of milk is estimated only about 54 percent.

India is a vast reservoir of animal genetic resources with large numbers of breeds of various animal species serving various purposes mainly under low input animal production system. More than 97 % of the total livestock are present in rural India. The ***pattern of distribution of animals*** of small (Sheep, goat, pigs and poultry) and large size (Cattle and buffaloes) is distinct depending on landholding size. Small animals are kept by small landholders for the reasons that these animals require less capital in terms of their purchase value and operational cost, being prolific with short generation interval, can be maintained on grazing on waste land (sheep and goat) or on kitchen waste or waste around human dwellings (pigs and poultry), require less amount of feeds and fodders, and well suited to poor resource farmers who have either no or small piece of land and less capital to invest.

The livestock and poultry have important position in socio-economic development and prosperity of a country like India being primarily an agricultural country. The livestock is playing a significant role in food security and to the livelihood of rural poor. The livestock in India is becoming more important sector of agriculture in view of the changing food consumption pattern of people for their preference to dairy, poultry and meat products and also for uncertainty of food security based on traditional agricultural practices.

Animal husbandry as a sub-sector of the agricultural economy, contributes 27 % in total agriculture output (GDP) with its increasing share consistently, plays an important role in the development process of our country, and is the most realistic and safe way to reduce poverty of rural India. The contribution of livestock and fisheries sector to the agricultural sector in India is 31.5 % of the value of the output, against the livestock contribution of about 40 % in the world agriculture. These sectors had contributed about 7.35 % to the total GDP in 2001 – 02. The milk had been the primary contributor to agricultural GDP and was higher than paddy, wheat and sugarcane.

It is expected that the growth in animal production sub-sector is faster than the crop sub-sector and reduces poverty more than a similar increase in crop production. Inspite of Green Revolution in the country, the ***annual growth rate of livestock sector*** has been higher than crop sector at national level as evidenced from the table below –

| **Years** | **1950-60** | **1960-70** | **1970-80** | **1980-90** | **1990-2000** | **2000-07** |
|---|---|---|---|---|---|---|
| Crop sector | 3.1 | 1.7 | 1.8 | 2.2 | 3.0 | 2.7 |
| Livestock sector | 1.4 | 0.4 | 3.9 | 4.9 | 3.8 | 3.7 |
| National GDP | 3.68 | 3.29 | 3.45 | 4.17 | 6.05 | 6.88 |

The milk and poultry production in India has increased since 1991 at an annual growth rate of 4.5% for milk, 12.3 % for poultry meat, and 5.7 % for eggs. The production and growth rate of different livestock products have been given in table below –

| **Livestock products** | **Prouction (Million tons)** | | **Annual growth rate (%)** |
|---|---|---|---|
| | **1982-83** | **2004-05** | **1982-83 to 2004-05** |
| Milk | 35.8 | 92.5 | 4.26 |
| Meat | 2.98 | 6.03 | 3.47 |
| Buffalo meat | 0.93 | 1.48 | 2.37 |
| Mutton | 0.52 | 0.71 | 1.55 |
| Poultry meat | 0.16 | 1.72 | 12.30 |
| Eggs (million) | 11454 | 45201 | 5.72 |

Many livestock products particularly the meat of different animal species (beef, mutton, and pork), and eggs are produced at much lower cost per unit of production in comparison to major exporting countries. Milk in India is also produced at lower cost than in Netherland, Germany, France, and Belgium.

The increasing demand of livestock products has made the Indian farmers to respond positively for livestock rearing and hence the animal production (milk, meat, eggs) have increased many fold in the country during the last 3 decades. This faster growth of animal industry in our country has become possible as a result of improvement in the genetic potential of animals of different species by selection and crossbreeding programmes, increase in feed and fodder supply and improvement in animal health services.

M.I. Darling said "Without cattle, fields remain unploughed, store and bins stand empty, and food and drink lose half their savour for in a vegetarian country what can be worse than to have no milk and ghee".

## 8.2 CHANGES IN FOOD HABBIT IN INDIA

There has been a change in food habbit in India during past 4 decades from low cost staple food grains to high cost animal food products (milk, meat, eggs, and fish), fruits, and vegetables. The consumption demand of high value food products including livestock products compared to staple foods had been more among rich people. There has been increase in per capita consumption of milk by 71%, meat by 30% and fish by 42%. The increase in per capita consumption of milk by poor and rich people had been nearly of the same magnitude (nearly 30 %) during 1983 to 2000 whereas the increase in consumption of other animal products (meat, eggs, and fish) had been little higher among rich people (121 %) than poors (100 %) and thus responsive to income changes of the people. The ***per capita consumption of livestock products (kg per annum)*** has been shown in table below –

| Animal products | Poor people | | | Rich people | | |
|---|---|---|---|---|---|---|
| | 1983 | 2000 | % change | 1983 | 2000 | % change |
| Milk | 15.7 | 20.5 | 30.6 | 89.7 | 117.2 | 30.7 |
| Meat, eggs & fish | 1.9 | 3.8 | 100.0 | 4.8 | 10.6 | 120.8 |

The share of animal food products in food expenditure (%) by rural and urban population showing increase in share of animal products has been shown in the following table (GOI, 2006) –

| Items of food basket | Rural people | | | Urban people | | |
|---|---|---|---|---|---|---|
| | 1983 | 1993 | 2004 | 1983 | 1993 | 2004 |
| Expenditure on food (% of total expenditure) | 65.6 | 63.2 | 55.0 | 58.7 | 54.7 | 42.5 |
| Milk & milk products | 11.5 | 15.0 | 15.4 | 15.7 | 17.9 | 18.6 |
| Meat, eggs, fish | 4.6 | 5.3 | 6.0 | 6.1 | 6.2 | 6.4 |

The dairying is a major component of livestock production in India. In absolute terms, dairying accounts for more than 2/3 rd of the livestock production.

The ***demand of animal food products*** has been increased locally as well as globally as a result of fast growing urbananization, sustained economic and income growth, and integration of global agri-food markets. This signifies a good potential for future growth of livestock sector, increase in exports of livestock products and hence better income opportunities to animal farmers particularly the small landholders and landless farmers who keep animals. There has been rapid increase in global demand for livestock products. The increase in consumption of total meat and milk after 1980 has been registered by 81 % and 29 %, respectively. The increase in annual per capita consumption of meat has doubled from 14 kg to 29 kg whereas the increase in milk consumption has been from 23 kg to 31 kg in developing countries but there was little change in developed countries.

## 8.3 LIVESTOCK AND POULTRY POPULATION

India has 11 % of the world livestock population, 17 % of human population and 28 % land. The livestock and poultry population in the world, the number of breeds of different species, and livestock & poultry population trend in India over the years since 1951 have been given here as under -

### 8.3.1 World population of livestock & poultry

Among the livestock in the world, India has 56 % buffaloes, 16 % cattle, 17 % goats, 5 % sheep, and 1.7 % pigs (Table 8.1).

**Table 8.1:** Total population (million) of different species of livestock and poultry in the world according to FAO Data bank (1998)

| Species | World | Asia | India | Indian share (%) | |
|---|---|---|---|---|---|
| | | | | World | Asia |
| Cattle | 1315 | 453 | 209 | 15.9 | 46.1 |
| Buffaloes | 164 | 159 | 92 | 56.1 | 57.9 |
| Sheep | 1061 | 402 | 56 | 5.3 | 14.0 |
| Goat | 693 | 437 | 121 | 17.4 | 27.6 |
| Pig | 957 | 577 | 16 | 1.7 | 2.8 |
| Horse | 61 | 17 | 1 | 1.6 | 5.8 |
| Camel | 19 | 5 | 1.5 | 7.7 | 30.5 |
| Chicken | 13683 | 6828 | 343 | 2.5 | 5.0 |

### 8.3.2 Livestock and poultry breeds in world

Total livestock and poultry population in the world is distributed in several breeds within each species, and their number.

**Table 8.2:** Number of livestock and poultry breeds in the world

| Species | World | Asia and Pacific | South Asia | Europe | America | Africa | India | |
|---|---|---|---|---|---|---|---|---|
| | | | | | | | FAO | ICAR |
| Cattle | 787 | 190 | 94 | 305 | 110 | 120 | 61 | 30 |
| Buffaloes | 72 | 57 | 3 | 2 | 3 | 2 | 19 | 15 |
| Sheep | 920 | 226 | 11 | 42 | 70 | 73 | 59 | 42 |
| Goat | 351 | 126 | 70 | 113 | 23 | 34 | 29 | 20 |

*Contd...*

*Contd...*

| | | | | | | | | |
|---|---|---|---|---|---|---|---|---|
| Pigs | 353 | 157 | 8 | 129 | 52 | 13 | 3 | 3 |
| Asses | 77 | 17 | 5 | 16 | 10 | 6 | 3 | 3 |
| Horses | 384 | 72 | 25 | 185 | 65 | 31 | 9 | 6 |
| Camel | 56 | 14 | 9 | - | - | - | - | 6 |
| Poultry | 606 | 72 | 27 | 406 | 45 | 57 | - | 18 |
| Total | 3606 | 931 | - | - | - | - | 183 | 145 |

### 8.3.3 Livestock Population Trend in India

India is one of the globally recognized 12 mega-biodiversity countries and bestowed with rich diversity of livestock germplasm, having uniquely distinguished breeds particularly of buffaloes. India ranks first in respect of cattle and buffaloes (5 7 % world buffaloes), second in goats, third in sheep and sixth in poultry population in the world. The 17 $^{th}$ Livestock Census (2007) indicated the total livestock population of 529.7 million and poultry birds of 648.8 million in India. The country with this population has 11 % of the world livestock population whereas the human population is 17 % of the world human population. There has been tremendous increase in both livestock and poultry population in India post independence, as shown below -

**Table 8.3:** Livestock and poultry population (million numbers) trend since 1951:

| **Species** | **1951** | **1961** | **1972** | **1982** | **1992** | **2003** | **2007** |
|---|---|---|---|---|---|---|---|
| Cattle | 153.3 | 175.6 | 178.3 | 192.45 | 204.58 | 185.18 | 199.1 |
| Buffaloes | 43.4 | 51.2 | 57.4 | 69.78 | 81.21 | 97.92 | 105.3 |
| Mithun | – | – | – | – | 0.15 | 0.264 | |
| Yak | – | 0.02 | 0.04 | 0.13 | 0.06 | 0.06 | 0.08 |
| Sheep | 39.1 | 40.2 | 40.0 | 48.76 | 50.78 | 61.47 | 71.5 |
| Goat | 47.2 | 60.9 | 67.5 | 92.25 | 115.28 | 124.36 | 140.5 |
| Camel | 0.6 | 0.9 | 1.1 | 1.08 | 1.03 | 0.63 | 0.52 |
| Horses & ponnies | 1.5 | 1.3 | 0.9 | 0.9 | 0.82 | 0.57 | 0.61 |
| Donkeys | 1.3 | 1.1 | 1.0 | 1.02 | 0.97 | 0.65 | 0.44 |
| Mules | 0.06 | 0.05 | 0.08 | 0.13 | 0.19 | 0.18 | 0.14 |
| Pigs | 4.4 | 5.2 | 6.9 | 10.7 | 12.79 | 13.52 | 11.13 |
| ***Total Livestock*** | **292.8** | **335.4** | **353.4** | **419.59** | **470.86** | **485.00** | **529.7** |
| Poultry | 73.5 | 114.2 | 138.5 | 207.74 | 307.70 | 489.01 | 648.8 |

During a period of 5 years from 2003 to 2007, the increase has been registered to 1.5 per cent per annum in cattle and buffalo population, 3 per cent in sheep, 2.5 per cent in goat, and 8.6 per cent in poultry. On the other hand, there has been registered a decrease in mithun, yak, camel, donkey, mule, pig, and duck population.

### 8.3.4 Land area and Stocking rate

India has 328.8 million hectares of total land area which is 2.29 % of the land in the world and has about 10.1 % of the world's livestock. Only 193 million ha land is under cultivation of which about 60 million ha is irrigated. According to 1982 census, the stocking rate per sq.km had been 58.5 cattle, 21.2 buffaloes, 14.8 sheep, 28.9 goats, 0.27 equines, 0.3 camels, 3.0 pigs, 63.2 poultry, and the overall livestock density was estimated as 127.64 animals.

## 8.4 LIVESTOCK CONTRIBUTION

The livestock is playing an important role to provide draft animal power (DAP) or agricultural operations and transport; animal's products as the source of food (milk, meat, eggs), cloths (wool/fibre, skins), dung (manure and fuel); generation of employment and supplementing income; as companion and source of entertainment to the rural people particularly of the landless, small and marginal farmers as well as of women. Therefore, livestock sector is important in poverty alleviation programmes. India has large livestock population to meet the demand of livestock products for human population like milk, eggs, meat and wool. However, the production and quality of these livestock products of Indian livestock are low and need improved technology.

### 8.4.1 Livestock Products

The different species of domestic animals produce different animal products which are marketed and earning money. These products are the sources of human diet, clothing and raw material for a number of industries *viz.* dairy industry, leather industry, woolen industry, bone industry, sport industry. These animal products are as under –

| Animal species | Animal products |
|---|---|
| 1. Dairy animals | Milk, meat, animal power, manure, hides, horns & (Cattle & buffaloes) hooves |
| 2. Sheep | Wool, meat, milk, hides, hooves, manure, skin |
| 3. Goat | Milk, meat, mohair, skin, manure |
| 4. Horses, Donkeys, Camel | Animal power, manure |
| 5. Poultry | Eggs, meat, feathers, manure |
| 6. Rabbits | Meat, fur, mohair |

### 1. Animals as Source of Food – Milk, Meat and Eggs

In ancient days before domestication men used to hunt and eat the animals. Even today, the animals provide essential foods of animal origin like milk from cattle, buffalo and goats; meat from pig (pork), sheep (mutton), goat (chevon) and poultry; and eggs from poultry. The milk and eggs are the main and common source of animal protein in human diet today particularly for vegetarian community.

During 2011 – 12, the livestock sector has contributed 121.84 million tones milk, 4.83 million tones meat, and 61.45 billion eggs. In 2012-13, the contribution increased to 132.4 MT milk, 5.95 million tones of meat, and 69.7 billion eggs. The milk is the output accounting about 67 % of the total output by livestock sector, meat share was 18.2 % and of eggs was 3.9 % of the value of livestock output.

## (i) Milk production

The milk is satistary diet, easily digestible and meets the requirements of many essential elements for all ages and categories of people (pregnant and nursing woman, labourers and the people engaged in mental job). All categories of people, particularly the vegetarians throughout the world consume milk and its products as a main source of protein and calcium in readily available form.

The Dairying has two components *viz.* dairy farming and dairy processing. The dairy farming is in the private unorganized sector but the dairy processing has 3 milk processing sectors viz. milk producer's co-operative unions with prominent large base federated onto 22 state co-operative federations, private and multi-nationalists sector as second number followed by Govt. owned dairy plants.

The Operation Flood programme was implemented successfully. There has been a significant increase in milk production after independence. India had about 17 % of the world human population and produced about 14 % of the total milk produced in the world (592 MT) during 2003 whereas the India's share was 9 % in world milk production in 1970.

**Table 8.4:** Milk production and per capita availability in India over the years

| Years | 1951 | 1968 | 1990 | 2006 | 2007 | 2008 | 2009 | 2010 | 2011 | 2012 |
|---|---|---|---|---|---|---|---|---|---|---|
| *Milk Prod. | 17.6 | 21.2 | 53.9 | 102.6 | 107.6 | 112.2 | 116.3 | 121.8 | 127.3 | 132.4 |
| **Availab. | 124.0 | 112.0 | 176.0 | 251.0 | 260.0 | 266.0 | 273.0 | 281.0 | 290.0 | – |

*Milk production (million tones).

** Availability of milk (gm)/capita/day.

In 1999, India became the largest milk producing country in the world producing 74.30 million tones of milk followed by USA with a milk production of 73.80 MT, Germany (28.33 MT), France (24.89 MT) and U.K. (15.01 MT).

In 2003, India continues to be the largest milk producer in the world producing 86.96 MT which was 14.7 %. The country wise milk production, India ranked first followed by USA (77.25 MT), Russia federation (32.8 MT) and Germany (28.35 MT).

During 2004, India achieved 91.5 MT milk produced followed by USA producing 77.2 MT milk.

The dairy cooperative movement initiated by NDDB (India) has done an appreciable job in increasing the growth rate of milk in India. The main reason of getting first position by the India goes to the increase in bovine population and improvement in the productivity of bovines during the past three decades.

The growth rate of milk production is about 4 % against the population growth of 2 % per annum, 2 % in crop production and 6 - 8 % in A.H. sector as a whole.

***Total milk production in the world*** *(2003)*: The total world milk production during 2003 was highest in Europe (36.03 %) followed by Asia (30.69 %), North America (16.79 %), South America (7.86 %), Africa (4.39 %) and Oceania (4.24 %).

*Milk from different species*: The world milk production was 592 MT of which 85.7 % was produced by cows, 12.3 % by buffaloes and 2.0 % by goats.

(*i*) The *buffalo milk production* was highest in Asia (96.94 %) followed by Africa (2.86 %) and Europe (0.20 %).

(*ii*) The *cow milk production* was highest in Europe (41.54%), followed by Asia (20.65%0, North America (19.55%), South America (9.14 %), Oceana 4.95 %) and Africa (4.17 %).

(*iii*) The *goat milk production* was highest in Asia (53.42%) followed by Africa (23.3 %), Europe (20.30 %), South America (1.53 %) and North America (1.45 %).

About 98 per cent of the milk in India is produced in rural sector and rest 2 per cent in urban areas.

*In India,* the bulk of the milk produced during 2003 is by buffaloes (54.5 %) followed by cows (41 %; out of which 25 % by indigenous, and 16 % by crossbreds) and goats (4.5 %).

**Table 8.5:** Milk production (Million tones) in India by different species over the years

| **Species/ year** | **1951** | **1961** | **1972** | **1982** | **1990** | **1995** | **2003** |
|---|---|---|---|---|---|---|---|
| Cows | 7.75 | 8.75 | 7.5 | 14.0 | 24.0 | 29.8 | 35.6 (41 %) |
| Buffaloes | 9.25 | 11.0 | 14.75 | 20.75 | 29.0 | 34.8 | 47.4 (54.5 %) |
| Goats | 0.4 | 0.65 | 0.75 | 1.05 | 1.9 | 1.7 | 3.9 (4.5 %) |
| Total | 17.4 | 20.4 | 23.0 | 35.8 | 54.9 | 66.3 | 86.9 |

***Growth in milk production in the world*** *(1994 – 2003)*: The world milk production is showing an increasing trend annually with a rate of 1.47 %. The growth rate was highest in Oceania (4.32 %) followed by Asia (3.68 %), Africa (3.52 %), South America (2.49 %), North America (1.45 %) and the trend in growth rate in milk production was decreasing in Europe (– 0.68 %).

Among the countries, Thailand remained the fastest growing milk producing country during 1994 – 2003 with an annual growth rate of 11.85 % followed by China (9.15 %), Jordan (7.74 %), Egypt (6.07 %), Algeria (5.28 %) and the growth rate in milk production has been 3.9 % in India. Most of the European countries had declining trend during this period.

**Table 8.6:** ***Average milk production*** **per cow (kg/year)** during 2000 and ***availability of milk/day/*** **capita (g)** during 2002 for different countries of all the six continents

| Continents | Productivity kg /Availability gm<br>Continent average | Countries & their average |
|---|---|---|
| **North America** | **4808/542** | (**USA 8388/727**, Canada 7324/697, Maxico **1393/263**); |
| **Oceana** | **4180/2199** | (**Australia 5153/1629,**<br>**New Zealand 3600/9877**); |
| **Europe** | **3987/807** | (Sweden **7717**/996, Denmark 7271/**2350**,<br>Netherland 7200/1848, U.K. 6190/687,<br>France 5630/1153, Germany 5880/926,<br>Italy 5499/547, Poland 3890/842,<br>Switzerland 5356/1506, Finland 6452/1334,<br>Spain 4820/**402**, Russian Fed. **2447**/621); |
| **South America** | **1564/356** | (**Argentina 4000**/630, Uruguay 1755/**1203**,<br>Brazil 1380/349, Chile 1350/382,<br>Venezuela 1278/**150**, Columbia **990**/378); |
| **Asia** | **1220/128** | (**Israel 9787/554**, Japan 6641/180, Jordan 3085/97,<br>China 1639/34, Thailand 2231/**28**, Turkey 1600/302,<br>Iran 1243/224, Pakisthan 1179/493, India 917/219,<br>Nepal 407/136, Srilanka, 319/43, Bangladesh **206**/40), |
| **Africa** | **486/85** | (South Africa **2667**/168, Algeria 1250/113,<br>Egypt 1089/74, Kenya 480/241, Sudan 480/**375**,<br>Zymbabwe 310/59, Tanjania 207/70, Ethopia 204/58,<br>**Burkino Faso 172/49**) |
| **World** | **2192/259** | |

During 2010-11, in India, the average annual milk production of indigenous cattle was estimated to be 1172 kg which is about 50 % of the world average. The daily milk production had been about 2.14 kg for indigenous cow, 6.87 kg for crossbred cows and 4.57 kg/day for buffaloes.

The ***per capita availability of milk*** **in India** has also increased from 124 g per day in 1950 – 51 to 195 g in 1994 – 55 and 281 g during 2010 – 11 compared to world average of 284 g per day. As per Economic Survey Statistics, 2012-13, the per capita availability of milk has increased from 176 grams per day in 1990-91 to 290 grams per day in 2011-12 and this is comparable with the world per capita availability of milk at 289 grams per day for 2011.

**Annual** ***consumption of liquid milk***: The per capita annual consumption pattern of liquid milk is maximum in Finland (177.9 kg) followed by Sweden (142.2 kg), Denmark (135.7 kg) and U.K. (112.5 kg). In India only 35.8 kg milk was consumed in liquid form annually based on 46 % liquid milk consumption of the total milk produced in this country. During 2010-11, the per capita milk consumption in India increased to 69 kg and meat to 3.7 kg which is still lower than world average of 85 and 40 kg, respectively.

***Milk utilization pattern*** **in India** during 2003 has shown that about 33 % of the total milk produced is retained by producers for home consumption and 51 % is sold to traditional un-organized sector of milk vendors, sweet makers and small dairy units which are engaged in loose and unprocessed liquid milk, traditional sweets, curd, khoa and chhana, paneer and loose ghee; and rest 16 % to organized sector (co-operative unions/federations) and private sector engaged in processed milk, milk powder, infant and malted milk foods, processed fresh dairy products, cheese, ice creams, butter and ghee.

**Table 8.7:** Milk utilization pattern (%) in India over the years

| Products/ Years | 1951 | 1961 | 1990 | 2003 |
|---|---|---|---|---|
| Liquid Milk | 39.3 | 45.1 | 45.7 | 33.0 |
| Ghee | 39.3 | 31.8 | 45.7 | – |
| Butter | 6.0 | 6.4 | 6.5 | – |
| Curd | 8.8 | 8.1 | 6.9 | – |
| Khoa | 4.0 | 4.7 | 6.5 | – |
| Milk powder | – | – | 3.7 | – |
| Paneer | – | – | 1.3 | – |
| Other | 2.6 | 3.9 | 1.9 | – |

***Milk marketing***: The marketing is the major constraint to the milk production in rural production system. It is in the hands of milk venders who collect the milk from rural households and do not pay remunerative price. The price paid to dairy farmers by milk venders is based on expedient disposal and cash payment rather than quality based. There is lacking of efficient channel of milk collection from rural areas to the milk processing plants, quality control measures, and remunerative price. The milk is a perishable commodity requiring expedient consumption otherwise spoiled and needs speedy transport and rapid chilling, pasteurization and hygienic storage till sent to sale outlets to reach consumption or converted in to milk products. The milk processed through organized milk industry is nearly 14 % of the milk produced and over 45 % is converted to products.

The quality control and evaluation system should not be under the scattered laboratories that are neither mandatory nor accredited for quality control and monitoring but there must be an efficient system at district, state and national level by delineated system of quality parameters based on natural constituents of milk (fat, protein, lactose, SNF and chlorides), presence of adulterants (urea, detergents, vegetable oils, mineral oil, sugar and other additives), microbial load (pathogenic, somatic cell counts) and other contamination and pollutants (antibiotics, drug residues, heavy metals, pesticides, fungal toxins, hormones, other anti-nutritional components and their metabolites).

## (ii) Meat production

The meat is a rich source of amino acids, vitamins and minerals (Iron and zinc) and has an importance in human diet. It has significant role in food security, livelihood and national economy. About 20 gm animal protein per person is required daily against the availability of only 14 gm in India whereas the world average is 107 g with maximum of 337 g in USA. India rank 5 $^{th}$ in meat production in the world having its 2.42 % share in global meat production of 269 m MT.

The distribution of 221 million tones of *meat production in the world* at the end of last century was:

| | |
|---|---|
| Beef and buffalo meat | 64.4 MT |
| Sheep & goats | 13.0 MT |
| Pigs | 92.6 MT |
| Poultry | 51.5 MT |

There has been faster growth in meat production than milk production. The total meat production in the world has increased to 269 million MT in the beginning of this century. This increase has been attributed to the increase in number of animals slaughtered, but the increase in yield growth was negligible in almost all the species. The animals slaughtered in India has been estimated by FAO as 12.4 m cattle, 10.86 m buffaloes, 19.54 m sheep, 54.3 m goats and 14.2 m pigs, though GOI has reported lower than this.

The ***meat yield in India*** during the first decade of this current century (2000-01 to 2009-10) has increased about 1.5 to 2.0 % for cattle, sheep, goat and poultry, and only about 0.8 % for pigs while the meat yield remained static for buffaloes. The meat yield of most of the species in India is 20 – 60 % lower than world average.

**Table 8.8:** M*eat production* (million tones) in India over the years

| Years | 1980 | 1998 | 2002 | 2006 | 2007 | 2008 | 2009 | 2010 | 2011 |
|---|---|---|---|---|---|---|---|---|---|
| Meat | 0.9 | 1.9 | 2.1 | 2.3 | 4.0 | 4.2 | 4.5 | 4.9 | 5.1 |

**Table 8.9:** Meat production from different species in India

| Species | Meat production (MT) | Average carcass weight (kg) | Slaughter % |
|---|---|---|---|
| Cattle | 1.3 | 103 | 6.5 |
| Buffalo | 1.2 | 138 | 11.1 |
| Sheep | 0.17 | 12 | 31.8 |
| Goat | 0.47 | 10 | 39.8 |
| Pig | 0.41 | 35 | 97.1 |
| Poultry | 0.44 | 1.3 | – |

In India about 30 % of sheep and 40 % of goats are slaughtered every year. The small ruminants had major share (44 %), followed by large ruminants (35 %) and poultry (19 %) in total meat production in India during 1980's. But the situation was changed during 1990's. The contribution of different species to total meat production had been as bovies (55 %), ovines (17.1 %), pigs (11.4 %) and poultry (16.3 %).

The *ovine meat* is mainly produced in A.P., W.B., MS, Bihar, Orissa and Rajasthan. About one-third of total *buffalo meat* is produced in U.P. alone, whereas about half of the total *poultry meat* is produced in A.P., MS, and T.N. About 70 % of the total *pig meat* (pork) is produced in Eastern and North-eastern states.

For meat industry in India about 11 million buffaloes are being slaughtered annually contributing 4.16 million tons (MT) of beef accounting for 32 % of total Indian meat production (USDA 2012). The buffalo meat has increased at the rate of 8 % per annum and doubling its export during the last 5 years

*Pork production* stood at 109.37 million tons in the World during 2010 with per annum growth (2000-2010) of 2.2 %. The share of pork to the total meat production globally is 38.9%. Pig meat is the largest consumed meat in the world with an average 15 kg/person/year followed by poultry (12.60 kg), beef (9.60 kg), and mutton and chevon (1.90 kg), according to FAO – Food Balance Sheets, 2007.

Per capita consumption of pork in India is 0.40 kg/head. Estimated pork production in the country during 2010-11 was 0.402 million tons resulting from slaughter of 10.677 million heads with an average yield of 38 kg per animal. Total export of pork from India was 180.63 MT which fetched Indian Rupees 215.42 lakhs during 2012-13, registering a decline over the years (APEDA).

The per capita consumption of poultry meat per annum in India was about 131 g against 15 kg in USA, and 2.4 to 5.9 kg in European countries during 1971.

### (iii) Egg production

Regarding the poultry population in 1971, there were only 21 chickens per 100 persons in India against the world average of 154 chickens per 100 persons with maximum in Belgium (446), Netherlands (445) and Canada (435) followed by UK (270) and USA (217). The states in India with largest poultry population during to 1972 census were A.P. (19 million), flowed by W.B., Bihar, T.N., MS and Kerala, Karnataka, Orissa, Assam, MP, UP, Pb, Gujarat (2.73 million).

Production of eggs per bird during 1971: The average annual laying capacity of a desi hen was only 60 eggs against the world average of 130 with maximum average of 227 eggs per laying bird in USA.

The egg production in India has increased @ 6 % per annum during the last 20 years from 1990. The numbers of improved layers to the total layers have increased from 45 to 70 % and in egg production has increased from 64 % in 1992 to about 85 % in 2010. At present, the average egg production in India is higher than world average and the average of developing countries though lower to average of developed countries. This has been attributed to genetic improvement. The share of improved layers to total egg production is about 2/3 with an average egg yield of improved year as about 305 eggs per year against about 100 eggs of an indigenous layer. This shows that there exists a scope of genetic improvement for increasing egg production.

The egg production has been more in A.P. and T.N. which have made a contribution of 32 and 18 % to the total eggs production while the contribution of Punjab, Haryana, MS. Karnataka, and W.B. had been about 5 – 6 % each.

The per capita availability of eggs per year in India during 1971 was only 11 against the world average of 115 eggs, with maximum in Belgium (541) followed by Netherlands (354), USA (352), Canada (320), UK (305), and West Germany (256).

**Table 8.10:** Increase in egg production and per capita availability after 1950-51, in the Country

| Years | 1951 | 1971 | 1980 | 1990 | 2003 | 2006 | 2007 | 2008 | 2009 | 2010 | 2011 |
|---|---|---|---|---|---|---|---|---|---|---|---|
| Eggs* | 1.8 | 6.0 | 10.0 | 21.1 | 40.4 | 50.7 | 53.6 | 55.6 | 60.3 | 63.0 | 65.5 |
| Available** | 5.2 | 11.0 | – | 25.0 | – | 42.0 | 47.0 | 49.0 | 51.0 | 53.0 | 51.0 |

* Number of egg production in billions

** Per capita availability of eggs/annum (numbers)

## 2. Source of Wool

The largest wool producing country is Australia. However, the wool production in Australia has a declining trend over the years: 814 m kg in 1992, 440 m kg in 2003 and 280 m kg in 2007. New Zealand is the second largest wool producer which is also showing the decreasing trend over the years: 200 m kg in 1993, 190 m kg in 2003, and 180 m kg in 2008. The China is important country for wool production and has shown an increasing trend over the years: 119 m kg in 1992, 151 m kg in 1999, and 185 m kg in 2007.

The wool is of three categories depending on the fineness. The fine wool called as *apparel wool* has fibre diameter below 24.5 μ, *carpet wool* has fibre diameter between 25 – 32 μ, and the wool having fibre diameter more than 35 μ is called *coarse wool.*

About 90 % of the wool produced in Australia is of fine quality (apparel) and 50 % of the fine wool in the world is produced in Australia. Of the total wool produced in the world about 48 % is of carpet type. New Zealand produces about 92 % coarse wool which is mostly used in carpet manufacturing.

The sheep population in India is 4.5 % of the world sheep population but the wool production is only 1.4 %. About 30 metric tones of pashmina is produced from goats reared at high altitude. Changthangi breed in Laddakh region of J & K and Chegu goats of H.P. and Uttrakhand produce pashmina of much finer speciality fiber.

The wool yield produced in India is low and that too of poor quality. Out of the wool produced in this country, about 85 % is of carpet quality with a fibre diameter of more than 30 microns and used for carpets and industrial fabrics, 10 % is of coarse grade and rest 5 % is of apparel quality (fine wool) with less than 25 microns diameter.

Pashmina fibre which is of fine quality is produced by goats reared at high altitudes of Himalayas. It is produced by Changthangi breed of Laddakh region (J & K) and Chegu goats of HP and Uttarakhand. This fine wool is used for making fine garments like shawls and fine quality fabric.

**Table 8.11:** *Wool Production* in India over the years

| Years | 1951 | 1968 | 1990 | 1995 | 2002 | 2006 | 2007 | 2008 | 2009 | 2010 | 2011 |
|---|---|---|---|---|---|---|---|---|---|---|---|
| Wool* | 27.5 | 29.8 | 41.2 | 45.6 | 51.0 | 45.2 | 44.0 | 41.0 | 43.1 | 43.0 | 44.4 |

* Wool production in Million kg.

India produced 42.99 million kg of wool in 2011-12 which increased to 46.05 million kg wool in 2012-13. India approached peak production of 51 million kg wool during 2002 and thereafter declining was observed. The cause of decrease in wool production is due to increase demand of meat (mutton) on high price and hence slaughter of lambs at early age.

The wool production is highest in Rajasthan (30 % of total wool) followed by Karnataka and J & K (17 % each) and A.P. (11 %).

**Table 8.12:** Wool production in India (Zone wise) and per sheep during 1992

| Traits/ Zones | North-west | North Temperate | South | East | Total |
|---|---|---|---|---|---|
| Sheep population | 21.45 | 3.63 | 20.86 | 4.85 | 50.79 |
| (millions) | (42.2%) | (7.2%) | (41.1%) | (9.5%) | |
| GFW (M kg) | 29.82 | 4.79 | 9.12 | 1.87 | 45.6 |
| | (65.4%) | (10.5%) | (20.2%) | (4.1%) | |
| Per sheep wool (kg) | 1.39 | 1.32 | 0.44 | 0.39 | 0.9 |
| (World Average wool prod./sheep = 2.4 kg) | | | | | |

Khadi sector and other rural based industries are utilizing about 15 m kg wool of the total wool produced in India, for hand spinning and weaving, 5 m kg raw wool is utilized by the organized sector and rest 25 m kg is utilized by the carpet sector. About 50 m kg wool is imported.

Ludhiana and Amritsar are the main centres of hosiery and depends totally on imported wool. The carpet yarn production centres are located at Panipat, Bikaner, Bhilwara, Beawar, Kekri and Bhadohi. The only dying centre of yarn produced is Bhadohi. The imported Australian wool is used in worsted industry located at Bombay, Bhilwara, Jamnagar, Baroda and Amritsar.

### 3. Source of bones, hides and skin

Large quantities of animal byproducts like bones, skin, hides and blood are produced. These byproducts are important from industry and business point of view.

### 4. Source of fuel

The livestock are indirect source of fuel. The dung from cattle and buffaloes is used as fuel in place of coal and wood to cook food, etc. The gobar gas plant is run by gobar (fresh dung) and is used to cook food, to enlighten bulbs, and to run chaff cutter, etc.

### 5. Source of livestock feed

The slaughter house byproducts and carcass wastes are processed for manufacturing of meat meal, blood meal, bone meal, feather meal for farm animals and poultry. The poultry droppings, cow dung and other waste of farm yard are used in the ration of pigs and poultry.

### 6. Converter of agricultural byproducts into valuable food

The animals convert roughages (cereal straw, bhusa, grasses, fodder tree leaves, rice and wheat bran) in to milk, meat, and eggs for human consumption.

## 8.4.2 Livestock in Agriculture

The contribution of livestock to the rural economy is enormous and can not be fully realized by counting only the income from sale of milk, meat, eggs, wool, hides, etc. The contribution in the form of animal draft power (ADP) and manure needs to be accounted besides the value related to social and cultural aspects.

The livestock particularly the cattle and buffaloes have important contribution in different agricultural operations, to maintain soil fertility, and to provide cash for purchase of various inputs used in crop production. The draft animal power (DAP) has an important contribution in agriculture. The various sources of DAP are the bullocks, buffaloes, camels, horses, mules, donkeys, yak and mithun. These animals are serving to the mankind in socio-economic services in carrying out various agricultural operations, transportation and as pack animals. Most of the share to DAP goes to cattle (70 % of total) followed by buffaloes (about 29 %). Whereas the camels, horses and ponies, mules, donkeys, yak and mithun are about 1 % of the total work animals in our country.

The agriculture requires three energy systems viz. manually operated hand tools, animal drawn and machine operated tools and systems. The machine operated tools and systems use the fossil fuel-based energy and the electric-based energy. All these three energy tools and systems are found to work simultaneously in agriculture operations even to day. The use of diesel energy and electrical energy in agriculture has increased progressively reaching 150 – 200 times during 2003-04 compared to base year1970. On the contrary, the use of animal as well as human energy has simultaneously come down to 1/9 $^{th}$ and almost ½, respectively during 2003-04 after 1970. The mechanical energy had about 10 % of its share of the total energy used in agriculture during 1970 which had progressively increased over the years (29 % in 1980, 58 % in 1990) and reached about 87 % in 2003-04.

In the present era of mechanization in agriculture through electro-mechanical power, about 60 % of the area in our country is still cultivable through draft animals and 20 per cent by use of tractors. The DAP is most economical source of power and easily available energy in rural India particularly for millions of small and marginal farmers. In India, there were estimated about 65 million drought animals which save fuel about 6 million tones of petroleum fuel worth 60 billion annually (2007). There are certain areas in hill regions of the country like mountain areas, highlands and undulated lands where DAP is feasible for carrying out the agricultural operations (field operations, transport and agro-processing) and it is difficult if not impossible to be substituted by fossil fuel-operated energy system. Thus, DAP based energy has key role for sustainable agriculture particularly the mountain agriculture. Mechanized agriculture needs to be supported by animal power and farmers can not stop using the hand tools but manual operations with hand tools are also required to be done as the supplement of mechanized agriculture. This is practically true for small and marginal farmers and in areas of mountain and highlands. Moreover, the mechanized farming sustains on fossil fuel of limited supply with certain life.

The DAP energy has many advantages over the fossil fuel and electric based energy.

- The DAP is easily available at cheaper cost on lower possible input,
- It is not possible to use machines for all the operations,
- The animal drawn equipments are of low cost, home made at village level and can be used for small and fragmented/scattered farms particularly in mountains,
- The animals utilizes the crop residues and byproducts as well as produce milk, meat, dung/ manure used to increase soil fertility,
- The mechanization reduces employment opportunities which is particularly very important in our country,
- The fossil energy has a limited life and it has also a negative impact on environment.

(i) ***Agricultural operations***: Livestock provide motive/tractive power to the agriculture. They are basic to all *agricultural operations* like ploughing, sowing of seeds, irrigation, inter-cultural operations, threshing, cane crushing, transportation of fodder and all kinds of agricultural produce from field to home and then to market.

(ii) ***Organic manure***: All classes of livestock and poultry as well produce *organic manure* in the form of faeces (dung0 and urine. The manure (dung) is the sole source to maintain soil fertility. The dung is a complete manure having all the major elements (N, P, K) required by crop plants. The cost of dung as manure was estimated Rs. 2156 crore in 1990-91. About 50 % of the total dung produced is used as farm yard manure and rest as fuel. The animal bones are used to manufacture the bone meal.

(iii) ***Income to purchase inputs for agriculture***: The sale of various livestock products (milk, and milk products, meat, egg and wool) is a good source of income for livestock owner particularly to the small farmers. The cash received is used for the *purchase of various inputs* like quality seeds, fertilizers, pesticides and agricultural implements, and other household materials for daily use, etc.

### 8.4.3 Income and Employment From Livestock

The dairying has an important economic role to play besides a way of life of the faming community in India and an adjunct (joined) to crop production by providing the required farm power and manure. The dairying and poultry farming has a good share to agricultural income and playing role in protecting the farmers from abject (miserable or degraded) poverty, overcomes the hardship of landless, small and marginal farmers and women in agriculture during natural calamities and agro-climatic uncertainties like drought leading to crop failure.

The milk and eggs provides regular income through their sale. About 16 % of the farm household's income in different parts of the country comes from livestock production. This was more in the states of Gujarat (24.4 %), Haryana (24.2 %), Punjab (20.2%) and Bihar (18.7%).

About 70 % of the total households in rural India are small and marginal farmers having their own livestock. During 2003, the marginal farmers (< 1ha land) were 48 % of rural households, had 24 % of the land and more than 50 % of the country's cattle and buffaloes, and 2/3 of small animals and poultry. Not only this, but the landless labourers are also benefited from livestock rearing. Medium and large farmers have major share of land holdings in rural India but the distribution of livestock is more equitable with no skewness among different categories of people.

A.H. sector provides large *self-employment* opportunities. About 19 million people worked in livestock sector during 1999-2000, out of which 11 million as principal status and 8 miilion as subsidiary status and it was 5 % of the total working population. About 73 % of the rural labour force and about 57 % of the total working population is engaged in agricultural sector for performing various activities while 8.8 % of the agriculture work force is involved in livestock sector with variation among the states from only 3 % in N-E states to 40 – 48 % in Punjab and Haryana. Further, the rural women are playing a major role in different activities of animal husbandry. About 75 % (70.5 % in 1993-94 to 76.6 % in 2004-05) of this labour requirement in livestock sector is met by village women and this figure has gone to about 90 % in important dairy states of Gujarat, Punjab and Haryana because most of the men force has diverted to cities for jobs.

Significant employment and income opportunities have been provided to different classes of people from animal sector:

(*i*) *Raising of animals*: Small and marginal farmers and landless labourers get opportunity to work in keeping of animals as the animal rearing involves intensive use of labour usually on the part of family members. This also provides permanent jobs to people in livestock farms maintaining cattle, buffaloes, sheep, goat, and poultry.

(*ii*) *Livestock feed supply:* A large number of people are engaged in fodder production and its supply, manufacturing and marketing of livestock and poultry feeds.

(*iii*) *Health care*: The animal husbandry Department of states and central government provide jobs to a number of veterinary graduates, stockmen, and other technical, ministerial and supporting staff. There is a scope for para-vets (A.I. workers, lab-technicians, veterinary pharmacists, meat inspectors) for supporting animal production and health services.

(*iv*) *Teaching and research activities:* A number of teachers, research scientists and extension workers get jobs for teaching to veterinary students, for conducting research and providing training and demonstration work on different aspects of animal husbandry.

There is a shortage of trained manpower (Veterinary & Animal Science Graduates) for animal health management, animal welfare, meat inspection, livestock rearing on farms, and to assist the farmers in advising them for sustainable animal production through the application of scientific breeding, feeding, management and health care. It is a real fact that veterinary colleges are lacking in faculty for teaching as per VCI guidelines. It is therefore the need of the day to admit and train more number of students and to establish more number of veterinary colleges. The minimum prescribed teaching staff strength of 84 (17 Professors, 23 Associate Professors and 44 Assistant Professors) for a veterinary college with an admission capacity of 60 students is required.

(*v*) *Medicine industry:* The work of manufacture of medicines, their distribution, advertisement and marketing have also generated employment.

(*vi*) *Livestock based industries:* A number of small and large scale industries are livestock based. These are -

- Dairy industry (production, collection, processing, transportation and distribution of milk and dairy plants in manufacturing milk products),
- Woolen industry in processing and manufacturing,

- Poultry farming in raising poultry and marketing the products,
- Leather industry in processing and manufacturing leather goods,
- Bone industry to manufacture buttons and bone meals,
- Sport industry (to manufacture of sport materials like football, volleyball, cricket ball, hand gloves from hides and skins of animals), and
- Handy craft industry.

### 8.4.4 National Economy

The agriculture is the backbone of India's economy. During 1980-81, the agriculture had contributed about 34.7 % to national GDP but its contribution has goes down to 12.3 % in 2010-11 based on price index of 2004-05, whereas the livestock share within agriculture has doubled from 14 to 27 %..

During 2001-02, the value of output from livestock sector was about 28.3 % of the value of the output of Rs. 635395 crores from Agriculture and allied sector (Central Statistical Organization) and the contribution of these sectors in the total GDP was 7.35 % and its contribution to national GDP has been 12.3 % during 2010-11.

During 2002-03, the *contribution of milk alone* (Rs. 103 804 crores) was higher than paddy (Rs. 73964 crores), wheat (Rs. 43815 crores) and sugarcane (Rs. 28205 crores).

During the last 10 years, the contribution of livestock sector has been between 3.5 to 4 % to the national GDP while its contribution to the GDP from agriculture has shown some increase from 25 % to more than 27 %.

| Contribution (%) | 1980 -81 | 2001 -02 | 2004 -05 | 2005 -06 | 2006 -07 | 2007 -08 | 2008 -09 | 2009 -10 | 2010 -11 |
|---|---|---|---|---|---|---|---|---|---|
| 1* | 34.7 | 25.9 | 16.04 | 15.46 | 14.69 | 14.29 | 13.36 | 12.41 | 12.34 |
| 2 | – | 7.35 | 4.02 | 3.9 | 3.74 | 3.63 | 3.68 | 3.58 | 3.37 |
| 3 | 14.0 | 28.3 | 25.04 | 25.02 | 25.48 | 25.39 | 27.59 | 28.86 | 27.28 |

*1 = Agr % to national GDP; 2 = AH % to national GDP; 3 = AH % to Agr.GDP.

During 2010-11, the livestock sector has contributed an output of Rs. 2075 billion, amounting 3.37 % to the total national GDP and 27.3 % to the agricultural GDP.

***Export and import of livestock products:*** India is poor in global livestock trade, sharing only 0.3 % of world export and 0.4 % imports. Earlier upto 1980's India used to import the livestock products. But the situation has changed in early 1990's. This has been attributed to a decrease in imports of milk and milk products and an increase in export of meat and meat products.

Among the livestock products export, the dairy export was just of nominal value of nearly 3.6 % compared to meat of 87.3 % and eggs of 8.4 % whereas live animals were of 0.5 % and others 0.2 % during 1996 - 2000. The dairy export increased to

13.5 % of the total export during 2001 – 07 while the export share of meat and its products had been of 76.5 %, eggs of 9.1 %, live animals of 0.8 % and others 0.1 % of the total export.

Regarding the *import* of livestock products, the major share goes to dairy products which remained around 95 % of the total import during 1961 – 85, but went down to 84.3 % during 1986-90 and 62.0 % during 2001 -07. The share of imports of other livestock products had been to 7.9 % of live animals, 4.3 % of meat and its products, 2.6 % of eggs and 23.2 % of other products of the total imports during 2001 – 07.

During 2002-03, total export earning from livestock, poultry and related products was Rs. 4226 crores. Out of the total export, leather sector accounted for Rs. 2470 crores in value terms.

During 2009-10, India made an export of 545731 tones of bovine meat worth Rs. 5436.1 crore and 52251 tones of sheep and goat meat worth Rs. 737.27 crore. The buffalo meat has registered an increase of 8% per annum during last 5 years and its export was doubled during 2008 – 2011, pushing out the US as the third largest exporter of beef. India has exported 1.28 MT of buffalo meat in 2011 – 12 and 1.5 MT in 2012 – 13 against 0.61 MT in 2009.

There is growing demand of meat and its product in neighbouring countries, especially Middle East. There are about 100 modern abattoirs and meat processing plants for hygienic meat production to meet the export demand of nearly 56 countries, besides 4000 registered and 25000 unregistered slaughter houses in the country to supply meat for domestic market.

USDA report (2012) and Bulletin of the International Dairy Foundation – 446 (March 2012), estimated Indian beef (buffalo meat) exports for 2013 at 2.15 million tons, 30 percent higher than 1.66 million tons in 2012 exports. Similarly, in 2011 exports was estimated 1.29 million tons, based on trade data.

The *export of animal products* indicate the importance of Buffalo in export earnings from animal products, the total export earnings in 2012-13 was Rs. 20,130.90 crore, in which, Rs. 17,400.60 crore was from buffalo meat and around Rs. 700 crore from buffalo dairy products.

**Table 8.13:** Animal Products Exported (APEDA Report 2012-13)

| Sr. No. | Items / Products | Rs. (crore) |
|---|---|---|
| 1 | Buffalo Meat | 17,400.90 |
| 2 | Sheep & Goat Meat | 425.60 |
| 3 | Poultry Products | 494.14 |
| 4 | Dairy Products | 1412.10 |
| 5 | Animal Casting | 18.37 |
| 6 | Processed Meat | 21.56 |

*Contd...*

*Contd...*

| | | |
|---|---|---|
| 7 | Swine Meat | 2.15 |
| 8 | Honey | 356.32 |
| 9 | Total Export: | 20,130.90 |

India is on top to export processed ***goat skin.***

### 8.4.5 Transportation

The livestock, besides the transportation of agricultural products, is the main source of rural transport in difficult, remote and hilly areas due to lack of the means and feasibility of other sources of transport to carry loads and passengers. In such cases the camels and equines (horses, mules and asses) are the only sources of transport.

### 8.4.6 Source of entertainment and social activities

Man also keeps the animals as a hobby (horse and poultry) and also uses them for riding like horse and elephants. The animals are main part of circus to play a number of games for recreation. The animals kept in zoo also attract people. Cultural life particularly in villages is intimately connected with cattle in many ways. The buffalo serve as a sacrificial animal in India, as a racing animal in some south east Asian countries, marriage and funeral procession in Indonesia and buffalo festival in Thailand.

❑❑❑

# 9

# Response to Selection

The different evolutionary processes together with artificial selection have resulted into many breeds of domestic animals that exist today. After domestication man has tried to change the genetic structure of animal population through judicious selection for bringing genetic improvement in performance potential. However, the aim has not been achieved so far and more genetic improvement is needed for milk, meat, egg, wool and other characters like race in horses, better show jumping and more intelligent guide dogs.

## 9.1 SELECTION OBJECTIVE AND SELECTION UNIT

The *selection,* in animal breeding, is an outcome of the process of differential reproduction and survival of animals which may be natural or artificial or both. Artificial selection is man's activity and depends on the choice of the breeder which is objective specific viz. to bring genetic improvement (to produce animals of high genetic merit) in certain traits of interest like production of milk, meat, wool, egg, pashmina fiber, pelt production or for any other purpose like, draft, race, load carrying capacity, etc. The objective of selection is to bring genetic improvement in a character of interest and hence the character under selection is known as *selection objective.*

The genetic selection is based on the breeding value of the individual(s) under selection. The individuals under selection may belong to the different families and selectd on the basis of certain criteria viz. individuals own performance or relatives performance. Secondly, the the individuals under selection may belong to the same family wherein all individuals of the family are selected based on the family mean as a deviation from population mean. Thus, whole family is considered as unit and is selected or rejected based on merit. The families may be half sib or full sib families whose breeding values are estimated based on family mean. Therefore, the individuals of different families under selection and the whole family under selection are called the *selection unit***.** Thus, the selection unit is either the individual or the family.

## 9.2 SELECTION AND CULLING CRITERIA

The individuals with superior performance are selected. The important point to consider is to identify the superior animal. The identification of superior animal is performance based and divides the population in two groups of better and poor performing animals. The better performing group is selected allowing them to reproduce and hence to produce the next generation with the expectation of better performance of their progeny while the poor performing group is culled not allowing them to reproduce and hence not to produce the next generation in the herd/flock. Therefore, the selection and culling are two faces of the same system of reproduction.

The two terms of selection and culling are used in different context for two sexes of animals. The term *culling* is used for the female animals of the population that are not allowed to reproduce because very few animals are culled and most of them are selected. Thus, only the process of culling is done for female animals on the basis of some criteria known as *culling criteria.* On the contrary, the term ***selection*** is used for male animals that are allowed to reproduce (selected) on the basis of some criteria known as *selection criteria.*

The selection is based on the breeding value of the selection objective (character under selection). The B.V. of the character for an individual under selection is estimated based on information from different sources. The various sources of information are called as *selection criteria* which are also known as ***basis of selection*** *or aids to selection* and these are as follows:

- Individual's own performance
- Relative's performance viz. performance of its ancestors (parents, grand parents), its collateral relatives (sibs, cousins) and its progeny.

The breeding value (B.V.) of the character for an individual is estimated based on individual's own phenotypic value of a character. Thus individual's own phenotypic value of a character is the source of selection (criteria of selection) and hence called the "individual selection" or *performance testing* because individual' own performance (phenotypic value) is used as a measure of its breeding value (genetic merit). This is also known as *mass selection* instead of individual selection if the individuals are kept together *en mass* for mating, *e.g.* Drosophila and Tribolium, in a bottle without keeping records of individual mating. But when the mating is controlled (records are maintained) as in case of large farm animals and mice etc., the term individual selection is used.

The B.V. of a character for an individual may also be estimated based on its ancestors performance (parents and grand parents) known as pedigree selection or based on the performance of its collateral relatives viz. family members (sibs) known as family selection or based on its progeny performance called as **progeny selection or** progeny testing.

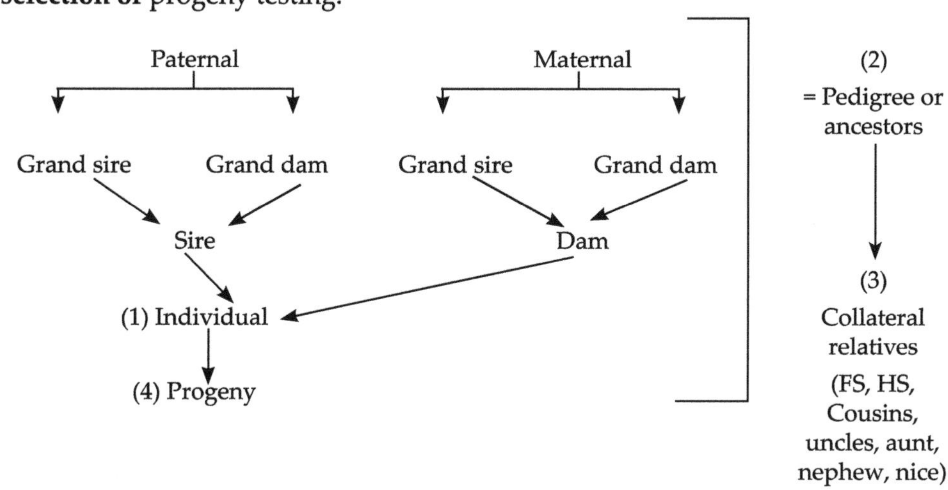

The breeding value obtained from various basis of selection is called as the *estimated B.V. (E.B.V.) or probable B.V. (P.B.V.) or predicted B.V.* This estimated B.V. is really the genetic worth or breeding worth of an animal which is due to the average effect of genes possessed by the animal for a character and half of this genetic worth is transmitted to the progeny. Therefore, the estimation of this breeding value must be accurate which affects the efficiency of selection and hence the genetic improvement.

## 9.3 IMPORTANCE OF SELECTION FROM RELATIVES PERFORMANCE

***Why and when the relative's performance is useful:*** The basis of estimating the B.V. of an individual from phenotypic value of its relatives lies in the fact that the relatives share the same genes possessed by the individual, to a certain extent depending on the degree of genetic relationship. The different selection criteria other than animal's own performance (relative's performance) are required under the following situations-

(*i*) When individual selection is not possible. The following are the traits for which individual selection is not possible:

- Sex-limited traits.
- Traits which can not be measured in living animals, like slaughter traits, longevity and lifetime production traits.

(*ii*) When selection is required at an early age but the trait under selection is expressed at later age of the animal. Under such conditions, the part records (part lactation milk yield to predict lactation milk yield, FLMY to predict milk yield in later lactations in dairy animals; part year egg production to predict the annual egg production in poultry; birth weight to gain in adult body weight may either be used or the information from relatives may be used to predict the B.V. of the trait.

(*iii*) When the heritability of the trait under selection is low, the individual selection is not effective. In this case, the information on relatives (family members like sibs) gives more reliable estimate of B.V. of the individual.

## 9.4 GENETIC EFFECT OF SELECTION

The change in gene frequency can be estimated only for qualitative characters and it is not possible for quantitative traits for the well known reason of their polygenic nature. However, the effect of selection on quantitative characters is reflected in progeny performance. The effect of selection is estimated by measuring the change in mean performance of progeny generation in comparison to the mean of parent generation.

The selection, without creating new genes, changes the genetic structure of the population by changing the frequency of genes and genotypes (Chapter 5). The frequency of desired genes is increased in the progeny generation through selection of superior breeding stock at the expense of the frequency of undesirable or less favourable genes by culling the genetically inferior animals. The increase in frequency of desirable genes due to selection improves the performance of progeny generation over the parental generation. This is the *genetic effect of selection.* The genetic effect of selection in terms of the change in mean performance of polygenic traits due to selection is now the subject matter to be discussed in next chapter. However, it is first better to know some aspects of selection considering the polygenic characters.

## 9.5 SEQUENTIAL SCHEMES OF SELECTION

The selection involves the identification of the individuals (males as well as females) which are superior. The selection is a complex process completed in different stages / ages of the animals. Broadly speaking, there are two stages of selection viz. selection based on physical attributes like birth defects (off-breed characteristics, anatomical and genetic defects), slow growth rate and late maturity. This is called as *initial selection* or *phenotypic selection*. The second stage of selection is the *genetic selection* which is based on the breeding value of animals under selection for various traits of economic importance.

The different stages of selection are taken as sequential schemes of selection. Such schemes of selection are known as *sequential selection.*

*Multi-stage selection* - The process of phenotypic selection starts with the birth of the animal.

The *first stage* of phenotypic selection is that the new borns are culled if they have off breed characteristics, anatomical or congenital and genetic defects.

The *second stage* of phenotypic selection starts when the animals are culled

(*i*) For slow growth rate and late maturity,

(*ii*) For non- functional reproductive system,

(*iii*) For poor libido, poor semen quality and poor semen freezability.

(*iv*) For any serious disease.

(*v*) From poor pedigree (dam performance).

- *Successive stage selection* – The selection of adult females is done for repeatable traits based on the value of the character measured at successive time viz. milk yield recorded in different lactations. The selection is done after completion of lactation.

  The adult females are also culled for a number of reasons viz. reproductive problems (anoestrus condition, repeat breeding), utero-vaginal disorders (prolapsed of vagina/uterus, retention of placenta, metritis), udder problems (mastitis, teat block, blood in milk), other diseases and also for low production (voluntary culling).

- *Selection based on correlated traits* – Some traits are expressed late in life of animal while others are observed after death of animal. Preliminary selection is done based on correlated traits expressed in early life. This reduces the space requirement, labour and cost in keeping all the animals.

- *Genetic selection* - This is the final selection which is based on the breeding value of the animal for the trait under selection.

The genetic selection can also be multi stage selection particularly in poultry viz. two stage selection. The first stage selection may be for 20 weeks body weight and age at first egg while the second stage selection may be for egg number and egg weight or it may be for all the four traits in the second stage.

## 9.6 SELECTION DIFFERENTIAL (PHENOTYPIC SUPERIORITY)

The selective breeding to produce superior progeny rests on the logic that the superiority of superior parents is transmitted and shown by the progeny. The superior animals are selected and the remaining ones are culled. Thus, selection divides the population in two groups viz. selected and culled (rejected). The selected group comprised the superior animals which have their mean performance ($\overline{P}s$) better than the mean of the whole population ($\overline{P}$).

*The difference in the mean performance of the selected group ($\overline{P}_S$) to be used as parents and the mean of the population ($\overline{P}$) before selection,* is called the ***selection differential***, denoted by **S**. Therefore, the selection differential is the superiority of the selected parents ($\overline{P}_S$) over the population average ($\overline{P}$). It is thus estimated as:

$$S = \overline{P}s - \overline{P} \qquad ...(9.1)$$

This selection differential is called the expected selection differential.

### 9.6.1 Standardized selection differential ($S/\sigma_P$) or Intensity of selection

The selection differential (S) may also be expressed in terms of standard deviation and called as the standardized selection differential. This is obtained by dividing the selection differential with the phenotypic standard deviation of the trait ($\sigma_P$). This standardized selection differential ($S/\sigma_P$) is also called as the intensity of selection, denoted by i. Thus,

$$\frac{S}{\sigma p} = \frac{(\overline{P}S - P)}{\sigma p} = i \qquad ...(9.2)$$

The intensity of selection, *i*, is the selection differential in standard deviation units of the trait and it is the mean deviation of the selected animals in units of σP of the trait. Therefore, the intensity of selection (i) is the numbers of σP of the trait by which the mean of the selected group (Ps) is above the population mean (***P***) before selection.

The use and importance of the standardized selection differential (intensity of selection) lies for the reasons -

- That the S can not be estimated unless the selection is actually done.
- It is also used to compare the different methods of selection and
- To compare the response to selection for different traits measured in different units.

The S in original units of measurement can be taken by multiplying i with standard deviation as:

$$S = i\,\sigma_P \qquad ...(9.3)$$

The intensity of selection, ***i***, is also taken from normal frequency distribution curve as:

$$i = \frac{Z}{p} \qquad ...(9.4)$$

**Where,** p is the proportion of animals selected or

It is the area of the truncated distribution

Z is the height of frequency distribution curve where the group of selected animals is represented, or

Z is the height of normal curve at the point where the selected individuals with the lowest record fall. This point of separation is called the *point of truncation (t).* The Z is thus the height of ordinate at the point of truncation.

The level at which the animals are culled at a point of truncation 't' is called the culling level. The culling level at 't' = $P + t\,\sigma_P$.

There are tables available showing the ordinate and area of normal frequency distribution curve to obtain the value of Z. This is known as the selection differential under truncation which indicates that all the animals are selected with records above a certain level of the trait. Therefore, the selection differential in standard deviation units may be taken from mathematical properties of the normal distribution as:

$$i = \frac{S}{\sigma_P} = \frac{Z}{p} = \text{intensity of selection}$$

The intensity of selection (***i***) is inversely proportional to the proportion selected (***p***). If the proportion selected (***p***) is given, the intensity of selection (***i***) can be estimated in terms of ***p*** the number of standard deviation of the trait ($\sigma_P$)

The intensity of selection (***i***) can be known without actually measuring the superiority of parents. The approximate value of ***i*** can be obtained from ***p*** for any character based on their relationship after consulting a table of values.

**Table 9.1:** i as a function of p.

| p | Intensity of selection (i) | | | | |
|---|---|---|---|---|---|
| | 10* | 20 | 50 | 100 | infinity |
| 1.0 | 0 | 0 | 0 | 0 | 0 |
| 0.9 | 0.17 | 0.18 | 0.18 | 0.19 | 0.20 |
| 0.8 | 0.32 | 0.33 | 0.34 | 0.35 | 0.35 |
| 0.7 | 0.46 | 0.48 | 0.49 | 0.49 | 0.50 |
| 0.6 | 0.74 | 0.77 | 0.79 | 0.79 | 0.80 |
| 0.5 | 0.74 | 0.77 | 0.79 | 0.79 | 0.80 |
| 0.4 | 0.89 | 0.93 | 0.95 | 0.96 | 0.97 |

*Contd...*

*Contd...*

| | | | | | |
|---|---|---|---|---|---|
| 0.3 | 1.07 | 1.11 | 1.14 | 1.15 | 1.16 |
| 0.2 | 1.27 | 1.33 | 1.37 | 1.39 | 1.40 |
| 0.1 | 1.54 | 1.64 | 1.71 | 1.73 | 1.76 |

* No. of individuals from which the parents are to be selected

### 9.6.2 Factors Affecting Selection Differential (S) or Intensity of Selection (i)

The following factors determine the intensity of selection **(i)** or selection differential **(S)** -

(*i*) *Proportion selected* (**p**)*:* There is inverse relation between proportion selected and selection differential. This implies higher the proportion of animals selected, the **S** will be low than when the ***p*** is low. The number or proportion of animals selected **(p)** depends on the breeding policy, population size and the capacity of the farm. The proportion selected (**p**) will be higher, if it is required to increase the herd size. This will decrease ***i*** and the **S.** On the contrary, the ***i*** and **S** will be increased in a herd of constant size because fewer replacements are required. However, the population size put a restriction on ***p***. In small herd size the chances of inbreeding are increased which have three fold effects viz. the inbreeding depression and random drift, reduction in genetic variability and thirdly reduction in ***i*** and **S**. On the contrary, in large size herd, the chances of inbreeding are less and ***i*** is increased. The increase in ***p*** can increase the herd size but it will reduce ***i.***

The ***p*** is affected by number of replacement stock being available. This is the function of the *fertility (conception rate and number of offspring in a birth), prenatal calf losses, sex ratio, post-natal calf mortality and culling (health, survivability and longevity)*, etc. The **S** is also larger in litter bearing species (pigs) compared to the species giving single birth (cattle, buffalo). The constraints of low breeding efficiency (late AFC and longer C.I.) and high calf mortality in genetic improvement can overcome by adapting the following ways –

- *Increase in herd size:* This requires an increase in the capacity of the farm. The intensity of selection (***i***) is increased in a large size herd and the level of inbreeding is reduced.
- *Increase in breeding efficiency:* This can be increased by providing better nutrition and better management. This will also reduce the calf losses. Further, the reproductive rate of superior animals can be increased by use of MOET technology.
- *Longevity or stayabilty of elite dams in the herd:* The dams of elite animals should be retained in the herd for longer time to get more calf crop of high genetic merit. A large number of progeny per adult elite female reach the breeding age with proper management. This will increase the selection differential and hence the genetic gain.

Conclusively, the **R** increases with increase in **S** which in turn increases with decrease in ***p***. However, more genetic variation is preserved in the population by higher p and results in greater cumulative **R**. The lower value of ***p*** does not result in maximum long term **R**. The chance to get lost the favourable alleles with small effect linked with unfavourable alleles is reduced with larger ***p*** compared to smaller p each generation.

(*ii*) *Phenotypic Variability* ($\sigma_p$): The phenotypic standard deviation of the selection objective is very important to influence the S and hence R. The S is small if the character is less variable and vice versa. The S increases with decrease in p and with increase in $\sigma_p$ of the character. The effect of the proportion selected (*p*) *and the* $\sigma_p$ on the magnitude of **S** is shown below –

| 50 % selected | 20 % selected | 20 % selected |
|---|---|---|
| $\sigma_p$ = 2 units | $\sigma_p$ = *2 units* | $\sigma_p$ = *1 unit* |
| *S* = *1.6 units* | *S* = *2.8 units* | *S* = *1.4 units* |

The above findings indicate that the **S** increases with the decrease in proportion selected and with the increase in $\sigma_p$ of the character.

The $\sigma_p$ depends on the breeding programme of the herd. The selection for many years reduces the variability in a trait under selection. The selective breeding increases the performance level of the herd and results fewer animals being available in the herd that exceed the herd average. This reduces the proportion of animals selected (p), $\sigma_p$ and selection differential. On the other hand, the **S** is larger in a herd with no selection for many years and it will result in low herd average and larger variability in the trait.

(*iii*) *Sex of animal*: The S is large in case of males. This is because fewer males are required for breeding and hence **i** will be higher in case of males. The use of A.I. and frozen semen technology has further increased the **S** for males. However, the effect of sex on **S** further depends on the species viz. strong selection can be carried out in poultry breeding among both males and cockerels.

(*iv*) *No. of character under selection:* The increase in number of traits under selection reduces the selection differential. This is because an animal may not be outstanding in all the characters. It is therefore advisable to include minimum number of traits in selection programme.

(*v*) *Accurate measurement:* Any error in recording a trait also influences the selection differential. A mistake in deciding the superior animal may result in small selection differential **(S)**.

### 9.6.3 Weighted Selection Differential

The expected selection differential is worked out considering that all the selected parents will have equal number of progeny in the next generation. However, it does not happen but the different parents do not have equal number of progeny due to differences in fertility and survivability of parents and the survivability of progeny. Therefore, it is essential to give weight to each parent by its number of progeny. The selection differential so estimated is called the weighted selection differential or observed or effective selection differential.

Weighted selection differential = $\left(\frac{\Sigma Xini}{\Sigma ni}\right) - \overline{X}$

Where, $X_i$ = observation on the $i^{th}$ parent

$n_i$ = number of progeny of $i^{th}$ parent

$\overline{X}$ = population mean

## 9.7 RESPONSE TO SELECTION

The phenotypic superiority of the selected parents called the selection differential ($S = P_S - P$) is due to the reason that

- The selected parents may have either better genes or they might have received better environment
- The selected parents may have better genotypes as well as better environment.

Therefore, the phenotypic superiority of selected parents (selection differential, S) corresponding to gene's effect and environmental effect has two main components *viz.*

S = Genetic superiority + Environmental superiority

= (Additive genetic + non-additive genetic) + environmental part

The genetic part of phenotypic superiority of selected parents (selection differential) is more important and of more concern. This is because it is only the genetic part of superiority which is transmitted to the progeny and environmental part of superiority is not transmitted. Therefore, the selection of superior animals will be effective to improve progeny generation if they are genetically superior rather than environmental.

Further, out of the total genetic superiority of selected parents it is only the additive genetic superiority which is fixed, transmitted to and shown by the progeny. This is because the genotypes are broken up in every generation due to halving nature of inheritance and crossing over. Hence the progeny do not get the same gene combinations (genotypes) as their parents have. Thus, the superiority due to gene combinations (gene interaction) corresponding to $\sigma^2_D$ + $\sigma^2_I$ is not inherited. The additive genetic variability which is due to individual effect of genes ($\sigma^2_A$) out of the total genetic variability ($\sigma^2_G$) is inherited. The additive genetic superiority, measured as additive genetic variance ($\sigma^2_A$) of the total phenotypic superiority (Ps – P), is thus responsive to selection. This additive genetic superiority or variability (($\sigma^2_A$) expressed as a ratio of total phenotypic superiority or variability ($\sigma^2_P$) is known as the heritability, denoted by $h^2$. The heritability expresses the extent to which the phenotypic superiority of a trait sh+own by selected parents is transmitted to the progeny.

Therefore, whole of the phenotypic superiority of selected parents (selection differential) is not transmitted but its portion equal to $h^2$ is transmitted. Thus, the process of selection results a change in performance of progeny generation in a trait under selection. The change in performance due to selection is known as ***response to selection*** or *genetic change* or *genetic gain* or *genetic improvement* or *genetic advancement.* This change in performance of progeny generation due to selection indicates the superiority of progeny generation over parent generation. Thus, ***response to selection is the difference between the mean of progeny of the selected parents and the mean of the parental generation.*** This is actually the predicted or expected response to selection (R) or direct response which equals to the expected inherited part of selection differential.

$$R = h^2 (Ps - P)$$

$$= h^2 S \quad ...(9.5)$$

The $h^2$ is thus an important concept. It is used to predict the response to selection and can be explained by taking the extreme values of $h^2$ as zero and one as under:

In the first situation of zero value of $h^2$ means that $V_A$ is zero. This indicates that there is no variation in breeding values of individuals of the population. In this case, the selection will not be effective because of no no difference in B.V. of selected and culled animals. This means that there is no genetic superiority of the selected parents to pass on to their progeny. Therefore, there will not be any difference in the mean performance of two successive generations. Thus, R = 0.0 when $h^2$ is zero.

In the second situation when $h^2$ is one, the phenotypic superiority is entirely due to the variation in breeding values (B.V.) of individuals. In this case, the gene interaction effects as well as environmental effects are not important to cause variation in phenotypic values. Thus, the ranking of individuals based on phenotypic values will be equivalent to the ranking based on B.V. Therefore, the superiority of selected parents being solely due to the superiority in B.V. will be passed on to their progeny and the progeny mean is expected to be equal to the mean of selected parents. Thus, R = S when $h^2 = 1.0$.

### 9.7.1 Different expression of predicted response to selection

The response to selection is predicted in different ways as under:

(*i*) R in terms of intensity of selection (***i***, standardized S) –

$$R = h^2 S$$

$$\frac{R}{\sigma_P} = \left(\frac{S}{\sigma_P}\right) h^2$$

$$= h^2 i$$

$$R = h^2 i \sigma_P \quad ...(9.6)$$

$$= b_{AP} i \sigma_P \quad ...(9.7)$$

Where, $i = \dfrac{S}{\sigma_P}$

$\sigma_P$ = phenotypic standard deviation of the trait

$b_{AP} = h^2$

(*ii*) R in terms of accuracy of selection ($r_{AP}$)-

$$R = h^2 i \sigma_P$$
$$= h\, h\, i\, \sigma_P$$
$$= h\, i\, \sigma_P \left(\frac{\sigma_A}{\sigma_P}\right)$$
$$= r_{AP}\, i\, \sigma_A \qquad ...(9.8)$$

Where, $\sigma_A$ = Standard deviation of the breeding value of the trait,

$r_{AP}$ = Correlation between B.V. and phenotypic value.

h = Accuracy of selection.

The response to selection (R) can be predicted from above formulae when the selection criterion is the individual's own record.

The above formula of predicting response to selection can be made generalized for other criterion of selection (selection based on performance records of relatives of the individual).

The genetic gain ($\Delta G$) can be predicted based on any criteria of selection (I) by using the formula:

$\Delta G = i\, \sigma_A\, r_{AI}$

where, $r_{AI}$ is correlation between the selection criteria (I) and B.V. of individual (A).

### 9.7.2 Direct and Correlated Response

The different characters are correlated to each other and hence a genetic change due to selection is expected to occur in the correlated trait. The trait which is under selection (X) is called as primary trait whereas the trait which is not under selection is called secondary trait (Y). The change in progeny generation due to selection in primary trait (X) which is under selection is called direct response whereas the response in secondary trait (Y: which is not under selection) due to selection in primary trait (X) is called correlated response, denoted as CRY.

The selection differential (S) of a trait under selection (X) is not only used to predict the direct response to selection (R or $\Delta G$) in X trait but it can also be used to predict the correlated response (CRY) in the secondary trait not under selection (Y).

The correlated response ($CR_Y$) was estimated by Falconer (1981) using the formula:

$$CR_Y = i\, h_X\, h_Y\, r_G\, \sigma_{P(Y)}$$

The $CR_Y$ can also be predicted from the following formula -

$$CR_Y = S_Y\, h_{XY}$$
$$= S_X\, b_{YX}\, h_{XY}$$
$$= \mathbf{i}\, \sigma_{\mathbf{P(X)}}\, \mathbf{b_{YX}}\, \mathbf{h_{XY}}$$

Where, $S_Y$ = selection differential in unselected (correlated) trait (Y),

$= S_X b_{YX}$

$S_X$ = selection differential in primary trait (X)

$= i\,\sigma_{P(X)}$

$$i = \frac{Z}{P}$$

$b_{YX}$ = phenotypic regression of Y trait on X trait

$h_{XY}$ = co-heritability

= ratio of additive genetic covariance to the phenotypic covariance of traits X and Y

$$= \frac{Cov_{A(XY)}}{Cov_{P(XY)}}$$

### 9.7.3 Factors Affecting Response To Selection

The equation of change in performance due to selection indicated that the response to selection depends on the following factors-

1. Additive genetic variability ($\sigma_A$) of trait
2. Intensity of selection (i)
3. Accuracy of selection ($r_{AP}$)
4. Population size
5. Generation interval.

(1) *Additive genetic variability in the trait* ($\sigma_A$): The additive genetic variability is the main stem and the raw material on which the selection acts. The magnitude of $\sigma_A$ has direct relation with R. The selection is not effective if the genetic differences among animals do not exist. However, the genetic variability of the trait ($\sigma_A$) is determined by the population and the character as well by the breeding programme of the herd, in terms of the selective breeding in the herd for that trait. Thus, it is beyond the control of breeder.

The nature and extent of genetic variability (additive genetic) present in the herd decides the choice of selection criteria for estimating the breeding value of a trait. The individual criterion of selection for within herd improvement is advocated for traits with high heritability and for those expressed in both the sexes at an early age (growth rate). The pedigree selection can be done for traits of high heritability. The selection based on performance of other sort of relatives viz. progeny and collateral relatives (sibs) is effective when heritability of trait is low, the character is sex limited and for the characters which are not expressed in the life of animal (carcass traits, lifetime production, life span, etc.).

(2) *Intensity of selection* (i): This depends on proportion of animals selected (*p*). When *p* is small, the selection is said to be more intense or rigorous. The **R** will be more when *p* will be small. This is because when few animals will be selected (small **p**), the selection differential will be high and hence **R** will be high. The **S**

is directly proportional to ***p***. If all animals are selected, the ***S*** will be zero and no change in progeny mean will occur. The genetic change occurs if some of the best animals are selected and hence ***p*** should be less which will increase ***i***.

(3) *Accuracy of selection* ($r_{AP}$): The B.V. for a trait can not be recorded directly but it is predicted and its prediction requires more sources of information for its accurate prediction. The information to estimate the true B.V. of an animal may be collected on performance of the animal itself and or of its any close relative. Thus, the estimate of true B.V. should be as much accurate as it can be to make the selection more accurate. The accuracy of selection is taken as the correlation between the true B.V. of an animal and the source of information (selection criterion) used to estimate the B.V. ($r_{AP}$), where A is the B.V. and P is the selection criteria. Thus, $r_{AP}$ is a measure of accuracy of selection. The selection criterion may be single record or average of repeated records of the animal itself or any relative.

The $r_{AP}$ is directly related to heritability because it is equal to square root of heritability ($r_{AP} = h$). Thus if $h^2$ is higher the $r_{AP}$ will be higher and hence accuracy of selection will be higher. The accuracy of selection should therefore be increased which is possible by increasing heritability. The heritability can be increased by reducing the environmental variance, use of multiple records, adjustment of data, accurate measurement of data and analyzing the data based on contemporary group means. The Accuracy can also be increased by use of combined selection to estimate B.V., making selection based on future performance (M.P.P.A.) of more number of records, and by keeping a population of large size.

(4) *Population size:* The population size influences the **R** by way of influencing the inbreeding and genetic drift.

(i) *Inbreeding:* The inbreeding is unavoidable in a population of small size. The inbreeding increases homozygosity and hence reduces the genetic variability and also the performance level due to inbreeding depression. The inbreeding is less when animals of both the sexes are equal which is not possible for the reason that fewer males are used than females. This needs to estimate the effective population size (Ne). The Ne is the number of individuals that would give rise to the same rate of inbreeding, if they bred in the manner of an idealized population in which the rate of inbreeding is $\Delta F = 1/(2Ne)$. The Ne reflects the amount of inbreeding expected.

$$Ne = \frac{4\ SDt}{S + D} \text{ because } \frac{1}{Ne} = \frac{1}{4S} + \frac{1}{4D}$$

$$= \frac{4\ SD}{S + D} \text{ and}$$

$$1/2Ne = \frac{1}{8S} + \frac{1}{8D}$$

Where, S and D are the number of sires and dams entering the population per year

The increase in inbreeding per year $(\Delta F) = \frac{1}{2Net}$ which becomes in case of using few males as $\Delta F = \frac{1}{8S\ t^2}$,

Where S = number of sires and

t = average generation interval for males and females in years.

The effect of inbreeding is to reduce the amount of genetic variability ($V_A$) and reduction in performance (inbreeding depression). This causes a decrease in response to selection (R). It is thus suggested that there should be sufficient Ne to reduce $\Delta F$ less than 1 % per generation.

(ii) *Genetic drift:* The genetic drift arises in small population due to sampling of genes. This may cause the loss of favourable alleles from the population and hence will reduce the **R**. Not only this, but it will affect the limit to selection which will reach more quickly.

(5) *Generation interval (t)*: The genetic response per unit of time (year) depends on the generation interval. This is because the genetic gain is observed in next generation, as the difference in mean performance of two generations. Thus, genetic gain will be higher in a population that breeds with younger animals than the population that breeds comparatively at later age. The genetic progress per unit of time (year) is important than genetic progress per generation.

## Dependence of factors affecting response

The factors that affect the response per unit of time ($h^2$, S, and t) are related in such a way that the change in magnitude of one factor affects the change in the magnitude of other factor in an adverse way. The R is more when the $h^2$ of the trait and the selection differential are more but the t is less. The $h^2$ of a trait may be increased based on average of more number of records. This will increase the generation interval (t) and decrease the selection differential (S) and hence reduces the response to selection. This is because the average of more records is more close to the genotypic value of the animal. Secondly, if the generation interval is minimized by culling the animals based on single progeny, the S is decreased. Therefore in order to influence the overall genetic gain, there should be a proper compromise between these factors.

## 9.8 ACCURACY OF SELECTION

The additive genetic relationship (a) between two individuals is taken to the extent to which they have breeding values (B.V.) or genes in common. There are two situations to understand this concept.

The first situation is that two animals are identical twins so as the additive genetic relationship **(a)** will be equal to one and hence a record on other animal will be as accurate of the animal's true B.V. as a record on the animal's own record. The accuracy of selection on a record of other animal's record is equal to the accuracy of selection based on animal's own record.

The second situation is that the additive relationship **(a)** between two animals is zero. This means that the two individuals are not relatives and hence their B.V. is independent of each other. Thus, the selection based on a record of unrelated animal will not give any indication of the animal's true B.V. This means that the accuracy of selection based on a record of other animal will be zero.

### 9.8.1 Estimation of accuracy of selection

It is thus clear that the accuracy of selection based on a single record on a relative will be equal to the extent to which the individual under selection and its relative have their genes in common **(a)** times the correlation between the B.V. of individual and selection criterion ($r_{AP}$). The $r_{AP}$ is the accuracy of relative's performance as a criterion to the relative's true B.V. Therefore, the accuracy of selection is indicated as:

*Accuracy of selection* = Additive relationship between individual and relative x $r_{AP}$

= a h

In view of the above, the *accuracy of selection* based on single record on individual itself or on a relative's record will be as –

- Based on Individual itself = a h = 1 x h = h
- Based on pedigree record
  - (*i*) One parent = a h = ½ h
    Both parents = √(½) h = 0.71 h
  - (*ii*) One grand parent = ah = ¼ h
    All 4 grand parents = √¼ h = 0.5 h
  - (*iii*) One great grand parent = a h = 0.125 h
    All 8 G.G. parents = $\sqrt{\frac{1}{8}}h$ = 0.35 h
- Based on family records
  - (*i*) Full sibs = ah = ½ h
  - (*ii*) Half sibs = ah = ¼ h
- Based on progeny record = ah = ½ h

The above relations have indicated that a single record on one relative is always less accurate than a single record of individual itself to predict its B.V.

The $h^2_n$ for multiple record of the individual is more than single record ($h^2$) and hence the accuracy of selection will be more. The $h^2$ based on single record as well as based on multiple records and the accuracy of selection on n records will be as shown below:

| Selection Criteria | Heritability | | Accuracy of selection based on n records |
|---|---|---|---|
| | Single record | Multiple records (n) | |
| Individual | $h^2$ | $\frac{nh^2}{[1 + (n - 1) r]}$ | $\sqrt{\left[\frac{nh^2}{1+(n-1)r}\right]}$ |
| Pedigree | $0.5 h^2$ | $\frac{0.50 nh^2}{[1 + (n - 1) r]}$ | $0.50\sqrt{\left[\frac{nh^2}{1+(n-1)r}\right]}$ |

*Contd...*

*Contd...*

| | | | |
|---|---|---|---|
| Progeny | $0.5\,h^2$ | $\frac{0.50\,nh^2}{[1+(n-1)\,t]} = \frac{2\,nh^2}{[4+(n-1)\,h^2]}$ | $0.50\sqrt{\left[\frac{nh^2}{1+(n-1)t}\right]}$ |
| Full sibs | $0.5\,h^2$ | $\frac{0.50\,nh^2}{[1+(n-1)\,t]} = \frac{nh^2}{[2+(n-1)\,h^2]}$ | $0.50\sqrt{\left[\frac{nh^2}{1+(n-1)t}\right]}$ |
| Half sibs | $0.25\,h^2$ | $\frac{0.25\,nh^2}{[1+(n-1)\,t]} = \frac{nh^2}{[4+(n-1)\,h^2]}$ | $0.25\sqrt{\left[\frac{nh^2}{1+(n-1)t}\right]}$ |

Where, r = repeatability (Correlation among records)

t = intra class correlation among FS/HS/Progeny

= $0.50\,h^2$ for FS and progeny

= $0.25\,h^2$ for HS

### 9.8.2 Increasing Accuracy of Selection

The efforts are made to make the selection more accurate and effective. The accuracy of selection can be increased by the following methods:

(*i*) **To minimize the environmental variance:** It can be minimized by providing uniform environment to all the animals, use of multiple records to estimate heritability, adjustment of data for environmental effects, accurate measurement of data, and analysis of data on contemporary group mean. See section 17.6 for details on ways to increase heritability.

(*ii*) **Combined selection:** When two or more criteria of selection are used to estimate an individual's B.V., it called as combined selection. The combined selection gives more accurate estimate of B.V. of the individual. It is therefore better to use combined selection by supplementing the individual's performance records with those of its relatives.

The family selection may be used to support individual selection when $h^2$ of a trait is low. It is better to select an individual with better record of a superior family compared to an individual with similar performance of a mediocre or inferior family. The half sib family selection is better than FS family selection. The reason being that family selection is more effective when the environmental variances common to all members of a family are as small as possible. This means that environmental similarities among family members should be low. The common environmental variance is less among half sibs than among full sibs. Thus, the environmental correlation among F.S. is more than among H.S.

Secondly, the F.S. family selection reduces the genetic variability in the population and also results in a certain amount of inbreeding. The selected parents based on F.S. family averages are more closely related. Thirdly, the effective selection intensity is also reduced for a given testing facility for F.S. family selection.

*(iii)* **Selection based on future performance:** The selection should be based on future performance (Most probable producing ability or expected producing ability) of an animal with more number of records. The MPPA or EPA is estimated as:

$$\text{MPPA} = H + \frac{nr}{1+(n-1)r}(e-H)$$

Where,
H = Herd average or contemporary average
C = Average of females based on all records
N = numbers of records
r = repeatability

$$\frac{nr}{[1+(n-1)\,r]} = \text{repeatability of multiple records}$$

The repeatability increases with increase in number of repeated records. Thus, the accuracy of evaluating an animal based on more records is increased when the temporary environmental effects are more important.

The accuracy of estimating MPPA or EPA of a cow is estimated as square root of heritability based on n records:

$$\text{Accuracy of MPPA} = \sqrt{\frac{nr}{1+(n-1)r}}$$

*(iv)* **Large population:** In small size population, the homozygosity is increased with consequent decrease in genetic variability and also the genetic drift occurs which leads to the fixation of alleles. This results in low $h^2$, hence low $r_{AP}$ and the low genetic gain. Therefore, the population size should not be allowed to decrease. The **S** is also low in small population.

## 9.9 GENERATION INTERVAL

The generation interval is the time period between two generations with respect to the same stage of life cycle. It is defined as the average age of parents at the time of the birth of their offspring which also become parents.

The generation interval is estimated species wise combined for both the sexes giving due weight to the proportion of offspring raised. Both the sires and the dams produce progeny of both the sexes but at different ages. The generation interval is estimated as the weighted average of the following four averages –

$$t = n_1\, t_{SS} + n_2\, t_{SD} + n_3\, t_{DS} + n_4\, t_{DD}$$

Where,

$t_{SS}$ = average age of sire to produce sons (males)
$t_{SD}$ = average age of sire to produce daughters
$t_{DS}$ = average age of dam to produce sons (males)
$t_{DD}$ = average age of dam to produce daughters

$n_i$ = weights being the proportion of offspring raised from four paths.

## Factors affecting generation interval (t)

The genetic progress per generation is converted in to the progress per unit of time (years). The conversion is essential to compare the progress for the same trait in different species like milk yield in dairy animals (exotic, native, buffalo, goat, etc).

The following factors are important to influence the generation interval:

(*i*) **Species:** The generation interval is different for different species. The generation interval (years) in cattle is 4.5 in males and 3.0 in females, 5.0 and 3.5 in buffaloes, 4.0 and 2.0 in sheep and goats, 0.7 and 0.6 in poultry.

(*ii*) **Breeding age:** The age at which the reproduction starts (age at first calving/ kidding) influence the generation interval (t). Breeding of animals at the earliest reduces the **t** and increase the response per unit of time. The breeding age may be reduced by providing balanced feeding and better care. This increases the number of replacements available to cull the old and low producing animals. It has bidirectional effect on intensity of selection (i). Increase in number of replacements increases the generation interval **(i),** whereas culling the old animals before their normal productive life increases the requirement of replacements per year and thus the **p** can be increased. This in turn increase the **i** but reduces **S.** The gain attained by reducing **t** will be cancelled by reduction in **S.** Therefore, more rapid population turn over (culling older animals) and reduction in **t** are the determining factors of relative change in **t** and **S** which in turn will affect response per year.

(*iii*) **Sex:** The generation interval (t) is also different for the two sexes depending on their use to produce the offspring. For example, in a progeny testing programme in dairy animals, many sires are used even after their daughters are tested for milk production and their genetic worth is evaluated based on more number of daughters. This increases the generation interval for male side. Therefore, the **t** should be taken as an average from both male and female side in order to predict the response per year. This is also done because there are overlapping generations in dairy animals. In species like poultry, the **t** is equal for both the sexes because the generations are distinct and do not overlap. This is because the selected progeny mostly replace the entire population at the same time.

(*iv*) **Selection schemes:** The selection schemes of varying numbers of offspring raised in any generation also affect the generation interval. There are two schemes depending on number of offspring or records. The first scheme involves one offspring with one record and second scheme involves more number of offspring or more records per animal. The second scheme increases the intensity of selection as well as accuracy of selection and thereby increases the genetic gain. But this also increases the generation interval and result comparatively less genetic progress per year than progress under first selection scheme. Thus, the two schemes of selection can be compared on the basis of genetic gain expected per unit of time and the scheme resulting maximum progress per unit of time will be optimum.

The generation interval is affected by the number of progeny born before the selection is done. A breeder has to wait until more offspring have been reared so as to increase the intensity of selection which will increase the gain. The generation interval is increased by including more offspring and thus response per unit of time is reduced. The increase in numbers of offspring is beneficial to certain limit beyond which it reduces the relative speed of progress (measured as 1/t). The **i** is certainly increased with additional numbers of offspring but the increase in **t** is greater and faster than increase in **i.** This results in low relative speed of progress.

**Conflict of Interest:** It is essential to minimize **t** to maximize the genetic gain per year. But on the contrary, there is an important and interesting aspect of **t** which does not necessitate its minimization and raises a conflict of interest between generation interval and the genetic gain per generation. The magnitude and direction to which the **t** should be influenced is decided by the trait of interest with respect to its variability ($\sigma_P$ and $h^2$) and its maximum production value (in terms of age of dam/female at which the peak production is reached, goals and objective of the breeder viz. commercial production or seed stock, market price of the animals, and the cost of rearing the replacement stock, maintaining the animal during production. These are the factors which raise a conflict of interest to influence the generation interval.

The main objective is to maximize the genetic gain per unit of time for those maintaining nucleus / breeding stock. On the contrary, the main aim is not the maximum rate of genetic change for those interested in commercial farming viz. production of animal products (milk, meat, wool, egg, etc) in a more economically efficient way. It is not necessary to minimize the **t** for economically efficient production of animal products particularly for females in order to maximize the genetic gain per unit of time. This is because the production in females is increased up to a certain age, remains constant for some period and then declines. Thus more production can be obtained by retaining older animals. Therefore, culling the females at an early age to minimize the **t** will affect adversely the productivity of the herd because the phenotypic potential of the new replacements is less than older though the genetic worth of young progeny may be better. Moreover, the depreciation per year (taken as selling price at early age minus selling price at old age divided by **t**) per female is decreased with the increase in **t** because the selling price at old age after a certain age is constant. Finally, lesser numbers of new replacements per year are required if the female stock is kept for longer in the herd and thus the new young progeny may be sold in more numbers. This will fetch more income because younger animals are more costly.

## 9.10 FACTORS HAMPERING THE GENETIC PROGRESS

Artificial selection had been effective in various species of animals to develop better breeds. However, the selection is not so easy to apply and to give desired results. The selection fails to give desired results and thus dampen the enthusiasm of breeders. There are some complications and constraints which hamper the response to selection. These can be grouped as genetic or theoretical and operational or practical complications.

**Genetic Complications**: The genetic complications which are responsible for the failure of selection effect are as:

(*i*) *Phenotypic value as the joint effect of genotype and environment*: The environment has a great effect to mask the genetic worth of an animal. The environment may lead to mistake in selection of animals given different environment. It is therefore better to keep the animal under selection in suitable and uniform environment so as to evaluate the real genetic differences to increase the accuracy f selection.

(*ii*) *Selection is based on phenotype*: It is not possible to identify the better genotype for the reason of polygenic nature of economic characters. The selection is thus based on phenotype. Moreover, the G – E interaction affects the phenotypic value for which it is changed over space and age of animal, etc.

(*iii*) *Heritability:* The heritability estimates of most of the economic traits are low and changes over space and time as well as by the mating systems. The phenotypic valu is a good indicator of the genotypic value only when the $h^2$ of a trait is high.

(*iv*) *Gene segregation*: The superiority of an individual is due to additive effect of favourable genes, better gene combination (heterosis), and better environment. The gene combinations are broken due to their segregation and independent assortment. Therefore, the progeny of better parents have a tendency to regress towards the population mean. The advantage of favourable combination of genes can be exploited by inbreeding to increase homozygosity and by crossing the inbred line to take the advantage of hybrid vigour. However, the superiority of heterozygote can not be used as breeding stock for the reason of gene segregation.

(*v*) *Gene action:* The genes when combined act in different ways. The selection favours the heterozygotes. This advocates the crossing of suitable inbred lines.

(*vi*) *Correlation among traits:* The selection is less effective due to to negative correlation between desirable traits (milk yield and fat %) as well as due to positive correlation between desirable and undesirable traits (milk yield and susceptibility to mastitis).

(*vii*) *Inbreeding:* The selection lines are small in size and hence inbreeding is unavoidable. Thus, there is some effect of inbreeding depression. Therefore, the depressing effect of inbreeding can be counteracted by adopting the mating system of the "mating best with the best" which is selective breeding.

**Operational Complications**: There are some practical complications in operating the effective selection. These are as:

(*i*) *Foundation stock:* The parental population may not be genetically superior and may have no genetic variability to make the selection effective.

(*ii*) Population Size: The population size of farm animals is generally low and hence the intensity of selection and selection differential is low. This makes the selection ineffective.

*(iii) Number of traits:* The ranking of animals for single trait is easy and becomes difficult as the number of traits under selection is increased. This is due to the reason that an animal may be superior in one trait but inferior in other trait. The intensity of selection is decreased for more than one trait becoming equal to $1/\sqrt{n}$ for each trait.

*(iv) Change of selection goal:* The goal of selection is changed without achieving the first goal and hence the primary objective of selection is forfeited.

*(v) Selection limit*: The genetic variability is reduced or completely vanished under long term selection. This makes selection ineffective. This requires the introduction of new genes from outside.

## 9.11 CONSTRAINTS IN GENETIC PROGRESS

Besides the above complications of selection there are some constraints in genetic improvement.

The population of farm animals are non-descript with low production performance and the animal rearing in India is mostly in the hands of farmers and poor people. However, some organized farms have been established in the past to maintain purebred animals and to produce genetically superior bulls for distribution to various agencies engaged in animal rearing. The constraints in genetic improvement through selection are little different in farmer's herd and organized herd.

The genetic improvement is possible by the following ways -

- Selection within improved breed
- Upgrading of non-descript and low producing stock with sires of high genetic merit of improved breed
- Crossbreeding between two or more breeds

The genetic progress through selection depends on genetic variability ($h^2$) in the trait and the selection differential. The generation interval is also important for the genetic improvement per unit of time.

***The following are the suggestions to overcome the constraints:***

The **heritability estimates** can be improved by reducing the environmental variance (see section 17.6). Secondly, in view of low $h^2$ estimate, the progeny testing as a criterion of selection is advocated.

The **accuracy of selection** ($r_{AP}$) is low due to low $h^2$ estimate. Therefore, efforts should be made to increase the accuracy of selection.

There is less scope of selection particularly among females and hence the **intensity of selection is low** due to low proportion of animals selected (p). Therefore efforts should be made to increase the p.

The **population size** is low which results in genetic drift and inbreeding as well as limits the intensity of selection. There is thus required to increase the capacity of the animal farms or to run P.T. programme in associated herd.

## NUMERICAL SOLVED EXAMPLES

**Example 9.1:** Find out the proportion of animals saved, the S of those saved and the culling level in a herd producing 2000 kg milk in a lactation with 400 kg $\sigma_P$ in case the cows with milk production above one $\sigma_P$ of mean are selected.

**Solution:**

(*i*) Proportion of cows saved = Taken from table of proportions, ordinates of normal curve and intensity of selection, the proportion saved (p) = 0.1587 for one $\sigma_P$ above mean (t = 1), and Z/p = 1.52. The t indicates the $\sigma_P$. Thus, 15.87 % of the cows are saved for breeding.

(*ii*) Mean of selected cows

$$(P_S) = Z/p\ \sigma_P + \overline{P}$$
$$= i\ \sigma_P + \overline{P}$$
$$= 1.52 \times 400 + 2000$$
$$= 2608 \text{ kg.}$$
$$S = \overline{P}_S - \overline{P}$$
$$= 2608 - 2000$$
$$= 608 \text{ kg}$$

(*iii*) Culling level

$$t = 1\ \sigma_P) = \overline{P} + i\ \sigma_P$$
$$= 2000 + 1 \times 400$$
$$= 2400 \text{ kg.}$$

**Example 9.2**: Find out the expected and effective selection differential, taking the population mean 27.0 from the following data when 6 individuals were selected.

| No. | 1 | 2 | 3 | 4 | 5 | 6 | Total |
|---|---|---|---|---|---|---|---|
| $X_i$ | 34 | 32 | 30 | 28 | 26 | 25 | 175 |
| $n_i$ | 16 | 36 | 50 | 38 | 35 | 45 | 220 |

**Solution:**

$$\overline{X}p = 27.0$$
$$\overline{X}s = 175 / 6 = 29.16$$
$$\text{Expected } S = 29.16 - 27.0 = 2.16 \text{ units}$$
$$\text{Weighted (effective) } S = \frac{\Sigma X_i n_i}{\Sigma n_i} - \overline{X}_p$$
$$= \frac{6327}{220} - 27.0$$
$$= 1.75 \text{ units}$$

The ratio of effective to expected S less is than one. This indicates that natural selection was operating in terms of differential fertility and viability.

**Example 9.3:** The selection for 6 months body weight in a flock of sheep was done. The mean body weight was 12 kg with $\sigma_P$ of 1 kg. Find out the mean of selected males and females in case 10 % females and 1 % males were selected.

**Solution:** The selection intensity (i) values were taken from table of proportions, ordinates of normal curve and intensity of selection:

(*i*) Selection of males:

$$i = \frac{Z}{p} = 2.64, \text{ since } p = 0.01 \text{ (1 \% selected males)}$$

$$S = i \times \sigma_P$$

$$= 2.64 \times 1 = 2.64 \text{ kg}$$

Thus, mean of selected males = 12 + 2.64 = 14.64 kg

(*ii*) Selection of females:

$$i = \frac{Z}{p} = 1.76, \text{ since } p = 0.10 \text{ (10 \% selected females)}$$

$$S = i \times \sigma_P$$

$$= 1.76 \times 1 = 1.76 \text{ kg}$$

Thus, mean of selected females = 12 + 1.76 = 13.76 kg.

**Example 9.4**: Clayton et al. (1957) worked out the expected and observed response to selection for abdominal bristle number in Drosophila. The $h^2$ from base population was 0.52 and the base population had an average of 35.3 bristle numbers whereas the selected parents had 40.6. The numbers of bristle in offspring generation were 37.9. Find out the realized response.

**Solution:**

$$S = 40.6 - 35.3$$

$$= 5.3 \text{ bristle}$$

$$R = h^2 S$$

$$= 0.52 \times 5.3 = 2.8 \quad \text{(expected response)}$$

$$\text{Realized } R = 37.9 - 35.3 = 2.6$$

**Example 9.5**: The standard deviation ($\sigma p$) of a trait was 3.35. The parents selected were 20 male and 20 females out of 100, giving the selection intensities as 1.40. Estimate the expected response.

**Solution:**

$$R = i\,\sigma p\, h^2$$

$$= 1.4 \times 3.35 \times 0.52 = 2.44.$$

**Example 9.6: A** herd of Sahiwal cows had the herd average of 2000 kg milk of first lactation and the average of the selected population was 2200 kg. The $h^2$ estimate of milk yield was 0.25 and generation interval was 5 years. Estimate the time required to double the milk production.

**Solution:**

$$S = 2200 - 2000 = 200\text{kg}$$

$$R = h^2 S = 0.25 \times 200 = 50\text{kg}$$

$$\Delta G/\text{year} = 50/5 = 10\text{kg}$$

No. of generation required to double the milk production = 2000/40 = 50 and hence 40 generation = 40 x 5 = 200 years.

**Example 9.7:** The average annual greasy fleece weight of a flock of sheep was 2.0 kg and the average increased to 2.5 kg, after culling the low producers. The EPA of 10 rams used in the flock was 3.0 kg and after culling the low producers the average increased to 4.0 kg. Assume the $h^2$ of fleece production as 0.4 and generation interval from dam to daughter as 2 years and that of sire to daughter as 3 years. Estimate the expected genetic gain per generation and per year

**Solution:**

$$S = 2.5 - 2.0 = 0.5 \text{ kg (females)}$$

$$S = 4.0 - 3.0 = 1.0 \text{ kg (male)}$$

$$G\text{ (F)} = 0.4 \times .05 = .02 \text{ kg}$$

$$\Delta G\text{ (M)} = .04 \times 1.0 = .04 \text{ kg}$$

$$\Delta G \text{ per generation} = \frac{0.2 + 0.4}{2} = 0.3 \text{ kg}$$

$$\Delta G \text{ per year (F)} = 0.2/2 = 0.05 \text{ kg}$$

$$\Delta G \text{ per year (M)} = 0.4/3 = 0.13 \text{ kg}$$

$$\text{Average } \Delta G \text{ per year} = \frac{0.05 + 0.13}{2} = 0.09 \text{ kg}$$

**Ans:** $\Delta G$ per generation = 0.3 kg = 300 gm

$\Delta G$ per generation = 0.09 kg = 90 gm

**Example 9.8:** Find out the observed selection differential, standardized and weighted (effective) selection differential from the data given below with $\overline{P}$ = 1800 units and $\sigma p$ = 110 units. Also find out the genetic gain (response to selection) taking the $h^2$ as 0.2.

| Sire no. | 1 | 2 | 3 | 4 | 5 | 6 | Total |
|---|---|---|---|---|---|---|---|
| Trait | 2000 | 1970 | 1920 | 1880 | 1850 | 1900 | 11520 |
| No. of progeny ($n_1$) | 10 | 15 | 12 | 14 | 11 | 8 | 70 |

**Solution:**

(*i*) Expected

$$S = \overline{P}s - \overline{P} = \left(\frac{11520}{6}\right) - 1800$$

$$= 1920 - 1800 = 120.0$$

(*ii*) Weighted (Effective)

$$S = \frac{\Sigma X_i n_i}{\Sigma n_i} - \overline{P} = \frac{134460}{70} - 1800$$

$$= 1920.8 - 1800 = 120.8$$

$$\text{(iii) Standardized S (Intensity)} = \frac{(\overline{P}s - \overline{P})}{\sigma p} = S/\sigma p = i$$

$$= 120/110 = 1.09$$

$$\text{Genetic gain } (\Delta G) = h^2 i \sigma p$$

$$= 0.2 \times 1.09 \times 110$$

$$= 23.98 \text{ units}$$

**Example 9.9:** The following are the dates of birth in years for sires and dams of 10 calves born in the year 1975. Estimate the generation interval for sire and dam.

| Calf no. | 1 | 2 | 3 | 4 | 5 | 6 | 7 | 8 | 9 | 10 |
|---|---|---|---|---|---|---|---|---|---|---|
| Sire | 1970 | 1968 | 1970 | 1968 | 1971 | 1970 | 1970 | 1972 | 1971 | 1970 |
| Dam | 1971 | 1967 | 1969 | 1970 | 1971 | 1968 | 1965 | 1970 | 1967 | 1967 |

**Solution**

G.I. = Year calf born – year sire or dam born

$$\text{G.I. for sire} = \frac{(5+7+5+7+4+5+5+3+4+5)}{10} = 5.0 \text{ years}$$

$$\text{G.I. for dams} = \frac{(4+8+6+5+7+7+10+5+8+8)}{10} = 7.0 \text{ years}$$

**Example 9.10:** The AFK of 500 goats is given here. Find out the G.I. for AFK.

| Age at kidding (years) | 2 | 3 | 4 | 5 |
|---|---|---|---|---|
| No. of goats | 140 | 170 | 100 | 90 |

**Solution:**

$$\text{G.I.} = \frac{(140 \times 2 + 170 \times 3 + 100 \times 4 + 90 \times 5)}{500}$$

$$= \frac{1640}{500}$$

$$= 3.28 \text{ years}$$

❑❑❑

# 10 Measurement of Response

The response to selection is also known as the genetic change. The comparison of the mean performance of two successive generations under selection programme provides an estimate of the genetic change or response to selection.

## 10.1 PURPOSE AND ASSUMPTIONS IN MEASUREMENT OF RESPONSE

The response to selection is measured for the following ***purposes:***

- To know whether selection is likely to result in genetic improvement.
- To compare the expected response to selection ($R = h^2S$) with that actually obtained in progeny generation (realized genetic change)
- To compare the rate of genetic change under different sire selection schemes and different breeding programme viz. inbreeding, recurrent selection, hybridization or strain/line crossing

The measurement of response to selection has certain ***assumptions*** which are as under:

- The genetic change is entirely due to selection and there is no change in environmental conditions.
- The generation means should be free from the effects of random genetic drift arising from sampling of gametes in each generation.

## 10.2 PROBLEMS IN MEASUREMENT OF RESPONSE

The R depends on the $h^2$ of the parental generation and hence the R in later generations can not be predicted without predetermining- the $h^2$ in each generation. The $h^2$ is expected to change in each generation under selection for the reasons that selection changes the gene frequencies and reduces the variance which in turn affects the $h^2$. However, the magnitude of change in gene frequencies and $h^2$ is very small unless only a few loci are involved. Thus the response is usually maintained with little change over several generations.

The measurement of response for more generation of selection is problematic. The response to selection varies from generation to generation and leads to the variability of generation means. The generation means do not progress in a regular fashion but fluctuate erratically and for this the response can not be measured with accuracy. The best is to measure the average response per generation.

**I. Variability of generation means:**

The different causes assigned to the variability of generation means are as:

- Population size which causes the random drift and the sampling errors in estimating generation means
- Variation in selection differential
- Effects of environmental factors

***(i) Population size:*** The selected lines are small in size. Thus there will be genetic drift and sampling effect which will cause the changes in gene frequencies at random. This will consequently lead to change of generation means. The random drift has cumulative effect because any change in gene frequency in one generation act as the starting point for the change in the next generation. The increase in population size can help to reduce the variability due to random drift and sampling errors.

***(ii) Selection differential***: The selection differential in different generations is different. This happens due to the reason that different parents do not contribute equally to the offspring generation in terms of their number of progeny. The unequal contribution of different parents is caused by natural selection and artificial selection.

The artificial selection of parents separates the parental population in two groups *viz.* selected and culled. These two groups differ in gene frequencies. On the other hand, the natural selection exert its influence on change of gene frequency in two stages viz. differences of fertility among the parents and the viability of parents as well as among their progeny. Therefore, the change in gene frequency due to selection occurs at three stages viz. artificial selection among the parents, natural differences of fertility among the parents, and natural differences of viability among the parents and progeny generation. Depending upon the choice of breeder some parents are culled without having reproduced at all, some are culled after getting one or two progeny and others are not culled as long as they can reproduce in the herd. The consequence of these three stages of selection is that some parents contribute more offspring than others.

*Weighting the selection differential:* The problem of unequal contribution of parents (selected to produce the progeny generation) in terms of their different number of offspring that are available for measurement in the progeny generation necessitates the weighting of the selection differential of the parents according to the number of offspring from different parents measured. This weighted selection differential is really the effective selection differential which is more relevant to the response observed in the progeny mean. Thus, it requires to distinguish between two types of selection differential viz. effective (weighted) and the expected (unweighted). The expected selection differential is the superiority of selected parents taken as deviation from population mean as defined earlier. Whereas the effective selection differential is obtained by weighting the expected selection differential (deviation of parents from population means) according to the proportionate contribution of the progeny measured from each parent. Therefore, the effective selection differential is the weighted mean deviation of parents. The weights are given to each parent (pair of parents) according to their proportionate contribution of individuals measured in next generation. The weighted selection differential takes into account the effect of natural selection also. Natural selection will be taken to opposing artificial selection if selected individuals with better phenotypes have low fertility and viability. The joint effects of natural and artificial selection together can be measured by weighting the selection differential. The effect of natural selection can be determined by comparing the effective with the expected selection differential. The ratio of effective to expected S, if less than 1 indicates that natural selection is operative. However the effective selection differential does not cover the effect of differences of viability among the offspring. Thus the whole effect of natural selection is not accounted for in the effective selection differential.

Falconer (1955) compared the effective and expected selection differential in a two way selection experiment for 6 weeks body weight in mice for 30 generations in upward selection and for 24 generations in downward lines. He could not find any vertical difference between the effective and expected S in upward selection which indicated that natural selection was not operating to influence the response. On the other hand, in downward selection, the effective S was found less than expected particularly with the increase in generation number beyond 18. This indicated that natural selection hindered the artificial selection and reduced the observed response by reducing the effective S. The natural selection could not influence the fertility of mice in upward selection because large mice produced large litters than small mice and litters were reduced to 8 young at birth for standardization. Litters having less than 8 young in upward selection were very few and so differential fertility had no effect, while in downward line the litters with 8 young were very few. Thus, the effective selection differential in downward lines was reduced to fewer-young and further the smallest mice after 18 generations did not breed at all which further reduced the effective selection differential.

***(iii) Environmental differences***: The variability of generation means is created by a number of environmental factors like differences in climatic conditions, nutrition level of animals and differences in general management practices which changes over time. With the passage of time the environmental conditions (climatic conditions, nutrition, and management practices) are changing. It is well known that the genotypic value of a quantitative character is modified and changed by the environmental factors. Therefore, any change in subsequent generation mean may also be due to change in environmental conditions. Thus, both the genetic and non-genetic factors contribute to the changes in performance traits in subsequent generations. Therefore, these changes in generation means reflect the phenotypic trend or phenotypic change in the character over time rather than only the genetic change. Thus, the direct comparison of generation means does not provide an estimate of the genetic change (genetic gain) but it provide an estimate of phenotypic change ($\Delta P$). Therefore, it requires partitioning the phenotypic change into its component parts viz. genetic change ($\Delta G$) and environmental change ($\Delta E$). The partitioning of $\Delta P$ is the main problem to measure the genetic change.

**II. Specific problems to measure response to selection in large livestock species:**

Besides the problem of variability of generation mean in measuring the genetic change, a number of other specific problems also occur which are specific to large livestock. These are:

(*i*) The generations are overlapping in large livestock

(*ii*) The sires are compared in different herds, years and regions

(*iii*) The selection intensities and also the selection criteria are different for the two sexes

***(i) Overlapping. Generation***: Large livestock species (dairy animals, sheep etc.) have overlapping generations. This means that the numbers of generations of selection are different for different animals. The older parents experience less cumulative selection than the younger parents. Consequently the progeny of younger parents are genetically superior to older parents. When the generation number of sire ($T_S$) and dam ($T_D$) of each offspring differ, the generation number of each offspring ($T_O$) is taken as:

$$T_0 = 1 + \frac{(T_S + T_D)}{2}$$

The estimate of the $\Delta G$ per generation is obtained as the intra year regression of offspring performance on offspring generation number. This method also needs to adjust the performance record of offspring for dam's effect.

Another method to measure the response from selection is to compare the contemporary progenies produced by sires belonging to two different generations (Smith, 1962). This method involves the equal allotment of dams of each age and generation to the sires of the two generations. The annual genetic change can be estimated as:

$$\Delta G = 2\sum_{i=1}^{i} \frac{\overline{X_1} - \overline{X_2}}{mk}$$

Where, k = is the number of years for which the comparison is done,

m = is the age difference in years between two groups of sires (older sires which are progeny tested, and younger sires which are non tested).

$\overline{X}_1$ and $\overline{X}_2$ are the average performance of progeny from younger and older sires, respectively.

***(ii) Comparison of sires in different herds:*** The ranking of different sires and their comparison can only be made if all of them are tested in one herd or population, and also in the same year. This is because the herd differences and year differences are caused by both genetic and environmental factors. To overcome this limitation, the sires are compared after making adjustment of data for the environment factors viz. herd differences, year differences etc. The sires which have been tested with large number of progeny in two herds, two years etc. are taken as reference sires and included with new sires in progeny testing programme. The use of reference sires provides a genetic control as an aid to compare the different sires which do not have progeny in the same herd. The concept of reference sire is clear from the following example:

Suppose the two sires X and Z are progeny tested in one herd. The sire X ranked first by a difference of 30 units than sire Z and in another herd, two sires Y and Z are progeny tested and the sire z ranked first with a difference of 20 units than sire Y as shown below:

| Herds | Sires and their ranking | | |
|---|---|---|---|
| | X | Y | Z |
| I | + 30* | Not used | – 30 |
| II | Not used | – 20 | + 20* |
| | Difference between X and Y = 30* + 20* = 50 units | | |

In the above PT programme, a direct comparison of sires X and Y is not possible but they can be compared in reference to sire Z which has been progeny tested in both the herds. The two sires X and Yare expected to differ in progeny performance by 50 units.

## 10.3 METHODS OF MEASUREMENT OF RESPONSE

Considering the partitioning of the phenotypic superiority of parents into its component parts (genetic and environmental), the phenotypic change ($\Delta P$) between two generations can be partitioned into the corresponding changes - the genetic change ($\Delta G$) and the environmental change ($\Delta E$). The $\Delta P$ equals $\Delta G$ only when environmental change between generations ($\Delta$ E) is zero. This is possible only when environmental effects on parents and progeny generations are exactly the same. However, the effects of environmental factors are beyond control and, therefore, response to selection of a population can be measured by partitioning the $\Delta P$ into its genetic and environmental parts:

$$\Delta P = \Delta G + \Delta E$$

### 10.3.1 Elimination of environment effect

The variation due to changes in environment can be reduced or eliminated by two methods in estimating the genetic change. The first method is the control system to estimate the environmental change using the control population whose genetic value remains unchanged over generations. The second method is to control the environmental change either in the experiment or in statistical analysis (phenotypic and genetic trend).

**1. Control population**: There are three ways of control system. These have been discussed here.

***(a) Unselected control population (Random bred control population):***

The unselected control population is one in which there is no selection and the parents in each generation are taken (selected) entirely at random. As no genetic change between generations ($\Delta$ Gc = 0) is expected in control population, so the phenotypic change occurred will be caused by the environmental factors ($\Delta Pc = \Delta E_c$). The effect of environment will be the same in control and in selected population ($\Delta Ec = \Delta E_S$). The change in phenotype of the selected population will be caused due to change in genetic structure as a result of selection plus due to change in environment corresponding to $\Delta G_S$ and $\Delta E_S$ respectively. Thus

$$\Delta Ps = \Delta G_S + \Delta E_S$$

Therefore, the genetic change ($\Delta G$) can be estimated by difference in the mean phenotypic merit of the control and selected population as:

$$\begin{aligned} \Delta G &= \Delta Ps - \Delta Pc \\ &= (\Delta Gs + \Delta Es) - \Delta E_c \\ &= \Delta G_S \quad (\text{since } \Delta E_S = \Delta E_C) \end{aligned}$$

***Assumptions:*** The expectation of control population for no genetic change ($\Delta G_C = 0$) is based on the following assumptions:

- The control and selected population are of the same genetic background so that both the populations respond equally to the same changes in environment.
- Both the populations (control and selected) should be raised under similar environments and management practices so that $\Delta E_c = \Delta E_s'$
- The size of control population should be large enough so that there is no genetic change due to sampling of gametes (genetic drift) or due to natural selection.
- There is no G x E interaction

***Limitations:*** This control system has certain limitations.

- It is mostly used in poultry because it is easy to maintain control population. On the other hand, to maintain an unselected control population of large livestock is costly of space and capital investment.
- The second limitation to use this control system is that it leads to a biased estimate of $\Delta$ G in later-generations. This is because the selected lines and control lines becomes genetically different after several generations of selection and hence they respond differently to the same change in environmental effects. In order to overcome this problem it is better to take new un selected control population every generation by randomly choosing the individuals from selected population. In the absence of control population which is the situation mostly occurred in the improvement of domestic animals, it is difficult to decide the magnitude of the improvement due to selection because of the progressive change in the management conditions.

The control population provides a genetic constant (no genetic change) and reflects only variation caused by the environment. The control populations are the herds, flocks or lines which are bred continuously at random as a base or reference.

The size of **control population** should be large enough to maintain effective random mating as well as to keep the inbreeding level low as far as possible. The increase in rate of inbreeding per generation depends upon the number of males and females in the population. The rate with which the homozygosity is increased is given by the following formula (Lush 1945, A.B. Plans):

$$\Delta F = \frac{1}{8M} + \frac{1}{8F}$$

Where, $\Delta F$ = the increase in inbreeding per generation

M and F = number of males and females in the population, respectively.

The factor $\frac{1}{8M}$ is more important because the males are less in numbers. This formula is used for the whole population regarding the rate of inbreeding in relation to population size. It is suggested that the population should contain about 25-50 males and 50-100 females. The control populations are mainly used in poultry. The control population size kept in Cornell (USA) is 50 cocks and 250 hens, for pigs the size is 12 boars and 12 gilts (USA) while it is 16 boars and 32 gilts in U.K., 10 rams and 250 ewes (UK) and 1 bull and 25 cows for Hereford cattle selection (USA).

Random change in gene frequency due to random genetic drift is the main problem in control population. The problem of random drift can be avoided to keep reconstituting the control population or to maintain at different locations so as to allow comparison between control populations and to have a check on random drift.

**(*b*) Two way selection or bidirectional selection:**

The entire population is divided in two groups - one for higher phenotypic merit and other for low phenotypic merit. The selection is done in both the lines for increased and decreased performance. Both the lines are maintained contemporaneously under the same environmental and management practices. Thus both the lines are affected simultaneously to the same extent by the environmental changes and each of the two lines provides control for the other. This reduces the variation between generation means because the environmental changes affect both the population equally. The response is measured from the difference between the two lines which is about twice as great as that in the two lines separately as:

$$\Delta P_u = \Delta G + \Delta E$$

$$\Delta P_D = \Delta G + \Delta E$$

The difference between two lines is:

$$\Delta P_u - \Delta P_D = (\Delta G + \Delta E) - (-\Delta G + \Delta E) = 2\Delta G$$

Therefore, the estimate of genetic change or genetic gain ($\Delta$ G) is half the difference.

This system has the following limitations:

- It has been observed that the response is not equal in the two lines (upward and downward lines) and it is called as asymmetry of response to selection.
- The response is unequal for various reasons discussed in chapter 27.
- This system is expensive and not feasible for large livestock because the selection for decrease performance is not justified.

**(*c*) Repeat mating**

This is another control system to estimate ΔG and ΔE. The same parents are repeatedly mated in adjacent years to get full sibs. This leads to an expectation that the full sibs born in adjacent years will not differ genetically and the difference in the progeny of the same sire is due to environment. However, single pair of sibs related to the extent of 50 % is expected to have only half of their genes in common and so will differ genetically. Therefore, the average of many pairs of full sibs should be taken with the expectation of their equal genetic merit. The average difference between the phenotypes of full sibs from adjacent years is taken as an estimate of environmental change (ΔE) between two years.

$$\Delta E = \Delta P_{\text{I year}} - \Delta P_{\text{II year}}$$

The phenotypic change between two adjacent years is also obtained for the entire population which contained both the component parts of the phenotypic change (ΔP = Δ G + ΔE). Therefore, an estimate of genetic change is obtained by substracting the environmental change (estimated from full sibs) from the phenotypic change estimated for the entire population between two years. Thus the genetic change (ΔG) is obtained from the difference of phenotypic change in entire population (ΔP = ΔG + ΔE) and the environmental change (ΔE) of repeated progeny:

$$\begin{aligned} \Delta G &= \Delta P \text{ in entire population} - \Delta E \text{ in repeated progeny} \\ &= \Delta P - \Delta E \\ &= (\Delta G + \Delta E) - \Delta E \end{aligned}$$

The method of repeat mating has the advantage relative to control population method that there is no accumulation of drift variance and that there is little chance of G x E interaction.

However, several records from several years need to be analysed. In this case, an estimate of average phenotypic change per year (ΔP/year) is obtained from the linear regression of phenotype on time and likewise an estimate of average ΔE per year is obtained from linear regression of cumulative ΔE on time. Now the estimate of average genetic change per year is obtained from the difference of the two regressions.

***Requirements***: This system has the requirements

- This control scheme of repeat mating requires a sufficient number of repeat mating for each pair of adjacent years.

- Secondly, it requires the *adjustment of age effect of dams,* at least for traits which are affected by dam's age. The dams in different adjacent years will differ in their "age and so the progeny of repeat mating (full sibs) will differ due to environmental differences plus due to the effect of differences in dams' age. Therefore, $\Delta E$ will be the biased estimate and so will be the $\Delta G$. To overcome this biasness, the records for dams' effect are adjusted, or like aged progeny from repeat mating in adjacent years (i.e. like-aged full sibs) are measured and compared.

***Advantages:*** The repeat mating is more useful and restricted to poultry and pigs, and is not practicable for large livestock. This control scheme is advantageous because

- It is less expensive even in a commercial herd/flock,
- It does not involve the problem of G x E interaction when there is divergence in the selected and control populations,
- The same genetic background is not an assumption and
- The population size is not a problem to avoid the effect of inbreeding.

### 10.3.2 Statistical Method (Phenotypic and genetic trend)

The statistical methods are applied to estimate the genetic, environmental and phenotypic trends as detailed below:

**(I) Estimation of phenotypic trend**

The change in mean over years is the phenotypic trend over time. It is estimated as regression of performance records (P) on time (years, T) as:

$$\Delta P = b_{P.T} = \frac{\sum Pt}{\sum t^2}$$

$$S.E.(b_{P.T}) = \left\{ \frac{[\Sigma P^2 - b_{P.T}(\Sigma Pt)]}{(\Sigma t^2)(N-2)} \right\}^{1/2}$$

Where, $b_{P.T}$ = linear regression of performance (P) on time (year) of calving (T)

$\sum Pt$ = Corrected sum of products of performance and time

$$= \sum Pt - \frac{(\sum P \sum Pt)}{N}$$

$\sum P^2$ = Corrected SS of performance records

$$= \sum P^2 - \frac{(\sum P^2)}{N}$$

$\sum t^2$ = Corrected SS for time (years)

$$= \sum t^2 - \frac{(\sum t^2)}{N}$$

N = Total numbers of records

**(II) Estimation of genetic trend**

The method of estimation of genetic change was developed by Smith (1962).

This was based on the comparison of mean performance of paternal half sisters in different years. Therefore, it requires the continuity of the genotypes over different years. However, the changes in performance of trait of paternal half sibs in successive years reflect the genetic plus environmental trend as:

$$E\,(b_{P.T}) = g + e$$

$$E\,[b\,(P - \overline{P}).T/s = ½\,g]$$

Where, $b_{P.T}$ = linear regression of performance (P) on time (T)

$b\,(P - \overline{P}).\,T/s$ = intra sire linear regression of P on T, each record being taken as deviation from contemporaries

g = genetic trend

e = environmental trend

The genetic trend (ΔG) is calculated as:

$$\Delta G = -2b(P - \overline{P}).\,T/s$$

$$S.E.\,\Delta G = \sqrt{4}\,V\,b(P - \overline{P}).T/s$$

This method eliminates year effect in environment and hence it is more precise. The' following expectations of regression have founded the method of Smith (1962) or some modification of it

$$E\,(b_{pT}) = g + e$$

$$E\,(b_{pT}/_{s}) = 0.5g + e$$

$$E[\,b(P - \overline{P}).\,T\,/\,s = -0.5\,g$$

$$E\,(b_{PT}/_{SD}) = e$$

The $b_{P.T/S}$ is the regression within sire which gives e + 0.5g because the sires did not vary and only dams contribute to genetic progress.

$$b\,(P - \overline{P}).\,T\,/\,s = (0.5\,g + e) - (g + e) = -0.5\,g$$

$b_{PT/SD}$ = Regression of performance on time within sire and dam = e

(because F.S. have same B.V.)

The followings are obtained by combining above 4 equations

$$g = 2(b_{P.T} - b_{p.T}/S)$$

$$g = b_{P.T} - b_{P.T}\,/SD$$

$$g = -2[b\,(P - \overline{P}).T/S]$$

Smith (1962) estimated the rate of genetic change as $2(b_{P.T} - b_{S.T})$

where, $b_{P.T}$ is the linear regression of population performance on time

$b_{S.T}$ is the within sire regression of progeny performance on time.

To avoid year effect, the expression 2b (P-S) T was suggested which is twice the pooled within sire regression on time of the difference between the population and individual sire means.

**(III). Environmental trend**

This is taken as a difference of Δ P and Δ G.

Therefore, $\Delta E = \Delta P - \Delta G$

$$S.E.\ \Delta E = \sqrt{[SE\ (\Delta P)2 + SE(\Delta G)^2]}$$

## 10.4 GENETIC CHANGE BY DIRECT SELECTION

The method given by Rendel and Robertson (1950) is used as a measure of genetic change by measuring the genetic superiority and generation interval from 4 paths of parents to offspring.

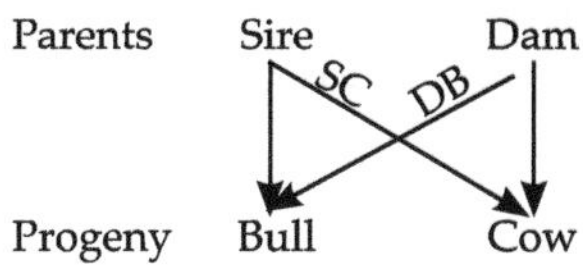


$$\Delta G/\ yr = \frac{G_{5C} + G_{5B} + G_{DC} + G_{DB}}{L_{5C} + L_{5B} + L_{DC} + L_{DB}}$$

Where, $L_{SC}$ = Average age of sires when their daughters were born which entered in herd

$L_{SB}$ = Average age of sires when their sons (bulls) were born and used as breeding bulls

$L_{DC}$ = Average age of dams when their daughters were born which entered in herd

$L_{DB}$ = Average age of dams when their sons (bulls) were born and used as breeding bulls

***1. Expected genetic gain by selection of sires of daughters ($G_{SC}$):*** The genetic superiority of i[th] sire having n daughters [$G_{SC(i)}$] is estimated as:

$$G_{SC(i)} = \frac{2nh^2}{4+(n-1)h^2}\left(\overline{X}_{Di} - \mu\right)$$

Where, $X_{Di}$ = Average record of n daughters of i[th] sire

$$G_{SC} = \frac{\Sigma n_i G_{sec(i)}}{\Sigma n_i}$$

Summation is done for all the sires and thus i = 1 to S.

***2. Expected genetic gain by selection of sires of bulls ($G_{SB}$):*** The genetic superiority and expected genetic gain due to selection of sires to produce bulls is measured as:

$$G_{SB(i)} = \left[\frac{2n_i h^2}{4+(n_i-1)h^2}\right](\overline{X}_D - \mu)$$

Where, $X_D$= Average record of n daughters of sire of i[th] bull

$$G_{SB} = \frac{\sum n_i G_{SB(j)}}{\sum n_j}$$

Where, $n_j$= Number of daughters of $i^{th}$ bull

***3. Expected genetic gain of dams of cows ($G_{DC}$):*** Expected Δ G by selection of cows to produce daughters is taken as:

The genetic superiority of $i^{th}$ dam of j $^{th}$ cow was estimated as:

$$G_{DC(i)} = \left[\frac{n_i h^2}{1+(n_1-1)r}\right](\overline{X}_D - \mu)$$

Where, $\overline{X}_D$ = Dam's record

$n_i$ = Number of repeated records of $i^{th}$ dam

r = Repeatability

$$G_{DC} = \sum m\, n_j G_{DC(ij)}$$

Where, $n_j$ = number of daughters of j $^{th}$ dam

m = total number of dams.

***4. Expected genetic gain of dam's of bulls ($G_{DB}$):*** Expected genetic gain due to selection of breeding bulls on the basis of their dam's performance.

The genetic superiority of dam of $i^{th}$ bull was computed as:

$$G_{DB(i)} = \left[\frac{nh^2}{1+(n-1)r}\right](\overline{X}_{Di} - \mu)$$

Where, $\overline{X}_D$ = Dam's average

$n_i$= Number of records of the dam of $i^{th}$ bull

$$G_{DB} = \sum n_i G_{DB(i)}$$

Summation is done for all the bulls and thus i = 1 to S.

The $G_{DB}$ was estimated by weighted sum of $G_{DB(i)}$, weights being the number of daughters ($n_i$) sired by the son of $i^{th}$ dam.

## 10.5 SELECTION BETWEEN POPULATIONS

The selection exploits the genetic variation whether exists between and or within the population. The different breeds or strains are kept by different breeders for their particular preference even for the same purpose. However, selection between populations requires the information on the relative merits of various populations under any set of conditions for making comparison. This task is difficult for single breeder as it requires relatively long time and large number of animals. Therefore, the various Government agencies, Universities, Breed societies (having the resources necessary for comparison) have been involved in this work.

***Requirements for comparing population:*** (i) The most important requirement is that all stock should be treated equally and hence comparison is made in a manner that minimizes the chances of any population receiving favoured treatment.

(*ii*) Each population should be represented by a random sample of animals or the method and extend of selection of animals should' be similar for each population. In case of P.T., the mates must be selected at random from the tester population. Moreover, in case of comparison at more than one place, the stock from each population should be allocated at random to each place.

(*iii*) The type of comparison is an important factor. There should be direct comparison which means that all populations under test for comparison should belong to the same location. Similarly, purebred progeny should not be compared with crossbred.

(*iv*) The numbers of measurements made on each population under test also influence the efficiency of a comparison. In this context, there are two important points. The first is that it is the number of sires represented in each population rather than number of progeny per sire, which determine the efficiency of comparison. Thus each population under test should be represented by many sires. The comparison based on two or three sires per population does not give the true representation of the entire population. Secondly, it is better to measure a different set of progeny each season or year than to take repeated measurements on the same set of progeny, unless the characters have low repeatability.

(*v*) In view of the possibility of G x E interaction, it is better to conduct the comparison in ail relevant environments.

## 10.6 RANDOM SAMPLE TEST

The random sample tests are conducted in poultry. The samples are taken from various strains at regular interval, usually annual. The random samples of eggs from the hatchery of a number of breeders are taken and raised the chickens in well defined set of conditions. An independent body conducts the tests on performance measurements and publishes the results. The random sample test has now been started for pigs and sheep. In India there are 4 centers of random sample test in poultry. These centres are at Banglore, Bombay Bhuvneswar and Gurgaon where 750 eggs for layers and 700 eggs for broilers are received for testing.

There are certain objectives of random sample tests; the first is that different breeders have different aims e.g. one strain may have been developed for a restricted feeding regime while another for one particular feeding regime, say *ad libitum*. Secondly, very large numbers of observations are often required to detect economically important differences between strains, especially for reproductive characters having high variability.

## Numerical Examples

**Example 10.1** The selection was practiced for 10 generations in a flock for egg number up to 300 days of age together with maintaining a random bred control population. Estimate the average phenotypic and genetic response per generation from the given data.

| Generation No. (X) | 1 | 2 | 3 | 4 | 5 | 6 | 7 | 8 | 9 | 10 |
|---|---|---|---|---|---|---|---|---|---|---|
| Selected line (S) | 62 | 64 | 64 | 66 | 66 | 68 | 71 | 71 | 76 | 75 |
| Control line (C) | 60 | 63 | 63 | 64 | 62 | 63 | 68 | 65 | 67 | 66 |
| Diff. (S - C) = d | 2 | 1 | I | 2 | 4 | 5 | 3 | 6 | 9 | 9 |

(*a*) Average ***phenotypic gain*** per generation ($\Delta Ps$) = b sx

$$= \text{Cov } sx / \sigma^2 x$$

This is estimated as the regression of generation mean of trait on generation number.

$$\sum S = 680; \sum X = 55$$

$$\sum X^2 = 385 - 302.5 = 82.5$$

$$\sum XS = 3856 - 3740 = 116$$

$$b_{SX} = 116 / 82.5$$

$$= 1.40 \text{ eggs}$$

$$= \Delta P_S$$

(*b*) Average ***genetic gain*** per generation can be estimated in two ways:

$$\Delta_G = b_{Xd}$$

or $$\Delta G = \Delta P_S - \Delta P_c$$

(*i*) This is estimated by the regression of deviation of generation mean of selected population from mean of control population (d) on generation number (X) = $b_{d.X}$.

$$\sum d = 39$$

$$\sum d^2 - \frac{(\sum d)^2}{N} = 213 - 152.1 = 60.9$$

$$\sum Xd - \frac{\sum X \sum d}{N} = 278 - 214.5 = 63.5$$

$$b_{d.X} = \Delta G = \frac{63.5}{82.5} = 0.77 \text{ eggs}$$

(ii) $\Delta G$ as the difference of $\Delta_{PS} - \Delta_{PC}$

$$\sum C = 641$$

$$\sum C^2 - \frac{(\sum C)^2}{N} = 41141 - 41088.1 = 52.9$$

$$\sum XC - \frac{\sum X \sum C}{N} = 3578 - 3525.5 = 52.5$$

$$b_{CX} = \Delta P_C = \frac{52.5}{82.5} = 0.63 \text{ eggs}$$

$$\Delta G_S = \Delta P_S - \Delta P_C$$
$$= 1.40 - 0.63 = 0.77 \text{ eggs}$$

Thus, $\Delta G$ is same by two methods.

**Example 10.2**: Estimate the genetic trend from the following data·

| Sire No. | Cow No. | Year | Milk yield | Cow temp. average | Deviation (d) |
|---|---|---|---|---|---|
| 1 | 1 | 1 | 2780 | 2202.8 | 577.2 |
| | 2 | 1 | 2720 | 2211.4 | 508.6 |
| | 3 | 2 | 2230 | 2466.3 | -236.3 |
| | 4 | 2 | 2624 | 2410.0 | 214.0 |
| 2 | 5 | 1 | 1580 | 2374.3 | -794.3 |
| | 6 | 1 | 1640 | 2365.7 | -725.7 |
| | 7 | 2 | 1540 | 2564.9 | -1024.9 |
| | 8 | 3 | 1600 | 2097.1 | -497.1 |
| | 9 | 3 | 1540 | 2105.7 | -565.7 |
| 3 | 10 | 2 | 2510 | 2426.3 | 83.7 |
| | 11 | 2 | 3070 | 2346.3 | 723.7 |
| | 12 | 3 | 1800 | 2068.6 | -286.6 |
| | 13 | 4 | 2420 | 2222.0 | 198.0 |
| | 14 | 4 | 2500 | 2206.0 | 294.0 |
| | 15 | 4 | 2100 | 2286.0 | -186.0 |
| | 16 | 4 | 1800 | 2346.0 | -546.0 |
| 4 | 17 | 1 | 1630 | 2367.1 | -737.1 |
| | 18 | 1 | 2850 | 2192.8 | 657.2 |
| | 19 | I | 3000 | 2171.4 | 828.6 |
| | 20 | 1 | 2000 | 2314.3 | -314.3 |
| | 21 | 2 | 1820 C′ | 2524.8 | -704.0 |
| | 22 | 2 | 3100 | 2342.0 | 758.0 |
| | 23 | 2 | 2600 | 2413.4 | 186.6 |
| | 24 | 3 | 2900 | 1911.4 | 988.6 |
| | 25 | 3 | 2450 | 1975.7 ~, | 474.3 |
| | 26 | 3 | 2520 | 1965.7 | 554.3 |
| 5 | 27 | 3 | 1870 | ~′o58.6 | -188.6 |
| | 28 | 3 | 1600 | 2097.1 | -497.1 |
| | 29 | 4 | 2310 | 2244.0 | 66.0 |
| | 30 | 4 | 2400 | 2226.0 | 174.0 |

**Solution:** Contemporary average = (year total – cow's yield) / n - 1

| Years (X) | 1 | 2 | 3 | 4 |
|---|---|---|---|---|
| $n_1$ | 8 | 8 | 8 | 6 |
| **Total M.Y.** | 18200 | 19494 | 16280 | 13530 |

Herd average = $\frac{67504}{30}$ = 2250.13 kg

Contemp. av. for cow No. 1 = $\frac{(18200 - 2780)}{7}$ = 2202.8 kg

Contemp. av. for other cows is calculated in the same way

Years S.S. (sire wise): $\sum X^2 - \frac{(\sum X)^2}{n}$

Sire No. 1 = 10 – 9 = 1.0

Sire No. 2 = 24 – 20 = 4.0

Sire No. 3 = 81 – 75.6 = 5.4

Sire No. 4 = 43 – 36.1 = 6.9

Sire No. 5 = 50 – 49 = 1.0

S.P. (Year x deviated yield) = $\sum Xd - \frac{\sum X d)}{N}$

Sire No. 1 = 1040.8 – 1595.25 = – 554.2

Sire No. 2 = –6758.2 – (–7215.4) = 457.2

Sire No. 3 = –151.0 – 981.7 = –1132.7

Sire No. 4 = 6967.2 – 5115.18 = 1852.02

Sire No. 5 = –1097.1 – (–1559.95) = 462.85

| Sire No. | N | SS years | SP (year x deviation) |
|---|---|---|---|
| 1 | 4 | 1.0 | -554.45 |
| 2 | 5 | 4.0 | 457.20 |
| 3 | 7 | 5.4 | -1132.70 |
| 4 | 10 | 6.9 | 1852.02 |
| 5 | 4 | 1.0 | 462.85 |
| Total | 30 | 18.3 | 1084.92 |

b = 1084.92/18.3 = 59.28 kg / year

Thus genetic progress ($\Delta$G) = – 2b (P – S).T = – 2(59.28)

= – 118.5 kg/year

❑❑❑

# 11

# Bases of Selection

## INDIVIDUAL SELECTION

The animal itself is the smallest unit of selection and improvement. The genetic selection requires the measurement of breeding value of the animal. The B.V. of an individual can not be measured directly from observation on the individual but it is estimated from the phenotypic value of the individual itself or from the relative's performance. The relatives of the animal under selection may be direct relatives (ancestors and progeny) or collateral relatives (full-sibs, half-sibs, aunts, niece, etc).

The breeding value (B.V.) of an individual based on its phenotypic value can be estimated for one character (single trait selection) or for more than one character (multi trait selection).

## 11.1 SINGLE TRAIT SELECTION

The phenotypic merit of the individuals under selection is determined in comparison to the average of other individuals of the population. The comparison is made with the average of other individuals kept under similar environmental conditions and that too at the same time. Therefore, the individuals are ranked relative to others under similar conditions. The breeding value (B.V.) of an individual is estimated from the phenotypic value of that individual as a deviation from the population mean (selection differential) times the heritability of the trait and it is known as the probable breeding value (PBV):

$$PBV = \overline{P} + h^2 (P_i - \overline{p})$$

Where, $\overline{P}$ = population mean

$P_i$ = phenotypic value of the individual

$h^2$ = heritability of the character

The PBV is the genetic superiority or inferiority of the individual in comparison to the population mean. The non- additive gene effects and the environmental factors bring the PBV nearer to the population mean rather than phenotypic value of the individual.

The accuracy of individual selection ($r_{AP}$) is the correlation of breeding value with phenotypic value which is equal to the square root of heritability ($r_{AP} = h$). Therefore, the accuracy of predicting the breeding value of an animal based on its own performance will be increased with the increase in $h^2$ of the trait and hence the progeny performance will come close to the phenotypic value of the selected parent rather than population mean.

Thus, efforts must be made to get accurate and higher estimate of heritability. This can be done by reducing the environmental variations. The accuracy of individual selection can further be increased by estimating the $h^2$ based on repeated records of the individual. The use of repeated records is helpful for traits of low $h^2$. The individual selection is applied when traits are expressed in the individual and when the $h^2$ of the trait is high.

***Advantages of individual selection:***

1. The information on individuals to be selected is easily available.
2. It can be applied earlier to progeny testing. This is used when pedigree information are not available.
3. The generation interval is shorter by this method compared to progeny testing.
4. This gives a direct estimate of B.V. rather than on the basis of relative's performance and hence it is more accurate.
5. This allows a greater selection differential because the B.V. of all the individuals can be estimated while in case of P.T. only a few individuals may be progeny tested as parents.
6. This method minimizes the environmental effects because the individuals for selection are tested in the same environmental conditions.

***Limitations of individual selection:*** The individual selection can not be applied under the following conditions:

1. When the traits are not expressed by the individual (sex limited traits).
2. When the traits are expressed in later life of the individual or after death of the animal.
3. When the traits have low heritability, the individual selection is not effective. This is because the accuracy of individual selection depends on $r_{AP}$ which is equal to square root of heritability. The phenotypic value is a poor indication of B.V. if $h^2$ is low. Thus, low $h^2$ reduces the accuracy of selection.

In view of these limitations, the individual selection particularly for dairy animals, the male selection is not applied because the most important economic characters are sex-limited (expressed only in females) and also have low heritability estimates. The individual selection is applied when traits are expressed in the individual and when the heritability of the trait is high.

## 11.2 AIDS TO INDIVIDUAL SELECTION

There are certain situations when the information of relatives is not available, and also the $h^2$ of character is medium to low. This embarrasses the situation making the selection more difficult based on individual information. Moreover, it may also happen that early selection is required before the final trait is expressed by the animal.

When the $h^2$ of a trait is low, the repeated records of same animal can be used to improve the estimate of real breeding value or real producing ability. The use of part records and of genetic correlation among the traits will be helpful when earlier selection is to be practiced for the traits expressed later.

**(1).** ***Additional records:***

Many traits are repeated several times in the life of the animal e.g. milk production, lactation length in dairy cattle and buffalo, egg production in poultry in poultry and fleece production in sheep. It is advantageous to predict the B.V. of an individual based on the mean of several records of that individual, if such records are available. The records on the same animal at different times show variation due to changes in the environment which prevail at the time the records are made, or sometimes due to error of measurement. The environment may include the internal conditions also e.g. endocrine functions of the animal, health of animal, as well as external conditions e.g. feeding, management and climate. As a result of environmental effects, the genetic potential of the animal is not fully expressed in a single record. The mean records will then give the better estimate of the real producing ability of the animal. The heritability based on n records is higher than based on single record. The heritability of n record is:

$$h^2_n = \frac{n\,h^2}{[1+(n-1)\,r]}$$

The relative efficiency of selection based on repeated records is always more efficient than single record. This can be obtained as the ratio of gain from selection on mean of n records over the gain from single record selection and it will be

$$\sqrt{\frac{\dfrac{nh^2}{1+(n-1)r}}{h^2}} = \sqrt{\frac{n}{1+(n-)r}}$$

(*i*) The above quantity increases with n for fixed value of r. Therefore, the more the numbers of repeated records are used, the greater is the efficiency in terms of gain per generation.

(*ii*) Secondly, efficiency increases with decreases in the value of r, for any given value of n. Therefore, as r increases the use of more than one record in selection is not advantageous. This is because high value of r indicates that the temporary environmental influences were not important and permanent effects are more important. As such if r is high there is no use waiting even for the second record.

(*iii*) The use of many records in selection is useful for the traits which are more variable from time to time in the individual's life i.e. more affected by temporary environmental effects. The use of mean of records in selection result much of the gain with the use of second record and further if r is small the gain from using third or fourth record may be considerable. The main advantage of this method is that it is used for the traits influenced more by temporary environmental effects which are cancelled out in taking the average of more records.

***Disadvantages*** of the method based on n records:

(*i*) It results in increased cost because the animal has to be kept till more than one record is available.

(*ii*) It increases the generation interval and consequently will decrease the response (genetic gain) per unit of time in comparison to that based on single record.

(*iii*) The selection differential is reduced because the variation decreases in calculating the mean.

(*iv*) The individuals with several records constitute a highly selected group, within which further selection will be comparatively ineffective, if continuous selection is carried out every time. Thus, the final relative efficiency of the two methods must be based on the annual genetic gain which is obtained by dividing the genetic gain ($\Delta G$) by the generation interval. Thus, the increased generation interval by the method of using several records will reduce the annual genetic gain. Therefore, using ***n*** records for selection is less efficient in terms of annual genetic gain.

**Most Probable Producing Ability (MPPA)**

When more numbers of records of animals are available and is desired to cull some animals and to retain others, it is then problematic to evaluate properly the animals with varying number of records. Female in a herd/flock often differ in age in terms of parity order. The real estimate of producing ability of a female to rank them in the herd is required if there is a difference in the number of records of different females. The MPPA or EPA is estimated as given under point (iii) of section 9.8.2.

**(2) *Use of part records*:** The part record selection uses the production performance for part period (viz. first 12 months growth, milk yield before lactation is completed). This method reduces the generation interval. The genetic gain ($\Delta G$) by use of part record will be:

$$\Delta G = b_{AP}(Xp - \overline{X})$$

Where, $b_{AP}$ is the regression of BV of total production (A) on part record (P),

$Xp - \overline{X}$ is the selection differential of part record.

Thus, for greater gain by use of part record, it is important that $r_{AP}$ be positive and greater for which the part production must be positively and highly correlated with total or full time production/record. If the genetic correlation is negative, net result of part record selection can be the reduction in total production.

## 11.3 METHODS OF SELECTION (Multi-trait selection)

The economic value of an animal depends upon several characters. This is known as overall performance (productivity) of total breeding value (net breeding merit) of animal. For example, a dairy cow will be more economical to maintain if she produces more milk for a longer period during lactation with higher fat content and remained dry for a shorter time between two successive calving. On the other hand, if a cow produces more milk daily of low fat content with shorter lactation length and goes dry earlier during the lactation, if will not be economical to maintain. Likewise, more profit from pigs depends on their fertility (litter size), mothering ability, growth rate, efficiency of food conversion, carcass quality, and also the resistance to various diseases. The economic production in poultry is affected by the annual yield of eggs, their weight, and age at egg production starts etc. The traits of direct economic importance in sheep are wool clip, fertility and lamb weight. Thus, the net economic value of an animal depends upon several traits. It is, therefore, essential to estimate the total breeding worth (net genetic merit) of an animal based on several characters. In practice, the breeding merit of an animal is often determined for several traits simultaneously and not for a single trait. It is required to improve the overall economic value based for several characters. This is known as *multi trait selection.*

### 11.3.1 Requirement of multi trait selection

The following information are required for improvement of several traits-

(*i*) The *economic value* of the traits. This is measured as the amount by which each unit of variation in it actually raises or lowers an individual's value, known as the relative economic value of the trait.

(*ii*) The *genetic significance* of the traits in terms of the heritability of the trait and genetic correlations among the traits.

(*iii*) The method of selection chosen and the *number of traits* to be included in selection criteria also influence the efficiency of simultaneous selection of several traits.

### 11.3.2 Methods of multi trait selection

Simultaneous selection for many traits can be applied based on individuals' own performance by adopting any of the procedure of selection. One may use tandem selection or independent culling or selection index. These procedures are known as methods of selection (Hazel and Lush, 1942).

## 11.4 TANDEM SELECTION

In this method of individual selection the individual traits are improved successively. The selection is practiced for one trait only at a time till the satisfactory improvement is achieved. After improvement in this trait the selection is started for improvement in the second trait till the goal is achieved to the desired level. Thus, trait A is improved first, then trait B, and so on.

(1) This method is less efficient than other methods. The genetic progress per unit of time method is less and that too for great efforts. The average genetic improvement per generation in each of n traits (which are independent and equally important) would be only 1/n times which would be made if it was the sole object of improvement over the entire period of selection.

(2) Secondly, it requires more time for improvement in all the traits because while selection is being done for one trait, the other traits must wait.

(3) The efficiency of this method depends on the genetic correlation among the traits under selection. A desirable genetic association would lead improvement in other trait while selection is made for one trait. The low genetic correlation will decrease the efficiency of this method because there will be on correlated response in other trait. An undesirable genetic correlation between two traits, in which selection for an increase in desirable direction in one trait results in a decrease in the other trait, would neutralize the genetic progress made in any one trait.

## 11.5 INDEPENDENT CULLING LEVELS (ICL) METHOD

This method of individual selection involves selection for two or more traits at a time. A minimum standard (level) for each trait is fixed and every animal to be selected must met this minimum standard fixed for each of the trait under selection. These minimum standards are called as culling levels which are different for different traits and are independent of each other. Thus these are called as independent culling levels. An animal fails to meet the minimum standard fixed for any one of the trait (culling level) will be rejected, irrespective of the fact that how good the animal was in other trait. Thus, an animal is rejected in spite of its being exceptionally good in one trait but if it was little poor in another trait.

The procedure of this method can be understood by the following table, taking an example in buffalo:

| Traits | Standard set | Buffalo No. | |
|---|---|---|---|
| **AFC (moths)** | 42 | 40 | 44 |
| **Milk yield (kg)** | 1900 | 1950 | 2100 |
| **Fat (per cent)** | 7 | 7 | 6.5 |

Under the use of ICL method of selection buffalo No. 1 will be selected and no. 2 will be rejected because no.2 fails to meet the standard for AFC and fat percentage, though it was much superior in milk production, thus, the buffalo superior in milk production will be lost from the herd.

### 11.5.1 Advantage of ICL

(i) *Selection for more than one trait*: This method is superior to tandem selection because the selection is practiced for more than one trait. (ii) *Culling at early age:* The practical and economic advantage of this method is that it allows culling of the animal earlier which are inferior in early expressed traits.

(ii) This method is simple, relatively efficient and practiced with arbitrary culling levels.

### 11.5.2 Disadvantages of ICL

(*i*) *No compensation for other traits:* The first disadvantage of this type of selection is that it does not permit superiority in some traits to compensate for deficiencies in other, and there is a possibility to cull genetically the superior animals.

(*ii*) *Culling level*: The method involve a tedious work to determine the optimum culling level.

(*iii*) *More emphasis to early expressed traits*: In case early selection, the early expressed traits get more emphasis rather than actually more important trait which are expressed later.

(*iv*) *Selection intensity*: It is reduced with increase in the number of traits.

The effectiveness of this method depends on the level kept for each of the traits. In case of keeping low levels, very few animals are culled and it affects the genetic progress. On the other hand, if the standards are kept high, it results in gradual extinction of population because it will find difficult to have adequate replacements.

This type of selection is used in selection of cattle for show purposes. The animal is selected with excellence in type, colour and conformation traits ignoring its performance for economic traits.

### 11.5.3 Selection intensity under ICL

The ICL method decreases the selection intensity of the individual traits and hence it decreases the S. The decrease is along the increases in number of traits. The genetic correlations among traits also influence the selection intensity. The negative genetic correlation among traits decreases more selection intensity in single trait because animal superior in one trait may be inferior in other trait. The effect of positive genetic correlation is less on the decreases of selections intensity because animal superior in one trait may also be superior in another trait. The decrease in selection intensity for including more number of traits suggests a restriction to a minimum number of traits. This is because animals superior in most of the economic traits are rate. It is, therefore that fancy traits like horn shape and coat colour should not be included in selection. Its effect is two fold.

- A decrease in intensity of selection for important traits.
- Secondly, if in dairy cattle fancy traits are considered in addition to economic traits, the work become more difficult and superior genotype for economic trait not associated with fancy trait of the breed will be rejected.

The selection intensity is decreased by selection for more than one trait. The intensity of selection for n traits which are equally important and independent is equal to $\frac{1}{\sqrt{n}}$ compared to selection intensity for only one trait. Heavy culling for one trait limits the amount of culling which can be practiced for others. For example, if selection is for a single trait and p = 0.25, the selection differential is 1.27

standard deviation. When two independent and economically equal important traits are to be selected with the restriction that p = 0.25, the p will obtained as $\sqrt{p} \times \sqrt{p} = \sqrt{0.25} \times \sqrt{0.25} = 0.50 \times 0.50 = 0.25$. Thus $P_1 = P_2 = 0.50$, instead of 0.25. Thus, selection can only be among the best 50 % of individuals for each trait instead among the 25%. The selection differential for p = 0.50 is 0.80 standard deviations instead of 1.27. Thus, the S is reduced. The simultaneous selection of equal intensity of n traits reduced the S from i = z/p to i '= z ' $\sqrt{p}$ where, z is the ordinate of point on the abscissa separating the best $\sqrt{p}$ per cent of the population. In this example, i = 1.27 and i′ = 0.80. Thus, $\Delta G$ per generation will be equal to $\sum i' h_j \sigma_{Aj}$ instead of $\sum ih_j \sigma_{Aj}$ Therefore, if the heritability and variances are equal for all n traits the $\Delta G = n\, i'\, h\, \sigma_A$.

When the traits are equally important, the same culling level ($P_1 = P_2 = \sqrt{p}$ ) for each trait is most efficient for every value of p. On the other hand, when one trait is more important than the other, the culling proportions differ for the two traits in such a way that the culling proportion for less important trait is less to maximize H. For example, when first trait is twice as important as second trait, the culling level are 19 and 1 for p = 0.80; 47 and 6 for p = 0.50; and it is 76 and 18 for p = 0.20.

## 11.6 TOTAL SCORE OR SELECTION INDEX

In this method of mass selection, the selection is practiced for several traits simultaneously based on an index. The selection index is an index of the net merit of an animal for many traits which vary in their relative economic values, having different amount of genetic variance expressed in terms of heritability estimates and are correlated with each other with varying size and sign of correlations (both genetic and phenotypic) among them. It is thus essential to evaluate each individual before selection for its overall genetic worth. The scores of credits and penalties are given to each animal according to the degree of superiority or inferiority in each trait. Some weight is given to each trait depending on the unequal economic importance together with the genetic variance and covariance among traits and thus allows equal attention to each trait. An index value is obtained by adding the score for each trait into one figure (total score for all the traits). The animal with the highest score is then selected for breeding. The amount of weight given to each trait in relation to other traits determines the influence of each trait on the final index. Thus, this method encompass all the advantage and disadvantage of an animal by permitting extra merit in one trait to off set slight defects in another.

The selection index method differs from ICL in following ways -

(*i*) The culling levels are flexible,

(*ii*) Each trait is weighed by a score for individual trait which are summed for each animal and this is taken as selection criterion,

(*iii*) Superiority in some traits can make up deficiency in others, unlike ICL which discards an animal failing to qualify in one trait regardless of its superiority in other trait.

### 11.6.1 Discriminant function to construct index selection

Selection index method involves the technique of discriminant function originally developed by Fisher (1936) and used by Smith (1936). They developed an optimum linear discriminant function for multiple trait selection in self perpetuating varieties of crops. Hazel (1943) extended the application of this technique in animal improvement discussing the genetic basis for constructing selection index (principle of selection index theory) when it is required to select individuals whose progeny will be of superior merit.

The discriminant function as a basis for making selection on several characters simultaneously is used to discriminate the desirable genotypes from undesirable ones on the basis of their phenotypic performance. It is used to construct a selection index based on the principle of fitting a multiple regression equation for predicting a dependent variable from two or more independent variables. Such a selection index should combine the different traits in a way that maximize the probability of progress in the aggregate economic value. The selection index will take the following from as:

$$I = b_1 X_1 + b_2 X_2 + \dots\dots\dots\dots\dots\dots + b_n X_n$$
$$= \sum bi\ xi$$

Where, $X_i$ = Phenotypic value for different traits

$b_i$ = Multiple regression coefficients for different traits

The $b_i$ and $X_i$ values are put up in the index to obtain the index values for each animal and the animals with top highest index value are selected.

### 11.6.2 Genetic basis of selection index

The selection indices are constructed to bring genetic improvement from phenotypic selection. The genetic improvement depends on the magnitude of genetic variability in terms of heritability of the trait. The overall net genetic improvement (H) for several traits should be the sum of genetic gains made for several traits.

All the traits are not equally important economically. Therefore, economic value of a trait to be included in selection should be given due emphasis. This is done by giving weight to the breeding (genotypic) values of all the traits ($G_i$) with their respective economic values ($a_i$). The economic value, denoted by ***a,*** is defined as the increment in profit occurring from improving the trait by one unit, independent of improvement in other trait.

Hazel and Lush (1942) had described the aggregate genetic worth (H) of the individual is a function of the B.V. of each trait ($G_i$'s) and their economic value ($a_i$'s). This is the average genetic superiority of a selected group or the genetic economic value (H) of the index represented as:

$$H = a_1 G_1 + a_2 G_2 + \dots\dots + a_n G_n$$
$$= \sum a_i G_i$$

Where, $\sum a_i G_i$ = Genetic worth of animal or, aggregate genotypic value.

$a_i$ = Relative economic values corresponding to one unit of $X_i$ which measure the amount by which profit is expected to increase for each unit change in $X_i$

$G_i$ = Expected values of $X_i$ due to additive gene effect (i.e. genetic value of $i^{th}$ trait).

The genetic worth or aggregate breeding value of an animal is H and the overall phenotype of an animal for multi traits determined by the discriminant function or index is I. Therefore, the index value (I) should be as closely correlated to the individual's composite breeding value (H) as possible. The $r_{HI}$ should be as high as possible to make index selection more efficient and will depend up on the correct estimation of $b_i$ values of the index. Therefore, the $b_i$ values (multiple regression coefficients) should be obtained so as to make $r_{HI}$ as high as possible. Thus, the $b_i$ should be such that index value (I) may best discriminate the individuals with greatest genotypic economic value (H). The selection thus based on I should be as accurate as based on H, if it was known.

### 11.6.3 Estimation of $b_i$ values

The ***$b_i$* values** are estimated by setting and solving the simultaneous equations whose number is equal to the number of traits under selection. The simultaneous equations use the economic value of each trait, genetic variances and co-variances as well as the phenotypic variances and co-variances of all the *n* traits.

$$b_1X_{11} + b_2X_{12} + \ldots + a_nX_{1n} = a_1G_{11} + a_2G_{12} + \ldots + a_nG_{1n}$$

.

.

$$b_1X_{n1} + b_2X_{n2} + \ldots + a_nX_{nn} = a_1G_{n1} + a_2G_{n2} + \ldots + a_nG_{nn}$$

The normal simultaneous equations for *n* traits under selection are set as follows:-

I. $b_1 V_{P1} + b_2 Cov_{P12} + \ldots + b_n Cov_{P1n} = a_1 V_{A1} + a_2 Cov_{A12} + \ldots + a_n Cov_{A1n}$

II. $b_1 Cov_{P21} + b_2 V_{P2} + \ldots + b_n Cov_{P2n} = a_1 Cov_{A21} + a_2 V_{A2} + \ldots + a_n Cov_{A2n}$

.

.

N. $b_1 Cov_{Pn1} + b_2 Cov_{Pn2} + \ldots + b_n V_{Pn} = a_1 Cov_{An1} + a_2 Cov_{An2} + \ldots + b_n V_{An}$

The above equations in matrix from can be written as:

$$bP = aG$$

Where, b = column vector of partial regression coefficients to be calculated,

$P^{-1}$ = inverse of phenotypic variance and covariance matrix of $X_i$,

G = genotypic variance and covariance matrix of $g_{i,}$

a = column vector of relative economic values.

Therefore, $$b = p^{-1}aG.$$

Thus, the $b_i$ values are obtained by multiplying a G with the inverse of P.

### 11.6.4 Selection index

The $b_i$ values (multiple regression coefficients) so obtained are put in the equation of the function (I) as under –

$$I = b_1 X_1 + b_2 X_2 + \ldots\ldots\ldots + b_n X_n.$$

Where, $X_{i's}$ are the phenotypic values for different traits.

This fitted equation (I) is called the selection index.

The $b_i$ and $X_i$ values are put up in the index to obtain the index value for each animal. This index value is the selection criterion based on several traits for each animal. The animals are then arranged in order of merit to select those with the highest index value.

### 11.6.5 Computation of different parameters of selection index

After the selection indices are constructed, the following phenotypic and genetic parameters of selection indices are compute as -

***(i) Variance of the selection index ($\sigma^2_I$) :***

$$\sigma^2_I = b'P\,b$$

***(ii) Variance of the aggregate genotype*** **($\sigma^2_H$)**:

$$\sigma^2_H = a'G\,a$$

Where, a' = row vector of relative economic values ($a_i$'s)

a= column vector of relative economic values ($a_i$'s)

***(iii) Correlation*** ($R_{IH}$) of index value (*I*) with aggregate genotype (H):

$$R_{IH} = Cov_{HI} / \sigma_H \sigma_I$$
$$= b'G\,a / \sqrt{(b'P\,b)\,(a'G\,a)}$$
$$= \sigma^2_I / \sigma_H \sigma_I$$
$$= \sigma_I / \sigma_H$$

***(iv) Genetic gain*** or ***gain in aggregate genotype ($\Delta H$) from selection:***

$$\Delta H = i\, r_{HI}\, \sigma_H$$
$$= i\,(\sigma_I / \sigma_H)\, \sigma_H$$
$$= i\, \sigma_I$$
$$= a'G\,b\,(i/\sigma_I)$$

Where, i = selection intensity

***(v) Response in i th trait of the index ($\Delta G_i$):***

$$\Delta G_i = g_i b\,(i/\sigma_I)$$

Where, $g_i$ is the $i^{th}$ row vector of G matrix

***(vi)* Response in $j^{th}$ trait not included in the index ($\Delta G_j$):**

$$\Delta G_j = g_j b\,(i/\sigma_I)$$

Where, $g_j$ is the row vector of additive genetic covariance of $j^{th}$ trait with trait included in the index

***(vii) Heritability of Index ($h^2_I$)***: The heritability of the index ($h^2_I$) is estimated as the ratio of genetic variance to the total variance of the index after Lin and Allaire (1977) as –

$$h^2_I = \sigma^2_g / \sigma^2_I$$
$$= b'\,G\,b / b'\,P\,b$$

Where, $b'$ = row vector of $b_i$ values

$b$ = column vector of $b_i$ values

### 11.6.6 Requirements to construct selection index

The requirements in constructing the selection index for the estimation of $b_i$ values are as under -

- The genetic and phenotypic variances of all the traits
- The genetic and phenotypic co-variances among of all the traits, and
- The economic values of all the traits under selection.

(i) *Estimation of the genetic and phenotypic parameters:* The genetic and phenotypic variances of all the traits as well as the genetic and phenotypic covariance among all traits are estimated by the standard procedure.

(ii) *Estimation of economic values* ($a_i$)**:** The $a_i$ values are needed to get $b_i$ values. It is important to know the method of estimating the relative economic value (a) of the trait, by taking the example of cattle. The most important economic traits are age at first calving (AFC), milk yield, lactation length, peak yield, and milk yield per day of calving interval etc.

The economic value of AFC is estimated by taking actual cost of raising the animal from birth till first calving. Actual expenditure incurred on feed and fodders per day per head along with labour and miscellaneous charges are considered for finding the economic value of each day of AFC. The minor expenditure on housing, depreciation on capital and medicine charges are not included as these are considered to be balanced out by the income from sale of dung, skin and hides etc. to find out the cost per head per day till AFC, the monthly expenditure and the number of animals are considered. It is better to divide the period from birth to AFC in at least 3 group viz. calves (up to 6 or 9 months), young stock (6 or 9 months to 2 years of age) and heiress (2years to AFC). The cost per head per day for each of the categories is calculated. The total cost per head per day of AFC is calculated by giving due weight to the length of age in days for calves, young stock and heifers but multiplying the cost per day in the category with the number of days in the category and summing up over the three categories, and then dividing by the total number of days from birth to AFC.

Economic value of milk yield is the selling price of milk during different month of the year. From this, the average cost of one kg. of milk is calculated. The economic value of lactation length, peak yield, and milk yield per day of calving interval may be calculated as the regression of milk yield on these traits. This results in the value of a unit of the traits in terms of kg. of milk the value is subsequently multiplied with the selling price of milk to get in terms of rupees the economic value of one unit of trait.

*Relative economic values:* Further in order to bring about similarity among different traits, the economic value of a standard unit of each trait is considered. A standard unit is taken as equal to one per cent standard deviation of the trait. In order to obtain the economic value of 1 % standard deviation of the trait, the respective standard unit of the trait equal to 1 % **σ** is multiplied by the economic value (cost per unit). Finally, keeping the economic value of milk as unity (one), relative values of other traits are calculated (See table below). This gives the vector of economic values of economic values when a proper sign is attached to the values. In this case positive sign to all the traits except AFC which is assigned a negative sign for the obvious reason to affect a decrease in AFC for economic reasons.

**Table 11.1:** Relative economic values of various traits of KF cattle (Singh and Tomar, 1988)

| Traits | Standard units (1 % σ) | Economic value | Economic value of 1 % σ of trait * | Relative economic value |
|---|---|---|---|---|
| AFC | 1.7037 month | 150.00 | – 255.55 | – 6.94 |
| FLL | 1.3357 days | 33.091 | 44.1996 | 1.2017 |
| FLMY | 10.5088 kg | 3.500 | 36.7808 | 1.000 |
| MY/CI | 0.0247 kg | 3.500 | 0.0865 | 0.0024 |

* E.V. of 1 % **σ** of trait = Standard units (1 % **σ**) × Economic value

### 11.6.7 Practical problems in constructing and applying selection index

There are certain practical problems in construction of selection index and to recommend in general for all herds and even in the same herd over time.

(*i*) The information required is not available or accurate enough.

(*ii*) The relative economic values ($a_i$) are very important in construction of selection indices. But, it is not so easy to assign these values in many cases. It requires information in the long time price average and cost of production of the traits. Sometimes the economic values changes due to change in market demand. Yamada et al (1975) and Lin (1978) had described the difficulties in assigning the $a_i$ values in the following situations -

(a) The standard to assign the $a_i$ values are not available for some traits like body confirmation traits,

(b) The assumption of constancy of $a_i$ values over the range of variation are not met in some traits,

(c) The $a_i$ values by imposing some restrictions are not the same to that of the selection index with no restriction,

(d) The $a_i$ values derived from multiple analysis vary according to the number traits considered in multiple regression equation, sampling variation, and the way the profit is defined,

(e) There is change in $a_i$ values with time and place.

(*iii*) To find the genetic and phenotypic parameters is not an easy job particularly when the number of traits included in the index exceeds three or more.

(*iv*) The genetic and phenotypic parameters differ in different populations due to different genetic constitution of the herds and different management practices. The genetic correlations have large sampling errors mainly due to small herd size. Because of these problems, an index constructed for one herd cannot be widely applied in general.

(*v*) The different management practices and environmental conditions at different places may cause the variability of the traits.

(*vi*) The net merit (aggregate genetic worth H) of an animal has been assumed as an additive linear combination of characters each weighted by its relative economic value. This assumption may not true always, and the economic value of a character either may not be linear or may not be independent of the other character. The example of milk yield and fat content may be taken. The net return for increate in milk production is not linear and fat content depends on milk yield of the animal. It is, therefore, advisable to use a model which covers both additive and non-additive functions.

### 11.6.8 Efficiency of S.I. over other methods of selection

The index method is more efficient method of selection than others methods –

(*i*) It allows the individuals which are superior in some traits to be selected regardless of their inferiority in other traits and thus superior genotypes are selected,

(*ii*) All traits do not have equal heritability and so the same selection intensity will not be expected to give the same proportionate improvement for each trait,

(*iii*) The traits are correlated to each other to which the index gives appropriate consideration. It also results in more genetic gain for the time and efforts made. The advantage of index method increases with the number of traits. Thus it requires less time to bring about an overall improvement.

## 11.7 COMPARATIVE EFFICIENCY OF DIFFERENT METHODS OF SELECTION

Hazel and Lush (1942) had mentioned that there is always the danger that selection will fall below its maximum efficiency because too much attention is paid to some characters and too little to others. In tandem method the selection may be for unimportant characters in the beginning of first few generation and more important characters may be for a brief period. In ICL method, the culling levels may be set too high for one trait and too low for other. In selection index method there may be mistakes in estimating genetic parameters and economic and value. Moreover, the value of Z/P (intensity of selection) is not properly taken because the value of i (selection intensity) is taken from statistical tables based on normal curve which are not possible always in biological data that show little skewness. If the curve is skewed with long tail of distribution towards low merit, the progress is faster with mild selection and less rapid with intense selection than in a normal curve. On the contrary, if the curve is skewed towards high merit, the progress is less with high selection but more with intense selection. Lastly, the selection progress is also affected by animal health which affects p, and also by the breeders' ability to record the observations properly.

The selection index method is a more efficient method of selection than tandem and ICL methods of selection.

(*i*) It allows the selection of animals superior in some traits regardless of their inferiority in other traits and thus superior genotypes are saved for breeding.

(*ii*) All traits do not have equal heritability and so the same intensity of selection will not be expected to give proportionate improvement for each trait.

(*iii*) The selection index gives appropriate consideration to the correlation among traits. It also results in more genetic gain for the time and efforts made. The advantage of index method increases with the numbers of traits. Thus, it requires less time for overall improvement.

Hazel and Lush (1942) compared the efficiency of 3 methods in terms of genetic gain, assuming that the traits under selection are independent, have equal heritability, economic weight and variances. Accurate estimation of genetic and economic parameters and low phenotypic correlation among the traits are also assumed for higher efficiency of S.I. method.

The tandem method of selection may be for unimportant traits in the beginning of first few generations but selection for more economically important traits may be for a short period. The ICL method may set the culling levels too high for one trait and too low for others. The selection index method may use the wrong estimates of genetic parameters and economic values.

The intensity of selection in terms of z/p is not properly taken because the value of *i* is taken from statistical tables based on normal curve which is not possible always showing little skewness. If the curve is skewed towards low merit the progress is faster with mild selection and less rapid with intense selection than in a truly normal curve. On the contrary, if the curve is skewed towards high merit the progress is less with high selection but more with intense selection.

The expected genetic gain in one generation of selection of one trait alone by three methods has been given as under –

$$H_{(TS)} = a\, h^2\, \sigma_P\, i$$

$$H_{(ICL)} = n\,(a\, h^2\, \sigma_P\, i)$$

$$H_{(SI)} = \sqrt{n}\,(ah^2\, \sigma_P\, i)$$

The ratio of gain from index to tandem selection = $H_{(SI)} / H_{(TS)} = \sqrt{n}$

The index selection based on equally important and uncorrelated traits is $\sqrt{n}$ times as efficient as the tandem method, regardless of intensity of selection. The progress in any one trait is only $\frac{1}{\sqrt{n}}$ times as efficient by the index selection as if selection was applied for that trait alone. The index is more efficient than tandem, the relative efficiency of the index increase with n (number of characters). The index is more efficient than ICL, its superiority rises with increasing n but falls with increasing i (intensity of selection).

The ICL method is more efficient than tandem selection for each trait at a time, the relative efficiency increases with the number of traits and intensity of culling. But ICL method is less efficient than index selection except in some cases of earlier selection. The ICL method is always intermediate in efficiency between tandem and index selection method.

However, the p value influences the efficiency of all the three methods. The lower value of p leads to more progress by all the 3 methods and the increase in progress is more rapid with ICL than other two methods. The relative efficiency of ICL method is closer to that of tandem method with mild selection but tends to be close to the total scores method when the selection is very intense.

Thus it was concluded by them that index method was most superior and tandem selection the least whereas the efficiency of ICL was intermediate –

(*i*) The index is more efficient than tandem, the relative efficiency of index increases with n.

(*ii*) The index is more efficient than ICL, its superiority increases with increasing n but decreases with increasing i.

(**iii**) The ICL is more efficient than tandem selection, its superiority increases with increase in n and i.

Young (1961) and Finney (1962) compared the efficiency of three methods of selection under general conditions of correlation among traits and their unequal importance.

Young (1961) concluded that:-

(*i*) The index is never less efficient than ICL in all the circumstances, though in some cases it is no more efficient; ICL is never less efficient than tandem but in some cases no more efficient than tandem selection.

(*ii*) An increase in number of traits increases the superiority of index over tandem and ICL but decreases with increasing differences in relative importance. Its superiority is more for equal importance of traits. An increase in number of traits also increases the superiority of ICL over tandem selection but decreases with increasing differences in relative importance of traits.

(*iii*) An increase in selection intensity decreases the superiority of index over ICL but the superiority of index over tandem is independent of intensity.

(*iv*) Since the change in correlation among traits affect the genetic gain, the phenotypic and genetic correlation may affect the relative efficiency of selection methods. The relative efficiency of the index over other methods is higher when $r_p$ is low or negative when the traits are of equal importance.

## 11.8 TYPES OF SELECTION INDICES:

The selection indices are of the following different type –

- General Selection Index,
- Restricted Selection Index,
- Multitrait-multisource Selection Index.
- Selection Index for desired gain.

### 11.8.1 Similarities in different types of indices

(*i*) The principle of theory of constructing selection indices of all the above types is the same as given by Hazel (1943),

(*ii*) The requirements in constructing the selection indices of all types are the same as given below in terms of phenotypic, genetic and economic parameters of the traits included in the indices,

(*iii*) The $b_i$ values (partial regression coefficients) are estimated by setting and solving the simultaneous equations,

(*iv*) The formulae for estimating the variance of index (I), variance of aggregate genotype (H), correlation of index value with aggregate genotype ($R_{IH}$), genetic gain in aggregate genotype (*ΔH*), response in a trait of the index ($\Delta G_i$), and heritability of index ($h^2_I$) for all type of indices are the same.

### 11.8.2 Dissimilarities in different types of indices

The different selection indices are applied under different situations. These situations have been mentioned separately for different selection indices.

## 11.9 RESTRICTED SELECTION INDEX

Under certain conditions, the breeder may like to effect change in the means of certain traits (say, r) out of p traits while keeping the means of p-r traits unchanged. This is the case of restricted selection which has been by Kamthorne and Nordskog (1959) with the constraints that some linear functions of genotypic values would remain constant (unchanged).

### 11.9.1 Conditions to impose restriction

There are some conditions which compels the breeder to impose restrictions. One trait may be at an optimum level of performance beyond which no improvement can be made because of physiological optima, or one does not want to reduce the means of a certain trait which is expected because of its negative genetic association with other economic traits under selection. Reports appeared in the literature (Salisbury and Van demark, 1961 and Swanson, 1967) suggested that the physiological optima to age at first calving and calving interval exist, below which the economic value of the animal decreases. Secondly, fat percentage is negatively genetically correlated with milk production and similarly egg production with egg weight and so an improvement in milk yield and egg weight, respectively, which will not be desirable. Therefore, the problem is, for example, to aloe a change in milk yield only and no change in the fat percentage.

### 11.9.2 Procedure to impose restriction

The problem of index construction with restriction requires to maximize the correlation between H and I such that $\Delta G_j = 0$, i. e., the genetic gain in a character ($X_j$) under restriction is zero. A solution to this problem was given by Kempthorne and Nordskog (1959) imposing restrictions equal to the number of traits that were not allowed to change their phenotypic value. They proposed that the usual index equation be solved subject to the simultaneous conditions that the covariance between the index and the linear function of the genotypic involved be zero, thus preventing selection on the index from causing any genetic change in this function.

For example, take two traits, $X_1$ and $X_2$, with a condition to impose restriction on $X_2$. To impose restriction that the new index shall produce no change in $G_2$, the index equation are solved subject to the condition that Cov ($G_2$, I) = 0. The aggregate genotype (H) for 2 traits is $H = a_1G_1+a_2G_2$ but when restricted to no change in $X_2$, the $H = a_1G_1$. Thus select for $H = a_1G_1$ by $I^* = b^*_I X_I + b^*_2X_2$ where * indicates the restricted selection index (RSI) – restricted in the index is to maximize H with the restricted selection that $\Delta G_2 = 0$. Mathematically, it can be shown that the covariance between $G_2$ and I is $b^*_1X_1\ G_2 + b^*_2X_2\ G_2$ which must be zero. In addition the equations for the $b_1$ to maximize $R_{HI}$ are

$$b^*_1X_1 + b^*_2X_1X_2 = X_1H$$

$$b^*_1X_1\ X_2 + b^*_2X_2 = X_2H$$

Thus, there are three equations including the restricting but only two unknowns. In order to find a solution we must add a dummy unknown, the so called Lagrange multiplier. The equation can now be solved and are:

$$\begin{array}{llll} b^*_1X_1 & +b^*_2X_1X_2 & +\lambda\ G_1G_2 & = X_1H \\ b^*_1X_1X_2 & +b^*_2X_2 & +\lambda\ G_2G_2 & = X_2H \\ b^*_1X_1G_2 & +b^*_2G_2G_2 & +0 & = 0 \end{array}$$

or,

$$\begin{vmatrix} b_1 \\ b_2 \\ \lambda \end{vmatrix} = \begin{vmatrix} V_{P1} & +Cov_{P12} & +Cov_{G12} \\ Cov_{P21} & +V_{P2} & +V_{G22} \\ Cov_{G12} & +V_{G2} & +0 \end{vmatrix}^{-1} = \begin{vmatrix} V_{G11} \\ Cov_{G12} \\ 0 \end{vmatrix}$$

These can be solved for $b_1^*$ to get the restricted selection index.

The $b_1$ estimates can also be obtained by the formula:

$$b = [I- P^{-1}\ GC\ (C'CP^{-1}GC) - C'GJP^{-1}Ga$$

Where, I is the identity (unit) matrix,

G is the genotypic variance covariance matrix,

$p^{-1}$ is the inverse of phenotypic variance and covariance matrix

G is the coefficient vector matrix, and

C′ is the transposed coefficient matrix, which is defined as:

(*i*) For $\Delta G_1 = 0$ and other ($\Delta G_2$, $\Delta G_3$ and $\Delta G_4$) unrestricted, C′ = (1 0 0 0) is a vector

(*ii*) For $\Delta G_2 = 0$ and other ($\Delta G_1$, $\Delta G_3$ etc.) unrestricted, c′ = (0 1 0 0) is a vector

(*iii*) For $\Delta G_1 = \Delta G_2 = 0$ and other ($\Delta G_3$ & $\Delta G_4$) unrestricted, C′ $\left[\frac{1000}{0100}\right]$

### 11.9.3 Genetic cost of the restrictions

Cunningham et al (1970) explained the method of calculating the genetic cost of the restrictions. There is percent reduction in rate of genetic gain in aggregate genotype if a particular trait is dropped from the index. This is the genetic cost of restriction. The response in selecting according to I* can be compared with the response in selecting directly without restriction. The percent reduction in rate of genetic gain in aggregate genotype is calculated as:

$$\text{Genetic cost\%} = 100\left[1-\sqrt{\left(b'Pb-b^2/w_{ii}\right)b'Pb}\right]$$

Where, $w_{ii}$ = corresponding diagonal element of $P^{-1}$ matrix,

$b^i$ = weighing factor (partial regression coefficient) for the $i^{th}$ trait, and

b′Pb = variance of the index.

### 11.9.4 Role (uses) of R.S.I.

The R.S.I. are of special use when the breeder wishes to keep the means of a certain trait unchanged which is otherwise expected because of association among the economic traits. In case the traits are positively correlated, the intermediate optimum in some traits(s) having arrived, the method of R.S.I. can help in improvement of some traits without affecting the average value of others, like age at first calving and weight of the calf. Likewise, if the traits are negatively correlated, viz., milk yield and its fat percentage egg size and production, the improvement in milk yield and egg production is possible without allowing a reduction in fat percentage and egg size, respectively, by the use of R.S.I.

The technique, of R.S.I. is not effective/ applicable when the traits are either uncorrelated of having perfect correlation. Considering that the population have reached plateau after intensive selection for a long time, high selection intensity for further improvement is no more desirable. The main reason is that the traits negatively correlated e.g. egg weight and production will be affected adversely. It is reasonable to reduce the selection intensity for such traits. In such R.S.I. methodology makes possible judicious reduction in selection pressure of such traits resulting into best desirable improvement.

### 11.9.5 Some selection experiments on R.S.I.

Very few reports are available in the literature regarding the use of R.S.I. Applanalp et al. (1963) selected turkeys for 7 generations using an index calculated to produce maximum gains in 9 week while holding 24 week body weigh constant. The results were in fair agreement of prediction, except a little negative response in 24 week body weight.

Okada and Harding (1967) designed R.S.I. to increase 14 day larval weight increasing 30 days adult weight of Tribolium, using the technique developed by Kampthorne and Nordskog (1959). They found that R.S.I. was effective for increasing larval weight while holding adult weight constant.

Scheinberg et al. (1967) also confirmed the biological validity of the R.S.I. model working with Tribolium castaneum for 10 generations. They compared unistage tandem selection with R.S.I. for high larval weight, short development time and high pupal weight imposing restriction on two traits at a time in each experiment.

Some selection experiments using the technique of R.S.I. have also been conducted with mice (Putledge et al., 1973 and Eisen and Bandy, 1977). They considered 6 week body weight and tail length at 6 week of age. They found that the index selection was effective and in agreement with theory.

Eisen (1977) also designed a R.S.I. on mice with the aim to maximize genetic change in post- weaning weight gain from 5 to 6 weeks of age while keeping genetic response in feed intake at zero. The result supported the validity of R.S.I. theory. He found the R.S.I. to be as:-

$$I = Pg - 0.067\ Pf$$

Where, Pg = phenotypic values for post gain and

Pf = feed intake from 3-6 weeks of age.

A number of selection indices with and without restriction have been given for selection of dairy cattle. However, they only utilized already available data of Tharparkar herd of NDRI – on the first and second lactation yield, the age at first calving and first calving interval. They observed that restriction on age at first calving did not reduce the gains while restricting the mean genotypic value of calving interval resulted in halving the gains.

### 11.9.6 Assumption and condition in construction and using R.S.I.

1. Relative weights are known without error. This can be achieved and would change over time and with region.
2. Relative economic weights remain constant over the range of variation likely to occur for the different traits. This is doubtful because of too good and too poor kind of cattle/buffaloes. These relative economic weights to not change through time, however, new indices may be constructed in due course.
3. Genetic and phenotypic variances are known without error.

4. The traits under consideration are heritable and the genetic correlation between the two traits is not perfect or absent.
5. Aggregate genotype and index are normally distributed.

### 11.9.7 Practical situation for using RSI in farm animals

The main selection trait in sheep is fleece weight, beside other features. The S.I. is required to increase the fleece weight of the flock as much as possible while stabilizing staple length and number of crimps at optimum level.

In cattle it is required to increase the birth weight of calf, increase the average monthly growth rate in the first year of life, increase the weight at first calving, no change in the age at first calving, increase the first lactation production, and no change in the fat test of milk

It is known that all these six traits, are heritable, more or less, and that there exist correlations, phenotypic as well as genetic, positive or negative amongst them. Thus any change in one trait would bring about an undesirable correlated change in other traits when classical selection index incorporating all the traits is used. It is here that the restricted selection index is of use and involves the imposition of restriction on the mathematical model that 2 traits do not change in value.

In poultry the situation arises when one wants to increase the egg production without affecting the egg size which is negatively correlated. Egg size is kept constant while using an index to maximize progress in genetic value based on egg weight, body weight and production.

## 11.10 MULTITRAIT – MULTISOURCE SELECTION INDEX

The selection index is obtained for individual selection considering many traits simultaneously. However, the different traits may not have equal heritability estimates but some may have low heritability. Thus it may not be possible to make improvement in the lowly heritable traits by individual selection. The accuracy of selection for such traits (with low $h^2$) can be increased by using the information from other sources (individual's relatives' records) for such traits. The selection index for two or more traits (AFC, FLMY, etc) is obtained based on the information from individual and its relative's records (FS and HS). Such a selection index of two or more traits based on information from two or more sources (Individual' performance + Relative's performance like FS and HS) is known as multi trait-multi source index.

This multitrait-multisource selection index is constructed using the principle of selection index theory (Hazel, 1943). The presentation given by Ronningen and Van Vleck (1990) of this selection index theory with practical applications, In: *General and Quantitative Genetics.* Ed. by A.B. Chapman may be referred by the interested readers for more details.

The theoretical outlines of constructing the multi trait – multisource index are as under -

$$\text{Aggregate genotype } \mathbf{(H)} = a_1 G_1 + a_2 G_2 + \ldots\ldots + a_n G_n$$
$$= \sum a_i G_i$$

$$\text{Index (I)} = b_{11} P_{11} + b_{12} P_{12} + \ldots + b_{1K} P_{1k} + b_{21} P_{21} + b_{22} P_{22} + \ldots + b_{2k} P_{2k} + b_{m1} P_{m1} + b_{m2} P_{m2} + \ldots + b_{mk} P_{mk}$$

$$= \sum b_{ij} X_{ij}$$

Where, $i = 1, 2 \ldots m$ traits,

$j = 1, 2 \ldots k$ sources of information,

$a = (a_1\ a_{2\ldots}\ a_n)$ row vector of known relative economic values,

$g = (g_1\ g_{2\ldots}\ g_n)$ column vector of unknown genetic values,

$b = (b_{11}\ b_{12\ldots}\ b_{mk})$ row vector of m k index coefficients, and

$X = (X_{11}\ X_{12\ldots}\ X_{mk})$ column vector of m k known phenotypic values.

The $b_{ij}$ are derived in such a manner so that $R_{IH}$ is maximum.

$$b = P^{-1} G a$$

Where, P is matrix of $P_{ij}$ elements of phenotypic values for $i^{th}$ trait of $j^{th}$ relative group and is a function of phenotypic and genotypic variance or covariance

G is matrix of genetic covariance between phenotypic values in I and genotypic values in H, and

a is row vector of economic values of the trait in H

$$a' = (a_1\ a_{1\ldots}\ a_1\ a_2\ a_{2\ldots}\ a_{2\ldots\ldots}\ a_n\ a_{n\ldots\ldots}\ a_n)$$

*Assumptions:* It is assumed that there is no inbreeding and the environment is not contributing to the covariance between records on different individuals/groups, the covariances can be expressed in terms of Wright's coefficient of relationship and additive genetic variance.

*Computation of the elements of P matrix:* The diagonal (variances) and off-diagonal (covariances) elements of P matrix are computed as under taking single record on individual –

**Diagonal elements** which are covariances of X's, ($\sigma^2_{Xij}$):

$$\sigma^2_{Xij} = [\sigma^2_{Pi} + (N-1)\, a_{jj}\, \sigma^2_{Ai}] / N$$

Where, $\sigma^2_{Pi}$ = phenotypic variances of $i^{th}$ trait,

$\sigma^2_{Ai}$ = additive genetic variance of $i^{th}$ trait,

$a_{jj}$ = additive relationship among individuals in group j, and

N = no. of individuals in the group.

**Off – diagonal elements** which are covariances among X's, ($\sigma_{Xij\,Xi'j'}$): There are 3 possible types of covariances –

(*i*) Different relative groups, same trait

$$\sigma_{Xij\,Xij'} = a_{jj'} \sigma^2_{Ai}$$

Where, $a_{jj'}$ = additive relationship between groups j and j'.

(*ii*) Different relative groups, different traits

$$\sigma_{Xij\,Xi'j'} = a_{jj'} \sigma_{Ai\,Ai'}$$

(*iii*) Same relative group, different traits

$$\sigma_{Xij\,Xi'j'} = [\sigma_{Pi\,Pi'} + (N-1)\, a_{jj}\, \sigma_{Ai\,Ai'}] / N$$

**Covariances** in G matrix ($\sigma_{Xij}\, g_i$): These will be of 2 types –

(*i*) When m and n are the same traits, m = n (in I and H)

$$\sigma_{Xij} A_i = \sigma_{Xmj} A_n = a_j\, \sigma^2_{An}$$

(*ii*) When m and n are two different traits (in I and H)

$$\sigma_{Xij} A_{i'} = \sigma_{Xmk} A_n = a_j\, \sigma_{AmAn}$$

## NUMERICAL EXAMPLES

The breeding value of an individual can be estimated based on single record in a herd, in many herds, as contemporary deviation and with average of multiple records of the individual.

**I. Single record in a herd**

Probable breeding value = $\overline{H} + h^2 (Pi - \overline{H})$

Where, $\overline{H}$ = herd average,

$p_i$ = individual's record

**Example 11.1:** A herd of Tharparkar cows had an average milk production of 6.0 kg with an $h^2$ estimate of 0.25. The average milk production of 4 cows is 9.0, 5.0, 8.5 and 6.0 kg. Estimate the B.V. of these cows.

**Solution:**

Cow no. 1 PBV = 6.0+0.25 (9-6) = 6.75kg

Cow no. 2 PBV = 6.0+0.25 (5-6) = 5.75kg

Cow no. 3 PBV = 6.0+0.25 (8.5-6) = 6.62kg

Cow no. 4 PBV = 6.0+0.25 (6-6) = 6.0kg

**II. Single record in many herds**

$$P.B.V. = \overline{H} + h^2 (P_{ij} - \overline{H}) + h^2_H (\overline{H}_i - \overline{H})$$

Where, H = population mean (all herds)

$h^2$ = $h^2$ of the character

$P_{ij}$ = $j^{th}$ individual of $i^{th}$ herd

$H_i$ = $i^{th}$ herd mean

$h^2_H$ = $h^2$ of the herd difference within population

**Example 11.2** A population comprising two herds had average milk production of 1900 kg whereas the herd average of two herds was 2000 kg and 1800 kg. Consider 3 cows in each herd with their milk production as 1900, 2000, 2100 for one herd and 1700, 1800, 1900 for second herd, $h^2$ of milk production as 0.25 and $h^2_H$ as 0.12. Find out the PBV of all the cows and rank them.

**Solution:** P.B.V. = $\overline{H} + h^2 (P_{ij} - \overline{H}) + h^2_H (\overline{H}_i - \overline{H})$

Cow no. 1 of herd 1 = 1900+ 0.25(1900-2000) +0.12(2000-1900)

$= 1900\text{-}25\text{+}12 = 1887$ kg

Likewise the P.B.V. of all other cows can be estimated and ranked them.

**III. Single record with contemporary deviation**

It is better to use the contemporary deviation to estimate the P.B.V. so as to remove the effects of year or season of calving. Thus

$$\text{P.B.V.} = \overline{H} + h^2 (P_i - P_c)$$

Where, $P_c$ is the contemporary average

The contemporary are those cows which calved within ± 2 months of the cow whose P.B.V. is to be estimated.

**IV. Average of multiple records**

When $h^2$ of a trait is low, it is better to select on the basis of family mean to increase accuracy but F.S. families are not available and thus the accuracy of individual selection can be increased by using multiple record on the same individual. This reduces the environmental effects and hence increases the accuracy of P.B.V.

$$\text{P.B.V.} = \overline{H} + \left\{ \frac{nh^2}{\left[1+(n-1)r\right]} \right\} (Pi - H)$$

Where, n = no. of records,

r = repeatability of the trait,

Pi = average of n records of $i^{th}$ individual,

$\overline{H}$ = herd average

**Example 11.3:** The average milk production of a flock of goats is 400 kg. The milk production of 4 outstanding goats in different lactations (L) is as under:-

| Goat no. | $L_1$ | $L_2$ | L3 |
|---|---|---|---|
| 1 | 460 | 480 | - |
| 2 | 390 | 400 | 450 |
| 3 | 480 | - | - |
| 4 | 400 | 380 | - |

Taking the $h^2$ as 0.25 and repeatability as 0.35, estimate the B.V. of all the 4 goats and rank them:

**Solution:**

$$\text{Goat no. 1: PBV} = 400 + \frac{2 \times 0.25}{1 + 0.35} (470 - 400) = 425.92 \text{ KG}$$

$$\text{Goat no. 2: PBV} = 400 + \frac{3 \times 0.25}{1 + (2 \times 0.35)} (413.3 - 400) = 404.8 \text{ kg}$$

$$\text{Goat no. 3: PBV} = 400 + \frac{1 \times 0.25}{1 + 0.0} (480 - 400) = 420 \text{ kg}$$

$$\text{Goat no. 4: PBV} = 400 + \frac{2 \times 0.25}{1 + 0.35} (390 - 400) = 396.2 \text{ kg}$$

Therefore, the ranking of 4 goats for their P.B.V. is

| | | | | |
|---|---|---|---|---|
| Goat no. | 1 | 2 | 3 | 4 |
| Rank | 1 | 3 | 2 | 4 |

**Estimation of MPPA:** Culling of low producers is essential to increase the productivity next year. This is done by estimating the 'most probable producing ability'. The estimation of MPPA is also essential under the situation where the $h^2$ estimate cannot be made for non availability of the information on pedigree of the individual. The MPPA is estimated as:

$$\text{MPPA} = \overline{\text{H}} + \left\{\frac{nh^2}{\left[1+(n-1)r\right]}\right\} (\text{Pi} - \text{H})$$

The contemporary deviation are used instead of herd average ($\overline{\text{H}}$) in order to reduce the seasonal variation

**Example 11.4:** The average milk production in a lactation of Tharparkar herd is 1800 kg. The milk production of 4 cows in different lactations is given as under:

| Cow no. | Lactation 1 | Lactation 2 | Lactation 3 | Lactation 4 |
|---|---|---|---|---|
| 1 | 1820 | 1860 | - | - |
| 2 | 1810 | 1836 | 1860 | 1840 |
| 3 | 1830 | 1850 | 1870 | - |
| 4 | 1840 | - | - | - |

Taking the repeatability of milk yield as 0.40, estimate the MPPA of all the cows and rank them.

**Solution**

$$\text{Cow no. 1: MPPA} = 1800 + \frac{2\times 0.40}{1+0.4}\ (1840 - 1800) = 1822\text{kg}$$

$$\text{Cow no. 2: MPPA} = 1800 + \frac{4\times 0.40}{1+(3\times 0.4)}\ (1816.5 - 1800) = 1812\text{kg}$$

$$\text{Cow no. 3: MPPA} = 1800 + \frac{4\times 0.40}{1+(2\times 0.4)}\ (1850 - 1800) = 1813\text{kg}$$

$$\text{Cow no. 4: MPPA} = 1800 + \frac{1\times 0.40}{1+0}\ (1840 - 1800) = 1816\text{kg}$$

Ranking of the cows is as under

| | | | | |
|---|---|---|---|---|
| Cow No. | 1 | 2 | 3 | 4 |
| Ranking | 1 | 4 | 3 | 2 |

**Example 11.5**: Construct an index to selection for fast growth and lean beef percentage from the information given below:

| Traits | $h^2$ | a | $\sigma_P$ | $r_g$ | $r_P$ |
|---|---|---|---|---|---|
| Average daily gain | 0.43 | 1.5 | 80.0 | 0.3 | - 0.1 |
| Lean beef % | 0.30 | 0.5 | 7.0 | | |

**Solution:** The genetic and phenotypic variance and covariance are estimated as:

$$\sigma^2_{P(1)} = (80)^2 = 6400;\ \sigma^2_{P(2)} = (7.0)^2 = 49$$

$$\sigma^2A\ (1) = h^2\sigma^2\ P(1) = 0.43 \times 6400 = 2752;\ \sigma A\ (1) = 52.46$$

$$\sigma^2A\ (2) = h^2\sigma^2P\ (2) = 0.30 \times 49 = 14.7;\ \sigma A\ (2) = 3.83$$

$$Cov_{P(12)} = rp\ \boldsymbol{\sigma}_P(1)\ \boldsymbol{\sigma}\ p(2) = -\ 0.1 \times 80 \times 7 = -\ 56$$

$$Cov_{A(12)}) = r_g\boldsymbol{\sigma}A(1)\boldsymbol{\sigma}A(2)=0.3x52.46x3.83=60.27$$

(*i*) The bi values will now be estimated as under

$$\begin{vmatrix} b_1 \\ b_2 \end{vmatrix} = \begin{vmatrix} V_{P(1)} & Cov_{P(12)} \\ Cov_{P(12)} & V_{P(2)} \end{vmatrix}^{-1} \begin{vmatrix} a_1V_{A(1)} + a_2Cov_{A(12)} \\ a_1Cov_{A(12)} + a_2V_{A(2)} \end{vmatrix}$$

Now putting the values, the equations will be –

$$\begin{vmatrix} b_1 \\ b_2 \end{vmatrix} = \begin{vmatrix} 6400. & -56.0 \\ -56.0 & 49.0 \end{vmatrix}^{-1} \begin{vmatrix} 1.5(2752) + 0.5(60.27) \\ 1.5(60.25) + 0.5(14.70) \end{vmatrix}$$

$b_1 = 0.63$ and $b_2 = 1.27$

(*ii*) The $b_1$ value can also be estimated by setting the two simultaneous equations and solving them as

$$b_1\ V_{p(I)} + b_2\,Cov_{p(12)} = a_1\ V_{A(I)} + a_2\,Cov_{A(12)}$$

$$b_1\ Cov_{p(12)} + b_2\ V_{p(2)} = a_1\ Cov_{A(12)} + a_2\ V_{A(2)}$$

Now putting the values, the equations will be

$$6400\ b_1 - 56\ b_2 = 1.5\ (2752) + 0.5(60.27)\ \ldots\ldots\ \text{Eq. 1}$$

$$-\ 56\ b_1 + 49\ b_2 = 1.5\ (60.27) + 0.5(14.7)\ \ldots\ldots\ \text{Eq. 2}$$

Now multiply eq. 1 with 49 and eq. (2) with -56 and adding two equations will cancell the $b_2$. Thus $b_1$ value can be obtained. Then the value of $b_2$ can be estimated by putting the value of $b_1$.

The selection index (I) will be:

$$I = 0.63X_1 + 1.27X_2$$

$$= 0.63(X_1 - \overline{X}_1) + 1.27(X_2 - \overline{X}_2)$$

$$I = X_1 + 2.015X_2$$

$$= (X_1 - \overline{X}_1) + 2.015(X_2 - \overline{X}_2 \text{ by diving with } 0.63$$

**Example 11.6:** Given the following information construct a selection index to maximize genetic gain I aggregate genetic rate. (ARS 1984)

$$h^2_X = 0.5,\ h^2_Y = 0.1,\ a_X = 1,\ a_Y = 4,\ r_{g(XY)} = 0.4,\ r_{P(XY)} = 0.4$$

**Solution**: The variances and covariances can be estimated with the given information:

$$h^2 = \frac{\sigma^2 A}{\sigma^2 p} \text{ and hence } \sigma^2_P = \frac{\sigma^2 A}{h^2} = 1.0$$

or the $h^2$ itself is an indicator of additive genetic variance ($\sigma^2_A$)

Therefore, $\sigma^2_X = \sigma^2_{A(X)}$ and $h^2_Y = \sigma^2_{A(Y)}$

$$\sigma^2_{P(X)} = 1.0 \text{ and also } \sigma^2_{P(Y)} = 1.0$$

$$r_{XY} = \frac{covxy}{\sqrt{\sigma^2 x \sigma^2 y}} \text{ and hence } Cov_{XY} = r_{XY}\sqrt{\sigma^2 x \sigma y}$$

$$Cov_{P(XY)} = r_{P(XY)} \sqrt{\sigma^2 p(x)\, \sigma^2 p(y)}$$

$$= 0.4\sqrt{IxI} = 0.4$$

$$Cov_{A(XY)} = r_{A(XY)} \sqrt{\sigma^2 A(x)\, \sigma^2 A(y)}$$

$$= -0.4\sqrt{0.5 \times 0.1} = -0.4 \times 0.2236 = 0.0894$$

The selection index (I) = $b_1 X_1 + b_2 X_2$

Where, $b_1$ are the partial regression coefficient estimated by setting and solving the following simultaneous as:-

$$b_1 V_{P(X)} + b_2 Cov_{P(XY)} = a_1 V_{A(X)} + a_2 Cov_{A(XY)}$$

$$b_1 Cov_{P(XY)} + b_2 V_{P(Y)} = a_1 Cov_{A(XY)} + {}^a2\, V_{A(Y)}$$

Now substituting the values of genetic and phenotypic variances, genetic and phenotypic covariance and economic values in the above equations, the following two equations will be obtained:-

$$b_1 + 0.4b_2 = 1 \times 0.5 + 4(-0.0894)$$

$$0.4\, b_1 + b_2 \times 1 = 1 \times (-0.0894) + 4(0.1)$$

Multiplying the equation (i) by o.4 and substracting equation (ii) from equation (i),

$$0.4b_1 + 0.16b_2 = -0.0569$$

$$\underline{0}.4b_1 + \underline{b}_2 = \underline{0}.3106$$

$$-0.84b_2 = -0.2537$$

$$b_2 = 0.302$$

Substituting the value of $b_2$ in equation (i)

$$b_1 + 0.1208 = 0.1424$$

$$b_1 = 0.1424 - 0.1208$$

$$= 0.0216$$

Therefore, $I = b_1 X_1 + b_2 X_2$

$= 0.0216\ X_1 + 0.302\ X_2$

$= X_1 + 13.98\ X_2$

**Example 11.7**: Based on first lactation milk yield (kg) and age at first calving (months) using the following statistics, construct the selection index.

| Traits | Economic value (a) | Genetic variance (VA) | Phenotypic variance (VP) | Cov. A | Cov.P |
|---|---|---|---|---|---|
| Milk yield(X) | 1.00 | 84687.21 | 125423.41 | -42.26 | 111.989 |
| AFC (X2) | -5.84 | 13.678 | 17.498 | | |

The following two simultaneous equations to be set are:-

$$b_1\ V_P\ (X_1) + b_2\ \text{Cov P}\ (X_1\ X_2) = a_1\ V_A\ (X_1) + a_2\ \text{Cov}_A\ (X_1\ X_2) \quad ...(1)$$

$$b_1\ \text{CovP}(X_1 X_2) + b_2\ V_P\ (X_2) = a_1 \text{Cov A}\ (X_1 X_2) + a_2\ V_A\ (X_2) \quad ...(2)$$

Putting the values given in the table we get the following two equations:-

$$125423.41\ b_1 + 111.989\ b_2 = 1.00(84687.21) + (-5.84)\ (-42.26 \quad ...(1)$$

$$111.989\ b_1 + 17.498\ b_2 = 1.00(-42.26) + (-5.48)\ (13.678) \quad ...(2)$$

Multiplying equation (i) with 17.498 and equation (ii) with 111.989, we get

$$219465.1\ b_1 + 1959.58\ b_2 = -122.139 \times 17.498 \quad ...(3)$$

$$12541.54\ b_1 + 1959.58\ b_2 = -8493.06 \times 111.989 \quad ...(4)$$

Substracting equation (4) from equation (3) we get

$$2182112.6\ b_1 = 1499853.6\ b_1 = 0.687$$

Substituting the value of $b_1$ in equation (2) we get

$$111.989 \text{ x } 0.687 + 17.498 = -122.139$$

$$76.97 + 17.498\ b_2 = -122.139$$

$$b_2 = \frac{122.139 - 76.97}{17.498}$$

The index, $I_1 = b_1\ X_1 + b_1 X_2$

$= 0.687\ X_1 - 11.378\ X_2$

$= X_1 - 16.56\ X_2$

**Example 11.8: Restricted S.I.**

Using the data of solved example 26.5, construct an index to improve the aggregate genotype for fast growing lean cattle with no change in lean percentage

**Solution:** The equations for the index are

$$\begin{vmatrix} b_1 \\ b_2 \\ \lambda \end{vmatrix} = \begin{vmatrix} P_{11} & P_{12} & P_{12} \\ P_{21} & \mathbf{P_{22}} & g_{22} \\ g_{21} & g_{22} & 0 \end{vmatrix}^{-1} \begin{vmatrix} g_{11} \\ g_{12} \\ 0 \end{vmatrix}$$

Now putting the values:

$$\begin{vmatrix} b_1 \\ b_2 \\ \lambda \end{vmatrix} = \begin{vmatrix} 6400 & -57.60 & 62.064 \\ 57.60 & 50.80 & 15.552 \\ 62.064 & 15.552 & 0 \end{vmatrix}^{-1} \begin{vmatrix} 2572 \\ 62.064 \\ 0 \end{vmatrix}$$

Solving the equations -

$$b_1 = 0.325\ b_2 = -\ 1.303$$

$$I = 0.325\ (X_1 - \overline{X}_2) + (-\ 1.303)\ (X_2 - \overline{X}_2)$$

$$r_{AI} = \frac{\sqrt{\sum bi\ gij}}{\sigma^2 a}$$

$$= \frac{\sqrt{[0.325 \times 2752 - (-\ 1.303)\ (62.064)]}}{2752}$$

$$= 0.544$$

= Accuracy of Index

The imposition of restriction on any trait in the index never increases the efficiency of index but usually reduces it.

❑❑❑

# 12

# Pedigree Selection

The pedigree is a list or record of ancestors in the past few generations of the individual. The ancestors are the parents, grand-parents, great grand- parents, etc. The pedigree having information on the economic characters of the ancestors is useful in selection of an individual. The B.V. of an individual is estimated on the performance of the ancestor(s). The selection criteria based on performance of ancestors (parents and grand-parents) is called as the *pedigree selection*.

## 12.1 BASIS OF PEDIGREE SELECTION

An individual receives genes from its ancestors. The percentage of ancestral genes is halved in each generation. This decides the genetic relationship between an individual and its ancestor. This relationship is reduced to half in each generation. It is important due to the halving nature of inheritance to consider more recent ancestors (parents) rather than distant ancestors (great grandparents) for pedigree selection. The inclusion of more remote ancestor results only in marginal gain in accuracy of selection due to halving nature and sampling nature of inheritance. The pedigree selection adds very little to the accuracy of estimating the B.V. of an individual if the information on individual is available. The significance of pedigree is decreased when information are available either on the individual or its family members (sibs and progeny).

## 12.2 GUIDES TO PEDIGREE SELECTION

The following factors determine that how much attention is to be given to pedigree selection.

- The degree of genetic relationship of the individual with its ancestor – more closely related ancestors should be given more emphasis.
- Heritability of the character – the pedigree selection is more accurate for traits of high heritability.
- Environmental correlation among animals used in the prediction.
- Information available on ancestors.

## 12.3 PRACTICAL DIFFICULTIES TO USE PEDIGREE SELECTION

- The ancestors' records are always not available.
- The records may be faulty due to stray mating.
- The pedigree records are destroyed with passage of time
- Most of the characters have low heritability

## 12.4 MERITS OF PEDIGREE SELECTION

- It is less costly as only compilation of pedigree is required.
- Allows selection at younger age and provides first hand information.
- It is helpful in multistage selection.
- It is useful for sex limited traits and those expressed in later life or after death of animal.
- It is helpful when two individuals have similar performance but one belongs to a better pedigree.

## 12.5 DEMERITS OF PEDIGREE SELECTION

- There is a disadvantage of using pedigree selection that all animals of similar pedigree are culled out in spite of the fact that an individual may be of good merit and free from recessive allele causing the defect.
- Some pedigree gets undue favour irrespective of the true merit of the individual. Better environment is provided to the progeny of favoured pedigree.
- It introduces non random biases because pedigree records are for different environmental conditions.
- Pedigree selection provides no basis of selection among individuals which are descendants of the same ancestor.
- The accuracy is usually low.

## 12.6 BREEDING VALUE BASED ON PEDIGREE RECORDS

The B.V of an individual from its pedigree records is estimated by the selection differential of its relative ($\overline{P}$) from their contemporary ($\overline{P}_C$) mean times the regression of the genotype of the individual on the phenotypic mean of the relatives ($b_{AP}$) and added to the mean of the contemporary group as:

$$P.B.V = \overline{P}_C + b_{AP}(\overline{P} - \overline{P}_C)$$

The $b_{AP}$ is taken as $rh^2$ where r is the coefficient of relationship between the individual and its relative. In estimating the PBV of individual, the record (s) of one parent or both parents or parents and grandparents can be used. The regression coefficients and accuracy of selection based on the record (s) of either parent or both parents for single and multiple records are given as under:

| Ancestors | Regression Coefficient | Accuracy of Selection |
|---|---|---|
| A. For single record: | | |
| Dam | $0.5\, h^2{}_D$ | $0.5\, h_D$ |
| Sire | $0.5\, h^2{}_S$ | $0.5\, h_S$ |
| Mean record of both parents | $\frac{1}{\sqrt{2}} h^2$ | $\frac{1}{\sqrt{2}} h$ |
| Parents and grand Parents | | $\sqrt{\frac{3-2h^2}{4-2h^4}}\, h$ |
| One grand parents | $0.25\, h^2$ | $0.25\, h$ |
| B. For more records: | | |
| Dam | $0.5h^2{}_D \frac{n}{1+(n-1)r}$ | $0.5\sqrt{\frac{nh^2{}_d}{1+(n-1)r}}$ |
| Sire | Same as for dam | Same as for dam |
| Both parents | $\frac{1}{\sqrt{2}} \frac{nh^2}{1+(n-1)r}$ | $\frac{1}{\sqrt{2}} \sqrt{\frac{nh^2}{1+(n-1)r}}$ |
| Parents and grand Parents | | $\sqrt{\frac{3-2h^2h^2{}_n}{2h^4h^2{}_n}}\, h_n$ |
| One grand parents | $0.25 \frac{nh^2}{1+(n-1)r}$ | $0.25\sqrt{\frac{nh^2}{1+(n-1)r}}$ |

The accuracy of selection is increased based on information of ancestors (parents and grandparents) combined with individual's own record than based on individual's own record.

The B.V of an individual can be estimated by combining information from all sort of relative whose records are available. This is done by giving weights to different relatives by the technique of multiple regression coefficients.

The records of different ancestors can also be used in combination of individual's own record to estimate the P.B.V of individual. In this situation, the partial regression coefficients are used because the records of two or more relatives are used. The partial regression coefficient for the individual is multiplied by the selection differential for the individual and this is added to the product of the partial regression coefficient and selection differential for each ancestor to predict the P.B.V. of individual, as shown below:

$$\text{PBV} = \overline{P}_{ic} + b_1 (p_i - p_{ic}) + b_2 (p_p - \overline{P}_{pc})$$

or $$= \overline{P}_{ic} + b_1 (p_i - \overline{P}_{ic}) + b_2 (p_p - \overline{P}_{pc}) + b_3 (p_{gp} - \overline{P}_{gpc})$$

or $$= \overline{P}_{ic} + b_1 (p_i - \overline{P}_{ic}) + b_2 (p_s - \overline{P}_{sc}) + b_3 (p_d - \overline{P}_{dc})$$

Where, $\overline{P}_{ic}$ = mean of individual's contemporaries

$\overline{P}_i$ = individual, s own record;

$\overline{P}_p$ = parents record

$\overline{P}_{pc}$ = parents' contemporaries mean

$\overline{P}_s$ and = $\overline{P}_d$ sire and dam's record

$\overline{P}_{sc}$ and $\overline{P}_{dc}$ = Average of sire and dam contemporaries

$b_i$ = partial regression coefficients

$P_{gp}$ = grandparent record;

$\overline{P}_{gpc}$ = mean of grandparent contemporaries

## 12.7 ACCURACY OF PEDIGREE SELECTION

The regression coefficients obtained indicated that the selection based on the performance of one parent is half as effective (accurate) as that based on individual's own performance. The values of b, are further reduced by half for each ancestor further back to the individual. The information on records of both the parents plus all of the four grandparents give lesser accuracy of selection which does not exceed 0.71 than based on individual's own record. However, the information on ancestors combined with individual's own performance give little higher accuracy of selection than based on individual's own record. The accuracy of selection based on single parent never exceeds to 0.50 and that based on either two parents or on parents and grandparents to 0.70.

**Table 12.1:** Accuracy of individual and pedigree selection (single record)

| Heritability | 0.1 | 0.2 | 0.3 | 0.4 | 0.5 | 0.6 | 0.7 | 0.8 | 0.9 | 1.0 |
|---|---|---|---|---|---|---|---|---|---|---|
| Individual | 0.31 | 0.44 | 0.54 | 0.63 | 0.70 | 0.77 | 0.84 | 0.89 | 0.95 | 1.0 |
| One parent Or all 4 g.p. | 0.16 | 0.22 | 0.27 | 0.32 | 0.35 | 0.39 | 0.42 | 0.45 | 0.47 | 0.50 |
| Two parents | 0.22 | 0.32 | 0.39 | 0.45 | 0.50 | 0.55 | 0.59 | 0.63 | 0.67 | 0.71 |
| Parents + Grand parent | 0.27 | 0.37 | 0.43 | 0.49 | 0.53 | 0.57 | 0.61 | 0.64 | 0.67 | 0.71 |
| Individual + One parent | 0.34 | – | 0.58 | – | 0.73 | – | 0.84 | – | 0.95 | 1.0 |
| Individual + One parent + One Grand parent + one GGP | 0.37 | 0.51 | 0.61 | 0.69 | 0.75 | 0.81 | 0.86 | 0.90 | 0.95 | 1.0 |

It can be concluded that pedigree selection is only useful to select the individual before its own record is available. The pedigree selection is less accurate than individual selection or P.T. based on 4 or more progeny

The accuracy of pedigree selection is expressed by the correlation of the genotype of the individual with the phenotypic mean of the ancestor used in selection. The ancestor records add little to the individual's own record for the traits of low heritability because the accuracy is increased with the increase of $h^2$ of the trait. Thus, for traits of low $h^2$ the information from ancestors without individual's own record results in low accuracy than individual selection. The accuracy of pedigree selection is generally low because of the followings:

(*i*) The different possibilities of combinations resulting from segregation of genes.

(*ii*) It is not possible to predict as to which half of the dam' s genes (better or inferior) are transmitted to the offspring because of the halving process and sampling nature of inheritance. Thus, the best lactation performance of the dam in sire evaluation is not a correct method.

(*iii*) The different environmental conditions prevail at the time of taking records of ancestors and the individual.

❑❑❑

# 13 Family Selection

The selection criteria to estimate the B.V. of an individual may be the information (performance records) of its collateral relatives. The more closely related collaterals to the individual are likely to have more common genes possessed by the individual and can provide more accurate information about the individual. The more closely collaterals can be grouped as full sib families and half sib families. The family means of these collateral relatives form the basis to select the superior individual. The procedure to estimate the B.V. of an individual on the basis of family mean is called the ***family selection or sib selection*** depending upon the inclusion or exclusion of individual's own record in estimating the family mean. The selection criteria is called as ***family selection*** when the individual's own record is also included to estimate the family mean but the selection criteria is called as the ***sib selection*** when the individual's own recorded is not included in estimating the family mean.

## 13.1 SIB SELECTION

It is the selection of an individual based on its sib performance. The sibs may be full sibs or maternal half sibs or paternal half sibs. Thus, sib selection is of two type viz. full sib selection and half sib selection. The sib selection is practiced for the following traits for which the measurements on the individual are not available or recorded-

- Slaughter traits
- Sex limited traits
- Threshold traits like disease resistance
- Traits with low heritability in species with high reproductive rate so as many sibs are measured in short time.

The full sib (F.S.) selection is more accurate than half sib (H.S.) selection. However, the HS selection is favoured for the following reasons

* The half sibs are easily available in more numbers than F.S.

- The rate of inbreeding is more for F.S. selection than H.S. selection. The inbreeding counter balances the effect of selection.

- The F.S. correlation is more likely to be increased by c-effects (common environmental shared by F.S.). The intra class correlation (t) is $rh^2$ for H.S. and $rh^2+c^2$ for F.S. where $c^2$ is the added contribution of maternal or common environmental effects. This reduces the accuracy of F.S. selection.

***Estimation of Breeding Value:*** The breeding value (B.V.) of individual based on sib's performance can be predicted from multiplying the regression coefficient of the genotype of the individual on the average phenotypes of the sibs by the selection differential of the sibs as:

$$\text{P.B.V.} = \overline{P}_c + rh^2 \frac{n}{1+(n-1)t} (\overline{P}_s - \overline{P}_c)$$

Where, $\overline{P}_c$ = average of contemporaries

$\overline{P}_s$ = average of sibs, n = number of sibs

r = coefficient of relationship between sibs and individual

= ½ for FS and ¼ for HS, if no inbreeding

t = Intra class correlation among sibs = $rh^2$

= ½ $h^2$ for and ¼ $h^2$ for H.S., if c-effects are not considered

$h^2$ = heritability of the trait

$$h^2 \frac{n}{1+(n-1)t} = h^2_s = \text{heritability of sibs}$$

The expected response to sib selection is:

$$R_s = i\sigma_p h^2 \frac{nr}{\sqrt{n[1+(n-1)t]}}$$

The accuracy of sib selection is taken as the correlation between the genotypes of the individual and the phenotypic average of sibs and estimated as:

$$\text{Accuracy} = r\sqrt{\frac{nh^2}{1+(n-1)t}}$$

## 13.2 FAMILY SELECTION

The selection criteria are known as family selection when based on the performance of the sibs plus the individual's own record. The family selection like the sib selection is of two types depending on the type of sib viz. full sib family selection and half sib family selection.

The B.V. of individual based on family mean is estimated as (Lasely, 1972)

$$\text{P.B.V.} = \overline{P}_s + h^2\left[\frac{1-r}{1-t}(P_1 - \overline{P}_s) + \frac{1+(n-1)r}{1+(n-1)t}(\overline{P}_s - P_{Cs})\right]$$

Where, $h^2 \dfrac{1+(n-1)r}{1+(n-1)t}$ = $h^2$ of family mean

$h^2 \dfrac{1-r}{1-e}$ = $h^2$ of within family mean

The expected response to family selection ($R_f$) is estimated as (Falconer, 1965)

$$R_F = i\sigma_P h^2 \frac{[1+(n-1)r]}{\sqrt{n[1+(n-1)t]}}$$

The accuracy of family selection (after Lasely, 1972) is estimated as:

$$\text{Accuracy} = \frac{h[1+(n-1)r]}{\sqrt{n[1+(n-1)t]}}$$

## COMMON ENVIRONMENT (C – EFFECTS)

The environment effects which are different for different families but same for all members of the same family are known as common environmental effects denoted as c- effects (Lerner, 1950). The family members share common environment during pre- and post-natal stage. The c- effects thus create resemblance within family members over and above the resemblance due to having common genes and this contributes to the variance between families. This increases the intra class correlation (t) among family members. The c-effects are more for F.S. than for H.S. The F.S. are affected by maternal effects viz. liter mates in swine and full sibs have the same intra uterine environment viz. all the daughters of a sire being born almost at the same time and being reared together are likely to be subjected to similar environment conditions like climatic conditions, feeding regime and management practices etc.

When environmental similarities (c-effects) are present among family members, the intra class correlation among the phenotypic values of family member (t) is increased equal to the amount of c-effects as ($t=c^2$), where $c^2$ is the portion of the total variation caused by difference in c-effects among families. This makes the denominator larger and hence the regression is deceased. Thus the c-effects decrease the accuracy of sib and family average

## 13.3 FACTORS AFFECTING ACCURACY OF SIB AND FAMILY SELECTION

From the perusal of the formulae of the accuracy of sib selection and family selection, it is clear that the accuracy depends on the $h^2$ of the trait, coefficient of relationship between sibs and individual (r) , number of sibs (n) and the degree of correlation (t) between the phenotypes of sibs. The values of r and t are affected by inbreeding and common environments, respectively.

(*i*) Magnitude of heritability: The accuracy of sib selection and family selection is more for traits of low heritability. The family selection is superior at low heritability. The relative efficiency of family and sib selection compared to individual selection decreases with increase in heritability. The accuracy of family and individual selection is equal at low $h^2$ values. The accuracy of sib selection compared to individual selection decreases along increase in $h^2$. When $h^2$ is greater than 0.5 the selection on any number of FS is never as accurate as individual selection and same is true for HS selection when $h^2$ is greater than 0.25.

(*ii*) Number of records: The accuracy of sib selection based on one FS is 0.5 h and based on one HS is 0.25 h. The accuracy of HS and FS selection as well as family selection is increased with increase in n but never exceeds more than 0.5 for HS ($\sqrt{0.25}$) and 0.71 for FS $\sqrt{0.50}$ selection irrespective of n (number of sibs) and $h^2$ of the trait. The family and sibs selection based on large families results in greater response. However, family size (n) is limited due to breeding space, available facilities and intensity of selection. The space and other facilities can be exhausted in two ways viz. more number of families with their small size and lesser number of families with their large size. The optimum family size for HS families has been proposed by Robertson (1957) as:

$$N = 0.56\sqrt{\frac{T}{Nh^2}}$$

Where, T = total number of individual to be measured (nN)

N = No. of families

n = family size (no. of sibs)

The large size family brings the mean phenotypic value close to the phenotypic value. The effect of family size (n) on the relative efficiency of family and sib selection compared to individual selection is more when t (correlation between phenotypes of sibs) is low. The accuracy of sib selection just equals that of individual selection based on 5 FS and 26 HS when $h^2$ is 0.1.

(*iii*) **Genetic correlation between sibs:** The type of family (HS and FS families) in terms of r between sibs also affects the accuracy of selection. The more close family members (FS and HS) provide more reliable information for selection compared to remote families. In other words, the accuracy of selection decreases when r (coefficient of relationship between sibs and individual) among family members decreases. Thus FS selection is more efficient than HS selection.

(*iv*) **Phenotypic correlation between sibs:** The degree of correlation between the phenotypes of sibs (t) affects the accuracy of selection. The family selection is better (even better than individual selection) when the phenotypic correlation (t) is low. When t is low, the $h^2$ will be low with little resemblance due to common environment. This makes the denominator smaller and consequently the regression is increased. Thus, with low values of t the accuracy of family selection is increased or say that c – effects decrease the accuracy of family average. Family selection is preferred to individual selection for characters of low $h^2$ and when c- effects are not important. Individual selection is better than family selection when t is larger than 0.25 for the reason that family selection acts only on the variance between family means while the individual selection operators on whole of the additive genetic variance.

## 13.4 ADVANTAGES OF FAMILY SELECTION

(*i*) The family selection can improve the characters of low heritability in species with high reproductive rates (pig, poultry) so as to get many sibs in a short time.

(*ii*) The family selection does not allow the generation interval to increase.

(*iii*) Family selection is a support to individual selection because it is better to select an individual from a superior family.

## 13.5 DISADVANTAGES OR LIMITATION OF FAMILY SELECTION

The family selection to be effective requires large family size and more number of families to avoid inbreeding as well as to increase the intensity of selection. In view of this it can be inferred that:

(*i*) The family selection is costly of space particularly when the breeding space and testing facilities are limited. The limited facilities reduce the intensity of selection.

(*ii*) The family selection as a unit of selection results in inbreeding and thus limits the genetic diversity. This is because only few families represent the next generation.

(*iii*) The F.S. family selection can only be applied in species with high reproductive rates to get large family size.

## 13.6 WITHIN FAMILY SELECTION

It is the selection criteria when individuals are selected on the basis of their performance expressed as deviation from their family mean. In this selection no weight is given to family mean. This criterion is preferred when C-effects are important i.e. when a large component of environmental variance is common to all members of a family. The C-effect arises when different families differ largely in the environment to which they are exposed viz. different families are getting either good or poor environment. The pre weaning growth in pigs and mice is the example. A large part of the variation in weaning weight is due to mothers and hence common to all members of the family (dam family). The selection within family eliminates the environmental differences among families.

The within family selection economizes the breeding space and minimizes the rate of inbreeding. Each family contributes equally to the parents of next generation, when single pair mating is made and two individuals from each family are selected as replacement of the parents.

The within family selection is better than individual selection when the sib correlation (t) is very high and caused largely by environment effects. The within family selection operates on a large amount of additive variance within families.

The within family selection is rarely applied to HS families because C-effects are not important for half sibs. This is more practicable for FS families because the FS share common environment.

The $h^2$ of within family selection is:

$$h^2_W = h^2 \left[\frac{(1-r)}{(1-t)}\right]$$

The accuracy of selection is the square root of this heritability.

The expected response ($R_W$) to this selection is:

$$R_W = i\,\sigma_P\, h^2 (1 - r) \frac{\sqrt{(n-1)}}{n\,(1-t)}$$

## 13.7 COMBINED SELECTION

The selection of an individual on the basis of two or more sources of information (selection criteria) is called as the **combined selection** or index selection. It is better to select on the basis of an index combining information from various relatives (dam, sibs and or progeny) with or without individual's own record. This is done by the technique of multiple regressions. These multiple regression coefficients are to be used as the weighting factors. All information available about each individual's breeding value combined into an index of merit is the optimal procedure for selection.

Two situations of combined selection have been discussed beloe:

***1. Combining family mean and within family deviation:*** It is obvious that the phenotypic value of an individual ($P_i$) is composed of two parts viz. $P_f$ which is the deviation of its family mean from the population mean and $P_w$ which is the deviation of the individual from the family mean called as the within family deviation. Thus

$$P_i = P_f + P_w$$

The weight given to these two parts decides the procedure of selection. When equal weight is given to both the parts it is known as *individual selection*. When family mean is considered as a basis of selection giving no weight to the within family deviation ($P_w$), it is known as *family selection* whereas the selection on the basis of the within family deviation ($P_w$) alone giving no weight to the family mean ($P_f$), is known as *within family selection* and the best individual from each family is selected. Lastly, when the two components ($P_f$ and $P_w$) are given different weights, the selection is known as ***combined selection*** or *index selection*.

The best estimate of an individual's breeding value on the basis of its phenotypic value is $h^2P$ where $h^2$ is the regression of breeding value on phenotypic value. Now considering the two parts of the phenotypic value (Pf and Pw) which are uncorrelated and give independent information about the B.V., the expected breeding value of an individual can be estimated from multiple regression equation as:

$$E.B.V. = h^2{}_f\, P_f + h^2{}_w\, P_w$$

Where, the $h^2{}_f$ and $h^2{}_w$ are the heritability of family mean and within family mean

The expected response to this combined selection is generally higher to about 10-20 % than individual selection, family selection and within family selection.

$$R_C = i\,\sigma_P\, h^2 \sqrt{1+\frac{(r-t)^2}{(1-t)}\times\frac{(n-1)}{1+(n-1)t}}$$

**2. *Combining information from individual and all kind of relatives:*** There are situation when information from the individual, its parents, FS, HS etc. are available and when the character on the individual is not measured (like sex limited characters). The individual can be selected to combine all information with an index. This involves the matrix method and takes the form of a multiple regression of B.V. on all the sources of information. The index so constructed is the best linear prediction of an individual's B.V. The index can be represented as:

$$I = b_1P_1 + b_2P_2 + b_3P_3 + \dots\dots\dots$$

Where, b's are the weighting factors (partial regression coefficients)

P's are the phenotypic values of different selection criteria.

Simultaneous equations equal to the number of sources of information are set. If the sources of information are three viz. the individual, the parents (dam) and paternal half sibs, then three simultaneous equations will be as follows.

$$b_1\, V_{P1} + b_2\, Cov\, p_{12} + b_3\, Cov\, p_{13} = V_{A\,1}$$

$$b_1\, Cov\, p_{21} + b_2\, Vp_2 + b_3\, Cov\, p_{25} = Cov_{\,A21}$$

$$b_1\, Cov\, p_{31} + b_2\, Cov\, p_{32} + b_3\, Vp_3 = Cov_{\,A31}$$

Where 1 = individual, 2 = parents and 3 = PHS

The b's can be estimated after putting the values of variances and co variances.

## NUMERICAL PROBLEMS

### Single record of individual and relatives

**Example 13.1:** The phenotypic value of a trait recorded on a calf, its sire and dam are 900, 800 and 450 units respectively. Taking the $h^2 = 0.43$, $\sigma P = 80.0$, estimate the B.V. of the calf.

**Solution:**

$$V_{P1} = V_{P2} = V_{P3} = (80)^2 = 6400$$

$$V_{A1} = h^2\sigma^2{}_{p1} = 0.43\times 6400 = 2752$$

$$Cov_{p\,12} = \sigma^2 \frac{1}{2}{}_P = \frac{1}{2}(2752) = 1376$$

$$CoV_{P\,13} = cov_{P\,12} = 1376$$

$$COV_{P\,23} = 0$$

$$Cov_{A\,12} = cov_{P\,13} = \frac{1}{2}\sigma^2{}_A = 1376$$

Now putting the values the equations will be:

$$\begin{vmatrix} b_1 \\ b_2 \\ b_2 \end{vmatrix} = \begin{vmatrix} V_{P1} & cov_{12} & cov_{13} \\ cov_{21} & V_{P3} & cov_{23} \\ cov_{31} & cov_{32} & V_{P3} \end{vmatrix} = \begin{vmatrix} V_{P12} \\ cov_{12} \\ cov_{13} \end{vmatrix}$$

Now the values are put up:

$$\begin{vmatrix} b_1 \\ b_2 \\ b_2 \end{vmatrix} = \begin{vmatrix} 6400 & 1376 & 1376 \\ 1376 & 6400 & 0 \\ 1376 & 0 & 6400 \end{vmatrix}^{-1} = \begin{vmatrix} 2752 \\ 1376 \\ 1376 \end{vmatrix}$$

After solving these equation we get

$$B_1 = 0.372,\ b_2 = 0.135 \text{ and } b_3 = 0.135$$

Therefore, $I = b_1 X_1 + b_2 X_2 + b_3 X_3$

$$= 0.372\, X_1 + 0.135 X_2 + 0.135\, X_3$$

$$= 0.372\,(900\text{-}\mu) + 0.135\,(800\text{-}\mu) + 0.135(450\text{-}\mu)$$

$$\text{Accuracy} = \sqrt{\frac{0.372(2757) + 0.135(1326) + 0.135(1326)}{2757}}$$

$$= 0.712$$

**Example 13.2:** mean records of individual and relative:

When mean of groups of individuals are taken, then n in the group is also required. The between group and within group variances are estimated as:

The intra class correlation between of families (t) = $\frac{\sigma^2{}_B}{\sigma^2{}_P}$ and hence between group component of variance ($\sigma^2{}_B$) = $t\sigma^2{}_P$

Likewise, within group component of variance ($\sigma^2{}_W$) = (1-t) $\sigma^2{}_P$

So that $\sigma^2{}_P = t\,\sigma^2{}_P + (1\text{-}t)\,\sigma^2{}_P$

$= V_P$ between families + $V_P$ within families

Now, if the average of first 2 lactation of a cow is 230 units and the mean of its 5 half sisters with 2 lactation is 300 units. Estimate the B.V. of the cow taking $h^2$ = 0.25, t = 0.5 and $\sigma_P$ = 34 units.

**Solution.** $V_{A1} = \sigma^2{}_A = h^2\sigma^2{}_P$

$$= 0.25(34^2) = 289$$

$$Cov_{A12} = cov._{HS} = \tfrac{1}{4}\,\sigma s^2{}_A$$

$$= \tfrac{1}{4}\,(289) = 72.25$$

$$V_{P1} = V_{\bar{p}1} = t\sigma^2_P + \frac{1}{n}(1\text{-}t)\,\sigma^2_P$$

$$= (t + \frac{(1-t)}{n})\,\sigma^2_P$$

$$= (0.5 + \frac{(1-0.5)}{2})\,(34)^2 = 867$$

$$V_{P2} = \text{Var.}_{\bar{p}2} = \sigma^2_B + \frac{1}{n}(\sigma^2_W)$$

Where, $\sigma^2_W = \frac{1}{4}\sigma^2_A$ = Between cow variance

$$\sigma^2_w = V_{\bar{p}2i} - \sigma^2_B$$

Taking $V_{p2i}$ = mean of first 2 lactation of cow i

Since all 5 cows each with 2 records, $= V_{\bar{p}2i} = V_{P1}$

Thus, $\sigma^2_W = (V_{P1} - ¼\,\sigma^2_A)$

$$\frac{1}{n}\sigma^2_W = \frac{1}{2}(V_{P1} - \frac{1}{4}\sigma^2_A)$$

So, $V_{P2} = \frac{1}{4}\sigma^2_A + \frac{1}{n}(V_{P1} - \frac{1}{4}\sigma^2_A)$

$$= \frac{1}{4}(289) + \frac{1}{5}[867 - \frac{1}{4}(289)] = 231.2$$

Now substituting the values in equation and solving them as

$$\begin{vmatrix} b_1 \\ b_1 \end{vmatrix} = \begin{vmatrix} 867 & 72.25 \\ 72.25 & 231.2 \end{vmatrix}^{-1} \begin{vmatrix} 289 \\ 72.25 \end{vmatrix}$$

$$b_1 = 0.316,\ b_2 = 0.213$$

Therefore, $I = b_1 X_1 + b_2 X_2$

$$\text{Accuracy of index} = \sqrt{\frac{0.316(289) + 0.213(72.5)}{289}} = 0.61$$

❑❑❑

# 14

# Progeny Selection

The selection criteria for evaluating an individual based on his progeny performance is known as progeny selection or progeny testing (P.T.). The P.T. is the most important basis of selection. The selection criterion is the mean phenotypic value of the individual's progenies compared to the mean phenotypic value of the contemporaries. The P.T. is regarded as a form of family selection because progenies are the family members of each other.

## 14.1 GENETIC PRINCIPLES OF P.T.

The P.T. stems on the principle of like begets like. The progeny of an individual is the index of parents' genetic worth. Each progeny inherits one-half of the genes from each parent due to the halving nature of inheritance. This is taken as the transmitting ability of the parent and hence the breeding value (B.V.) of the parent is twice the mean deviation of the progenies from the population mean.

Second principle is the sampling nature of inheritance. Chance at segregation may result in any one or few progenies receiving a good or poor than average sample of genes from its parent. Different genes are transmitting by a parent to its different progenies due to segregation of genes and the probability to receive exactly the same set of genes by its different progenies is very low for polygenic traits. This segregation is called the Mendelian error. Therefore, the estimation of B.V. of an individual (parent) based on one or few progenies may be misleading. Moreover, the polygenic traits are influenced by environmental factors and the different progenies of the same parent do not get same environment. Thus, the environmental factors cause differences in the performance of different progenies. Therefore, the chance at segregation and environmental factors causes greater deviation from true B.V. of a parent. The effects of these two factors are balanced out by estimating the B.V. with the increase in progeny size.

## 14.2 SUPERIORITY OF PROGENY SELECTION OVER OTHER CRITERIA

The *individual selection* is not possible for traits of low $h^2$, sex limited traits and the traits expressed after death of the individual. Most of traits in farm animals have low $h^2$ and are sex limited. Thus these traits cannot be improved through individual selection and hence the B.V. is estimated based on the performance of relative (dam, sister and progeny).

The *pedigree selection* based on the performance of dam is not a good selection criterion for the reasons of halving process and sampling nature of inheritance. It is not possible to know as to which half (better or inferior) of dam's genes are transmitted to the progeny. Moreover, the pedigree selection is effective for traits of high $h^2$. In pedigree selection, only one sample is tested.

The *sib selection* has also low accuracy due to sampling nature of inheritance. Every individual receives different set of genes from its parent and the probability to set same get of genes by the individual and its sibs is very less.

Finally, the *progeny testing* is the most important and best criterion of selection. The P.T. gives the best and most reliable information about the genetic merit (B.V.) of parent (individual under selection). This is because the P.T. is based on mean performance of many progenies. This overcomes the limitation of Mendelian error of gene segregation and hence provides the true estimate of B.V. of an individual.

## 14.3 ADVANTAGES OF P.T.

- The P.T. is a better method for sex limited traits, the traits with low $h^2$, and slaughter traits in meat animals.
- The P.T. is useful to prove a sire whether he is free from any recessive gene. To identify the individual for harmful gene, it is mated to its sibs or progeny.
- The main feature and principal advantage of P.T. is the increased selection intensity. This is because the progeny are also used a parent and await selection at the time of testing their own parent. Thus P.T. is a modified from of family selection.
- Its accuracy increases with increase in progeny number.

## 14.4 LIMITATIONS OF P.T.

- The time and cost required is the main limitation. It increases the generation interval.
- Due to larger generation interval the genetic gain per year is low.
- P.T. is effective on adequate number of progeny and hence requires high reproductive rate
- It requires to progeny test and compares many sires to find which one is superior. Thus selection is needed among sires under test. Thus, there is low selection intensity.

## 14.5 SOURCES OF ERROR IN P.T.

The various sources of error in P.T. and the ways to correct them are as under:

- Randomly distributed sources of variation – caused by Mendelian segregation, environmental effects and the error in number of data.

These errors can be reduced by increasing the number of progenies to estimate the B.V.

- Systematic error – arises by age and weight effect of cows, parity number, lactation length, days open, dams belonging to selected group, herd differences and period effect. These can be reduced by adjustment of data for these effects.

## 14.6 PRECAUTIONS IN PROGENY TESTING

In view of the various sources of error in P.T., the following **precautions** need attention:

- The adult females should be allotted randomly to all the sires under test to avoid the effect of production level of the dam on the progeny. Allotment of superior dams to a particular sire overestimates the B.V. of that sire.
- Different sire group should be raised and compared under similar environmental conditions viz. same nutrition regime and management practices. The progeny should be born in the same and year as well as should be recorded at the same age/lactation.
- All the healthy progeny should be included and excluded those which could not complete the lactation either due to their death or culling during lactation or due to disease or abnormal calving.
- The data recorded should be adjusted for environmental effects such as age, year, season, lactation length etc.
- More number of progeny per sire should be available and recorded to cancel out the randomly distributed sources of error.

## 14.7 REQUIREMENTS FOR P.T.

In addition to the above precautions, there are several requirements for an effective large scale P.T. programme.

- It is always better that more than required number of males must be kept for P.T. so as to have an opportunity of selection among the P.T. sires. About 4-5 males should be progeny tested for each required progeny tested sire for future breeding.
- The facilities should exist for frozen semen, and to record information on progeny. Moreover, a large population is also required.
- The progeny should be tested in several herds and stations and comparison should be made with contemporaries from other sires. This increases the accuracy.
- No progeny should be culled until the end of the test.
- The P.T. should be completed as early as possible in the life of the males.
- Similar environment is required for all the progeny groups.

## 14.8 CONSTRAINTS IN P.T. AND THEIR SOLUTIONS

There had been the following constraints for effective result of P.T. in the past.

(*a*) *Small population size:* Most of the organized farms have small size population due to the reason of having very limited capacity to maintain large population. This leaves no scope of culling the low producing females. Secondly, due to small herd size and close breeding, the inbreeding with its adverse effect on production and reproduction are inevitable. The genetic improvement is less in a herd of small size. Therefore, it is better to progeny test the sires in associated herds or famer's herd.

The problem of small population size in single herd testing and of progeny testing large numbers of bulls can be solved if the sires are tested simultaneously in different herds (known as associated herd P.T.) and also involving the farmer's herd (known as field progeny testing). Thus, these are 3 types of P.T. programmes. Such P.T. programme (associated herd or multiple herd P.T. and field P.T.) allows testing more number of sires and that too in different herds based on more number of progenies. This will result higher intensity and accuracy of selection. Under these programmes the progenies are raised in different conditions and hence lead to more accuracy of sire evaluation taking into account the G-E interaction.

(*b*) *Unplanned mating:* There had also been unplanned mating in these herds to get required number of daughters from all the bulls under test. The constraints of small population and unplanned mating do not allow to progeny test the bulls based on large numbers of daughters with greater accuracy and to progeny test the large numbers of bulls required for large scale breed improvement programme.

The unplanned mating results in wastage of the breeding data and also the non-orthgonality of data. The non-orthgonality in terms of unequal number of daughters for different sires had been the intense at most of the animal farms in India, varying from one to several hundred daughters per site. This result in loss of breeding data and that all the sires under test are not evaluated, besides creating the problem in data analysis and biasness in predicting the breeding value. Based on a number of studies at NDRI Karnal during 1985-2001 by Tomar and his students, the data of about 40 per cent of the total sires under test could be used in estimating the B.V. and the data on rest of the sires were wasted for want of adequate number of daughters. This puts restriction to P.T. of a large number of sires under test. Therefore, it is proper to prepare a calculative mating plan to obtain nearly equal progenies from different sires. The studies conducted on this aspect by Tomar and his students (1985-2001) have indicated that 5 pregnancies are required to get one replacement daughter. Therefore, to run a P.T. programme effectively, it is required to get 100 inseminations or 50 pregnancies from each bull to get about 10 - 15 daughters to be performance recorded.

(*c*) *Infrastructure:* There is a problem in associating the farmer's herd that the farmers do not keep the records. The reasons are their illiteracy and they find no immediate use of record keeing. Moreover, the facilities of A.I. and treatment of sick animals are not available at farmer's door.

(*d*) *Generation interval:* The generation interval in dairy cattle and buffaloes is very long for the reasons of slow growth rate, delayed maturity and poor breeding efficiency. The generation interval can be reduced by using young bulls, replacing older cows by heifers, improving the plane of nutrition and proper management. The use of part lactation record will also help in proving the sire at an earlier age.

In progeny testing programme the sires have been used for long to get sufficient progenies for evaluation. This increases the environmental variance in progeny test and reduces the genetic gain per unit of time. The reduction in test mating duration of bulls to one year, optimum feeding and management of young stock and using advanced reproductive technologies, such as in vitro fertilization (**IVF**) etc might be the best strategies to reduce the generation interval. Secondly, the genomic selection of young males at a very early age rather than traditional progeny testing is better option. This will reduce the generation interval and could lead to doubling of the rate of genetic gain.

(*e*) *Bull disposal:* A large number of sires remain idle over the years between test mating to the time the P.T. results are available. Keeping the bulls in store for such a long time is an expensive and moreover, some of the bulls either die or become sterile during this period. This problem has been solved to some extent by the frozen semen technology. It is now possible to dispose of the bulls after collecting and freezing the semen.

## 14.9 METHODOLOGY P.T. IN ORGANIZED HERD

In view of the requirements and constraints experienced earlier, a sound technical programme of P.T. is required to be formulated for organized herd.

- In order to overcome the problem of population size the P.T. programme should be carried out in the associated herds.
- About 4-5 males should be kept under test for each required progeny tested bull for future breeding.
- About 50 pregnancies should be settled from each bull so as about 10 daughters of each bull must be performance recorded.
- It is better to complete one set of bulls within a period of 2 years. This is known as test cycle.
- In each set of bulls, about 8-10 males should be tested so that at least 2 top ranking bulls are selected on the basis of their B.V. estimated from their daughter's performance.
- Taking a set of 10 males, 2 progeny tested and 8 under test (to be tested), it will require 500 pregnancies in a period of 2 years. Now taking into consideration that only 66% of the breeding females will calve each year, 250 pregnancies will be settled in one year from 380 breedable females. Therefore, the breeding herd should have at least 380 – 400 breedable females. The herd of 380 breedable females is divided in two groups viz. *elite cows* (80) used for mating with proven bulls (already progeny tested) to

produce elite heifers and test bulls used for breeding in next set and second group of breedable females is the *test cows* (300) used for test mating.

(*i*) *Test mating*: Test cows (300) x test bulls (8). These test bulls should be produced from elite mating. The test mating will produce about 200 pregnancies from 8 test bulls, thus 25 pregnancies from each bull will be obtained in one year. About 10-15 daughters from each bull in a test cycle of 2 years will be available for performance recording (first lactation completion). The male calves produced from test mating will not be used for breeding but will be disposed off. The heifer calves produced from test mating will be used for test mating of next set.

(*ii*) *Elite mating:* Elite cows (80) x proven bulls (2). These proven bulls are progeny tested and will be used on elite cows. The male calves produced will be used as test bulls. The elite female population for next batch will be taken from the herd (progeny of elite cows and test cows).

- The 2 best proven bulls based on P.T. results should be used on about 80 best yielding cows (elite cows) to get a set of young males for testing in next batch (test males) and remaining 8 which are the sons of 2 best proven bulls of an earlier batch should be used on the remaining cows (test cows).
- ***B.V. Estimation:*** The B.V. of bulls based on its progeny performance is estimated by multiplying the regression of B.V. of parent on the phenotypic performance of its progeny with the selection differential of the progeny of $i^{th}$ sire from the mean of the contemporaries (Pi -Pc). The regression (b) is taken as

$$b = \frac{r\,nh^2}{[1 + (n - 1)\,t]}$$

$$= \frac{0.5nh^2}{[1 + (n - 1)\,t]}$$

$$= \frac{2nh^2}{4 + (n-1)h^2} \quad \text{since } t = 0.25h^2$$

Where, r = coefficient of relationship between individual and his progeny

t = Intra class correlation among progeny which is ¼ for H.S.

n = No. of progenies

Thus, the B.V. is estimated as: $\text{B.V.} = \bar{P}_C + \frac{2nh^2}{4 + (n-1)h^2}(\bar{P}_i - \bar{P}_c)$

- The quality semen doses of about 3000 – 4000 from each of progeny tested as well as from each test bull should be frozen and stored for future use in the institutional as well as in the farmer's herd.

## 14.10 ACCURACY OF P.T.

The accuracy of P.T. is expressed as the correlation of genotype of parent with the average phenotype of its progeny (half sibs) and hence it is estimated as $r_{G\bar{P}}$ - which is

$$r_{G\bar{P}} = r_{A\bar{P}} = 0.5\sqrt{\frac{nh^2}{1+(n-1)t}}$$

The accuracy of selection based on one progeny is the same as that based on one parent or one FS and it is 0.5h. The accuracy increases with increases in number of progeny and the increase is much larger than for sib selection. The accuracy of P.T. with about 5 progeny gives as high as that of individual selection for traits of moderate heritability and it is more effective if n exceeds 5 in number. The accuracy of P.T. quickly reaches a level higher than that of individual selection and tends towards 100% for any value of $h^2$. Thus P.T. gives a higher accuracy than any other selection criteria.

The relative accuracy of P.T. and individual selection is estimated as:

$$0.5\sqrt{\frac{n}{1+(n-1)t}}/h = \sqrt{\frac{n}{4[1+(n-1)t]}} = \sqrt{\frac{n}{4+(n-1)h^2}}$$

The P.T. compared to individual selection is relatively more accurate at lower values of $h^2$. In comparison to family selection the P.T. is almost twice as effective as that selection based on H.S. families and about 1.4 times to that based on F.S. family mean for any value of $h^2$.

There are a number of factors which influence the result of P.T. like selection of dams mated to sires under test (contemporary dam average) and common environment (c-effects) which cause similarity between progeny. The presence of c-effects reduces the accuracy of P.T. The c-effects can be caused by season differences in lactation milk yield and when cows are from different herds. The $C^2$ is the environmently determined correlation which exists between daughters of a bull\. The $C^2 = t = 0.25h^2$ for season differences and $C^2 = h^2 = 4t$ for cows belonging to different herds. The estimation of the true B.V. of a male based on P.T. requires some correction in order to minimize the effects of level of dams production etc. these corrections have been discussed under sire evaluation.

## 14.11 FIELD PROGENY TESTING

***Necessity of field progeny testing:*** The population size of organized herds is small. Thus, progeny selection in single herd is based on limited progenies per bull. This leads to low intensity of selection and low accuracy of selection. Secondly, lesser number of bulls can be put under progeny testing which are not sufficient to meet the programme of genetic improvement of large population. Thirdly, the bulls based on single herd are tested under one type of environmental conditions and hence the genetic improvement made under farm conditions may not be relevant in other conditions in case of genotype-environment interaction.

Further, the organized herd of any breed is a very small part of total cattle and buffaloe population on country basis and hence the genetic improvement of these organized herds does not make a significant contribution to increase the productivity of the population as a whole. Moreover, a large number of progeny tested bulls or large numbers of quality doses of semen are required for breed improvement programme including selective breeding of well defined breeds and grading up of non-descript population at national level.

The single herd P.T. due to small population size of organized herds is thus a constraint in progeny testing of bulls based on few numbers of daughters in single herd. Therefore, the alternate breed improvement programme to bring sustainable genetic improvement of the breed through bull selection should be undertaken. In view of these, it is suggested to plan and execute the network programme of progeny testing involving various organized herd of the same breed (multi-herd P.T.) and also the farmers' herd (known as field P.T.). The multi herd P.T. improves intensity and accuracy of selection to a certain extent but with a limitation of testing of lesser numbers of bulls which are not sufficient to bring genetic improvement of entire population of the breed as a whole. Thus, the genetic merit of large number of bulls can not be evaluated through associate herd P.T. programme. In order to meet the large number of superior progeny tested bulls based on large numbers of daughters, it requires large scale P.T. of bulls involving farmer's herd. Such a large scale programme requires the collaborative approach of research and development organizations like, State Agricultural Universities, State Animal Husbandry Departments, State Milk Producers Cooperative Union, State Livestock Development Board, Research Institute of the breeding tract, etc.

***Requirements of field P.T.:*** The *organizational setup* is the main requirements to carry out the P.T. in field condition. This should include a central/ national coordinating agency, participating units in breeding tract, bull rearing centers, and extension units.

There should be a coordinating agency at national level. This will be responsible for formulating the technical programe, to monitor proper and uniform implementation through participating units in the breeding tract, and to create facilities for central data, its evaluation and follow up programme. In addition it will be responsible to create the infrastructure facilities, extension services, performance recording facilities, and providing incentives to farmers.

The participating unit breeding tract should be made responsible for execution of technical programe and to generate the data on animal identification (its number, its sire and dam number) and data collection on various production and reproduction traits.

The bull rearing centers are required. These centers will be established by selecting the bulls from various sources on pedigree performance or based on milk competition in field. These centers will rear the male calves in sufficient numbers to meet out the requirement of breeding bulls which should be free from communicable diseases and genetic defects.

The extension services should include the A.I. staff for providing the A.I. facilities at the doorstep of farmers through establishing its own A.I. centers as well as through State Govt. A.H. Deptt. The staff will also follow up the pregnancy diagnosis, calving records, milk recording, data collection and will also provide the treatment to animals for various disorders/diseases

The farmers should be given sufficient incentives so as they keep the heifers born to bulls under test as well as to progeny tested bulls at least up to the completion of first lactation. The incentives may be in the form of free treatment of their animals, free A.I. services, and rewards for better performing animals and their better coordination in the programme.

***Constraints in field P.T.:*** It requires large investment for creating infrastructure in the field in terms of scientific and technical staff, physical facilities and to give sufficient incentives to the farmers. The great constraints at farmer's level are their illiteracy and compulsion to sell their animals before completion of first lactation. There is also a technical constraint of inaccurate information due to casual milk recording. In addition to these, the poor conception rate and high mortality of male calves in field condition are also the constraints.

## 14.12 SIRE EVALUATION

The result of P.T. are expressed in the form of an index which is the index of the genetic worth of the sire and such an index is know a ***sire index***. Based on sire index a numerical value is obtained which indicates the production ability of the sire. Since a number of sires are progeny tested, it requires ranking all the sires for their genetic worth so as to select the best sire.

### 14.12.1 Biasness in sire evaluation

The biases in obtaining the accurate estimate of sire's B.V. arise due to genetic and environmental differences. The genetic biasness is created by genetic differences among herds and due to the differences in genetic merit (production level) of the mates of the sire. The major environmental biases are introduced by the different environmental conditions (feeding, management, climatic factors etc.) of different herds where the year and season of calving create minor biasness. In order to eliminate or reduce these biases, it is better to adjust the data before estimating the B.V of sires. Considering all these points, a number of sire indices have been proposed by different workers.

### 14.12.2 Sire Indexing Methods

The different indices developed are for two purposes viz. indices which simply rank the sires and the indices which provide the estimates of breeding value of sires. The B.V. is estimated for indexing in a single herd a as well as for indexing in many herds.

The following notations have been used in the construction of various sire indices:

$\overline{D}$ = Average of all daughter of a sire under test

$C_D$ = Average of contemporary daughters of the sire

$M$ = Average of mate's of the sire or dams average

$C_M$ = Average of contemporary mates

$H$ = Herd average

$n$ = Number of daughters

$b = 0.5h^2$, intra-sire regression of daughters on dam

***(i) Simple daughter average index:*** Edwards (1932) proposed an index as-

$$I = D$$

This index is the simplest measure in a single herd under similar environment. It does not take into account of the production level of mates of the bull (dams of he daughters) and hence subjected to bias.

***(ii) Equi-parent or intermediate index:*** Hansson (1913) proposed this index which is also known as Yapp's index (Yapp, 1925) also known as Mount Hope Index because it was first used at Mount Hope Farm in 1928. This index is

$$I = 2D - M = D + (D - M)$$

This index makes adjustment of the variation in production level of the dam (mates). However, it over corrects the production level of mates allotted to different sires. This index is based on the assumption that the dams and daughters are raised under similar conditions so that the daughter dam difference reflects the sire effect otherwise it will reflect managment practices in the herd.

***(iii) Corrected daughter average index:*** Krishnan (1956) gave the following index-

$$I = D - b\,(M - H)$$
$$= D - 0.5\,h^2\,(M - H)$$

This index is superior to first two indices because it corrects for the production level of dams allotted to the sire over herd average. Thus, daughter's average is corrected for the differences in the production level of their dams. The correction factor is $b\,(M - H)$.

Searle (1964) compared the records of daughters with those of their herd mates in the same year and referred to as contemporary comparison or stable mate comparison. The sire's merit was estimated as-

$$I = H + \frac{0.5\,h^2\,n}{[1 + (n - 1)\,0.25\,h^2]}\,(D - H)$$
$$= H + \frac{2n\,h^2}{[4 + (n - 1)\,h^2]}\,(D - H)$$

***(iv) Contemporary daughter average index:*** The contemporary comparison reduces the environmental variation de to herd, year and season. The records of the daughters of a sire are compared with the daughters of all other sires in the same herd calved in the same month or within two months.

The sire index proposed by Sunderasan et al (1965) is-

$$I = H + \left[\frac{n}{(n+k)}\right] (D - C_D)$$

Where, k = constant based on sire error variance

$$I = H + \left[\frac{n}{(n+12)}\right] (D - C_D) - b\,(M - C_M)$$

The following sire index also takes into account the number of daughters and environmental conditions-

$$I = H + \frac{0.5\,h^2\,n}{[1 + (n-1)\,t]} (D - C_D)$$

***(v) Corrected contemporary daughter average index*:** The following sire index as an extension of the contemporary daughter average index has been proposed. This index besides adjusting the number of progeny and period variation also adjust for the differences in production level of dams allotted to different sires. The index is-

$$I = H + \frac{0.5\,h^2\,n}{[1 + (n-1)\,0.25\,h^2]} [(D - C_D) - b\,(M - C_M)]$$

***(vi) Least square constants:*** The sire constants are obtained by least square technique which adjust the data for environmental effects including the non-orthogonality of data. The index obtained is

$$I = \frac{2nh^2}{4+(n-1)h^2}(S_i)$$

Where, $s_i$ is the sire constant for $i^{th}$ sire

***(vii) Maximum likelihood method and REML:*** The least squares method minimizes the error variance whereas the maximum likelihood (chance) method estimates the parameters by maximizing the logarithm of the likelihood function. The likelihood function is the likelihood of simultaneous occurrence of observations (variables). However, the ML estimates are biased because no account is taken of the degree of freedom in estimating the variance components. The ML method was improved by a method known as restricted maximum likelihood (REML) which takes care of the bias in estimates as well as avoids negative estimates of component of variance (Searle et al, 1992). The variance components by REML are estimated based on residuals calculated after fitting by ordinary least squares from the fixed effects part of the model. This maximizes a marginal maximum likelihood. This is also called as the residual or marginal maximum likelihood

***(viii) Best Linear unbiased prediction (BLUP):*** The BLUP method of estimation was developed by Henderson (1949, 1973). This method is more powerful than the conventional selection index approach. This provides directly comparable estimates of the average breeding value of groups of animals born in different years. It takes into account the complications of non random mating, sires from more than one herd, environmental trends over time, herd differences for B.V. of dams and bias due to selection. Thus BLUP takes account of the fixed effects such as herd–year- season and it is applicable when mixed models are used whereas BLUE is applicable when all effects are fixed. The BLUP eliminates the non genetic biases in estimating B.V Not only this but it also removes the genetic biases taking into account the effects of non –random mating, genetic merit of dams and selection.

***(ix) Other methods for multi-herds:*** Some methods for evaluation of sires used in more than one herd have also been proposed. These are Stable-Mate Daughter Average Index, Contemporary Daughter Average Index and Corrected Stable-Mate Daughter Average Index.

### 14.12.3 Efficiency of sire indices

The following criteria may be used to judge the efficiency or accuracy of various methods:

(*i*) Repeatability of the index which is taken as correlation between an index based on n daughters and an index based on n′ future daughters.

(*ii*) Regression of a sire's future daughters on the average performance of number of present daughters (Robertson and Rendel, 1950)as:

$$b = \frac{0.25\,nh^2}{1+(n-1)0.25h^2} = \frac{nh^2}{4+(n-1)h^2}$$

(*iii*) Error variance (or within sire variance)

(*iv*) Rank correlation

(*v*) Coefficient of determination ($R^2$) of fitting the model:

$$R^2\,(\%) = \frac{\text{s.s of fitting the model}}{\text{TSS}} \times 100$$

## NUMERICAL PROBLEMS - PROGENY TESTING

**Example 14.1:** The average milk production of Sahiwal herd comprising 150 progenies of 6 sires is 2000 kg with $h^2$ of yield as 0.30. The average milk production and number of progenies per sire is given below. Rank all the sires based on their breeding value.

| **Sire no.** | 1 | 2 | 3 | 4 | 5 | 6 |
|---|---|---|---|---|---|---|
| N | 30 | 20 | 25 | 15 | 10 | 50 |
| $x$ | 1600 | 2250 | 1840 | 2200 | 2300 | 2100 |

**Solution:** (i) Mean of contemporaries of all sires will be

1. $\frac{300000-48000}{120} = 2100$ kg  2. $\frac{300000-45000}{130} = 1961.5$kg

3. $\frac{300000-46000}{125} = 2032$ kg  4. $\frac{300000-33000}{135} = 1977.7$ kg

5. $\frac{300000 - 23000}{100} = 1978.5$ kg   6. $\frac{300000 - 105000}{100} = 1950$ kg

(ii) Regression of B.V. of sire on his progeny ($b_{AP}$)

$$b_{AP} = \frac{2nh^2}{4 + (n-1)h^2}$$

| Sire. | N | h² | $\frac{2nh^2}{4+(n-1)h^2}$ |
|---|---|---|---|
| 1. | 30 | 0.3 | 18/12.7 = 1.41 |
| 2. | 20 | 0.3 | 12/9.7 = 1.24 |
| 3. | 25 | 0.3 | 15/11.2 = 1.34 |
| 4. | 15 | 0.3 | 9/8.2 = 1.09 |
| 5. | 10 | 0.3 | 6/6.7 = 0.89 |
| 6. | 50 | 0.3 | 30/18.7 = 1.60 |

(*i*) Breeding value of sires = $\left(\bar{H} + \frac{2nh^2}{4+(n-1)h^2}(\bar{p}_i - \bar{p}_C)\right)$

Sire 1. 2000 +1.41 (1600-2100) =1295.0 kg

2. 2000 +1.24 (2250-1961.5) =2354.7 kg

Like wise, the B.V. of other sires can be estimated.

**Example 14.2**: From the following data on milk yield construct the sire indices by different possible methods, taking $h^2$ of milk yield = 0.28, population mean = 1660.

| Sire No. | Daughters records | Dam's Average | Contemporary Daughters | Dams |
|---|---|---|---|---|
| 1 | 1252, 1858, 2343, 1880, 1532, 1743, 1274, 1806, | 1600 | 1930 | 1550 |
| 2 | 1780, 1510, 1781, 1561, 1744 | 1340 | 1710 | 1310 |
| 3 | 2455, 1498 | 1680 | 2200 | 1590 |
| 4 | 1581, 2051, 2100, 1583, 2155, 2178 | 1500 | 2000 | 1490 |
| 5 | 1258, 1616, 1424 | 1315 | 1510 | 1300 |

**Solution: (i)** Simple daughter average index ($I_1$) = $\bar{D}$

| Sire no. | 1 | 2 | 3 | 4 | 5 |
|---|---|---|---|---|---|
| Index | 1711 | 1675.2 | 1976.5 | 1941.3 | 1432.6 |
| $n_j$ | 8 | 5 | 2 | 6 | 3 |

(ii) Equi-parent index ($I_2$)= $\bar{D}$ $\bar{M}$

| Sire No. | 1 | 2 | 3 | 4 | 5 |
|---|---|---|---|---|---|
| | 1822 | 2010.4 | 2273.0 | 2382.6 | 1550.2 |

Corrected daughter average index ($I_3$) = $\bar{D}$ -0.5h $_2$ ($\bar{M} - \bar{H}$)
(Krishnan index)

Sire No. 1: 1711.0 – 0.5(0.28) (1600 – 1660) = 1719.4

Sire No. 2: 1675.2 – 0.5(0.28) (1340 – 1660) = 1720.0

Sire No. 3: 1976.5 – 0.5(0.28) (1680 – 1660) = 1973.7

Sire No. 4: 1941.3 – 0.5(0.28) (1500 – 1660) = 1741.3

Sire No. 5: 1432.6 – 0.5(0.28) (1315 – 1660) =1720.0

(iv) $I_4 = \bar{H} + \frac{2nh^2}{4+(n-1)^2}(\bar{D} - \bar{H})$

Sire No. 1: 1660 + $\frac{2\times9\times0.28}{4+(7)0.28}$ (1711 – 1660) =1698.3

Sire No. 2: 1660 + $\frac{2.8}{5.12}$ (1675.2 – 1660) = 1668.3

Sire No. 3: 1660 + $\frac{1.12}{4.28}$ (1976.5 – 1660) = 1742.8

Sire No. 4: 1660 + $\frac{3.36}{5.4}$ (1941.3 – 1660) =1835.0

Sire No. 5: 1660 + $\frac{1.68}{4.56}$ (1432.6 – 1660) = 1576.0

(v) Dairy search Index ($I_5$ ) = $\bar{H} + \frac{n}{n+12}(\bar{D} - C_D) - B(\bar{M} - C_M)$

Sire No. 1: 1660 + $\frac{8}{20}$ (1711 – 1930) — 0.5 × 0.28 (1600 -1550)

=1600- 87.6 -7.0 = **1565.4**

Sire No. 2: 1660 + $\frac{5}{17}$ (1675.2 - 1710) – 0.14 (1340 – 1310)

=1660 - 10.2- 4.2 = **1645.6**

Sire No. 3: 1660 + $\frac{2}{14}$ (1976.5 – 2200) - 0.14 (1680 -1590)

=1660 - 31.9 – 12.6 = **1615.5**

Sire No. 4: 1660+ 6/18(1941.3 - 2000) – 0.14(1500 - 1490)

= 1660 – 19.5 – 1.4 = **1639.1**

Sire No. 5: 1660 + 3/15 (1432.6 - 1510) – 0.14 (1315 - 1300)

= 1660 – 15.5 – 2.1= **1642.4**

❑❑❑

# 15 Selection for Different Kinds of Gene Action

The genotypic value of a quantitative trait has two component values *viz.* additive genetic value (breeding value) caused by additive effect of genes and non-additive genetic value known as gene combination value (GCV) caused by interaction effect of genes, i.e. deviation caused by dominance (D) and epistasis (I). Some traits are more affected by additive gene action while others by non-additive. It is more important to identify the type of gene action for influencing a particular trait under selection. The type of gene action can be identified by the population parameters (mean, variance and covariance).

It is well known that the genes carried by an individual and influencing a particular trait act in different ways to produce their effect viz. additive and non-additive effect. The sole purpose of selection is to increase the frequency of the desirable genes which have additive effect in particular but at the same time the genes with non additive effects can not be ignored. The selection theory for each type of gene action is different. Thus, once the kind of gene action is known the selection system to exploit the particular types of gene action may be followed. The type of gene action can be identified by the population performance for a metric trait in terms of mean, variance and covariance.

## 15.1 SELECTION FOR ADDITIVE GENE ACTION

The nature and extent of genetic variability (additive genetic) present in the herd decides the ***choice of selection criteria*** for estimating the breeding value of a trait. The selection, in a random mating population, increases the frequency of desirable genes controlling the trait with additive effects. Thus, the selection of the breeding individuals is done on the basis of breeding value of the character under selection. The bases of estimating the B.V. of individuals are the individual selection, pedigree selection, family selection and progeny selection.

The *individual selection* for within herd improvement is advocated for traits with high heritability and for those expressed in both the sexes at an early age (growth rate). The *pedigree selection* can be done for traits of high heritability. The selection based on performance of other relatives viz. *progeny* and *collateral relatives* (sibs) is effective when heritability of trait is low, the character is sex limited and for the characters which are not expressed in the life of animal (carcass traits, lifetime production, life span, etc.).

Further the individual selection may for single trait or it may be multi trait selection for which tandem selection, ICL, and selection index method are used.

The amount of genetic progress ($\Delta G$) in traits controlled by additive gene action depends on the phenotypic superiority of selected parents (selection differential) and the extent to which this superiority is fixed and transmitted. It is only the amount of additive genetic variability ($\sigma^2_A$) which is fixed and transmitted. This amount of additive genetic variability is expressed as heritability of the trait ($h^2$). Therefore, heritability and superiority of selected parents directly affect the genetic progress. Thus, $\Delta G = h^2 S$.

## 15.2 SELECTION FOR NON-ADDITIVE GENE ACTION

The selection and breeding methods for traits affected by non additive gene action are crossbreeding, recurrent selection and reciprocal recurrent selection.

The **crossbreeding** is practiced to take advantage of heterosis and of good qualities of both the parental breeds in complementary combination.

The ***recurrent selection*** includes the crossing of a large number of individuals with a tester line (highly inbred) and to evaluate the progeny. The individuals with best performance are selected and inter mated. This process is repeated over and over. Consequently the line being selected becomes homozygous for genes different from tester line provided the heterosis is due to over dominance.

In the ***reciprocal recurrent selection,*** the randomly selected individuals from each of two non inbred strains are progeny tested in crosses with each other. The individuals with best performance in crosses from the two strains are inter-mated to continue their respective lines.

Both recurrent and reciprocal recurrent selection result in improved performance irrespective of gene action (dominance, over dominance, additive effects). However, these are effective only for those traits affected by non additive gene action.

## 15.3 CONCEPT OF GCA AND SCA

Observational components of genetic variance between crosses:

The genotypic value of a quantitative trait has its two components viz. additive genetic value also known as breeding value (B.V.) caused by additive effect of genes and non additive genetic value known as gene combination value (G.C.V.) caused by interaction effect of genes. The variation in these two components of genotypic values among the individuals of a population gives rise to the corresponding two components of total genetic variation (additive genetic variance and non-additive genetic variance) in a random mating population. This total genetic variance present in a random mating population is distributed if the mating is non-random (inbreeding and out breeding). The distribution of total genetic variation is between lines and within line genetic variance on inbreeding whereas it is distributed between crosses and within cross genetic variance on crossing the lines.

The total genetic variance between crosses ($\sigma^2_X$) is obtained considering that each line is used in one cross only. But when a line is crossed with several other lines, the total genetic variance between crosses can be partitioned into and estimated as two observational components expressed as combining ability viz. *general combining ability* (GCA) of the lines and *specific combining ability* (SCA) of the cross. The concept of GCA and SCA was introduced by Sprague and Tatum (1942).

***Diallel cross:*** The estimation of the observational components of variance in term of G.C.A. and S.C.A. requires the crossing of a number of inbred lines in all possible combinations. The crossing of a number of lines among themselves in all possible combinations is called as the *diallel cross*. Thus in diallel cross, each line is crossed with every other line. This is the commonly used experimental design for crossing inbred lines. The analysis of a diallel cross provides the information on the nature and amount of genetic parameters as well as on the GCA of parents and the SCA of the crosses.

***GCA:*** The crossing of a line with several lines gives an estimate of the mean performance of the line based on all crosses made by crossing this line with all other lines in all possible combinations. The mean performance of a line in all of its crosses with other lines is expressed as a deviation from mean of all crosses ($\mu$). This deviation of the line from $\mu$ is called as the *general combining ability (G.C.A.)* of the line. Thus, each line will have its G.C.A. Therefore, the GCA of a line is the average value of the $F_1$'s produced by crossing a line with several other lines and expressing it as a deviation from the over all mean of all the possible crosses ($\mu$).

***SCA:*** The cross of any two lines should have its expected mean value equal to the sum of the G.C.A. of two lines. However, it may not be true but the cross may have its expected mean value different from the sum of the G.C.A. of two lines. This difference in the mean value of a particular cross from the sum of G.C.A. of two lines is called the *specific combining ability* of the cross (S.C.A.). Thus, each cross will have its S.C.A. The SCA is thus the mean performance of a particular cross of two lines expressed as deviation of this cross from the GCA of the two lines.

The GCA's are the main effects due to genic values which cause differences in GCA. The SCA is due to the interaction effect (dominance, over dominance and epistasis) which causes differences in SCA. This is the genetic basis of SCA. The deviation of true mean of a cross ($X_{ij}$) between any two lines ($i^{th}$ and $j^{th}$) will contain the sum of their GCA's and SCA. Thus,

$$X_{ij} - \mu = GCA_i + GCA_j + SCA_{ij}$$

Where, $X_{ij}$ = Genotypic value of a cross

$\mu$ = overall mean of all crosses (involving all lines in all possible combinations, general mean of all crosses in a diallele cross)

$$GCA_i = X_i - \mu$$

$$GCA_j = X_j - \mu$$

$$SCA_{ij} = X_{ij} - (\mu + GCA_i + GCA_j)$$

The GCA's and SCA's are uncorrelated. The between cross total genetic variance will thus be as:

$$\sigma^2_X = \sigma^2_{GCA\,i} + \sigma^2_{GCA\,j} + \sigma^2_{SCA\,ij}$$
$$= 2\sigma^2_{GCA} + \sigma^2_{SCA}$$

The variance of GCA is the additive genetic variance ($V_A + V_{AA}$) in the base population which is due to differences in additive gene effects whereas the variance of SCA ($V_D + V_{AD} + V_{DD}$) is due to the non-additive gene effects

Further the variance of GCA increases linearly with F while the variance of SCA is expected to increase with higher powers of F and so the variance of SCA is more important as a cause of variation among crosses at higher degree of inbreeding (F). Therefore, it is advisable to use highly inbred lines for selection among crosses to be most effective. This means that when non-additive gene action has more influence on a trait, the most effective procedure of improvement is to produce highly inbred lines and then crossing them to find that which lines combine best.

The variance of GCA and the variance of SCA are the two components of the total genetic variance between crosses. These two components can be estimated directly from observations on phenotypic values and hence these are the observational components.

The measurement of GCA is important because the utilization of variance of GCA is applied as a method of improvement by crossing. By comparing the GCA of different lines, the best lines are selected which are most likely yield the best cross among all the crosses of the available lines. The correlation between performance of parental inbred lines and the crossbred progeny is generally high for traits with high GCA and large additive gene effects. Thus prediction of cross performance with reasonable accuracy for such traits can be made from performance of the parental lines.

## 15.4 GENETIC PRINCIPLE TO EXPLOIT NON ADDITIVE GENE ACTION (HETEROSIS)

The selection response based on B.V. gradually declines and finally ceases due to continuous selection for several generations in closed population. This is explained due to exhaustion of additive genetic variance. However, the genetic variation of non-additive nature still exists in traits like viability and fertility. It is therefore important to exploit the non-additive genetic variance. An individual has a special breeding value due to the non- additive gene effects. Some procedures of producing of inbred lines and their crossing combined with selection have been suggested to exploit non-additive gene action (heterosis) for improvement. Such procedures have been applied in improvement of maize and have also been used in pigs and poultry. Therefore, a breeder is interested to produce crossbreds or $F_1$ individuals which are superior to the inbred lines and the base population. The production of inbred lines and their crossing without selection or selection alone can not bring desired improvement. The crossing of inbred lines originating from base population produce no genotype that could not occur in the base population and therefore the mean of all crosses will not differ from the mean of base population.

However, the line crosses highlight some valuable gene combinations which can be multiplied. The selection of lines that cross well is thus practiced. The combination of both methods (developing inbred lines, their evaluation and selection) is advantageous over straight forward selection applied to base population. Through inbreeding and crossing the desirable and best genotypes present only in certain individuals of base population can be replicated to the required number by repeating the cross. This is the merit of the method and principle to take advantage of the existing genetic variability in base population (*genetic principle to exploit heterosis*). However, there is a practical difficulty to find those lines which combine best and, therefore, it is necessary to produce many lines to cross them with each other in order to find which one give rise to superior genetic combinations.

The replication of the genotype in a cross has an additional advantage to allow the genotypic value to be measured with little error. The measurement of genotypic value by its phenotypic value in the base population in not much accurate. It is the breeding value which can be reproduced by selection of individuals in a base population (non-inbred), whereas in the cross the genotypic value can be measured and reproduced till inbred lines exist. Therefore, selection alone, without inbreeding, to the base population can take advantage of the additive genetic variance to improve the character, while inbreeding and crossing combined with selection provide a better means of improvement when non additive variance in a character exist in substantial amount. Therefore, the selection is made for combining ability of the lines.

Highly inbred lines of farm animals are not available due to reduction in fertility and survivability among inbred. Therefore, crosses are made between different breeds or mildly inbred lines of farm animals.

## 15.5 SELECTION METHODS FOR COMBINING ABILITY

### *(Additive and non-additive gene action)*

The causal component of additive genetic variance is measured in terms of GCA as observational component whereas the causal component of the non additive genetic variance is measured in terms of SCA as the observational component. The relative magnitude of improvement due to GCA and SCA is proportional to their variances which are respectively equal to the additive genetic variance and non-additive genetic variance in the base population. The selection methods for GCA and SCA are different.

### 15.5.1 Selection for GCA

#### Breeding plans for additive genetic differences:

The additive gene action (AGA) is responsible for differences of GCA. The AGA is indicated when the trait has high $h^2$ and show little or no heterosis or inbreeding depression. This means that average values for the parental, $F_1$ and $F_2$ generation are all the same with some individuals being superior over others. These best individual are best because they have more contributing or plus genes that act additively and not because of heterozygosity. It is therefore, essential to find the genetically superior individuals and select them for breeding. Thus, selection should be based on the individual's breeding value, because the correlation between the performance records of parental inbred lines and crossbred progeny is generally high for traits with large additive gene effects. To make selection effective for AGA, the environment variance should be minimized by keeping the environment standard as much as possible equal for all individuals. The superior individuals (progeny of superior parents) then must be mated with superior. Thus, the selection for the improvement of GCA is through individual selection with pedigrees, progeny tests or family selection without inbreeding as a supplement to make selection more accurate. Mass selection alone is more reliable when $h^2$ is higher. It is usually more important than any other basis of selection if $h^2$ is much higher than 0.25 and if the trait can be measured early enough on the individuals which are to be selected or rejected. But when $h^2$ is low or when the trait is sex limited or when the trait can not be measured early, and when the trait can not be measured on living animal, the other basis of selection are more important. The selection without inbreeding is effective because the variation in GCA is attributable to additive variance in the base population from which the lines or families are derived. The GCA of all the available lines can be measured and some selection can be applied so that a portion of lines can be discarded on the basis of their own performance before crosses are made. This is possible because there is some degree of correlation between a line's performance as an inbred and its GCA. The selected individuals are mated to individuals of their own population to produce the next generation. The selection for GCA is repeated and thus it is knows as recurrent selection for GCA (Jenkins, 1940).

### 15.5.2 Selection for SCA

#### Breeding plans to exploit heterosis (NAGA):

The selection for SCA means to take advantage of hybrid vigour when NAGA is important. Selection on the basis of individuality is not effective when traits are largely affected by NAGA but the improvement in such traits depends upon heterozygosity through cross breeding resulting in expression of heterosis. Thus, the ultimate goal is to produce breeds or varieties each of which is homozygous which will produce the desired phenotypic merit when crossed with each other. This is important when epistatic effects and overdominace is important. The seed stock lines which are homozygous (produced by inbreeding) are selected and crossed with each other to get some special combination of genes which produce the desired phenotypic merit. The non-additive genetic variation is very important for viability and fertility in all species of farm animals.

**1. *When the SCA is caused by dominance*:** The dominance gene action can be exploited by two ways. One way is by producing homozygous dominant individuals for the pairs of genes which affect the trait, and secondly by producing heterozygous individuals. There are two difficulties to use the first method viz. to distinguish homozygous dominants from heterozygotes, and that the quantitative traits are polygenic which reduces the probability of getting all dominant genes into one homozygous strain. The second method uses the production of heterozygous individuals which can be achieved by producing two complementary homozygous lines. Though the production of two complementary homozygous lines is problematic but can be achieved by directed process of selection during inbreeding. This will create maximum differences of gene frequency in two complementary lines. These lines are then crossed to obtain hybrids. However, for polygenic traits it is not possible to determine the genotype from phenotype i.e. which individuals are complementary homozygous. To identify the complementary homozygous individuals (i.e. those which are homozygous in opposite ways) is by the method of finding the lines or breeds that combine and nick best when crossed. This method is used for production of hybrid seed corn as:

* The homozygosity of all pairs of genes is increased by inbreeding and forming several inbred lines.

* These lines are crossed and tested for their combining ability that which produce the best line cross progeny.

* The lines producing most superior progeny will be homozygous in opposite ways for many pairs of genes and will give greater heterozygosity in the progeny.

* The progeny are sold and not kept for further breeding, and these lines are kept pure to cross again and again to produce the line cross progeny.

**2. When *SCA is caused by over dominance and epistasis*:** The most practical procedure is the formation of different families, or lines and crossing them to find those which result in greatest hybrid vigour. Two breeding systems have been proposed when heterosis is largely dependent on over-dominance. The first is the recurrent selection proposed by Hull (1945) and the second is reciprocal recurrent selection proposed by Comstock et al. (1949). Both these methods involve progeny testing to take advantage of over dominance.

***(i) Recurrent selection:*** This procedure was proposed by Jenkins (1940) for improving GCA in plant breeding and was modified by Hull (1945) to utilize SCA as well as GCA. This is thus called as Hull's modification of recurrent selection. This is the recurrent selection to inbred tester lines because it produces stock which combines best with inbred tester line or recurrent selection for SCA. This method has been used in poultry breeding. In this method a highly inbred line is selected as a tester (tester line) known to have good GCA. A large number of individuals (females) of a line under testing are crossed with this tester line known to have good GCA and the resulting progenies are evaluated. The female individuals giving best results are selected based on performance of their test cross progeny and inter-mated. These selected females are then mated with males of their own line to produce next generation of parents to be tested with inbred tester line. The crossbred progeny are not used for breeding. The selected individuals from source population are mated again and again among themselves to produce next generation of complementary stock and also with the tester inbred line. The cycle is repeated to several generations. The complementary stock will be homozygous recessive and hence their test cross progeny will show greater heterosis. By this method the GCA of only one line and the SCA of cross is improved while the GCA of inbred tester is already determined. This method has also a practical difficulty to produce and maintain the highly tester line. Therefore, usually less inbred tester line is used which goes on changing genetically in subsequent generations and thus optimum gain is not achieved. Therefore, another method of selection is required.

***(ii) Reciprocal recurrent selection:*** It is important to exploit the heterosis that the lines to be crossed should differ widely in gene frequencies at all loci affecting the trait and showing the dominance. The differences can be produced by selection or by inbreeding. The directed selection is more effective and leads simultaneous selection for both GCA and SCA. The procedure of selection for combining ability to make maximum use of both GCA and SCA is known as reciprocal recurrent selection. This procedure was proposed by Comstock et al (1949). The use of RRS in animals was first discussed by Heisdorf (1950) and Bell et al (1952). The RRS has become the modern breeding system in poultry to improve performance of the crossbreds.

This procedure implies progeny testing of each of the two lines (A and B) by crossing with each other. The crosses are made reciprocally (AB and BA) and the parents of pure lines are evaluated on crossbred progeny performance (AB and/or BA hybrids). The best parents of both lines are selected whereas the rest parents as well as all the crossbred progeny are culled. The crossbred progeny are used only to test the combining ability of parents. The selected parent are re-mated to members of their own line to produce the next generation of individual which are to be tested by mating them with individuals of the second line produced in the same way. They are evaluated based on their cross`bred progeny and the best parents of both lines are selected to produce next generation of parents to be tested. Thus, RRS is a kind of progeny testing in evaluating the parents based on the performance of their progeny and producing the next generation based on selective breeding. This cycle is repeated over and over.

***RS vs RRS:*** The start is made with the two breeds, lines or varieties differing in gene frequencies. Secondly, deliberate inbreeding is avoided as far as possible. his can be done by using all of the females as parents in their own lines and to intensify the selection among males. The disadvantage of this method is that the performance of the lines for the character under selection is reduced. This is because this method is used for the characters which respond to inbreeding depression and heterosis under directional dominance. Thus, the change of gene frequency by selection is towards the extremes and so the mean value of the lines declines. That is to say the selection is made to make both the lines homozygous for individual genes for loci which exhibit over dominance. However, with reciprocal recurrent selection the decline in performance is not so deleterious as that of the effect of deliberate inbreeding because the inbreeding affects all loci and so mean value of all characters showing directional dominance declines whereas in reciprocal selection only the character under selection declines.

Comparing the recurrent selection and reciprocal recurrent selection, the recurrent selection is superior at the beginning. This is because of more differences in gene frequencies of inbred tester and reference population to be tested compared to that between the two segregating population used in reciprocal selection. But the recurrent selection is inferior if the loci with additive or semi-dominant gene action prevail. This is because it is difficult to differentiate between homozygous dominant and heterozygous in test cross progeny if one allele is dominant over others. Thus the desirable dominant gene can not be accumulated in the segregating population.

Both these selection systems viz. RS and RRS are inferior to pure line selection or *within line selection* (***WLS)*** for the traits governed by additive gene effects with high heritability.

***RRS vs WLS for crossbred response:*** The present day is the era of crossbreds for the reason that they show heterosis due to the existence of non-additive gene effects. The RRS improves the performance of crossbred. However, the magnitude and direction of genetic correlation between purebred (P) and crossbred (C) half sib's performance ($r_{g\,(PC)}$) decides the merit of RRS and WLS regarding the relative emphasis to be givn to the purebred and crossbred information in selecting for crossbred performance. A moderate to high positive $r_{g\,(PC)}$ favours WLS but RRS is said to be more efficient over WLS when $r_{g\,(PC)}$ is low and negative (Bell, 1982). This is for two reasons viz. existence of non-additive gene effects and secondly the two populations under RRS do not have identical gene frequencies which cause the covariance to be similar or negative.

The effectiveness of RRS and WLS depends on the relative importance of additive and non-additive genetic variance. The RRS is more efficient than WLS to improve the heterotic traits. The response to selection under RRS is highest than WLS in the presence of over dominant genes.

At later stage of using reciprocal method of breeding, the performance of purebreds shows negative correlation with the crosses. This is because the continuous selection for combining ability improves the cross but the parent lines deteriorate due to increased homozygosity at heterotic loci. It is only the heterotic loci which show further segregation after most of the loci showing additive gene effects become fixed.

The RRS is used for egg production in poultry and also encouraging results were obtained with maize.

The comparison of the two methods of selection has been shown diagrammatically here:

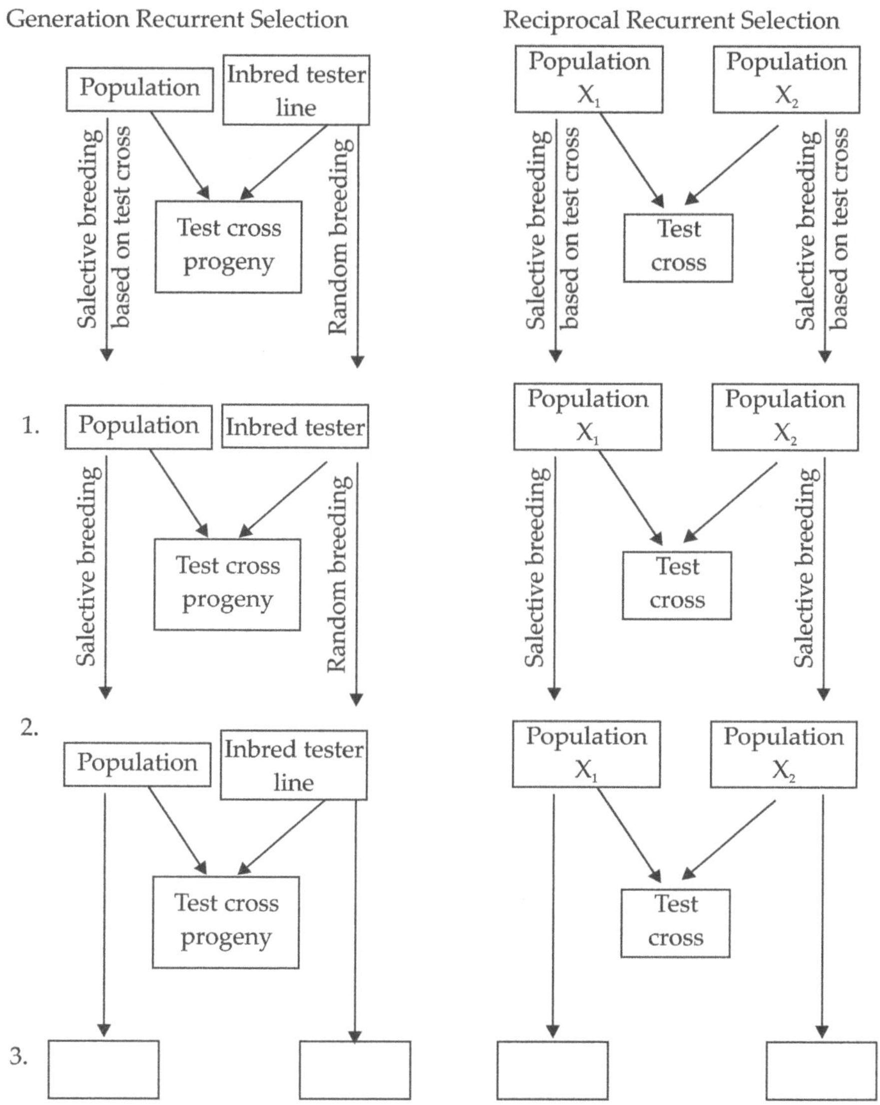


## 15.6 EFFECTIVENESS OF RRS

The merit of RRS and WLS (Within Line Selection) regarding the relative emphasis to be givn either to the information on purebred or on crossbred in selecting for crossbred performance depends on the magnitude and direction of genetic correlation between purebred (P) and crossbred (C) half sib's performance ($r_{g\,(PC)}$). Thus it is essential to know the procedure to estimate $r_{PC}$. The effectiveness of this breeding system (RRS) in poultry can be evaluated by magnitude and sign of the genetic correlation between purebred and crossbred half sib's performance ($r_{g\ (PC)}$) and by the expected genetic response in crossbreds from RRS.

The breeding structure is assumed such that S sires of each line (A and B) are individually mated to D dams from his own line and D dams of other companion line to reproduce n progeny of both purebred (AA and BB) and crossbred (AB and BA), respectively. For example, both purebreds (AA and BB) and two reciprocal crosses (AB and BA) are reproduced by mating each sire with D females of its own strain and D females of other strains. The purebred progenies are saved for selection but the crossbreds are culled after performance recording. The purebred parental half sib families with highest performance of both the lines are selected based on their crossbred half sib's performance.

### 15.6.1 Estimation of $r_{g\ (PC)}$

The genetic correlation between purebred and crossbred half sib's performance ($r_{g\ (PC)}$) can be estimated under two conditions viz. when each dam has one progeny and when the dams have different number of progenies.

***(i)* One progeny per dam:** The half sib families exists within which the observation on $j^{th}$ half sib of the $i^{th}$ sire family can be represented as –

$$Y_{ij} = \mu + S_i + e_{ij}$$

The HS analysis of variance is conducted to estimate the genetic and phenotypic parameters of performance record (egg production) separately for the 4 genetic groups (AA, BB, AB, and BA) using the following models –

$Y_{ij\ (P)} = \mu_{(P)} + S_{i\ (P)} + e_{ij\ (P)}$ for purebred (P) progeny (AA and BB) and

$Y_{ij\ (C)} = \mu_{(C)} + S_{i\ (C)} + e_{ij\ (C)}$ for crossbred (C) progeny (AB and BA)

The covariance [$\boldsymbol{\sigma}_{S\ (PC)}$] between $S_{i\ (P)}$ and $S_{i\ (C)}$ is estimated as:

$$\boldsymbol{\sigma}_{S\ (PC)} = \frac{1}{(S-1)} \Sigma[y_{i.\ (P)}\, y_{i.\ (C)}]$$

Where, $y_{i.\ (P)}$ = Sire constant for strain A sires producing AA purebred progenies and also sire constant for strain B sires producing BB purebred progenies

$y_{i.\ (C)}$ = Sire constant for strain B sires producing AB crossbred progenies and also sire constant for strain B sires producing BA crossbred progenies

The least means and sire constants of all the sire families of all the four different genetic groups (AA and AB; BB and BA) are estimated.

The $r_{g\ (PC)}$ is estimated from sire component of variance and covariance for strain A sires and for strain B sires, separately as:

**A strain sires:**

$$r_{g\ P(AA)\ C(AB)} = \frac{\boldsymbol{\sigma}_{S(AA)}\, \boldsymbol{\sigma}_{S(AB)}}{\sqrt{\boldsymbol{\sigma}^2_{S\ (AA)}\, \boldsymbol{\sigma}^2_{S\ (AB)}}}$$

Where, $\sigma_{S(AA)}\,\sigma_{S(AB)}$ = Covariance for A strain

$$= \frac{1}{(S-1)}[\,\Sigma\,(S_{i\,(AA)} \times S_{i\,(AB)})]$$

**B strain sires:**

$$r_{g\,P(BB)\,C(BA)} = \frac{\sigma_{S(BB)}\,\sigma_{S(BA)}}{\sqrt{\sigma^2_{S\,(BB)}\,\sigma^2_{S\,(BA)}}}$$

Where, $\sigma_{S(BB)}\,\sigma_{S(BA)}$ = Covariance for B strain

$$= \frac{1}{(S-1)}[\,\Sigma\,(Si_{(BB)} \times Si_{(BA)})]$$

$S_{i\,(AA)}$ and $S_{i\,(AB)}$ = Sire constant for $i^{th}$ sire of A strain used to produce AA and AB progenies

$S_{i\,(BB)}$ and $S_{i\,(BA)}$ = Sire constant for $i^{th}$ sire of B strain used to produce BB and BA progenies.

The covariance between paternal purebred and crossbred ($Cov_{PC}$) half sibs can also be estimated by using the sire family means as –

$$Cov_{PC} = \frac{1}{(S-1)}\left[\frac{\Sigma PC - (\Sigma P\;\Sigma C)}{S}\right]$$

Where, P and C indicate the purebred and crossbred, and

S = number of sires

***Pooled estimate of*** $\mathbf{r}_{g\,(PC)}$: Further, the ANOVA is conducted to obtain the pooled estimate over the two purebreds and two reciprocal crossbreds, by using the following model –

$$Y_{ijk} = \mu + g_i + S_{ij} + e_{ijk}$$

Where, $Y_{ijk}$ is the observation on $k^{th}$ progeny of $j^{th}$ sire within $i^{th}$ genetic group (purebreds or crossbreds).

The ANOVA is conducted to partition the variance into between groups, between sires within groups and residual, separately for purebreds (2 groups) and for crossbreds (2 groups).

The heritability and genetic correlation (rg (PC)) are worked out based on the variance components and least squares constants for sires of purebreds (P) and crossbreds (C) as:

$$r_{g\,(PC)} = \frac{Cov_{PC\,(Sires)}}{\sqrt{\sigma^2_{S\,(P)}\,\sigma^2_{S\,(C)}}}$$

*(ii) Dams having different number of progenies* (Hierarchal system of mating): The genetic correlation between the performance of paternal purebred and crossbred HS (rg (PC)) produced under hierarchal mating system when the dams have different number of female progeny for performance recording (egg production) can be estimatd by the procedure givn by Biswas and Craig (1969). Sib analysis is conducted to estimate different components of variance of the performance of

purebred progeny and cossbred progeny. The sire wise means of purebred and crossbred progenies are computed to estimate the variance among sire progeny means and covariance between purebred and crossbred progeny means. The rg (PC) is estimated as –

$$r_{g\,(PC)} = \sigma_{S\,(PC)}\left[\left(\sigma^2_{S\,(P)} - \frac{\sigma^2_{D(P)}}{H_{DP}} - \frac{\sigma^2_{W(P)}}{H_{WP}}\right)\left(\sigma^2_{S\,(C)} - \frac{\sigma^2_{D(C)}}{H_{DC}} - \frac{\sigma^2_{W(C)}}{H_{WC}}\right)\right]^{-1/2}$$

Where, $Cov_{PC} = \sigma_{S\,(PC)}$

$$= \frac{1}{(S-1)}\left[\Sigma X^2 - \frac{(\Sigma X)^2}{S}\right]$$

$\sigma^2_{S\,(P)}$ = Variance among sire's purebred progeny means or

$\sigma^2_{S\,(C)}$ = Variance among sire's crossbred progeny means

$$\sigma^2_{S\,(P)} \text{ and } \sigma^2_{S\,(C)} = \frac{1}{(S-1)}\frac{\left[\Sigma X^2 - (\Sigma X)^2\right]}{S}$$

Where, X = urebred or crossbred mean

$\sigma^2_{D(P)}$ or $\sigma^2_{D(C)}$ = Dam component of variance for purebred or crossbred (value estimated from ANOVA separately for purebred and crossbred)

$\sigma^2_{W(P)}$ or $\sigma^2_{W(P)}$ = Within component of variance for purebred or crossbred (value estimated from ANOVA separately for purebred and crossbred)

$H_D$ = Harmonic mean of number of dams mated to a sire for purebred or crossbred.

(Generally equal numbers of dams are mated to each sire to produce purebred and crossbred populations and hence no need to calculate the harmonic man).

$H_W$ = Harmonic mean of number of progeny per sire for purebred or crossbred

$$= \frac{1}{H_W}$$

$$= \frac{1}{\frac{1}{9}\left(\Sigma \frac{1}{n_i}\right)}$$

### 15.6.2 Expected genetic response in crossbreds from RRS

The ***breeding structure*** of the population for RRS in estimating $r_{PC}$ has already been given above.

***Expected response***: The average change in the crossbred B.V. of the purelines resulting from RRS indicates the genetic gain in cross performance by one generation of selection. This genetic gain can be estimated from the following formula given by (Calhoon and Boren, 1974):

$$\Delta G_T = ½ [\Delta G_{AA} + \Delta G_{BB}]$$

Where,

$$\Delta G_{AA} = i_{AB}\ \sigma P_{S\,(AB)}\ \beta\ G_T\, P_{S\,(AB)}$$

$$\Delta G_{BB} = i_{BA}\ \sigma P_{S\,(BA)}\ \beta\ G_T\, P_{S\,(BA)}$$

The regression of the B.V. in the cross (T) of a single individual in the pureline on the sire mean of the crossbred progeny (S) is -

$$\beta\ G_T\, P_{S\,(AB)} = \frac{h^2_C}{4}\sqrt{\left[\frac{n}{\{1+(n-1)rh^2_C\}}\right]}$$

Where, the subscript c indicates the heritability based on cross performance.

The correlation between the two variables ($r_{GT\ PS}$) is computed as

$$r_{G(T)\,P(S)} = \beta\ G_T\, P_S \frac{(\sigma P_S}{\sigma G_S)}$$

$$= \frac{h^2_C}{4}\sqrt{\left[\frac{nd}{\{1+(nd-1)rh^2_C\}}\right]}$$

This is the correlation between the family average phenotype and the genotype of an individual not included in family average.

***Expected correlated response to WLS in crossbred:*** This can be estimated as given by McNew and Bell (1976):

$$CR_{GTi} = [\frac{Cov}{2\,\sigma^2_{PC}}\, i_C\ \sigma P_{S\,(C)}]\ i$$

Where,

$Cov$ = Covariance of additive effects of alleles in purebreds and crossbreds

= Twice the cov of sire's purebred and crossbred half sib family means.

$\sigma^2_{PC}$ = phenotypic variance of crossbred half sib family mans

$i_C$ = selection intensity based on crossbred performance

$\sigma P_{S\,(C)}$ = phenotypic standard deviation of the trait (T) in crossbreds

$i$ = $i^{th}$ sire line.

## 15.7 COMPARISON OF RRS AND WLS FOR CROSSBRED RESPONSE

It has been explained/interpreted above that the effectiveness of RRS and WLS depends on the following genetic parameters -

(a) $r_{g\,(PC)}$ which is the genetic correlation as discussed above and

(b) Relative amount of additive and non-additive genetic variance

The comparison of direct response in crossbreds due to RRS and correlated response in crossbreds due to WLS is done by two methods:

1. Using the genetic and phenotypic parameters of both the purebreds and both the reciprocal crossbred genetic groups assuming equal selection intensity in purebred and crossbred lines as suggested by McNew and Bell (1976):

$$\text{RRS/WLS} = \frac{[(\sigma^2_{AC})i + (\sigma^2_{AC})j\ (\sigma_{P\,HS\,(AA)} + \sigma_{P\,HS\,(BB)})\ nd]}{(\text{Cov } i + \text{Cov } j)\ (\sigma_{P\,HS\,(AB)} + \sigma_{P\,HS\,(BA)})\ (nd + n + 2)}$$

Where, RRS/WLS = Ratio of direct response in crossbred from RRS to direct response from WLS

$(\sigma^2_{AC})i$ = additive genetic variance in crossbred due to line i

= Twice the sire component of variance

Cov i = additive genetic covariance between paternal purebred and crossbred HS of $i^{th}$ line

= Twice of covariance of sire's purebred and crossbred HS family means

$\sigma_{P\,HS\,(P\text{ or }C)}$ = phenotypic $\sigma$ of HS family means of purebred or crossbred, respectively

n = numbers of offspring in purebred FS family

d = number of purebred dams mated to a sire

i = purebred population 1 or 2

The covariance between purebred and crossbred is important with respect to the type of gene action. For example, when there is additive gene action, 2 Cov = $\sigma^2_{AP}$ and the WLS for purebred response exceeds to crossbred response to RRS. But in case of non-additive gene effects and the difference in gene frequencies between two populations (1 and 2), Cov becomes smaller. It is clear that the Cov should not be negative, as for overdominance for RRS to excel over WLS for crossbred response. But when Cov is negative, RRS certainly excels over WLS (Bell, 1982).

**(1)** *Using the pooled estimates of the purebreds and the reciprocal crossbreds, the comparison of correlatd response in crossbred from WLS to direct response from RRS per year is estimated as suggested by Sellier (1982):*

The correlated response in crossbred from WLS to direct response from RRS per year (WLS/RRS) as:

$$\text{WLS/RRS} = r_{g\,(PC)} \left(\frac{i_P}{i_C}\right)\left(\frac{t_C}{t_P}\right)\frac{\alpha h_P}{\beta h_C}$$

Where, $i_P$ and $i_C$ = selection intensities in WLS and RRS, respectively

$t_C$ and $t_P$ = generation intervals in WLS and RRS, respectively

$h_P$ and $h_C$ = square root of heritability in purebred (WLS) and crossbred RRS), respectively

$\alpha$ = coefficients when WLS is applied

= 1 when mass selection is applied

$\beta$ = coefficient when RSS is applied based on PT of males with its ***n*** crossbred offsprings

$$= \sqrt{\frac{n}{\left[4+(n-1)h^2{}_C\right]}}$$

The $r_{g\ (PC)}$ greatly affects the WLS/RRS. When $r_{g\ (PC)}$ is nealy 1, WLS/RRS is larger than 1 and WLS is usually better than RRS. When $r_{g\ (PC)}$ is low or negative, RRS is more efficient.

### *Modification in RRS*

Several modifications proposed in RRS have their own advantages, modified RRS being more efficient than RRS. The RRS is a family selection based on crossbred performance alone. In modified RRS the selection of purebred is based on their own performance as well as on the performance of their crossbred sibs. The selection index is constructed by giving proper weightage to the purebred and crossbred's performance. The information in the selection index may be incorporated on the individual, dam, purebred full and half sibs, and crossbred paternal and maternal half sibs. Thus, the selection of sires or dams can be done on the basis of the general index incorporating the available information as –

$$I = b_1\, I + b_2\, I_{FS} + b_3\, I_{HS} + b_4\, I_{CHS}$$

Where, $b_i$ = weights given to a particular source of information,

**I**, $I_{FS}$, $I_{HS}$ = records on individual (I), purebred FS and HS, respectively,

$I_{CHS}$ = record on the individual's paternal crossbred HS.

## 15.8 SELECTION AND BREEDING FOR THRESHOLD CHARACTERS

### - *Disease resistance*

Animal breeders have given main emphasis to select animals for higher milk yield, meat, wool and egg etc. On the other hand, the genetic differences in resistance to various pathogens to develop highly productive and disease resistance strains have not been utilized. This field has been totally neglected, except very few attempts, for the inability to recognize its potentialities and lack of research rather than lack of resistance.

Hutt (1958), working on respiratory disease of White Leghorn, had mentioned that some birds need protection but others do not, while some hens die of these diseases or stop laying, others in the same pen of the same flock are completely unaffected and continue to lay well. It is not that the unaffected birds are not exposed but it is much more probable that they are able to tolerate the infectious organisms, to resist them in some way and thus to withstand a bit of environment that may be disastrous to other birds. Two strains of a breed housed together can differ significantly in resistance to a disease and, moreover, remarkable differences may be observed among sire families within a strain. There are many examples of genetic variation to many diseases and parasites in different species of farm animals, viz., mastitis, TB, FMD, John's disease, protozoan disease in cattle; respiratory disease in poultry; internal parasites in sheep. Thus, it is obvious that heredity play an important role in determining whether an individual will withstand a disease or succumb to it. Now the important point to be valued is about the identification/selection and multiplication of resistant individuals so as to produce a resistant strain.

Tomar (1984) also found genetic variability in the occurrence of certain diseases in Murrah buffaloes.

There are three possible ***methods to identify and select the individuals*** for differences in resistance/susceptibility to a certain disease/parasite. These are as under:

- Occurrence of the disease under normal environmental conditions,
- Inoculation of animals with pathogen causing the disease, and
- Use of indicators of resistance.

The first method to identify the resistant animals on the basis of the incidence of the disease is not effective in the sense that under the changed environmental conditions the resistant become susceptible. The second method of inoculation is not likely to be applied because of the objections to deliberate infection with pathogenic organisms. The third one requires very great efforts and research to find a better indicator of genetic resistance; however, which do not require deliberate exposure to pathogens. Some of the good indicators are available in literature as given below:

(*i*) Adams and Rickard (1963) tested the bactericidal potencies of the lipid, sebum-like material from the teat canal of the udder that differed in susceptibility to *Str. Agalactiae* along with its composition for fatty acids. They found that materials from udder quarters resistant to infection contained higher proportion of some fatty acids (myristic, palmitic and linoleic acids) than the material from susceptible quarters which contained significantly more palmitic acid compared to resistant quarters. Stovikova and Sojda (1975) found that of the 60 fatty acids constituting the neutral lipid fraction of the lactosebum, the levels oflauric and myristic acids were highest in cows resistant to mastitis and the variation in lipid composition of the lactosebum has genetic component. These evidences show that resistance is associated with a high level in the lactosebum of fatty acids that excel in bactericidal potency.

(*ii*) An indicator of resistance to *eye cancer* in Hereford cattle has been identified.

The animals with pigmented eyelids have far less eye cancer than those with white faces and lids not pigmented. The incidence of lesions decreases steadily with the increase in pigmentation and a solid ring of pigment (at least half an inch wide) around the eye ensure pigmentation of the lids. It has been shown that the amount of coloured skin around the eye is influenced by heredity and hence is amenable to selection.

(*iii*) The differences among the chicks that are resistant and susceptible to *Salmonella pullorum* causing the pullorum disease were found that the resistant chicks have better control to maintain their body temperature. This was measured by determining the number of days the newly hatched chicks with their body temperature of 102°F (when they hatch) raise their temperature up to 105-1 07°F which is normal at ten days of age. The differences among them in this ability appeared by 6 days after hatching, the chicks most efficient in making transition are more resistant to pullorum. This ability is believed to accelerate the production of antibodies and other defensive mechanisms.

(*iv*) Briles et al. (1977) found that birds resistant for Marek's disease have RBC antigen $B^{21}$ whereas the susceptible one lacked this antigen but carry another antigen $B^{19}$.

(*v*) Crittenden et at. (1970) reported embryos lacking the antigen R] are resistant to infection by certain viruses causing avian leucosis and those carry the dominant allele for R] are all susceptible.

Falconer (1989) has discussed the efficiency of individual selection and family selection on the basis of the incidence of the disease with two visible classes. The effectiveness of selection depends on the selection differential which, in turn, depends mainly on the proportion selected. But in case of threshold characters, the selection differential depends on the incidence rather than on the proportion selected. Selecting all the resistant individuals or only a portion of them will give their mean as the mean of the resistant group. Thus, selecting a smaller proportion than the incidence, does not give an added advantage. It is, thus, evident that maximum selection differential is achieved when the proportion selected is equal to the proportion of resistant individuals. The individual selection is less effective when the difference between the two (proportion of resistant and proportion selected) is more. Further, the individual selection is also less effective when the proportion of resistant is very low but the family selection is more effective. An individual's phenotype as resistant or susceptible is not precisely known whereas the family is more precisely known for the proportion of affected members of the family. However, the precision depends on the family size and on the incidence of a disease in the family.

## NUMERICAL EXAMPLES: Reciprocal Recurrent Selection (RRS)

***Example 15.1*** The data on egg production of the progeny of 9 sires of two strains (A and B) were recorded on a poultry farm. Each of 9 sires of each strain was mated with females of their own strain to produce purebred progenies (AA and BB) and also with another strain to produce crossbred progenies (AB and BA). The females had different number of female progeny to record egg production. Analysis of variance for egg production was conducted separately on 4 genetic groups as well as on 2 groups (purebreds and crossbreds) after adjusting the data for environmental effects. The information obtained after analysis of data are given below –

Table 1: ANOVA for egg production and the phenotypic and genetic parameters

| Source of Variation | D F | Mean squares for genetic groups | | | |
|---|---|---|---|---|---|
| | | AA | BB | AB | BA |
| Between sires | 8 | 41.23 | 44.15 | 80.18 | 59.05 |
| Within sires | | 29.86 | 29.00 | 51.96 | 34.39 |
| | | (61) | (64) | (59) | (57)* |

* Figures in parenthesis are error degree of freedom.

| | | | | |
|---|---|---|---|---|
| n = HS family size: | 6.77 | 7.11 | 6.55 | 6.33 |
| $\sigma^2_S$ = Sire variance | 1.68 | 2.13 | 4.31 | 3.89 |
| $\sigma^2_A = 4\,\sigma^2_S$ | 6.72 | 8.52 | 17.24 | 15.56 |
| $\sigma^2_P\ (\sigma^2_S + \sigma^2_e)$ | 31.54 | 31.13 | 56.27 | 38.28 |
| $h^2$ | 0.213 | 0.273 | 0.306 | 0.406 |
| Overall mean ($\mu$) and Sire Constants: | | | | |
| $\mu$ | 68.76 | 71.00 | 72.25 | 76.34 |
| $S_1$ | 1.78 | – 0.09 | 3.50 | – 2.34 |
| $S_2$ | 2.34 | 2.37 | – 1.10 | 2.50 |
| $S_3$ | – 1.51 | 0.69 | – 1.31 | 1.95 |
| $S_4$ | – 1.12 | – 2.31 | – 0.06 | 0.34 |
| $S_5$ | – 1.00 | – 1.73 | 0.39 | – 2.90 |
| $S_6$ | – 0.58 | 2.63 | 2.08 | – 2.82 |
| $S_7$ | 0.54 | 1.72 | – 2.82 | 3.32 |
| $S_8$ | – 2.36 | – 2.97 | – 1.52 | 0.19 |
| $S_9$ | 1.91 | – 0.31 | 0.84 | – 0.24 |

Estimate the $r_{g\,(PC)}$ for strain A sires, strain B sires and pooled estimate of $r_{g\,(PC)}$.

**Solution:**

***(i)*** $r_{g(PC)}$ ***for strain A sires:***

$$\sigma_{S(P)}\,\sigma_{S(C)} = \frac{1}{(S-1)}\Sigma[y_{i.(P)}\,y_{i.(C)}]$$

$$= \frac{1}{8}[(1.78 \times 3.5) + (2.34 \times -1.10) + \ldots .. \ldots + (-2.36 \times -1.52) + (1.91 \times 0.84)]$$

$$= \frac{1}{8}(10.8193) = 1.324$$

$$\sqrt{\sigma^2_{S(P)}\sigma^2_{S(C)}} = \sqrt{1.68 \times 4.31}$$

$$= 2.6908$$

$$r_{g(PC)} = \frac{\sigma_{S(P)}\,\sigma_{S(C)}}{\sqrt{\sigma^2_{SP}\sigma^2_{SC}}}$$

$$= 1.324/2.6908 = \mathbf{0.49}$$

***(ii)*** $r_{gPC}$ ***for strain B sires:***

$$\sigma_{S(P)}\,\sigma_{S(C)} = \frac{1}{(S-1)}\Sigma[y_{i.(P)}\,y_{i.(C)}]$$

$$= \frac{1}{8}[(-0.09 \times -2.34) + 2.37 \times 2.50 + \ldots .. \ldots + (-2.97 \times 0.19) + (-0.31 \times -0.24)]$$

$$= \frac{1}{8}(12.216) = 1.527$$

$$\sqrt{\sigma^2_{SP}\sigma^2_{SC}} = \sqrt{2.13 \times 3.89}$$

$$= 2.8784$$

$$r_{g(PC)} = \frac{\sigma_{SP}\,\sigma_{SC}}{\sqrt{\sigma^2_{SP}\sigma^2_{SC}}}$$

$$= 1.527/2.8784 = \mathbf{0.53}$$

***(iii) Pooled Estimate of*** $r_{g(PC)}$: This pooled estimate of $r_{g(PC)}$ and heritability for purebred and crossbred are computed based on the variance components and least square constants for sires of purebred (P) and crossbred (C) after conducting the ANOVA. The sire constants are used to estimate the covariance ($Cov_{PC}$) as explained above:

***Table 2:*** ANOVA for egg production of purebreds and crossbreds

| Sources of variation | D F | Mean squares | Variance components |
|---|---|---|---|
| ***Purebreds*** | | | |
| Between groups | 1 | 24.21 | - |
| Between sires within group | 16 | 40.45 | 2.045 ($\sigma^2_S$) |
| Residual (Error) | 107 | 26.23 | 26.23 |

*Contd...*

*Contd...*

| | | | |
|---|---|---|---|
| ***Crossbreds*** | | | |
| Between groups | 1 | 117.32 | - |
| Between sires within group | 16 | 68.05 | 4.15 ($\sigma^2_S$) |
| Residual (Error) | 98 | 41.27 | 41.27 |

| | **Purebreds** | **Crossbreds** |
|---|---|---|
| K value = | 6.94 | 6.44 |
| $\sigma^2_A = 4\,\sigma^2_S$ | 8.18 | 16.60 |
| $\sigma^2_P = (\sigma^2_S + \sigma^2_e)$ | 28.275 | 45.42 |
| $h^2 =$ | 0.289 | 0.365 |

$$Cov_{PC} = 1.436$$

$$r_{g\,(PC)} = \frac{1.436}{\sqrt{2.045 \times 4.15}} = \frac{1.436}{2.913} = \mathbf{0.493}$$

***Example 15.2*** In an experiment in poultry 8 sires each was mated to 6 dam of its own strain to produce purebred and also to 6 dam of other strain to produce crossbred. The sires had different number of progenies for recording egg production. The ANOVA was conducted which yielded the estimated parameters: $\sigma^2_{S\,(P)} = 1.05$ and $\sigma^2_{S\,(C)} = 2.51$; $\sigma^2_{D\,(P)} = 1.47$ and $\sigma^2_{D\,(C)} = 3.49$; $\sigma^2_{W\,(P)} = 22.21$ and $\sigma^2_{W\,(C)} = 38.24$. The sire wise mean egg production and number of progenies under each sire for purebred and crossbred were found as –

| Sire no. | 1 | 2 | 3 | 4 | 5 | 6 | 7 | 8 |
|---|---|---|---|---|---|---|---|---|
| Purebreds | 69 (27) | 72 (32) | 69 (34) | 70 (35) | 65 (28) | 73 (31) | 71 (31) | 70 (32) |
| Crossbreds | 70 (30) | 76 (28) | 72 (35) | 73 (33) | 70 (32) | 73 (32) | 73 (33) | 74 (32) |

(Figures in parenthesis are the number of progenies)

Estimate the genetic correlation ($r_{g\,(PC)}$) between purebred and crossbred HS progeny performance.

***Solution:***

$$r_{g\,(PC)} = \left[\left(\sigma^2_{S(P)} - \frac{\sigma^2_{D(P)}}{H_{DP}} - \frac{\sigma^2_{W(P)}}{H_{WC}}\right)\left(\sigma^2_{S(C)} - \frac{\sigma^2_{D(C)}}{H_{DC}} - \frac{\sigma^2_{W(C)}}{H_{WC}}\right)\right]^{-1/2}$$

Variance among sire's purebred progeny means ($\sigma^2_{S\,(P)}$):

$$\sigma^2_{S\,(C)} = \frac{1}{S-1}\left[\Sigma X^2 - \frac{(\Sigma X)^2}{S}\right]$$

$$= \frac{1}{7}\left[40622 - \frac{(559)^2}{8}\right]$$

$$= 5.839$$

Variance among sire's crossbred progeny means ($\sigma^2_{S(C)}$):

$$\sigma^2_{S(C)} = \frac{1}{7}\left[42223 - \frac{(551)^2}{8}\right]$$
$$= 3.982$$

Covariance between purebred and crossbred progeny means ($\sigma_{S(PC)}$):

$$\sigma_{S(PC)} = \frac{1}{(S-1)}\left[\Sigma PC - \frac{\Sigma P \Sigma C}{S}\right]$$
$$= \frac{1}{7}\left[40622 - \frac{(559 \times 581)}{8}\right]$$
$$= 3.517$$

Harmonic mean of no. of progeny per sire

$$(H_{WP}) = \frac{1}{\left[\frac{1}{8}(0.2577)\right]}$$
$$= \frac{1}{0.0322} = 31.06$$

Harmonic mean of no. of progeny per sire

$$(H_{WC}) = \frac{1}{\left[\frac{1}{8}(0.2519)\right]}$$
$$= \frac{1}{0.0315} = 31.75$$

$$r_{g\,(PC)} = 3.517\left[\left(5.839 - \frac{1.47}{6} - \frac{12.21}{31.06}\right)\left(3.982 - \frac{2.49}{6} - \frac{18.24}{31.75}\right)\right]^{-\frac{1}{2}}$$
$$= 3.517\,[(5.839 - 0.245 - 0.393)\,(3.982 - 0.415 - 0.574)]^{-½}$$
$$= 3.517\,(5.201 \times 2.993)^{-½}$$
$$= 0.89$$

❑❑❑

# 16 Selection Experiments

Improvement of livestock to meet the human needs started since the days **of** domestication without the knowledge of genetic principles. The science of genetics helped the animal breeder for improvement of livestock more rapidly and economically.

## 16.1 CHOICE OF MODEL ANIMAL FOR SELECTION EXPERIMENT

Long term selection experiments have been conducted in order to determine the effects of selection on quantitative traits particularly with laboratory animals like fruit fly (Drosophila melanogaster) and mice. The selection experiments with farm animals have been mostly carried out on pigs and fowl. Conducting selection experiments with large domestic animals is difficult because of low reproduction rate, long generation interval, more breeding space and labour requirement and high economic value of the animal, and requiring more capital investment. Therefore, laboratory animals· are mostly used in selection experiments. This is because of their rapid high reproduction rate, short generation interval, ease of handling more numbers, and less breeding space, labour and capital requirement. These factors make the Drosophila, Tribolium and mice favourable for test material. The generation interval in fruit fly and mice is about 2 weeks and 8-12 weeks, respectively.

Recent studies have shown that the science of genetics has no species barrier and the basic genetic principles are applied to all. Therefore any experiment carried out to know the basic genetic principles is relevant to animal breeding. There are reasons to undertake the experiments with lab animals on quantitative genetics. The relevance or contribution of lab experiments to large animals can be viewed for two reasons viz. how far the conclusions drawn in such lab experiments are true for large animals and the usefulness of lab experiment to large animals improvement.

Correct selection or **choice of** model lab **animal** is more important to maximize the relevance of lab experiments for farm animals. rt should be kept in mind that choice of model lab animal should be based on the basis of the genetic theory involved rather than the availability of the model animal or the available facilities to conduct the experiment or the availability of trained personal. Conducting experiment on Drosophila is advocated for genetic studies like genetic basis of chromosomal theory, as it contained only 4 pairs of chromosomes. Tribolium is the best species/ model animal to test the existing breeding methods for non additive genetic variance or to find the genotype environment interaction because the Tribolium can sustain the changes in temperature and culture etc. Mice being closer to domestic mammals are more suitable for studying the specific problems of animal breeding and particularly the lactation studies, and maternal effects etc. Likewise, the Japanese quail is mostly useful for breeding experiment related to poultry.

## 16.2 RELEVANCE OF LAB EXPERIMENTS TO LARGE LIVESTOCK

(*i*) The lab experiments provide prior information which can be tested with large livestock-. Any attempt to try some idea like to explore the causes of heterosis, genotype-environment interaction, and efficiency of breeding methods (reciprocal recurrent selection etc.) on large animals without sound evidence based on lab experiments will be merely wastage of time and resources. Any genetic theory can be further checked on large animals after it is established on lab animals. But testing of the adequacy of theory developed on lab. animals posses a problem. The results obtained on lab experiments are not straight way applied to large animals. There may be some different factors working on lab animals and farm animals. However, based on the results of lab experiments ·it may lead to develop refined theories. It requires an experimental test of the population which will indicate whether the theory is adequate for the population under test. For example, Hammond (1947) recommended that' animals should be best selected for a trait under the environment which favours its fullest expression and that once a superior line is developed it can be used in other environment. This theory of Hammond was later on criticized and a new 'theory was developed by Falconer (1952) suggesting that the breed should be selected in the environment in which it has to live.

(*ii*) The genetic theory developed on lab animals fu1ly supported with experimental proof creates confidence, thinking, an aptitude to learning and understanding the problem.

(*iii*) The lab experiments help to know the correct population size for estimation of a particular parameter and accuracy of its estimation. The genetic control is essentially required in selection experiment to know the comparative effect of selection and genetic drift.

(*iv*) Lab experiments help to examine, un5!erstand and in establishment of basic genetic concept like Mendelian inheritance, sex linkage, gene frequencies, number of genes affecting a trait, origin of new variation by mutants, rate of inbreeding depression etc. This help to understand the genetic structure of population and properties of a population which can not be from a general theory.

(*v*) The lab experiments are also helpful in testing the validity of the theories of quantitative genetics like theory of additive and non-additive gene action, correlated response theory, asymmetry of response, limit of response to selection, theory of response to selection, theory of mate mal effects, genotype environment interaction etc.

Roberts (1965) has reviewed the contribution of the lab mouse to animal breeding research, citing the various Jab experiments conducted by various workers which demonstrated the followings:

- Additive genetic nature of variation in body weight;
- Correlated response by shortening ears, feet and tail, and higher ovulation rate as a result of selection for weight;
- Asymmetry of response in body weight due to maternal effect, directional dominance and inbreeding;
- Limit to selection for body weight due to loss or exhaustion of additive genetic variance
- Genotype - environment interaction for body weight;
- The significance of non additive genetic variance in body weight, and
- The existence of maternal effects

The results of selection for thorax length in Drosophila indicated that the response was asymmetrical and that selection ceased to have an effect (Robertson and Reeve, 1952). Experiment for increased shank length In the mature fowl was conducted by Lerner (1958) with making additional observations on the correlated changes in vigour and body proportions. The selection resulted in a rapid increase in shank length in the first seven generations after which there was no consistent progress. The selection for increased shank length had a detrimental effect on the vigour of the birds, and after the selection was ceased and the shank length decreased, there was an improvement in vigour. Body conformation, length of spinal column and posture of the birds were also altered as a result of increase in shank length.

(*vi*) The extrapolation of results of genetic studies from any experiment of a population is not definitely justified for other population or species. But when the traits of different species are governed by the same kind of gene action, the results obtained with one species can be extrapolated to another. The additive gene action largely determines the bristle number in Drosophila, body size in mice and pupal weight in Tribolium and responds to selection, and hence these can be easily extrapolated from one species to another. It is, therefore, not important that the species have analogies with respect to phenotype, economy and taxonomy but the extrapolation should be based on similar genetic parameters.

The findings of selection experiments with Drosophila, mice and fowl can be concluded in general the followings:

(*i*) Selection can rapidly change the traits which show additive genetic variation.

(*ii*) The continued selection over many generations showed that response is more in the first few generations, then declining and finally ceases reaching limit to selection.

(*iii*) Two way selection experiments indicated asymmetry of responses in two lines and that the difference between the means of the two lines at the selection limit considerably exceeds to that of the original population.

## 16.3 CONSEQUENCES OF SELECTION EXPERIMENTS

The basic effect of selection is to change the gene frequencies and consequently the genetic properties of population which determine the effect of further selection. The population mean is changed as a result of selection without having the idea of the effect of selection on individual loci affecting the quantitative trait. Therefore, it is not possible to predict the changes of gene frequencies. Thus the response to selection can be predicted only when the genetic properties remain unchanged. The selection experiments can only discover and explain the consequences of selection.

The selection theory states that:

1. The realized or observed response to selection in progeny generation should agree with that expected theoretically on the basis of equation $R = h^2$. Sd. Obtained from information on parental population.

2. Long term selection should lead to fixation of alleles with the consequent loss/exhaustion of genetic variance at selection limit. The reduction in genetic variance will also reduce the phenotypic variance and at last selection will not result in any response.

The validity of selection theory can be tested from experimental studies of selection. The observations that have emerged from selection experiments regarding unexpected consequence of selection theory are:

***Insufficient response***: The observed or realized response obtained from the selection experiment should agree to the expectation based on the $h^2$ estimate and selection differential. But it has been observed that the realized response deviates and generally falls short than expected based on selection theory. This disagreement between expected and observed (realized) response to selection is called as the insufficient response.

***Asymmetry of response:*** The selection in opposite direction, known as bidirectional selection or two way selection, should lead to equal response in the two lines selected for increased performance (upward selection) and selected for decreased performance (downward selection). But it has been observed that the response is not equal in upward and downward lines. This is called as inequality of selection progress or the asymmetry of response to selection.

### 16.3.1 Insufficient Response

The causes of variation in theoretical (predicted or expected) response and actual obtained (realized) response to selection (genetic progress) or the factors which potentially reduce the net response to selection can best be explained from the component factors affecting the response. These are $\Delta G = h^2.S$

***1. Heritability***: The heritability is the function of the trait and population, besides being affected by some other factors.

*(a) Low and wrong $h^2$ estimates*: The maternal effects and other common environmental effects called as C-effects such as herd differences, seasonal influences etc. prevent an accurate estimate of $h^2$• The other factors which cause wrong $h^2$ estimates are inaccurate measurement of data, scale effect, un adjustment of data, violation of the assumption in $h^2$ estimation. In view of these factors the actual genetic merit can not be known. As a result the accuracy of B.V. falls short of expectation and the realized genetic progress is less than expectations.

Similarly when $h^2$ of a trait is low, the $r_{AP}$ (accuracy of selection) is low and hence most of the variation in the character is of environmental in nature. This will lead more deviations from expected phenotypic merit of the progeny. It has been observed that the $h^2$ estimates of most of the economic traits of farm animals are of low magnitude.

*(b) Gene interaction*: The long term prospects of selection can be seriously over estimated in the presence of gene interaction (additive x additive). This is due to the over estimate of $h^2$ in the presence of gene interaction. The contribution of genes, which interacts, towards genetic progress can be much larger if such genes are linked. The epistatic interactions disintegrate if selection is discontinued and gain in performance due to interaction effect disappears.

*(c) G-E interaction*: The G x E interaction may also decrease genetic progress. Such interactions change the ranking of bulls in different environments and thus the genetic gains obtained in one environment will not be consistent but will only be partly or not at all repeatable in another environment depending on the magnitude and direction of genetic correlation between performances in different environments. The interaction between breeding value and the environment in different years was suggested in poultry by Dickerson (1955). When the correlation ($r_G$) between performances in two environments is zero the selection progress is zero for which the term genetic slippage was given by Dickerson for the reduction of the breeding value when populations perform in environment other than that in which they were selected.

In this context the G-t: interaction is important because the environmental conditions at the time of predicting the response on the basis of information on parent generation are different when the response to selection is measured on the performance of the progeny generation. This time gap creates the environmental differences. Thus there is a possibility of G-E interaction which if exist will influence the genetic progress.

**2. *Selection differential***: This influences the R in the following ways -

*(a)* The selection progress is also influenced by the ***artificial and natural counter selection.*** The accidental loss of selected parents (through viability, fertility or longevity) reduces the selection differential and thus it results in low progress i.e. the progress falls short than expected. This is because the R depends on S of the selected parents and if the selected parents will differ in fertility and viability then they will not contribute equally to the progeny of next generations and hence due to unequal number of progeny of different parents, the contribution of their genes to next generation will differ. The R depends on the realized (effective/weighted)

selection differential rather than the attempted selection differential. Thus the actual response obtained will not be equal to the predicted response. The situation arises for the traits which genetically correlated to fitness i.e. the traits under selection which influence viability or fertility. The selection experiment for change in shank length (Lerner, 1959) in fowl was found to have a detrimental effect on vigour of the birds. The selection resulted in an increase in shank length in early (up to 7) generations after which the progress was not consistent. The ratio of actual to expected response decreased. But after the selection was stopped, the shank length decreased and there was an improvement in vigour and reproduction.

The natural selection may aid or hamper the artificial selection and it depends upon the genetic correlation between the traits under selection and fitness. The natural selection hampers the artificial selection when attempts are made to change the optimum performance level. The reproductive rates of selected individuals with extreme phenotypes are lower and thus the realized selection differential is lower than expected. This makes the actual response to fall short than expected.

(*b*) The female progeny of culled animals or of the low rank P.T. bulls remain in the breeding herd and cause insufficient response.

(*c*) The bulls are selected on pedigree performance before P.T. The bull from superior dam may not be superior due to sampling nature of inheritance. However, if the bull is superior, it may not transmit its genetic superiority to his progeny because of sampling nature of inheritance.

(*d*) Practically, the ~n directed towards the characters other than milk yield (primary character) also causes the insufficient response to selection. (e) The realized selection differential is reduced if additional traits are included in the selection goal. This reduces the actual superiority than the S if the selection is only for one trait.

**3. Population** size: It is an important factor, when selection lines are small the sampling errors will be more frequent and larger in magnitude. This will contribute to the variability among lines for response to selection. This can be explained as:

(*a*) The true genetic merit may not be reflected from the phenotypic merit of the selected individuals when the sample size is small and it will result low $r_{AP}$. Thus the observed response will fall short of predicted. The low $r_{AP}$ means low $h^2$ which indicate that most of the differences among individuals are caused by environmental effects and therefore the deviations from expected phenotypic merit of the progeny from selection errors as well as from environmental variations will be larger.

(*b*) Second way of causing the variability in R by small number of parents per generation is through the ***inbreeding***. The inbreeding causes fixation of alleles at random which means that different lines may undergo fixation for different alleles having different effects. The inbreeding depression counterbalances the effect of selection.

(*c*) Thirdly the effect of sampling variation depends upon the ***initial gene frequencies*** in the base population as well as during the course of selection. The effect of sampling variation is very less when the genes affecting a trait under selection are at intermediate frequencies than when they are at low frequencies. The effect with low gene frequencies is more because in some lines the genes will be present while in others may be absent. Thus it will affect the response because of the presence or absence of genes with large effect.

A substantial amount of difference in response between lines starts appearing in later generations than in early generations. The response in early generations is mainly due to the genes which have major effects and are at intermediate frequencies. The response is consistent in all lines till these genes are fixed after which these genes do not contribute to genetic differences among individuals. ]n later generations the selection is emphasized for the numerous loci with initially low frequencies and with small effect on the trait.

### 16.3.2 Asymmetry of Response

The different magnitudes of response in opposite direction have been observed in many two way selection experiments. It is the second surprising feature of the unexpected result of selection experiment. This disappoints a breeder when the selection progress falls short than expected. In a series of experiments, Falconer observed the phenomenon of asymmetrical response for most of the traits in mice when selected in opposite direction, particularly for body weight of mice. The response was more in downward line than in upward line. Similar observations were made by Robertson and Reeve (1950) in thorax length of fruit fly. The total change was greater for decreased length.

Falconer mentioned that the realized heritability in the two directions must be significantly different if the asymmetry is real. Many factors may cause selection response to differ in two directions. These main causes which may create asymmetrical response are discussed below:

***1. Population size***: The population size is an important factor. The lines under selection are not very large in size. This causes the random drift, inbreeding and reduction in genetic variance.

The ***random drift*** can easily result the asymmetry of response if there is only one selection line in each direction. The effect of random drift can be assessed only by replication of the selection.

The inbreeding is unavoidable in small population and hence the ***inbreeding depression*** occurs which reduces the mean performance. This results into less response in upward line but higher response in downward line. The progress for increased performance is counteracted by the effect of inbreeding for the traits affected by inbreeding depression and hence the net genetic gain is less than expected. On the other hand, the progress due to selection is more in downward line by the effect of inbreeding depression. Therefore, the effect of selection and inbreeding is antagonistic in upward line but synergetic in downward line. This results the genetic gain to fall short in upward line because of opposite (antagonistic» effect of two forces while the inbreeding depression augment the effect of selection in downward line and hence asymmetry of response occurs.

**2. *Correlated characters*:** The milk production often called as mothering ability particularly in small mammals (mice) affects the growth of young one up to weaning age. Selection for low weaning weight at 6 weeks in mice brings simultaneous change in mothering ability. The mothering ability, for which there is no selection, is reduced when selecting for low weaning weight. The selection for weaning weight has thus different effect in two lines. The decrease in weaning weight due to selection is more in small mice due to reduction in their mothering ability and hence the rate of genetic progress is more compared to an increase in weaning weight in larger line (upward selection).

**3. *Selection favouring heterozygous:*** The heterozygotes are superior to either homozygote in case of over dominance. In this case the heterozygotes are selected in upward line. This results the equilibrium to be reached gradually in the upward line because heterozygotes produce two alleles in equal frequencies to the next generation. At equilibrium state the genetic variation still exists but there will be no response to selection. Thus as the gene frequencies approach equilibrium when selection favours heterozygotes, the response is slow. In downward line the response is more till the favoured alleles approach fixation.

**4. *Genetic differences:*** The asymmetry of response to selection may be due to any following form of genetic differences in the initial population. These are:

***(i) Directional gene frequency*** - Some alleles affects the trait in one direction while others in other direction. The frequencies of these two types of alleles may be different. This is called as the directional gene frequency which indicates the differences in the frequencies of the alleles affecting the trait in two opposite directions. The frequencies of the alleles affecting the trait in one direction may be high, which expects to cause a more rapid response in the direction of the less frequent alleles and low response in the direction of the most frequent alleles. The frequencies of less favourable alleles are low due to natural selection, particularly the alleles related to fitness. This causes a more rapid response in the direction of greater loss of fitness when the selection in one direction reduces fitness more than selection in other direction.

***(ii) Directional dominance of genes*** - There may be unequal distribution of genes which are increasing and decreasing the value of a trait. The directional dominance occurs when most of the genes are dominant in one direction (either to decrease or increase the value of the trait). It is a situation when genes dominant in one direction are more in number or have large effect than those which are dominant in other direction. When the frequencies of alleles acting in one direction are already high, further progress will be slow than the progress in opposite direction.

The dominance of alleles masks the true B. V. of heterozygotes and of dominant homozygotes. Thus selection is not effective or accurate when made to increase the frequency of dominant alleles. On the other hand, the response to selection for recessive alleles is not affected by the masking effect of dominant alleles. However, the' effect of neither of the two genetic symmetry is immediately observed in first few generations but after the gene frequencies become differentiated in two lines.

***5. Selection differential***: The selection differential of the two lines may be different for the following reasons - (a) Antagonistic or synergetic effects of the natural and artificial selection, (b) differences in fertility of two lines and c) scale effect. When the S will be different, the R will also be different in two lines.

The use of wrong scale may be a cause of asymmetry. The scale effect causes differences in skewness, of the genetic and environmental variation, in opposite direction. The difference in skewness may also be due to G-E interaction. The individuals of upward line are of high genetic value and receive better environment than the individuals of downward line. The genetic variation is fully exploited or expressed in better environment than in poor environment and hence the genetic variation is less among individuals receiving better environment. The example of school education may be cited here. Secondly, the individuals with high genetic value are more prone to environmental variation compared to those having low breeding value. The example is the intelligence. Both result in low $h^2$ of upward line than of the downward line.

### 16.3.3 Limits to Artificial Selection

**Concept:** The selection acts on additive genetic variance which goes on diminishing in future generations as a result of continuous selection. The response to selection, therefore, diminishes. The quantitative characters are controlled by polygenes which segregate in the population. These genes are brought to fixation by the selection or inbreeding or random drift due to small population size or also they may reach equilibrium in case of over dominance. Thus due to continuous selection there will be no genetic variance and hence there will be no response to further selection. When there is no response to selection it is said that selection limit has reached and such a population is called at selection limit or plateaued population. This is also known as selection plateau. Thus the phenotypic level of a trait at which the response ceases is termed the selection limit. The selection limits have been noted under long term selection experiments in laboratory animals (mice, Drosophila) and poultry.

*Examples of selection limit due to long term selection experiments:*

- **Drosophila -** Thorax length, wing length and abdominal bristle numbers obtained by Robertson and Reeve (1952), Reeve and Robertson (1953), Robertson (1955), Clayton and Robertson (1957).
- **Mice -** Body weight and litter size obtained by Falconer (1955, 1965, 1971, 1973), Roberts (1966), Falconer and King (1953)
- **Poultry -** Egg production and shank length obtained by Lerner and Dempster (1951), Dickerson (1955).

Most of the above selection experiments were bidirectional and the selection limits reached between 20-30 generations. The selection resulted to the mean at the limits far beyond the range of variation present in base population.

### No Selection Limit

Some long term selection experiments have shown no selection limits and the response continued over many generations of selection. The examples are:

- Oil content and protein content in maize. The high line did not reach the selection limit up to 78 generations of selection.
- Pupal weight of Tribolium increased continuously up to 75 generations.
- One of the 3 replicates for increased abdominal bristle number in Drosophila continued responding up to 88 generations.
- The important livestock species also do not seem to be at selection limit because there is continuous improvements in traits of most economic importance viz. milk yield and body weight in cattle, body weight and litter size in pigs and sheep, weight gain in broilers and egg number in layers are showing improvement though these species have been under continuous selection for many generation. The rate of genetic improvement ranged from I to 2% per year of the mean.

## 16.4 TOTAL RESPONSE AND ITS DETERMINANTS

The observed range of response or total response ($R_T$) is taken as the difference between the upper and lower limits and expressed in relation to the initial genetic variance or phenotypic variance as $R_T/\sigma_A$ or $R_T/\sigma_P$. The determinants of total response are the number of gene loci affecting a trait, population size and mutation.

(i) The numbers of genes affecting a character have inverse relation with the magnitude of their effect but have direct relationship with the response to selection that can be achieved. The genes few in number affecting a character have large effects but if the genes are many in number controlling the character their effects will be small. The $R_T$ in relation to $\sigma_A$ depends on the number of gene loci affecting a trait and hence contributing the genetic variation in the trait. The larger the number of loci, the larger will be the response in relation to the original variance. This is because the number of genotypes formed depends upon the number of loci affecting a trait and the frequency of the most extreme genotype is getting reduced with the increase in number of loci viz. 1/4, 1/16, 1/64, 1/256, 1/1024 and 1/4096 for one through six loci affecting a trait The most extreme genotype represents the selection limit. Therefore, selection limit is expected within the range of variation in base population. Since extreme genotype is rarer in the base population for larger number of loci, so the selection limits are far away from the original mean.

The total response is related to the additive variance in the base population.

The additive effect of a gene is taken as the average of the difference of two homozygotes and represented by **'a'**. The total difference between homozygotes is thus 2a and by summing over all loci is $\sum$ **2a**. Assuming the equal effect of all loci on the character, the range of difference which also represents the total response ($R_T$) becomes as $R_T = 2\,n\,a$ with n as the number of loci each having a homozygote difference of **2a**. Further assuming the equal gene frequencies of 0.5, the additive variance in the base population is

$$\sigma^2_A = 2\,pq\,[a + d\,(q - p)]^2$$
$$= 2pqa^2 = 0.5\sum a^2 = 0.5na^2$$

The degree of dominance does not make any contribution for the case of equal gene frequencies. Squaring the range of response to develop the relation with additive variance as:

$$R^2_T / \sigma^2_A = \frac{4n^2a^2}{0.5na^2} = 8n$$

In the presence of large number of genes, the response that could be achieved in the theoretical maximum response which sets an upper limit to expectation. This is the total response what would be attained when genetic variance is caused by an infinite number of loci. Robertson (1960) has shown the theoretical maximum response ($R_{max}$) as:

$$R_{max} = 2n_e\, i\, h^2 \sigma_P$$

This may be more with favourable recessive genes of low initial frequencies. The ratio of observed to theoretical maximum response gives an indication of whether the character is influenced by very large number of genes having equal effects. The long term experiments on thorax length and abdominal bristle number of Drosophila and the 6-week body weight and litter size in mice for which selection limits have reached, showed the ratio of observed to theoretical maximum response ranging from 0.21 to 0.36. This means that observed response had fallen short of the maximum response which indicated that none of these characters were controlled by a very large number of genes having equal effects.

(*ii*) The ***population size*** also influenced the total response. Selected lines are often small in size and hence random drift causes fixation of some unfavourable alleles in spite of selection against them. This results the response to fall short than expected. The limit to selection is then decided by the chance of fixation of favourable alleles. The selection and inbreeding interacting together decide this chance of fixation. The inbreeding depends on population size whereas the selection on the selection coefficient (s) or intensity of selection (*i*). Therefore, the chance of fixation can be taken as a function of effective population size (Ne) and the selection coefficient as NeS or Ne i. The coefficient of selection depends on the gene effect ($2a/\sigma_P$) and intensity of selection as $S = i\,(2a/\sigma_P)$. The quantity $2a/\sigma_P$ is the difference of value between two homozygous relative to phenotypic standard deviation and hence called as the standardized effect of a locus. In view of the above, the total response is expected to be more in a population of large size under more intense selection. This was confirmed by Jones et al. (1968) who found that total response was greater for increased abdominal bristle number in drosophila in the selected lines of larger size with high intensity of selection.

The fixation of alleles either by inbreeding or selection can be tested from the results of repeated experiments i.e. different replicate lines are subjected to selection. The fixation due to inbreeding is indicated by reaching the different replicate lines to different selection limits because different alleles at many loci are fixed in different replicates. This causes the response to show on crossing these lines. In case of fixation of alleles by selection, the different lines have the same selection limits because the same alleles at most loci are fixed and these lines will not respond to selection.

(*iii*) The **mutation** seems to be important in affecting the selection limits in long term response. The mutation give rise to new genetic variation, though new mutants have very small effect in next or in next few generation for their low frequencies which is 1/2N where N is the population size. The frequency of new mutants increases in case of their favourable effect on the character under selection. The continued increase in the frequency of favourable mutant increases the genetic variance and hence contributes to response to selection. The new mutants arise in every generation and have their cumulative effect to the response but after about 20 generations. The response should then continue without approaching selection limit when there is mutation.

## 16.5 NATURE (EXPECTATION) OF SELECTION LIMIT

### (Genetic properties of lines at plateau)

The quantitative characters are controlled by many genes. The response to selection continues over a number of generations with gradual diminution of the rate till a limit is reached. The selection acts on additive genetic variance. The effect of long term selection experiment on variance can be explained as under

(*i*) Theoretically it is expected that selection should lead to fixation of alleles (homozygosity) with the consequent loss or exhaustion of genetic variance. Consequent upon Lie absence of genetic (additive) variance, the heritability estimate should be zero and relaxation or reversed selection will bring no change in the level of the performance i.e. response should cease.

(*ii*) As a consequence of the decline in genetic variance as the limit is reached there should be a decline in phenotypic variance.

**Evidence from selection experiments:** The evidence from selection experiments on these expectations of selection limit are given here:

Selection experiment on abdominal bristle number in Drosophila indicated that the phenotypic variance in the large line remained at the same level up to about 25 generations when the limit is reached. The comparison with the control line indicated that the variance did not decrease at all, whereas in the small line a sudden and large increase was observed (Clayton and Robertson, 1957)

Long term selection experiments on abdominal bristle number in Drosophila conducted by Clayton and Robertson (1957) and Jones et al. (1968) have shown that the phenotypic variance increased very much during selection in both directions (high and low lines) than in base population.

Selection for thorax length in Drosophila (Robertson 1955) showed that the population at selection limit respond to selection in reversed direction, respond rapidly and returned to the unselected level when the reverse selection was made in upward line but the low line (for reduced thorax length) did not respond to reversed selection. Similar results were reported by Robertson and Reeve (1952) for two way selection of thorax length of fruit flies following selection for increased and decreased length (selection in opposite direction) after the selection limit had been reached, resulted in a rapid return in. the lines selected for longer thorax but this .did not happen in the lines selected for shorter thorax. Therefore the two lines behaved differently to reverse selection and indicated the presence of genetic variance after the selection ceased to have an effect. It may be probably that the genes for shorter thorax had been fixed in the homozygous condition but in the larger thorax lines the genes must have been still quite heterozygous.

The evidences from selection experiments about the nature of the selection limit do not confirm the simple theoretical expectation of reduction in phenotypic and genetic variance. The phenotypic variance fails to decline and the genetic variance of some kind still remains after the selection limit reached.

The ***increase in phenotypic variance*** instead of its declining as expected may be due to the following reasons:

(*i*) The selection limits represents the fixation of alleles concerned leading to an increase of the frequency of homozygotes which are more sensitive to environmental effects as compared to heterozygotes and it will increase the environmental variance. This increase of environmental variance not only compensates the decrease in genetic variance but the variance is increased.

(*ii*) This depends on the character under selection. Scale effect may be a cause for the characters for which the variance changes along change in mean through selection (the variance is not independent of the mean for some characters). The selection changes the mean and the variance is changed with it automatically.

(*iii*) The genetic variance may not be completely exhausted. The existence of additive genetic variance at selection limit may be tested by two way selection experiment and applying reversed selection. The presence of additive genetic variance is indicated if the selected line responds to reversed selection. Inbreeding of selected line provides the evidence of the presence of non-additive genetic variance. However, some lines lose all additive genetic variance as per expectations but others retain at selection limit.

The ***existence of additive genetic variance at limit*** may be due to

(i) The selection limit is not reached in reality. This may be due to -

- reduction in fertility in selected lines which reduces the selection differential,
- low frequencies of unwanted recessive alleles which may remain unfixed. Further reduction of frequencies of such alleles is very low by selection for they contribute the non additive genetic variance.

(ii) When the selection favours heterozygous, there is genetic variation due to genes at selection limit. Under this situation the natural selection prevents the fixation of the alleles because the homozygous are sterile for some traits e.g. pygmy gene in mice which control the body size (Falconer, 1960). Reversed selection shows immediate and rapid response. This situation is very common, when the natural selection acts through lethality or sterility.

(iii) The nature of the trait or its scale of measurement may impose the limit. There are measurement limits in score traits like 0 to 5, 0-10 etc. which are attained through selection before the genetic variance is fully exhausted. Thus a false limit is reached.

## 16.6 FACTORS AFFECTING SELECTION LIMITS

It is essential to understand the genetic basis of limits to selection and the methods to overcome the limits to selection when planning for improvement of economic traits. The selective breeding has been in practice since long for the improvement of farm animals. As a consequence of this, the limit to selection may have reached and it is not possible to get further response to selection. Al Murrani (1974) published a review on the limits to artificial selection regarding the genetics of limit and attempts to overcome the limits to selection. There are many factors which cause the cessation of response and leads to the limits to selection.

***(1) Exhaustion of genetic variance***: Many loci control the quantitative traits. As a result of continuous selection for long time, it is expected that the genes will get fixed and there will remain no genetic variance. This will lead zero $h^2$ estimates and I thus no change in population mean for the selected trait is expected even with relax selection or reversed selection.. Experimental evidences for body weight in mice (Falconer and King, 1953; Roberts 1966), for thorax length in Drosophila (Robertson, 1955) and for fecundity in Drosophila (Brown and. Bell, 1961) have shown that the exhaustion of additive genetic variance was the cause of selection limit. The loss of genetic variance as a cause of limit to selection is based on the theory that new variance and renewed response should occur in crossbred population produced by the two lines selected in two directions (high and low expression of the trait) and have reached limit to selection.

The absence of additive variance and the fixation of all loci affecting a character are distinct. One is over dominance when selection favour heterozygotes for their superiority over both homozygotes and some loci segregate without showing any additive variance. This results in gene frequency equilibrium at an intermediate level. The second is the opposition of artificial selection by natural selection. One homozygote may be rejected for having low value of the character while other homozygote may not contribute due to its inviability or infertility under natural selection and thus heterozygote becomes superior. The additive variance will be exhausted for overall fitness but not for the character under selection.

***(2) Reasons of failure to respond in the presence of genetic variance:*** The following factors may be responsible for this -

*(i) Overdominance gene action:* Some form of overdominance when heterozygotes are at advantage may be a barrier to further advance. The heterozygotes may be superior compared to the homozygotes or the homozygotes may be less fit than heterozygotes and thus the genetic variance exists at the limit. Heterozygote superiority may lead to an equilibrium gene frequencies and $h^2$ at equilibrium should be zero because all the genetic variance is non additive type. Under the

situation the reverse selection will be effective to decrease the performance level of the population for trait under selection.

The overdominance as a cause of plateauing for wing length (Reeve and Robertson, 1953) and bristle number (Clayton and Robertson, 1957) in Drosophila has been observed. There was an increase in variance as the response diminished at the limit and response to reverse selection was observed in bristle number. For wing length also the variance and $h^2$ estimate was greater, at the limit, than in the unselected stock. Natural selection may favour heterozygotes. The recessive lethal genes may be the factor in the maintenance of heterozygosity and the selected individual may be heterozygous carrier of lethal genes. Thus the heterozygotes have the selection advantage.

The pygmy gene in mice, selected for small size, appeared due to mutation (McArthur, 1949). The mice homozygous for pygmy gene are small but sterile and hence the gene could not be fixed due to natural selection against homozygotes. The mice heterozygous for pygmy genes were favoured in selection because they are smaller in size. Thus there existed the genetic variance due to the gene but without further progress. In case of reversed selection there is a joint synergetic effect of artificial and natural selection in the same direction and hence there is immediate and rapid response.

*(ii) Low frequency of recessive genes*: The selection reduces and eliminates the recessive genes until they become rare in the population. Thus the unfavourable recessive genes are brought to low frequency. The variance caused by recessive genes at low frequencies is non additive. In this situation, the mean will change in unfavourable direction on inbreeding. Therefore the genetic variance may persist at the selection limit due to the segregation of recessive genes which are at low frequency. This makes the further selection ineffective. Litter size and body weight in mice threw some light on this matter.

*(iii) Negative genetic correlation*: There are some traits in farm and lab animals showing negative genetic correlation. The continued selection may develop the negative genetic correlation. The examples are: egg production and egg weight in poultry, milk production and fat % in milk, shank length and hatchability in poultry. As a result of continuous selection for two traits simultaneously, there will be no response due to the high negative genetic correlation (Lerner and Dempester, 1951)

*(iv) Genotype-environmental interaction*: The combined effect of genes for maximum value of the character are changed in different environments and also in subsequent generation. Dickerson (1955) pointed that the combination of gene effects changes under different environments or successive generations. The loss in performance (genetic slippage) occurs if the environment is changed. The slippage between generations may occur simply because the response obtained in one environment is irrelevant to the changed environment. For example, a strain may be resistant to some disease in one environment but not in other.

*(v) Natural selection*: The natural selection through differential viability may oppose artificial selection. In this situation, the effective S is less than expected due

to the effect of fertility and viability of selected parents. In mice selected for low body weight (small mice), at selection limit, it was observed that $h^2$ was positive and the reverse selection for increased weight showed the response (Roberts, 1966). Thus the additive genetic variance existed at the limit. The fertility and viability of the small mice was affected adversely naturally. Low fertility in selected line reduces the S in later generation and so there was no further response. On the other hand, the large mice reached at selection limit due to exhaustion of additive genetic variance (Roberts 1966, Falconer and King 1953). In case of Drosophila the situation' was found reverse by Robertson (1955).

*(vi) Nature and scale of character*: The selection limit is imposed by the nature of the character and its unit of measurement. There are certain characters measured in scores or percentage or in number like bristle numbers in Drosophila which have their lower limit of zero. The limits of measurement of the character may be reached without fixation of genes concerned.

*(vii) Linkage:* The linkage of loci is not an important cause of limit to selection and it works in combination with some other causes of limits. The linkage affects the limit only when the linked loci have equal effects on the character and maximum effect will be when the chance of fixation of these loci was between 0.7 and 0.8. The linkage is not an important cause of limits to selection for quantitative traits because their inheritance is polygenic and the polygenes are distributed among many pairs of chromosomes. This makes the linkage effect very small. However, when the two loci are linked they will be inherited together without producing recombination. The progress under selection with no recombination (i.e. no crossing over or with linkage) is less (about 70 to 80%) than with usual amount of recombination. The linked genes may have opposite effect on the trait under selection and thus the response obtained due to selection of favourable gene at one locus will be offset by the loss associated due to selection of undesirable allele at the linked locus. Under the situation of linkage the population mean should not change by relaxed selection but the reverse selection will bring a change. The relaxed selection for several generations may allow crossovers to break up the unfavourable linkage and show further response to upward selection.

## 16.7 METHODS TO OVERCOME THE LIMITS TO SELECTION

It is difficult to overcome the limits to selection for the reason that it is difficult to know the actual reason of limit and secondly it is not feasible in terms of time required and the expenses to meet the testing facilities. However, the limit to selection reached due to exhaustion of additive genetic variance can overcome by introducing new genetic material either through mutation or through out crossing.

To induce mutation is not advantageous. This is because the effects of mutagenic agents are not selective but random. That is to say that there is no surety that the induced mutation will be favourable. Moreover in case of farm animals the dose of mutagenic agents is beyond the animals' tolerance. However, attempts have been made to induce mutation as a source of ne«r genetic variance but the results are contradicting and inconsistent. Some advance was achieved after crossing the irradiated male mice at selection limit without the establishment of the effectiveness of the method (Roberts, 1967). This method is thus not promising.

The out crossing may be done in two ways viz. crossing of two selected st'rams at limit and crossing of selected strain at limit to the unselected strain.

The first system includes the crossing of two selected strains which attained selection limits and then to select the crossbred progeny. This method resulted heterosis for large body weight in mice (Falconer and King, 1953). The cause of heterosis was the dominance effect of genes controlling the skeletal dimensions and fatness - one strain being long bodied but not v.ery fat and the other strain was not large but was very fat. Both these strains were selected for large size over a period of more than 30 generations and attained the limits due to the exhaustion of the additive variance. The crossbreds of these two strains had significant rate of response to selection in both directions. In other experiment this method of crossing selected strains in mice was also found successful to increase the body weight but not to decrease it (Roberts, 1967). However, the large mice developed fertility problem.

The crossing of a selected strain which approached limit to selection with an un selected strain and then selecting the crossbred progeny to break the limit in body weight in mice was found successful. It required many generations of selection first to recover the selected level and then some more generations of selection to obtain the progress. Thus the period is very long. Therefore, crossing of two selected strain is superior compared to crossing with an unselected strain.

In both these mating systems, the superiority of the crossbred progeny is due to over dominance which results heterosis. In order to exploit maximum heterosis in a trait and to utilize the non additive genetic variation, two methods are used. These are recurrent selection to a tester line (RST) for specific combining ability and reciprocal recurrent selection (RRS) for specific combining ability. Both these methods are efficient only when over dominance is important. It is better that all the selected strains are tested in all possible combinations including reciprocals to find out the superior progeny.

The above methods to break the limit to selection may not be applied in farm animals because mainly one strain is only available. The inbreeding combined **with** selection was tried for improvement in litter size in a strain of mice at a selection limit (Falconer, 1971). In all 9 lines of independent origin from the plateaued strain were chosen and inbreeding by full sib mating was used. Five lines were eliminated up to seventh generation. The selection within each of 4 remaining line was practiced. The inbreeding reduced the litter size. All the 4 lines were crossed when F = 89% and repeated cross when F = 99 %. Heterosis in F I and F 2 of the first cross was above the selected level of the strain and thus the improvement gained from inbreeding and crossing proved the success of the system in breaking the limit for litter size. This method was tried for body weight in mice for a strain which reached to a selection limit due to exhaustion of most of the additive variance due to fixation of favourable alleles affecting body weight. The breeding method with relaxation, selection and inbreeding was not proved successful to change the 6-week body weight. The inbreeding with selection and the subsequent crossing of the best inbred lines when F = 0.50 did not produce any improvement in body weight. This failure was due to lack of directional dominance and also of the lack of differentiation in gene frequencies between the lines when they were crossed. Thus no method will be effective, unless it introduces more genetic variance, in the absence of directional dominance and the lack of differentiation of gene.

Selection within an inbred line is effective to minimize the inbreeding depression and also to increase the combining ability of the line. But the results are inconsistent regarding the effect of selection in inbred line on the rate of inbreeding depression. However, the increased genetic variability of the lines crossed leads to increased heterosis.

## 16.8 RELEVANCE OF LIMIT TO SELECTION

The farm animal breeders remain interested with genetic progress in short time and hence the long term selection experiments are not relevant. Selection is practiced since long in various breeds of livestock. The limit to selection in broiler chickens under intense selection has not been reached so far.

The important species of livestock do not seem to attain the selection limit as they are responding to selection and continued improvement is being made. Therefore, the understanding of the nature of selection limits and methods of overcoming those to make further progress is not relevant at present in case of large livestock species.

❑❑❑

# 17 Mating Systems I. Inbreeding

The random mating does not change the genetic structure of population and hence it does not fulfill the objectives of genetic improvement of livestock. The mating systems based on the genetic relationship among mated pairs, are effective to change the genetic structure of population and are better tools for genetic improvement of farm animals.

## 17.1 GENETIC RELATIONSHIP

The concept of relatives and genetic relationship stems on the common origin of individuals of a population. The individuals of a population have been descended (originated) from their common ancestor and hence share common genes to some extent. The individuals sharing common genes directly or through their common ancestor(s) are said to be ***relatives.***

The common genes (genes in common) shared by relatives are the duplicate copies of the ancestral genes and hence identical in origin, known as *genes identical by descent*. The common genes (identical genes) received by two or more individuals from their common ancestor(s) cause the genetic relationship between them. Therefore, the basis of genetic relationship between any two or more related individuals is that they have the common genes that are identical by descent.

The ***ancestor*** of an individual is one that appears in the pedigree of the individual (X) and hence inherited some of its genes to the total genetic make up of the individual. The ancestor is also called as ***descent.*** The individual (X) of the succeeding generation which received genes from the ancestor (descent) is called as ***descendant***. The ***common ancestor*** of an individual is one that appears in both the sire's and dam's pedigree of the inbred individual (X). Therefore, the common ancestor contributes (inherits) its genes to the inbred individual (X) through its both the parents (dam and sire).

***Types of relationship***: The proportion of the identical genes shared by the relatives decides the *degree of relationship* between them. The relatives and relationship are of two types depending upon the way they receive the common genes.

1. *Direct relationship:* This is the relationship when an ancestor and its descendents are related to each other on direct genes donor-recipient basis. The

ancestor transmits the genes while the descendent (progeny) receives. Thus, there exists a relationship of an individual with its ancestors (parents, grand - parents) and with its progeny (son, daughter, grand-son, grand-daughter). This is called as the direct relationship because the genes are transmitted directly from ancestor to the descendents. The relatives with this relationship are called direct relatives and they belong to different generations.

2. *Collateral relationship*: The two individuals may not be related directly but they are related through their one or more common parents (ancestors) in the pedigree, due to genetic contribution of a common ancestor. The relationship between two animals due to genetic contribution of common parent (ancestor) in their pedigree is called the collateral relationship and the individuals having this relationship are called the collateral relatives. The collateral relatives may belong to the same generation (full sibs, half sibs, cousins) or to the different generations (uncle or aunt and nephew or niece).

## 17.2 CONCEPT OF GENE'S IDENTITY

Receiving and sharing of genes in common by relatives has the primary genetic consequence. This results the progeny of relatives (inbred individual) to inherit the same genes from each parent to some extent and hence leads to an increase in the probability of the inbred progeny to become homozygous. Therefore, the inbreeding increases the homozygosiy and this increase is at the expense of heterozygosity. The increase in homozygosity is due to increase in identical homozygotes that carry genes identical by descent and this leads to gene fixation.

The concept of *genes identical by descent* is basic to study the genetic relationship. The two alleles at a locus in an individual may be different (heterozygous state) or similar (homozygous condition) in structure (nucleotide sequence), function and origin. The similarity of allelic forms of gene at a locus is of two type viz. *identity by descent* and *identity by state and function.*

Two related individuals (having common ancestor) may receive replicate (duplicate) copies of the ancestral genes and transmit these replicates to their progeny, if they are mated. Thus, the inbred progeny produced from mating of such relatives may receive both genes at a locus that are replicated copies of one and the same gene of the ancestor and hence both alleles at a locus carried by inbred progeny have same origin. Two alleles that have originated by replication of gene in the ancestor are called the *genes identical by descent* or simply identical or *autozygous genes* (Cotterman, 1940). The individuals carrying the identical genes by descent at a locus are called the *identical homozygotes or autozygotes.*

The two alleles at a locus in homozygous state may not be identical by descent but they are of independent origin, though they are identical in structure and function. Such alleles in homozygous state at a locus are called the *genes identical in state* if they are identical in structure and function but not the replicates of the same allele of an ancestor within the depth of the pedigree. Such identity of alleles is called as *functional identity.* The homozygotes carrying alleles identical by state are called as the *independent homozygotes or allozygotes.* Cotterman (1940) called the alleles of allozygous individual as *allozygous* or *genes independent by descent* but identical in state (structure and function).

The genes identical by state (allozygous) are also identical by descent but through remote ancestor because they are also the replicates of the ancestral genes of the several past generations beyond normal pedigree depth which is considered to define the inbreeding. Thus, pedigree depth is taken as a base to define the genes identical by descent. The two alleles at a locus are identical by descent is a probability statement.

The concept of genes identical by descent is useful to measure the degree of relationship between mated pair and to measure the degree of inbreeding so as to avoid or advocate the inbreeding in future generation.

## 17.3 MEASUREMENT OF RELATIONSHIP

The relationship between the individuals of a population is measured in terms of the common genes they share. The relationship is expressed as the proportion of identical genes possessed by the relatives. This is the extra similarity in genes the relatives have over and above that present in the base population. Wright (1921) developed a measure for expressing the degree of relationship between two individuals that are related to each other directly or collaterally. He called this measure as the coefficient of relationship.

***Coefficient of relationship:*** It is defined as the *probable proportion of genes that are same (identical by descent) in two relatives due to their common ancestry, over and above that present in the base population.*

The coefficient of relationship is a measure of the similarity of genotypes in two relatives (X and Y) and denoted by $R_{XY}$. The $R_{XY}$ is the correlation between breeding values of two individuals. This is a numerical value which indicates that how much more alike are the breeding values in two related individuals compared to any other individuals in the same population. Therefore, $R_{XY}$ is a measure of the common genes the two related individuals have due to their common ancestry, over and above that present in the base population. This is expressed either in percentage or proportion (decimal fraction) of common genes ranging from 0 to 1.0.

***$R_{XY}$ of Direct relatives:*** The halving process of inheritance determines the genetic relationship between direct relatives. This means that each parent transmit only half of its total genes to each of the progeny. Therefore, the genetic relationship between a parent and its progeny is 50 percent. This means that the parent and offspring contain half of their total genes in common (genes identical by descent). This is the genetic relationship between a parent and its progeny. Due to the halving process of inheritance the similarity in genes (genetic relationship) is reduced to half every generation between an ancestor and its descendents. Thus, the genetic relationship between an individual and its direct relatives is as given here –

| Generation | Relatives | Genetic relationship* | |
|---|---|---|---|
| 1 | Parent (Father/mother) | = ½ | = 50 percent |
| 2 | Grand-parent | = ½ x ½ | = ¼ or 25 percent |
| 3 | Great grand-parent | = ½ x ½ x ½ | = 1/6 or 12.5 percent |
| 4 | Great-great-grand parent | = $(½)^4$ | = 1/ 16 or 6.25 percent. |

- The genetic relationship indicates the proportion of ancestral genes to the total genes of progeny of subsequent generation. In this way, the genetic contribution of an ancestor is only 1.5 percent of the total genes of the descendant of $6^{th}$ generation, and the contribution is less than one percent (0.78) of ancestral gene to the total gene pool of the descendant of $7^{th}$ generation.

Based on halving process of inheritance, ½ is taken as the base term and the number of generations between any two direct relatives is taken as the exponent term (*n*) to estimate the genetic relationship between them. Therefore, the genetic relationship between any two direct relatives (X and Y) is obtained as:

$$R_{XY} = (½)^n$$

Thus, the extent of genetic contribution of the ancestor to the total genetic constitution of descendent depends on the number of generations between them.

***$R_{XY}$ for Collateral relatives:*** The collateral relatives have common ancestor in their pedigree through which they receive same genes (common genes or genes identical by descent) from their common ancestor. The common genes possessed by any two collateral relatives through their common ancestor cause the genetic relationship between them. Now it needs explanation that any two collateral relatives are related to what extent.

***(i) Half sibs:*** Take an example of the genetic relationship between half sibs which is 25 percent. The half sibs have one common parent (either mother or father). The common parent transmits only half of its gene to each of its progeny. For example, the individuals X and Y are two half sibs having the same father. The two half sibs (X and Y) may receive from their father the same genes or different genes. The probability of having received the same genes by two half sibs is 50 percent and the probability of receiving different genes is also 50 percent. The father is heterozygous for many gene loci and hence will produce two types of gametes carrying different genes. Due to sampling nature of inheritance, it is expected that one-half of the common genes received by X from its father would have also been received by Y from its father. Therefore, due to halving process and sampling nature of inheritance, the combined probability of receiving the same genes by two half sibs (X and Y) from their father is ½ x ½ = ¼ or 25 percent. Thus, $R_{XY}$

***(ii) Full sibs:*** Now, take that X and Y are full sibs, having both the parents (father and mother) common. As a result of one common parent (father) the similarity of genes between two half sibs was 25 percent. In this case of full sibs another 25 percent of their genes should be identical. Thus, 25 percent of genes in the body cells of full sibs are identical from sire side and likewise another 25 percent would be identical from mother side. Therefore, the total probability of having received the identical genes by two full sibs is 25 + 25 = 50 percent.

The coefficient of relationship between an individual and its collateral relatives are as given here –

| Relatives | Coefficient of relationship (%) |
|---|---|
| Full sibs (full brother or sister) | 50.0 |
| Half sibs (half brother or sister) | 25.0 |
| Uncle / aunt, Niece/nephew (from FS family) | 25.0 |
| Do - (from HS family) | 12.5 |
| Progeny of FS from two families | |
| (Double first cousin, 4 common grandparents) | 25.0 |
| Progeny of FS from single family (2 common grandparents) | |
| (Single first cousin/first cousin/full cousin) | 12.5 |
| Half cousin (Progeny of half sibs, one common grandparent) | 6.25 |
| Second cousin (progeny of full cousins) | 3.125 |
| Third cousin (progeny of second cousins) | 0.78 |

## 17.4 METHODS TO COMPUTE $R_{XY}$

The correlation coefficient between two variables ($r_{XY}$) is a measure of the mutual relationship and it is computed as a ratio of the covariance between them to the geometric mean of their variances as –

$$r_{XY} = \frac{\text{Cov}_{XY}}{\sqrt{(\text{Var X})(\text{Var Y})}}$$

The coefficient of relationship between two individuals ($R_{XY}$) which is a measure of the genetic correlation between them is computed similarly from genic covariance and variances as –

$$R_{XY} = \frac{\text{Genic Cov.}_{XY}}{\sqrt{(\text{Genic Var. X})(\text{Genic Var. Y})}}$$

The $R_{XY}$ is measured by two methods viz. path coefficient and coancestry method.

### 17.4.1 Wright's Method of path coefficient

The genic covariance between two individuals (Cov $_{XY}$) is equal to the sum of all pathways connecting them through a common ancestor. Thus, the genic covariance is measured from path coefficient as-

$$\text{Genic Cov.}_{XY} = \sum (½)^n (1+F_A)$$

Where, n = number of generations connecting X and Y through common ancestor (A)

$F_A$ = Inbreeding coefficient of common ancestor

$R_{XY} = \sum (½)^n$ If X and Y are not inbred.

The $R_{XY}$ is computed for collateral and direct relatives by path coefficient as under -

**(*i*) Collateral relatives**

$$Rxy = \sum(½)^n \frac{(1+F_A)}{\sqrt{(1+Fx)(1+F_Y)}}$$

Where, $1+F_X$ = Genic $V_X$

$1+F_Y$ = Genic $V_Y$

$Rxy = \sum(½)^n$ If two individuals (X and Y) and their common ancestor are not inbred.

**(*ii*) Direct relatives:**

The following formula to compute the coefficient of relationship for direct relatives has been derived from the formula used for collateral relatives by multiplying both the numerator and denominator of the formula of collateral relatives with $\sqrt{(1+F_A)}$. This is done because in case of direct relatives one of them (say X) is the ancestor (A) of the second individuals (say Y).

$$Rxy = \sum(½)^n \frac{(1+F_A)}{\sqrt{[(1+F_A)(1+F_Y)]}}$$

$$= \sum(½)^n \sqrt{\frac{[(1+F_A)}{(1+F_Y)]}}$$

Where, Fx = inbreeding coefficient of individual X,

$F_Y$ = inbreeding coefficient of individual Y,

$F_A$ = inbreeding coefficient of common ancestor

$R_{XA} = \sum(½)^n$ If X and A are not inbred.

The above two methods of computing the coefficient of relationship between two individuals give the similar results.

***Rules for tracing the paths:*** There are certain rules to trace the paths connecting two relatives through common ancestor. These rules are helpful to compute the contribution of each path when the relatives have more than one common ancestor and hence connected by more than one path through each common ancestor.

(*i*) The path should connect the two related individuals either directly or through common ancestor. The path starts from one of the relative and ended to the second relative.

(*ii*) The path starting from one relative goes back to the common ancestor through various intermediate ancestors in the chain of segregation, if any and then the path takes turn to forward from common ancestor going to the second relative. The number of generations falling between one relative and common ancestor are denoted by n whereas those between common ancestor and second relative are denoted by n′. Therefore, the path first goes backward from one relative to the common ancestor and then comes forward (reverse) to the second relative.

(*iii*) Once the path started its reverse journey from the common ancestor, it will not go backward again. Thus, no reversal of direction in path is allowed except at the common ancestor.

(*iv*) No individual in the path is counted more than once. Thus, a path can not pass through the same individual twice.

***Rules for considering the inbreeding coefficient of common ancestor as*** $(1 + F_A)$:

There may be some pedigree in which the common ancestor is inbred. In this case, the term $(1 + F_A)$ is multiplied with the contribution of each path in which the inbred common ancestor falls as: $(\frac{1}{2})^{n+1}(1+F_A)$. Thus, $(\frac{1}{2})^{n+1}(1+F_A)$ is the contribution of each path.

(*i*) The inbred ancestor in the pedigree should be the common ancestor. There are some pedigrees which had certain intermediate inbred ancestor (s) but they are not common ancestors. In this case, the inbreeding coefficient of intermediate inbred ancestor (s) is not considered. Therefore, unless the inbred individual in the pedigree is the common ancestor, its inbreeding coefficient is not multiplied with the contribution of the path.

(*ii*) The inbreeding coefficient of the common ancestor ($F_A$) gets multiplied with the contribution of each path and with all the paths in which the inbred common ancestor falls as the common ancestor and not as the intermediate inbred ancestor.

(*iii*) When the common ancestor is not inbred, $F_A = 0$ and hence $1+F_A = 1.0$. Therefore, it is then not multiplied.

(*iv*) The inbreeding coefficient of the parents (sire and dam) of the individual is not considered unless they have direct relationship and one become the common ancestor.

### 17.4.2 Coancestry method

The coancestry of two individuals is the probability that two gametes taken at random, one from each individual, carry allelic genes identical by descent.

***Synonyms of coancestry:*** Coancestry has the following synonyms -

***(i) Coefficient deparent or kinship:*** Malecot (1948) called the coefficient of relationship of two individuals as the *"coefficient deparent"* or as *kinship*. He defined it as the probability that two homologous genes drawn at random, one from each of the two parents of the individual are autozygous (identical by descent) and denoted it by $\mathbf{\O}_{XY}$ as:

$$\mathbf{\O}_{XY} = \frac{1}{4}\,[P\,(a{=}c) + P\,(a{=}d) + P\,(b{=}c) + P\,(b{=}d)]$$

Where, a and b are two homologous genes in X individual,

c and d are two homologous genes in Y individual.

***(ii) Coefficient of parentage:*** Kempthorne (1957) called it as *coefficient of parentage.*

***(iii) Coancestry:*** Falconer (1960) called to the degree of relationship by descent between two parents as *coancestry* or *coefficient of kinship* or of *consanguinity* and symbolized by f.

The coefficient of relationship denoted as $R_{XY}$ is the correlation between the genic values of two individuals and it is exactly twice the coefficient of consanguinity. Thus $R_{XY} = 2\,\mathbf{\O}_{XY}$. The inbreeding coefficient of the progeny ($F_1$) is equal to the coefficient of coancestry of its parents ($f_{XY}$).

$$\text{Thus } Fz = f_{XY} = \mathbf{\O}_{XY} = \frac{1}{2} R_{XY}$$

***Computation of Coancestry:*** For the procedure to compute coancestries, and basic coancestries, see section 10.7 of chapter 10 in "Genetic Structure of Livestock Population" by Tomar and Tomar (2014).

The important required results of coancestry can be given as –

The F of an individual is equal to the coancestry between its parents

$$F_X = r_{PQ}$$

Coancestry between two individuals is the average of 4 coancestries between their parents

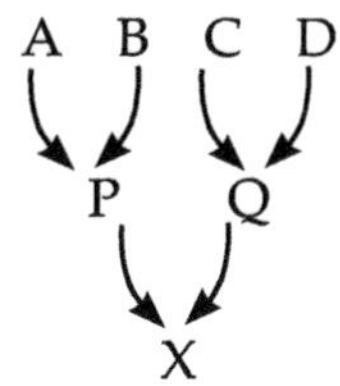


$$r_{PQ} = \tfrac{1}{4}\,(r_{AC} + r_{AD} + r_{BC} + r_{BD})$$

(*iii*) Coancestry of an individual with itself

$$r_{XX} = \tfrac{1}{2}\,(1 + F_X)$$

(*iv*) Coancestry of an individual (A, which belongs to previous generation) with the progeny (Q, which belongs to next generation) of two individuals (C and D). It is equal to the average of two coancestries of the individual of previous generation (A) with the parents (C and D) of another individual of next generation (Q):

$$r_{AQ} = r_{A.(CD)} = \tfrac{1}{2}\,(r_{AC} + r_{AD}) \text{ since, } Q = \tfrac{1}{2}C + \tfrac{1}{2}D$$

*Use of genic variance and covariance to compute coefficient of relationship ($R_{XY}$):*

(*a*) When no inbreeding

$$r = R_{XY}$$

$$= \text{Genic Cov}_{XY}$$

$$= 2\,f_{XY}$$

(*b*) When relatives are inbred

$$r = \frac{\text{Genic Cov}_{XY}}{\sqrt{(\text{Genic var X})\,(\text{Genic var Y})}}$$

$$= \frac{\text{Genic Cov}_{XY}}{\sqrt{(1 + F_X)\,(1 + F_Y)}}$$

$$= \frac{2\,f_{XY}}{\sqrt{(1 + F_X)\,(1 + F_Y)}}$$

The genic variance and covariance are computed from coancestries. The coefficient of coancestry ($f_{XY}$) is computed as –

$$f_{XY} = \tfrac{1}{2}\text{ genic cov}_{XY}$$

$$= \text{F of progeny of X and Y}$$

$$r = 2\,f_{XY}$$

$$= 2\,F \text{ and hence}$$

$$F = \tfrac{1}{2}\,r$$

The genic covariance between two individuals equals their coancestry. Thus,

Genic Cov $_{XY}$ = $f_{XY}$

= Average covariance of either individual with parents of other individual

= ½ (Sum of cov. of X with parents of Y)

= ½ ($Cov._{XC}$ + $Cov._{XD}$), if C & D are parents of Y

or ½ (sum of cov. of Y with parents of X)

= ½ ($Cov._{YA}$ + $Cov._{YB}$), if A & B are parents of X

Genic Var(X) = $Cov._{XX}$

= 1.0 + ½ (genic cov. of its parents)

= 1.0 + ½ ($Cov._{AB}$) since A & B are parents of X

= 1.0 + 0 (if X is not inbred)

= 1 + F x (if X is inbred)

Genic Var(Y) = 1 + $F_Y$

***Genetic covariance*** is composed of the fractions of the causal components of variance ($V_A$ and $V_D$) as:

Genetic cov = r $V_A$ + u $V_D$

Where, r = fraction (coefficient) of additive variance, or

= coefficient of relationship, or

= genetic correlation between relatives, or

= correlation between B.V. of two relatives, or

= probability of having identical genes by two relatives due to their common ancestry

u = fraction (coefficient) of dominance variance, or

= probability of both alleles at a locus in an individual being identical to those in another relative, or

= probability of having same genotype by relatives through ancestors.

= zero, unless the relatives have path of coancestry through both of their respective parents like FS and double first cousins.

The two coefficients (r and u) are related to the coancestries of the relatives and their parents in a random mating population as:

$$r = 2 f_{XY}$$

$$= 2 [¼ (f_{AC} + f_{AD} + f_{BC} + f_{BD})]$$

$$= ½ (f_{AC} + f_{AD} + f_{BC} + f_{BD})$$

or

$$= f_{YA} + f_{YB} = f_{XC} + f_{XD}$$

$$u = f_{AC} f_{BD} + f_{AD} f_{BC}$$

Where, f = coefficient of consanguinity or coancestry.

The size of these two coefficients among different kinds of relatives is as –

| Relatives | r | u |
|---|---|---|
| Monozygotic twins | 1.0 | 1.0 |
| Individual itself | 1.0 | 1.0 |
| Full Sibs | 0.5 | 0.25 |
| Parent-offspring | 0.5 | 0 |
| HS, uncle-nephew, offspring – GP | 0.25 | 0 |
| Double first cousins | 0.25 | 0.0625 |
| Single first cousins | 0.125 | 0 |
| Offspring great GP | 0.125 | 0 |
| Unrelated individuals | 0 0 | |

## 17.5 FACTORS AFFECTING THE DEGREE OF RELATIONSHIP

It is evident from the formula of the coefficient of relationship that the degree of relationship between two individuals depends upon:

- Number of generations between two individuals,
- Number of paths connecting two individuals,
- The inbreeding coefficient of the two individuals among whom the coefficient of relationship is to be computed, and
- The inbreeding coefficient of common ancestor.

### 1. Number of generations between two individuals

The inheritance is halved in each generation. Therefore, the contribution of the common ancestor to transmit identical genes is reduced with the increase in number of generations connecting two individuals among whom the degree of relationship is to be measured.

In case of direct relationship, the coefficient of relationship between parent and offspring is ½ ($R_{PO}$ = ½) whereas between the progeny and grand parent is ¼, and between the progeny and great-grand parent is 1/8. This is because the number of generations connecting progeny with the parent, grand- parent and great-grand parent are 1, 2, 3, respectively. Thus with the increase in number of generations between direct relatives, the coefficient of relationship is reduced to half in each generation.

Similarly, the coefficient of relationship between two collateral relatives is reduced with the increase in number of generations connecting them.

- The examples are given below:

| Relatives | No. of generations connecting them | Coefficient of relationship between them |
|---|---|---|
| Half sibs | 1 + 1 = 2 | 0.25 |
| Half cousins | 2 + 2 = 2 | 0.0625 |
| Full sibs | 1 + I =2 Through each of two common ancestor | 0.50 |
| First cousins (single) | 2 + 2 = 4 Through each of two common ancestor | 0.125 |

## 2. Number of paths connecting two individuals

The number of paths connecting two individuals depends upon the number of common ancestors and the repeated appearance of common ancestor in the pedigree. Both these factors increase the relationship between two related individuals (collaterally or directly related).

*(a) No. of common ancestors*: The increase in the number of common on ancestors increases the relationship between two individuals. If they have more than one common ancestor in their pedigree they receive similar genes from all the common ancestors. Two half sibs have one common ancestor and the relationship between them is 25% whereas two full sibs have two common ancestors for which their relationship is 50%.

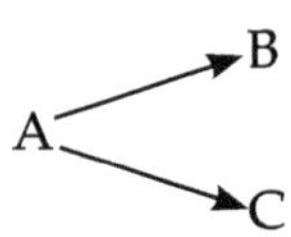


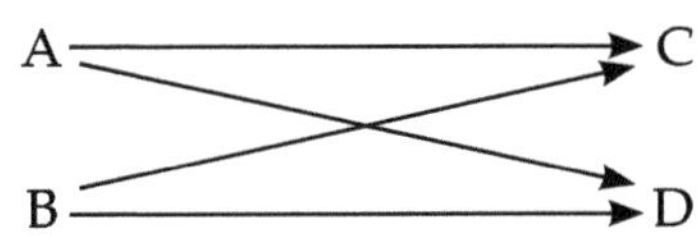


$R_{BC}$ = $(½)^2$ for CAB path
= ¼ = 0.25

$R_{CD}$ = $(½)^2$ = ¼ for DAC path
= $(½)^2$ = ¼ for DBC path
Sum = ½ = 0.50

*(b) Repeated appearance of common ancestor in the pedigree:*

- ***Inbreeding of individuals***: This increases the numbers of paths connecting two individuals and hence increases the relatedness between them. The increase in relatedness leads inbreeding in two ways.

The first way is that two related individuals receive common genes from the common ancestor through more than one path. The contribution of each path is added.

The second way is the consequent effect of the first effect (increase in relatedness). As a result of the increase in relatedness, either one or both the related individuals (descendents of related parents) become inbred.

The inbreeding creates variability between lines or families (sire family, dam family) and hence the similarity of genes between two related individuals is reduced. Thus, inbreeding of one or both the related individuals tends to reduce the relationship between them. Inbreeding causes the inbred individuals to become more homozygous and less variability in genotype. This is turn reduces the probability of two individuals to have the same genes by about the average of their intensity of inbreeding (inbreeding coefficient).

However, the net effect of the repeated appearance of the common ancestor in the pedigree of two related individuals is the increase in their relationship. This is because the first effect (the parents of the related individuals becoming relatives) is greater in magnitude which increases the relationship between them compared to the second effect which makes them inbred and reduces the relationship. This reduction can be seen from the following 2 pedigrees connecting two related individuals by more than one path with common ancestor (s).

*The first pedigree* takes the following form -

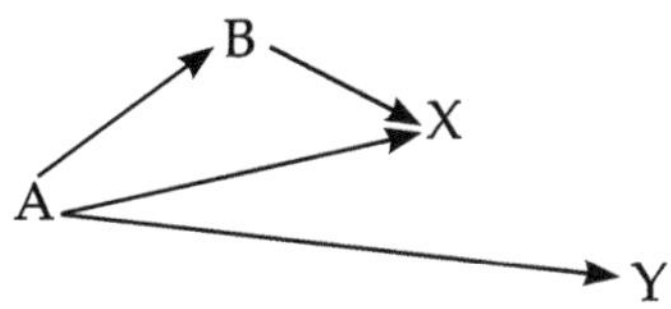


***Pedigree 17.1:*** Half sibs one of them produces by sire-daughter mating.

In this pedigree the two individuals X and Y are connected by 2 paths with their common ancestor A. Had they been really the half sibs *(collateral relatives)* they would have been connected by only one path (Y A X) and their relationship would have been $(½)^2$ = ¼ =. 0.25 but they are also connected by an additional path (Y A B X) as one of the half sib (X) has been produced by sire-daughter mating. The contribution of the additional path is $(½)^3$= 1/8 = 0.125. Thus the total relationship between them has become 0.2.5 + 0.125 = 0.375. Besides this, one of the two individuals, X, has become inbred because it is the progeny of sire-daughter mating (A and B). Thus X has become more homozygous than Y because the individual X has gene pairs homozygous which were heterozygous in its common ancestor, A. This has reduced the similarity of genotypes between X and Y. This is taken into account by dividing the sum of contribution of all the paths with the average inbreeding coefficient of two individuals (X and Y) to compute $R_{XY}$ as:

$$R_{XY} = \frac{\Sigma(½)^n}{\sqrt{1 + (F_X)\,(1 + F_Y)}}$$

The individual X is the inbred progeny from sire-daughter mating and hence $F_X = 0.25$.

Therefore,

$$R_{XY} = \frac{0.375}{\sqrt{(1 + 0.25)\,(1.0)}} = \frac{0.375}{1.118} = 0.3354$$

This means that X and Y have 33.54% similar genes and hence the *coefficient of relationship expresses the similarity of genotypes between two related Individuals instead of taking it in terms of the probability of genes having received by Iwo individuals from the common ancestor.*

In totality, the repeated appearance of the common ancestor (A) and connecting two individuals (X and Y) by more than one path had increased the relationship from 0.25 to 0.375 but consequently the inbreeding of X has increased the percentage of homozygous genes in X which it received from A whereas the second individual Y remained heterozygous. This reduced the relationship between X and Y from 0.375 to 0.3354. However, the net result of the repeated appearance of the common ancestor in the pedigree is the increase in the relationship between two individuals related collaterally.

Similar is the effect of the repeated appearance of the common ancestor on the relationship between two *direct relatives.* The parent-offspring relationship is 0.50 ($R_{PO}$ = 0.50) but when the ancestor appears more than once in the pedigree of the descendent, the relationship between them is increased But this makes the descendent inbred-which become more homozygous than its ancestor and hence the relationship between them is reduced. However, the net effect is the increase in relationship between two direct relatives. The $R_{AX}$ in the above pedigree is computed as:

$$R_{AX} = \sum \tfrac{1}{2}^{n} \sqrt{\frac{(1+F_X)}{(1+F_Y)}}$$

Where, $\sum \frac{1}{2}^{n} = (\frac{1}{2})^1 + (\frac{1}{2})^2$ for X A and XBA paths, respectively.

$= \frac{1}{2} + \frac{1}{4} = 0.75$

$F_A = 0$

$F_X = (\frac{1}{2})^2$ for A B path

$= \frac{1}{4} = 0.25$

Therefore, $R_{AX} = 0.75\sqrt{\frac{1.0}{1.25}} = 0.75 \times 0.8944 = 0.6708$

Thus, the $R_{AX}$ has increased from 0.50 to 0.75 due to the repeated appearance of A in the pedigree of X but due to the inbreeding of X, the $R_{AX}$ has been reduced from 0.75 to 0.6708 with the net result in an increase in the relationship between two direct relatives from 0.50 to 0.6708 due to the repeated appearance of common ancestor.

The *second pedigree* takes the following form:

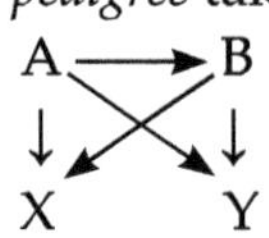


**Pedigree 17.2**: Full sibs from sire-daughter mating

In this pedigree the two individuals X and Y are a combination of direct and collateral relationship because B is the parent of both X and Y whereas A is the parent as well as the grand parent of both X and Y. Had they (X and Y) been simply the *collateral relatives* as full sibs, their relationship would have been 0.50 but in this pedigree the common ancestor A has repeated more than once and therefore X and Y are connected by more than one path through each of their two common ancestors (A and B). The sum of contribution of all the paths toward $R_{XY}$ is taken as:

$$\Sigma (½)^{n} = (½)^2 \text{ for YAX path}$$
$$+ (½)^3 \text{ for YA B X path}$$
$$+ (½)^2 \text{ for YBX path}$$
$$+ (½)^3 \text{ for YBAX path sum}$$

However, both X and Y are inbred being both the progeny of sire-daughter mating and hence the $F_X = 0.25$; $F_Y = 0.25$. Now, the $R_{XY}$ is computed as:

$$R_{XY} = \frac{\Sigma (½)^{n}}{\sqrt{(1 + F_X)(1 + F_Y)}}$$

$$= \frac{0.75}{\sqrt{(1 + 0.25)(1.0 + 0.25)}} = \frac{0.75}{1.25} = 0.60$$

Therefore,. X and Y have 60% similar genes. Due to repeated appearance of both the common ancestor (A and B) in the pedigree of X and Y, the relationship increased between X and Y from 0.50 to 0.75 but the inbreeding of both X and Y made them homozygous probably for different gene pairs due to crossing over and recombination of genes and hence the relationship has reduced from 0.75 to 0.60. Thus finally, $R_{XY} = 0.60$.

In case of the relationship between two *direct relatives* in this pedigree, similar is the effect of repeated appearance of the ancestor on their coefficient of relationship. This can be seen by computing the relationship among 4 pairs of direct relatives which is 0.75. This is the genetic covariance among them. Now the coefficient of relationship among all of them can be computed as:

$$R_{AX} = \frac{\Sigma (½)^{n} (1 + F_A)}{\sqrt{(1 + F_X)(1 + F_Y)}}$$

$$= \frac{\Sigma (½)^{n}}{\sqrt{(1 + F_A)(1 + F_Y)}}$$

$$= \frac{0.75}{\sqrt{1.25}} = \frac{0.75}{1.18} = 0.6708$$

Or,

$$= \frac{\Sigma(½)^{n} \sqrt{(1 + F_A)}}{\sqrt{(1 + F_X)}}$$

$$= \frac{0.75}{\sqrt{1.0/1.25}} = 0.75 \times 0.8944 = 0.6708$$

Similarly,

$$R_{BX} = 0.75\sqrt{\frac{1 + F_B}{1 + F_X}}$$

$$= \frac{0.75\sqrt{1.0}}{\sqrt{1.25}} = 0.6708$$

$$R_{AY} = 0.75\sqrt{\frac{1+F_A}{1+F_Y}}$$

$$= \frac{0.75\sqrt{1.0}}{\sqrt{1.25}} = 0.6708$$

$$R_{BY} = 0.75\sqrt{\frac{1+F_B}{1+F_Y}}$$

$$= \frac{0.75\sqrt{1.0}}{\sqrt{1.25}} = 0.6708$$

Thus, the relationship between all the 4 pairs of direct relatives has increased from 0.50 to 0.75 for repeated appearance of their ancestor and finally decreased from 0.75to 0.67 due to their inbreeding. But the net effect is the increase in relationship.

## 3. Inbreeding of common ancestor

The inbreeding of the ancestor increases the relationship between two related individuals either related directly or collaterally. The relationship is increased with the rate of inbreeding of the ancestor. The inbreeding increases homozygosity and hence the inbred ancestor will transmit more frequently its identical genes to its descendents and this will increase the similarity of genotypes between two collaterally related individuals. Similarly, the relationship between two direct relatives with inbred ancestor will be greater compared to when the ancestor is not inbred. Conclusively, the inbred ancestor contributes more identical genes to its progeny. This additional contribution of the inbred ancestor is taken care of by multiplying the contribution of each inbred ancestor from each of its path with the inbreeding coefficient of the ancestor as $(1 + F_A)$. Therefore,

$$R_{XY} = \sum (½)^n (1 + F_A)$$

The effect of inbreeding coefficient of the ancestor can be seen from the following 2 pedigrees- -

The *first pedigree* for two collaterally and directly related individuals with inbred ancestor may take the form of half sibs from inbred sire-

In this pedigree, the two individuals X and Y are half sibs *(collateral relatives)* from their inbred sire, S. The inbreeding coefficient of S is 0.125 as it was produced form half sib mating. Therefore,

$$R_{AX} = \sum (½)^n (1 + F_A)$$

$= (½)^2 (1 + 0.125)$ for YSX path

$= 0.25 \times 1.125 = 0.28125$

The S being inbred is expected to transmit more identical genes to both its progeny (X and Y) which has increased the coefficient of relationship between X and Y from 0.25 to 0.28125.

Now, take the relationship of *direct relatives* (say between S and X) which are related as parent-offspring and hence $R_{57}$ should be 0.50. However, the parent S is inbred and it has transmitted its more of the identical genes to X, resulting in an increase of their relationship. The $R_{XS}$ is computed as:

$$R_{XS} = \sum (½)^n (1 + F_A)$$

$= (½)^1 (1 + 0.125)$ for XS path

$= 0.5 \times 1.125 = 0.5303$

Or, $$R_{XS} = \sum (½)^n \frac{\sqrt{(1 + F_A)}}{\sqrt{(1 + F_X)}}$$

Therefore, the inbreeding of ancestor has increased the relationship between two direct relatives.

The *second pedigree* of two collaterally and directly related individuals may take the following form of full sib mating with inbred sire. -

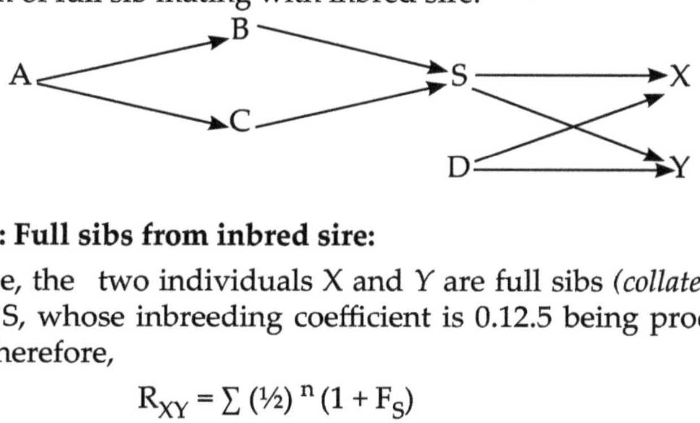


**Pedigree No. 17.3: Full sibs from inbred sire:**

In this pedigree, the two individuals X and Y are full sibs *(collateral relatives)* from inbred sire, S, whose inbreeding coefficient is 0.12.5 being produced from half sib mating. Therefore,

$$R_{XY} = \sum (½)^n (1 + F_S)$$

$= (½)^2 = ¼ = 0.25$ for Y **B** X path

$+ (½)^2 (1 + F_S) = ¼ (1.125) = 0.28125$ for Y **S** X path

Sum = 0.53125

Thus, the individuals X and Y have 53.125% similar genes which is greater than 0.50, if they had not been the progeny of inbred sire.

For *direct relatives,* say for S and X, their relationship will be same as obtained under half sibs from inbred sire ($R_{SX} = 0.5303$).

## 4. Combined effect of inbreeding of the common ancestor and the individual

The inbred individual is produced by the mating of its related parents. The followings are two pedigrees in which either one or both the individuals have been produced as inbred (from related parents) from their inbred ancestor.

The *first pedigree* may take the following form:

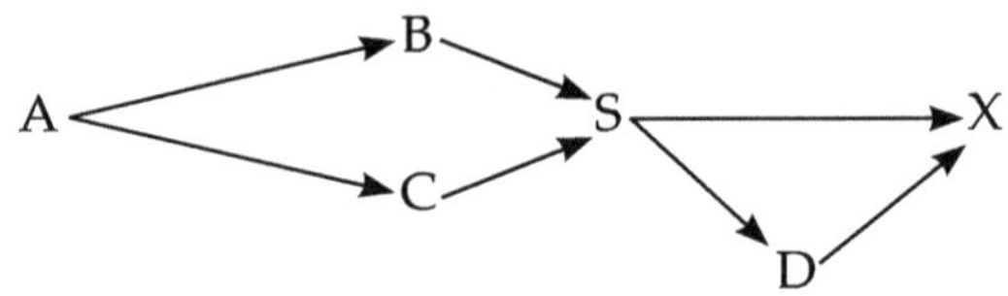


**Pedigree No. 17.4**: Half sibs from sire-daughter mating with inbred sire:

In this pedigree, the D and X are the progeny of S and thus S is the common ancestor of both X and D for which they are half sibs but the relationship between X and S is direct being parent offspring. Thus the relationship between D and X is a combination of both direct and collateral relatives. Therefore, $R_{DX}$ will be computed by using the following formula:

$$R_{DX} = \sum (½)^n \frac{(1 + F_A)}{\sqrt{(1 + F_X)(1 + F_Y)}}$$

$$= \sum (½)^n \frac{(1 + F_S)}{\sqrt{(1 + F_D)(1 + F_X)}}$$

Where, $\sum (½)^n (1 + F_S) = (½)^1 = 0.50$ for X D path

$+ (½)^2 (1 + F_5) = 0.28125$ for X S D path

Sum = 0.78125

$F_D = 0.28125$

Therefore, $R_{DX} = 0.78125/\sqrt{(1 + 0)(1 + 0.28125)} = 0.78125/1.1319 = 0.6902$

Now, the coefficient of relationship between two direct relatives (S and X) will be computed as:

$$R_{SX} = \Sigma\left(\frac{1}{2}\right)^n \frac{(1+F_S)}{\sqrt{(1+F_S)(1+F_X)}}$$

Where, $\sum (½)^n (1 + F_S) = (½)^1 (1.125) = 0.5626$ for XS path

$+ (½)^2 (1.125) = 0.28125$ for XDS path

Sum = 0.84375

$$R_{SX} = \frac{0.84375}{\sqrt{1.125 \times 1.28125}} = \frac{0.84375}{1.20058} = 0.7028$$

The S has contributed its 75% genes to X out of which 50% directly and 25% through its daughter D. The S is inbred and hence its contribution has increased from 0.75 to 0.84375 due to its inbreeding of 12.5%. This sire-daughter mating has reduced the relationship between the two directly related individuals (S and X) because the X is highly inbred (0.28125) than the ancestor S *(0.125)*. Thus, the descendent X has some more homozygous gene pairs that were heterozygous in its ancestor, S and hence the similarity of genotypes between S and X has been reduced from 0.84375 to 0.7028. The two individuals thus have 70.28% of similar genes.

The *second pedigree* may take the following form -

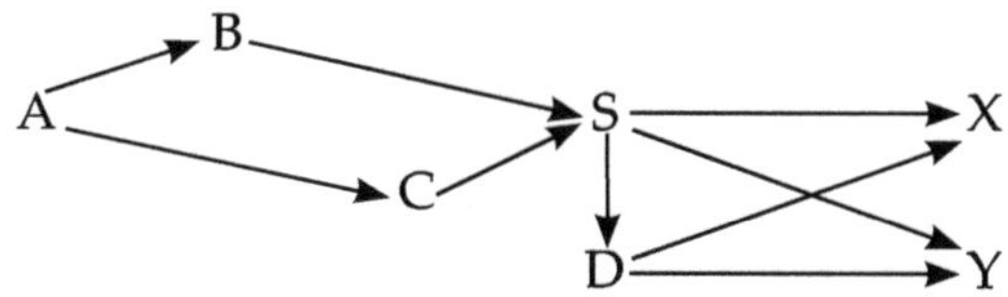


**Pedigree No. 17.5**: Full sibs from sire-daughter mating with inbred sire.

The relationship between full sibs is 0.50 and it increased to 0.53125 when sire was inbred, and to 0.60 when both X and Y were inbred produced from sire daughter mating. Now the next situation is a pedigree of full sibs produced by sire-daughter mating with inbred sire. The coefficient of relationship is computed as:

$$R_{XY} = \sum (\tfrac{1}{2})^n \frac{(1 + F_S)}{\sqrt{(1 + F_X)(1 + F_Y)}}$$

Where, $F_S = 0.125$

$F_X = (\tfrac{1}{2})^{n+1} (1 + F_S)$

$= (\tfrac{1}{2})^{1+1} (1.125)$ for SD path

$= 0.25 (1.125) = 0.28125$

$F_Y$ is also 0.28125

Now, contribution of all the paths $\sum (\tfrac{1}{2})^n (1 + F_S)$

$= (\tfrac{1}{2})^2 (1.125) = 0.28125$ for Y**S**X path

$= (\tfrac{1}{2})^3 (1.125) = 0.140625$ for Y**S**DX path

$= (\tfrac{1}{2})^2 = 0.25$ for Y**D**X path

$= (\tfrac{1}{2})^3 (1.125) = 0.140625$ for Y**D**SX path

Sum $= 0.8125$

$$R_{XY} = \frac{0.8125}{\sqrt{(1 + F_X)(1 + F_Y)}} = \frac{0.8125}{\sqrt{1.28125 \times 1.28125}} = 0.6341$$

## 17.6 USE OF DEGREE OF RELATIONSHIP

The knowledge of the degree of relationship is of practical importance, in the following ways:

(*i*) The relationship is used to estimate the probable breeding value (PBV), in case if the individual's own performance is not available either due to sex limited nature of traits or when the traits are expressed in later life of individual or after death of the animal such as carcass traits and longevity. The slaughter of a relative of the individual gives an indication of the carcass quality of the individual. Likewise after the death of a relative, the relationship gives an indication of the longevity of the animal in question. The relative's values are proportional to the degree of their relationship.

(*ii*) The degree of relationship (closeness) between two related individuals is also required at the time of purchasing the animals from outside the herd. Both the parties involved in sale and purchase of the animal (owner and customer) are benefitted by the knowledge of the degree of relationship. The knowledge of relationship and sale price of earlier relative (which was sold earlier) of the individual, put to sale, the buyer may think for maximum bid (how much maximum should be paid) whereas the owner may guess that how much he should demand.

The breeder may also use the knowledge of the degree of relationship in another way to buy animals, belonging to an excellent pedigree, lacking in one trait but not to the extent of rejecting the animal. One of the two animals in this pedigree is excellent in performance and the owner is demanding high price whereas the sale price is less for another animal of this pedigree which lacked in one trait. If these two animals have high degree of relationship, it is better to buy the second animal at low price (whose PBV is not lesser than the high priced one) for genetic improvement of the herd. It is therefore, essential to know the degree of relationship for its above mentioned advantages.

(*iii*) The knowledge of the relationship between two individuals is also useful in formulating the mating plan either to avoid inbreeding or to produce inbred progeny.

## 17.7 INBREEDING

The mating between relatives is known as inbreeding or genetic assortative mating. The inbreeding is a mating system when the mates are more closely related, up to 4-6 generations, than the average members of their population. The limit of relationship has been kept up to 6 generations, for the reason that the contribution of ancestral genes to the total genetic constitution of the descendent of $7^{th}$ generation is less than one percent on the basis of halving nature of inheritance. Therefore, inbreeding is considered as the mating of animals related up to 4-6 generations. The progeny produced by mating of two relatives is called *inbred individual*. The extent to which an inbred individual carry the genes identical by descent is the degree of inbreeding which depends upon the degree of relationship between the parents of the inbred individual.

## 17.8 TYPES OF INBREEDING

The inbreeding is of two type viz. close breeding and line breeding.

### 17.8.1 Close breeding

This is the mating of more closely related individuals like parent-offspring mating (sire – daughter, son-dam), full sib mating. Sometimes, the half sib mating is also taken as close breeding. The close breeding has some ***advantages*** and hence it is done for some specific purposes which are –

* To develop highly inbred lines.

- To discover undesirable and eliminate the recessive genes with the help of parent-offspring mating.
- To get more uniform progeny

The close breeding has following ***disadvantages***:

* Intensification of undesirable characteristics in the progeny

* Inbreeding depression, increase in reproductive failure and breeding problems

* Inbred are more susceptible to various diseases and adverse environment.

### 17.8.2 Line breeding

It is the mating of animals in which the relationship of an individual is kept as close as possible to particular ancestor in the pedigree, preferably male ancestor. This is the mating of distant relationship than those for close breeding. The descendants are mated to outstanding animals (sire) up to 3-4 generation. In this system, the mating may be between a fixed sire and his daughters and grand daughters. This develops the sire line as shown below-

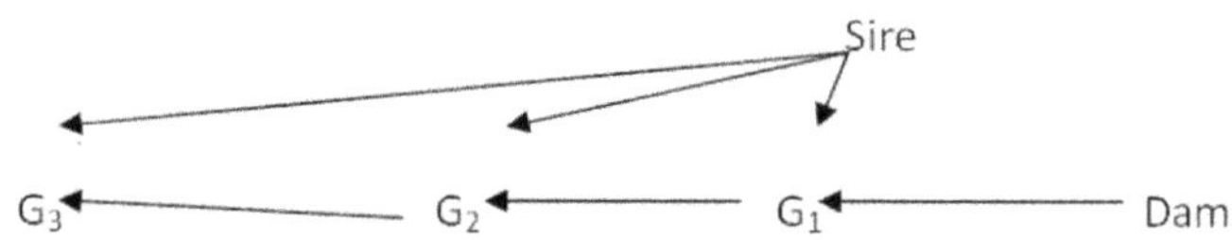


**Fig. 17.1:** Line breeding (mating of a sire to his offspring in successive generations).

The prime aim and advantage of line breeding is to maintain a high relationship to an outstanding ancestor for retaining its good proportion of genes. The relatives for their mating are selected because of a particular superior ancestor in their pedigree. The linebreeding thus incorporate into the linebred progeny the genes from the superior ancestor. The line breeding based on this logic can take various forms like parent-offspring mating in one generation or in subsequent generations continuously or after one-two generation as shown below-

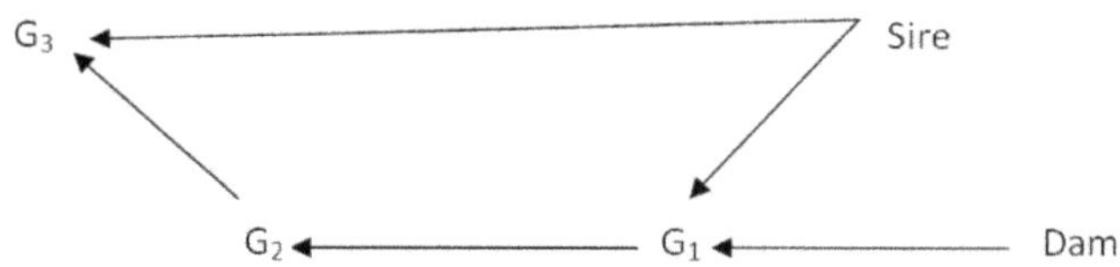


**Fig. 17.2** Line breeding (mating of sire to grand daughter)

In view of various forms of line breeding, it is difficult to classify some mating either as close breeding or line breeding, *e.g.* sire-daughter mating is taken as close breeding but it concentrates effectively the hereditary material of the sire and hence logically should be called as line breeding.

***Half sib mating as line-breeding***: The HS mating is practiced to avoid intense inbreeding and used as a system of line breeding when the ancestor is either dead or not available for breeding. In this situation half sib mating is the closest possible mating for line breeding. Below are given some of the pedigree to illustrate line breeding:

| Sr. No. | Pedigree | $F_x$ | $R_{AX}$ | Parent A inheritance |
|---|---|---|---|---|
| 1 | 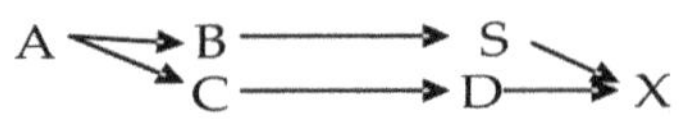 | 0.031 | 0.06 | 25 % |
| 2 | 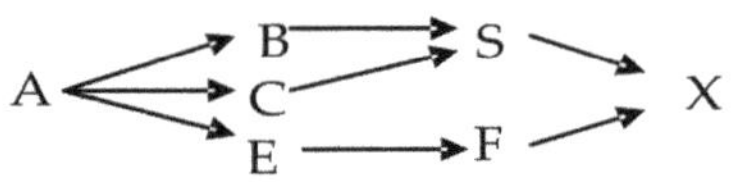 | 0.0625 | 0.36 | 37.5 % |
| 3 | 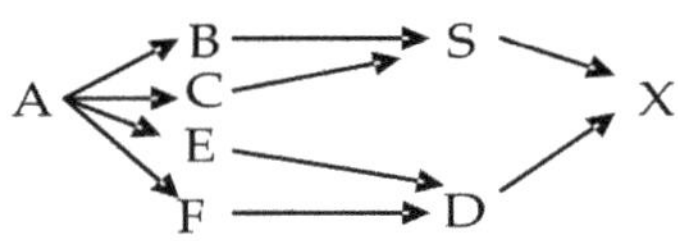 | 0.125 | 0.47 | 50 % |
| 4 | 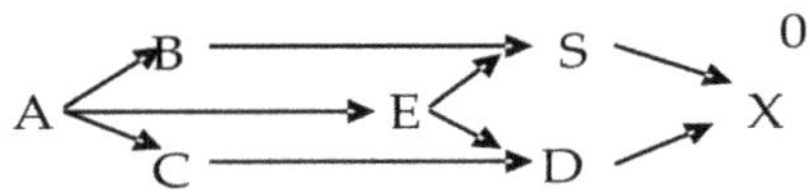 | 0.219 | 0.45 | 50 % |
| 5 | 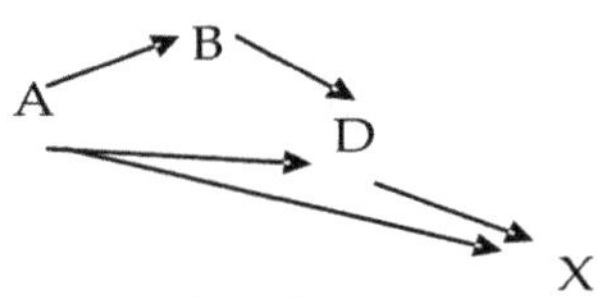 | 0.375 | 0.78 | 87.5 % |

In pedigree 1, the sire and dam of X are related due to their sires (B and C) are half brothers; in pedigree 2, the sire was produced from HS mating but not the dam; in pedigree 3, HS mating have been made in the pedigrees of sire and dam of X (4 individuals B, C, E, F are half sibs). The individual X is not highly inbred in these 3 pedigrees and $R_{AX}$ is reduced with reducing the number of paths between them (A and X). In pedigree 4, the line-breeding is through only one male (E) rather than 2 males (C and E as in pedigree 3) and the $R_{AX}$ is nearly the same in pedigree 4 (0.45) and pedigree 3 (0.47) but the $F_X$ is nearly doubled in pedigree 4 than in pedigree 3. This is the disadvantage of pedigree 4 wherein the line-breeding to ancestor A is through a single male (E) and inbreeding is more intense.

***Line-breeding differ from ordinary inbreeding:*** The ordinary inbreeding and line-breeding differ with respect to the relationship of an individual (kept close in line-breeding) to a particular ancestor. Secondly, in ordinary inbreeding (other than line-breeding), the related parents are mated with no attempt to increase the relationship of the inbred progeny to any particular ancestor. Thirdly, the line breeding is practiced to retain or concentrate the genes of an outstanding animal.

***When to use line-breeding:*** The line-breeding concentrates the inheritance of any one ancestor and hence it should be practiced in better than average herd, particularly when it is not possible to get good quality sires from outside. Secondly, there should be at least two sires in the herd in which the line breeding is practiced, to keep low the $F_X$. Thirdly, the line-breeding should be used in a purebred population of high genetic merit after identifying an outstanding individual based on progeny test. It should not be used in grade or commercial herds producing animals for the market. Line breeding is useful when the outstanding individual is dead or not available. Line-breeding must be directed through one or more of the sons of outstanding ancestor.

***Why line-breeding is favoured:*** The line-breeding is favoured because it is less intense than close breeding and hence results in low ID as well as other dangers of close-breeding. Secondly, the line breeding perpetuates the special character of an outstanding animal or strain. This system is followed by selecting the best and culling the poor animals. In line-breeding, the sire daughter mating is not recommended but HS mating are done among the progeny of a particular sire. This is done to avoid intense inbreeding.

***17.8.3. Development of inbred lines*:** Inbreeding is useful for certain purposes. It is then important to know the method of development and also the problems faced to develop them.

Inbreeding should be more intense to observe the effect of recessive genes hidden in parental stock. Thus, it is important to know the method of developing inbred lines. For maximum intensity of inbreeding in first generation, two types of mating are done.

First is parent offspring mating which includes mating of a sire to his own daughter, or mating a dam to her own son. Such mating produce progeny having about 75% inheritance of the parent mated to its own offspring. This will give an indication of the number and kind of recessive genes that a parent carries when it is mated to several off-springs.

The second kind of mating is the mating of full sibs which results in 25% inbreeding in the first generation. The parent-offspring and full sib mating both produce 25% inbred progeny. In FS mating two different individuals (both grand-parents) instead of one contribute to the inbreeding but in parent-offspring mating only one individual, that is parent as well as grand- parents of the inbred individual, is responsible for all the increase in homozygosity which is 25%. Thus a specific recessive trait is more likely to be expressed in the parent due to a single parent. However, the more intense form of inbreeding is more likely to express traits regardless of the kind of mating because more intense inbreeding identify more quickly the lines having detrimental genes. The lines with more recessive genes may be discarded and the superior inbred lines may be developed.

The highly inbred lines developed posses problems besides that development of inbred lines in animals is expensive, time taking and show inbreeding depression in performance traits. The main problem is the loss of their fertility and survivability.

The inbred lines have poor fertility and survivability and thus such lines become incapable to reproduce and maintain. Most of the inbred animals which die before weaning age probably have defective recessive genes. However, the lines with better performance and normal vigour should be propagated further.

### 17.8.4 Use and improvement of inbred lines

The inbred lines are developed mainly for crossing them to take advantage of hybrid viogour. This is the main use of inbred lines. The best combining lines based on the performance of their line bred progeny are selected. The parents whose progeny show better performance are said to have superior combining ability. In case of animals, the inbreeding is not possible to the extent of 100% as in case of plants. It is therefore advisable to start crossing lines when inbreeding is 40 to 50%. However, higher degree of homozygosity increases the repeatability of performance in recurrent crosses of the same lines.

As has been mentioned repeatedly about the inbreeding depression shown by inbred lines, it is therefore needed to correct for the inbreeding depression i.e. to say how to improve the inbred lines. The improvement of inbred lines of corn is done by several methods, some of which can also be used to improve inbred lines of animals. The improvement of lines may be by crossing two lines which are superior in one trait sand inferior in other trait, e.g. one lines is superior in growth rate and inferior or mediocre in trait whereas the reverse is true for second line. In this situation, these two lines are crossed to combine the two qualities, the line cross progeny is backcrossed to the first as well as to the second line. The back cross progeny are selected for better growth rate and carcass trait, and inbreeding in each line produced by back crossing is crossing is continued. The back crossing with selection is continued within each line for each trait as shown below taking two lines A and B.

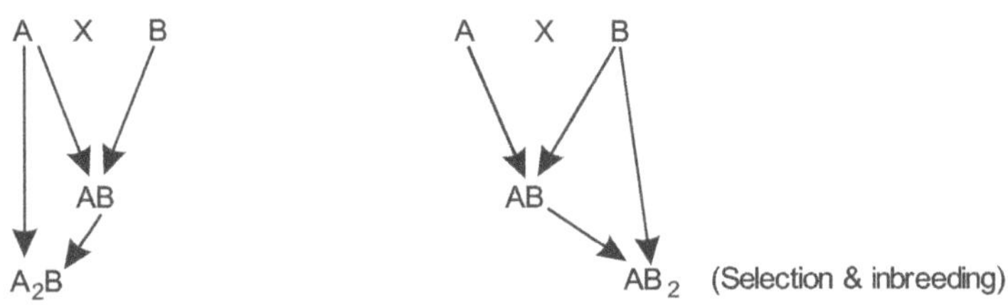


Back crossing and selection within each line is continued. This method incorporates new desirable additive genes into an inbred line.

It is not always necessary to cross two inbred lines to combine their best qualities. This can be done by mating of several out bred females having desirable quality with males of an inbred line. The inheritance of some of these females producing superior progeny can be incorporated into the inbred line by continued back crossing of her descendants with the original inbred line. Similarly, out bred males may also be used to introduce new genes into an inbred line.

The development of superior inbred lines and crossing those having the best combining ability is one of the methods to increase the efficiency of livestock production. But this method is not suitable for small herds. In large purebred herds the inbred lines may be developed for commercial use. The use of inbreeding for commercial production is being exploited in poultry and pigs.

## 17.9 DISADVANTAGES OF INBREEDING (GENERAL)

* The frequency of undesirable recessive genes is increased

* Inbreeding results in inbreeding depression in growth, reproduction and vigour in terms of vitality

* Inbred animals are more susceptible to environmental changes

## 17.10 APPLICATIONS OF INBREEDING

The inbreeding is a useful tool in animal breeding. Though inbreeding results in inbreeding depression but there are certain cases when inbreeding is advantageous-

* Inbreeding is used to determine the genetic worth of an animal by mating a sire with his about 30 daughters. It will ***test the sire for recessive alleles***.

* The inbreeding brings the undesirable recessive alleles in homozygous condition and it may thus help to uncover and eliminate recessive genes through selection.

* Inbreeding is used to ***produce distinct families*** within a breed. The selection between families for traits of low $h^2$ is more effective than individual selection.

* Inbreeding is used to ***develop inbred lines*** as a seed stock which can be crossed according to their combining ability. Thus, inbred lines can be used in crossbreeding programme. The highly inbred lines are also used in bioassay and other research work related to genetics, nutrition and physiology.

* Inbreeding should be used only for production of seed stock, to increase the purity of breeding animals.

* Inbreeding with selection had helped in the past to ***develop several breeds*** of livestock viz. Long horn cattle, Leicester sheep, Shire horse of Robert Bakewell and later on Merino, Rambouillet and South Down breeds of sheep were developed. Some strains of lab.animals (mice, guinea pigs, rabbit) had also been developed.

* Inbreeding also helps to know the ***type of gene action*** affecting a trait as the traits affected by additive gene action do not show inbreeding depression.

* Inbreeding also increases the homozygosity of favourable genes. The homozygotes with better performance can be selected for further breeding and these can pass on their superiority (own characteristics) to their progeny more successfully. This is called as the ***prepotency*** which is defined as the ability of an animal to stamp its own characteristics on the progeny. The prepotency depends on the homozygosity of dominant genes. Thus, the inbred are more potent.

However, the inbreeding should be practiced with selection. Secondly, inbreeding should be practiced only for research work and for the production of seed stock, particularly the males. The inbred females should not be used in commercial herd because of their low performance than outbreds.

## 17.11 WAYS TO MINIMIZE INBREEDING

The level of inbreeding primarily depends on the mating plan, number of sires used for breeding and the population size.

***Mating plan:*** The mating should be allowed among unrelated animals to avoid inbreeding. This requires the estimation of the coefficient of relationship of a bull with all the breedable females in the herd.

***Population size and number of sires:*** The inbreeding is unavoidable in a closed herd of limited size having about 100 females and 5 – 10 sires. An increase in herd size is not possible for the reason of limited resources, etc. Further, it is the effective population size (Ne) which decides the level of inbreeding. The rate of inbreeding ($\Delta F$) with respect to Ne for unequal number of animals of two sexes is:

$$\Delta F = \frac{1}{2Ne} = \frac{1}{8Nm} + \frac{1}{8Nf}$$

The Ne depends more on the sex with smaller number and hence on the number of bulls used. The Ne is reduced with decrease in the number of males used for breeding and consequently the $\Delta F$ is increased. The Ne should be such that $\Delta F$ is not more than 1.0 % per generation. It is thus essentially required to use more number of bulls for breeding.

As the $\Delta F$ is increased by use of limited number of sires in a single herd, it is better to introduce new breeding bulls in the herd from outside to reduce the $\Delta F$. The associated herd progeny testing programme and frozen semen technology have made it easy to introduce new germ plasm in the herd.

## 17.12 GENETIC EFFECT OF INBREEDING

See section 10.1 and 10.2 of chapter 10 in "Genetic Structure of Livestock Population" by Tomar and Tomar (2014).

## 17.13 PHENOTYPIC EFFECT OF INBREEDING

The inbreeding has adverse effect particularly on fitness traits like reproduction, vigour (vitality) and growth characters.

*1.* The inbreeding tends to depress the ***growth rate*** in farm animals. This results in small margin of profit when the live weight decides the profit in meat animals. The growth rate also affects the production and reproduction of animals.

**2.** The ***reproductive efficiency*** is reduced with inbreeding in farm animals. The reproductive efficiency is adversely affected in terms of delay in puberty and testicular development, reduction in gametogenesis (number of ova shed by females) and increase in embryonic losses, though with variable results.

3. The inbreeding into an increase in death rate and hence loss in ***vigour.*** The inbred animals are more susceptible to environmental conditions.

4. The inbreeding also causes the appearance of ***genetic defects*** which are mostly recessive in inheritance and remain hidden in out-bred animals. The inbreeding brings them in homozygous state and hence abnormalities are expressed and identified.

5. The ***phenotypic uniformity*** among inbred animals is increased for the characters which are mono-factorial. The variations within inbred line are more due to environment than to genetic causes.

**6. *Change in Population Mean on Inbreeding*:** The population mean of productive traits is reduced due to inbreeding and it is known as ***inbreeding depression.***

See point 5 section 10.3 of chapter 10 in "Genetic Structure of Livestock Population" by Tomar and Tomar (2014).

**7. Change of variance on inbreeding:** The inbreeding besides reducing the mean value also changes the variance, both genetic as well as environmental variance.

See point 6 section 10.3 of chapter 10 in "Genetic Structure of Livestock Population" by Tomar and Tomar (2014).

## 17.14 MEASURE OF DEGREE OF INBREEDING

The primary consequence of inbreeding is the increase in homozygosity. The inbreeding increases the homozygosity due to the genes identical by descent (autozygosity). The genes identical by descent are the result of genetic similarly of the two gametes produced by two relatives mated to produce the inbred progeny.

### 17.14.1 Definition

The extent to which an inbred individual carry the genes identical by descent is the degree of inbreeding and measured by *coefficient of inbreeding*. The degree of inbreeding of an individual (degree of genes identical by descent) depends upon the degree of relationship between the parents of the inbred individual. The coefficient of inbreeding is a measure of the increase in homozygosity in an individual over the parents, due to inbreeding. The inbreeding coefficient also indicates the degree to which the parents of inbred individual are related.

Wright (1921) developed the procedure of path coefficient to measure the degree of inbreeding as the correlation between the genic values of the two uniting gametes and coined the term ***"coefficient of inbreeding"***. The inbreeding coefficient of an individual (F) is a function of the common ancestry of its parents and expresses the extent to which the two parents of the inbred individual are related through their common ancestor. This is the degree of relationship by descent between two parents of the inbred individual. The common ancestry increases the probability of two alleles at a locus that they are identical by descent.

Another approach based on probability rules was developed by Haldane and Moshinsky (1939), Cotterman (1940) and Malecot (1948). The inbreeding coefficient was defined by Malecot (1948) as *the probability that two genes at a locus in an inbred individual are identical by descent (autozygous).* Thus, an individual with inbreding coefficient F has a probability equal to F that the two genes at a particular locus are identical by descent and a probability equal to 1-F is that they are not identical but independent in origin.

### 17.14.2 Inbreeding coefficient is a relative concept

It is relative to a population at specific time in its history. The degree of inbreeding in terms of the coefficient of inbreeding is measured from the reference of base population in which F was zero (F = 0). Thus, the base population is assumed to be random mating in which inbreeding was started and hence from which the degree of inbreeding (F) is measured. The F measures the amount of autozygosity relative to that in the base population in which the alleles were not identical by descent. The F of an individual in the present population is then the probability that the two allelic genes had arisen from replication of a single allele in subsequent generations after the inbreeding had started in the base (reference) population. The reference population is that existed a few generations back to the present generation and the inbreeding coefficient in the present generation (population) then measures the degree of inbreeding occurred during the span of few generations between reference population (generation) and the present one. The value of F may range from 0 to 1 or from 0 to 100 percent.

### 17.14.3 Interpretations of F

The inbreeding coefficient is interpreted in the following different ways.

*(i)* **Population level:** The F is a measure of the proportional reduction in heterozygosity (or the proportional increment in homozygosity) in an inbred population relative to a random mating population with same allele frequency. This is useful to know the distribution of genotype frequencies in an inbred population with different degree of inbreeding and the fixation of alleles in the population.

*(ii)* **Single individual - single locus:** The F is the probability that the two allelic genes in an inbred individual are identical by descent (autozygous). The F represents the probability that a random locus that was heterozygous or allozygous has become homozygous (autozygous) due to inbreeding. The amount 1-F represents the probability that allelic genes are not identical.

*(iii)* **Single individual - all loci:** Inbreeding affects all loci and hence increases homozygosity embracing all loci. Therefore, considering all loci, the F indicates the percentage of all loci (equal to F) in an individual's genotype expected to become homozygous for the genes identical by descent that were heterozygous in the base population. The amount 1-F represents the probable proportion of loci that are still heterozygous and allozygous. The F, however, measures the additional increase in homozygosity (autozygosity due to genes identical by descent) due to inbreeding over and above what was present before inbreeding was started.

*(iv)* **Alleles of different individuals:** The idea of probability can also he extended for alleles that are not necessarily in the same individual. The F is defined

as the probability that any two randomly chosen alleles in a population (not necessarily in the same individual) are identical by descent. This concept is used to measure the genetic relationship within and between populations.

## 17.15 METHODS TO COMPUTE INBREEDING COEFFICIENT ($F_X$)

The following two methods to compute the inbreeding coefficient are in use

### 17.15.1 Wrigh's method of path coefficient to compute $F_X$

Wright (1921) applied the approach of path coefficient analysis to compute the coefficient of relationship between two individuals ($R_{XY}$) and the inbreeding coefficient of an individual ($F_X$):

$$Fx = \sum (½)^{n+1} (1+F_A)$$
$$= ½ \sum (½)^{n} (1+F_A)$$

Where, n = No. of generations between sire and dam of individual through common ancestor.

$\sum$ = Summation over all the paths connecting sire and dam of the individual through the common ancestor and over all the common ancestors, if the common ancestors are more than one.

$F_A$ = Inbreeding coefficient of the common ancestor for each path.

***Rules for tracing the paths:*** The rules to trace the paths and the rules for considering the inbreeding coefficient of common ancestor ($F_A$) are the same as outlined earlier in estimating the coefficient of relationship. In estimating the $F_X$ of inbred individual, the numbers of generations are counted connecting the sire and dam of the inbred individual (X) through common ancestor (A).

***Steps involved in estimating inbreeding coefficient:***

(*i*) To find out the common ancestor and the common inbred ancestor

(*ii*) Tracing the different paths connecting the sire and dam of inbred individual through common ancestor

(*iii*) Contribution of each path is $(½)^{n+1}$ or $(½)^{n+1} (1+F_A)$ if the common ancestor is inbred

(*iv*) Summation of all the paths.

### 17.15.2 Coancestry method to compute F

There is another way to consider the inbreeding coefficient of the progeny produced by mating of two relatives. The inbreeding coefficient of an individual depends on the amount of common ancestry (co-ancestry) of its parents, i.e. the common genes they have. Therefore, instead of inbreeding coefficient of the progeny, the degree of relationship by descent between two parents (coancestry) of an inbred individual is considered.

The relationship by descent between two parents is called the coancestry. In other words, the *coancestry* of two individuals *is the probability that two gametes taken at random, one from each individual, carry alleles identical by descent.* Thus, the coancestry of any two individuals (parents) is identical with the F of their progeny. The coancestry is particular to two parents whereas the inbreeding coefficient is particular to an individual at a particular locus. The coancestry is symbolized by f or r.

## SOLVED EXAMPLE

*Coancestry method to compute $F_X$.*

The method to compute the inbreeding coefficient of individual X by coancestry method has been illustrated from the following pedigree no.17.6 -

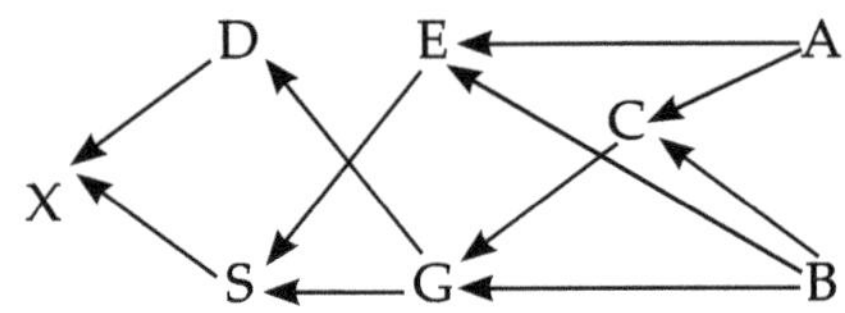


**Pedigree 17.6:**

***Solution:*** The F of an individual ($F_X$) is equal to the coancestry between its parents ($r_{SD}$)

$$F_X = r_{SD}$$

Coancestry between two individuals ($r_{SD}$) is the average of 4 coancestries between their parents (G and E) -

$$r_{SD} = \frac{1}{4}(r_{GG} + 2r_{EG} + r_{EE})$$

(*i*) $r_{GG} = \frac{1}{2}(1 + F_G)$

$F_G = r_{AB} = 0$

Thus, $r_{GG} = \frac{1}{2}(1 + 0) = \frac{1}{2}$

(*ii*) $r_{EE} = \frac{1}{2}(1 + F_E)$

$F_E = r_{BC} = r_{B(AB)} = \frac{1}{2}(r_{AB} + r_{BB})$

$r_{AB} = 0$ since, parents of A & B are unknown

$r_{BB} = \frac{1}{2}(1 + F_B) = \frac{1}{2}$ since, $F_B = 0$

$r_{BC} = \frac{1}{2}(r_{AB} + r_{BB}) = \frac{1}{2}(0 + \frac{1}{2}) = \frac{1}{4}$

Thus, $r_{EE} = \frac{1}{2}(1 + F_E) = \frac{1}{2}(1 + \frac{1}{4}) = 5/8$

(*iii*) $r_{EG} = r_{(B+C)(A+B)} = r_{BC}\, r_{AB}$

$= \frac{1}{4}(r_{BA} + r_{BB} + r_{CA} + r_{CB})$

$= \frac{1}{4}(0 + \frac{1}{2} + \frac{1}{4} + \frac{1}{4})$

$r_{CA} = r_{(A+B)A} = \frac{1}{2}(r_{AA} + r_{AB})$

$= \frac{1}{2}(\frac{1}{2} + 0) = \frac{1}{4}$

$r_{AA} = \frac{1}{2}(1 + F_A) = \frac{1}{2}(1 + 0) = \frac{1}{2}$

Thus, $r_{EG} = \frac{1}{4}(0 + \frac{1}{2} + \frac{1}{4} + \frac{1}{4}) = \frac{1}{4}$

Now, $r_{SD} = \frac{1}{4}(r_{GG} + 2r_{EG} + r_{EE})$

$= \frac{1}{4} (\frac{1}{2} + 2 \times \frac{1}{4} + 5/8)$

$= \frac{1}{4} (1 + 5/8) = \frac{1}{4} + 5/32 = 13/32 = 0.4062$

Thus, $F_X = r_{SD} = f_{SD} =$ **0.4062 Answer.**

### 17.15.3 Coancesry Vs Path Coefficient

(*i*) *Advantages:* The coancestry method is more convenient and more easily used particularly to compute F under regular systems of close inbreeding. This method gives correct results and leaves no scope of error which occurs in counting paths.

The method of coancestry is useful in preparing mating plans to avoid inbreeding because it computes F generation by generation and hence convenient to compute coancestries so as to compute F of the progeny to be produced. Therefore, this method keeps a running tally generation by generation to estimate the degree of inbreeding.

(*ii*) *Difference:* The coancestry method differs from path coefficient in a way that this method works forward from present generation to the next in contrast to path coefficient method which works back from present generation to the common ancestor.

(*iii*) *Similarity:* The two methods are identical in principle based on the halving nature of inheritance and random union of gametes from two parents. Therefore, the coancestry of two parents ($f_{XY}$) is identical (equal) to the inbreeding coefficient of their progeny ($F_Z$).

(*iv*) *Coancestry and the genic variance and covariance:* The coancestry is translated into genetic variance and covariance as:

***(a) Genic variance***: The genic variance of an individual is the genic covariance of the individual with itself. For example, the genic variance of an individual X will be:

$$\text{Var.}_X = \text{Cov.}_{XX}$$

The genetic relationship of an individual with itself is 1.0. This is the genic variance of an individual which is 1.0 for non-inbred population but it is equal to $I + F_X$ when the individual is inbred ($F_X$ = inbreeding coefficient of the individual X). Thus,

$$\text{Genic Var(X)} = \text{Cov.}_{XX}$$

$$= 1.0 + \tfrac{1}{2} \text{ (genic cov. of its parents)}$$

$$= 1.0 + \tfrac{1}{2} (\text{Cov.}_{AB}) \text{ since A \& B are parents of X}$$

$$= 1.0 + 0 \text{ (if X is not inbred)}$$

$$= 1 + F\,x \text{ (if X is inbred)}$$

Thus, $$F_X = \text{Var.}_X - 1.0$$

$$= f_{AB}$$

$$= \tfrac{1}{2} \text{Cov}_{AB} \text{ (A \& B are parents of X)}$$

Likwise, $$\text{Genic Var(Y)} = 1 + F_Y$$

***(b) Genic covariance***: The genetic covariance of two individuals (X and Y) is equal to their coancestry. This is taken as the average covariance of individual with the parents of the other individual. Therefore,

Genic Cov $_{XY} = f_{XY}$

= Average covariance of either individual with parents of other individual

= ½ (sum of Cov. of X with parents of Y)

= ½ (Cov. $_{CX}$ + Cov.$_{DX}$), where, C and D are parents of Y.

or = ½ (sum of Cov. of Y with parents of X)

= ½ ($Cov_{AY} + Cov_{BY}$) where, A and B are parents of X.

### 17.15.4 Regular Systems of Inbreeding And Their F

The regular systems of inbreeding are parent offspring mating and the mating of full sibs as well as half sibs. The selfing does not occur in case of farm animals.

**Pedigree 17.7:** ***Parent offspring mating:*** This is done in the following two ways–

(*a*) Line breeding: When the same parent is repeatedly mated to the progenies of subsequent generations. The inbreeding coefficient of this system is computed as –

$$F_t = ¼\,(1 + 2\,F_{t-1} + F_A)$$

Where, $F_t$ = inbreeding coefficient in $t^{th}$ generation,

$F_{t-1}$ = inbreeding coefficient in one generation back, and

$F_A$ = inbreeding coefficient of the parent used repeatedly.

The pedigree of this type of mating takes the form as –

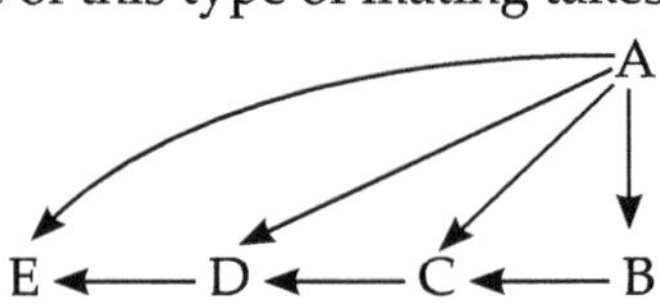


***Solution:*** The F of progeny of different generations are calculated as –

$$F_C = ¼\,(1 + 2 \times 0 + 0) = ¼ = 0.25$$

$$F_D = ¼\,(1 + 2 \times 0.25 + 0) = 0.375$$

$$F_E = ¼\,(1 + 2 \times 0.375 + 0) = 0.437$$

(*b*) Mating of offspring with younger parent: In this system old parent is replaced by younger parent. The F under this system is calculated as –

$$F_t = ¼\,(1 + 2\,F_{t-1} + F_{t-2})$$

The pedigree may take the following form –

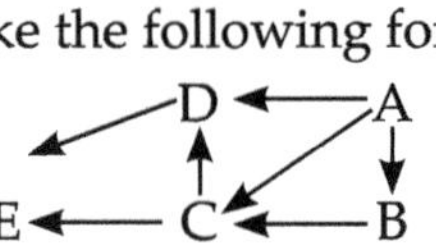


***Solution:*** *The* F of progeny of different generations is calculated as:

$$F_C = ¼\,(1 + 2 \times 0 + 0) = ¼ = 0.25$$

$$F_D = \tfrac{1}{4}(1 + 2 \times 0.25 + 0) = 0.375$$

$$F_E = \tfrac{1}{4}(1 + 2 \times 0.375 + 0.25) = 0.50$$

**Pedigree 17.8** ***Full sib mating:*** When there is full sib mating in each subsequent generation, the F of the progenies is calculated as –

$$F_t = \tfrac{1}{4}(1 + 2F_{t-1} + F_{t-2})$$

The pedigree under regular FS mating can be shown as under –

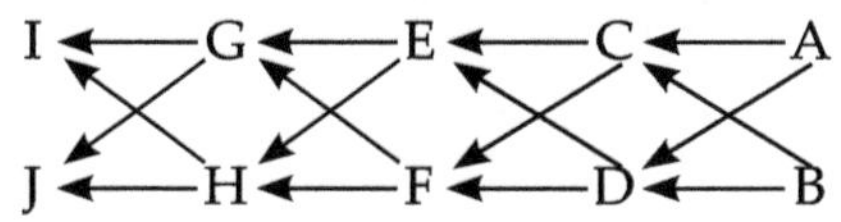


***Solution:*** The parents A and B of first generation are not inbred and hence

$$F_C = F_D = 0$$

$$F_E = F_F = \tfrac{1}{4}(1 + 2 \times 0 + 0) = \tfrac{1}{4} = 0.25$$

$$F_G = F_H = \tfrac{1}{4}(1 + 2 \times 0.25 + 0) = 0.375$$

$$F_I = F_J = \tfrac{1}{4}(1 + 2 \times 0.375 + 0.25) = 0.50$$

**Pedigree 17.9** ***Half sib mating:*** When there is half sib mating regularly in subsequent generations.

**(*a*)** ***HS mating:*** When the dams are half sibs to each other as well as to the sires. The recurrence equation to calculate the inbreeding coefficient of the progeny in $t^{th}$ generation in this case is:

$$F_t = 1/8\,(1 + 6F_{t-1} + F_{t-2})$$

The pedigree may take the following form -

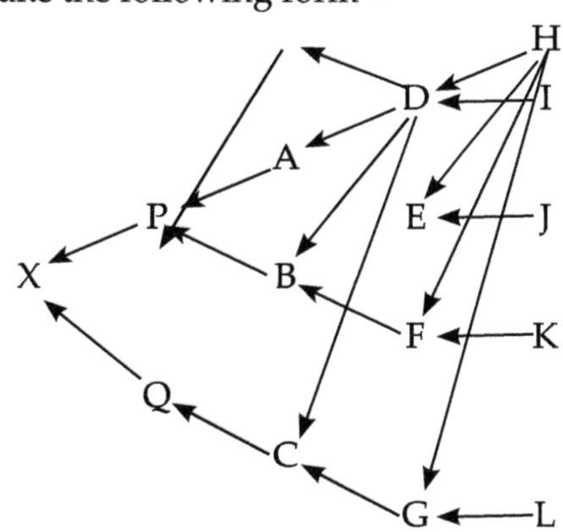


Solution:

$$F_{(1)} = F_A = F_B = F_C = 1/8\,(1 + 6 \times 0 + 0) = 0.125$$

$$F_{(2)} = F_P = F_Q = 1/8\,(1 + 6 \times 0.125 + 0) = 0.219$$

$$F_{(3)} = F_X = 1/8\,(1 + 6 \times 0.218 + 0.125) = 0.305$$

**(b) *HS mating:*** When dams are full sibs among themselves but they are half sisters of sires: In this case the recurrence equation for calculating $F_t$ is as under –

$$F_t = 1/16\,(3 + 8\,F_{t-1} + 4\,F_{t-2} + F_{t-3})$$

The pedigree may take the following form -

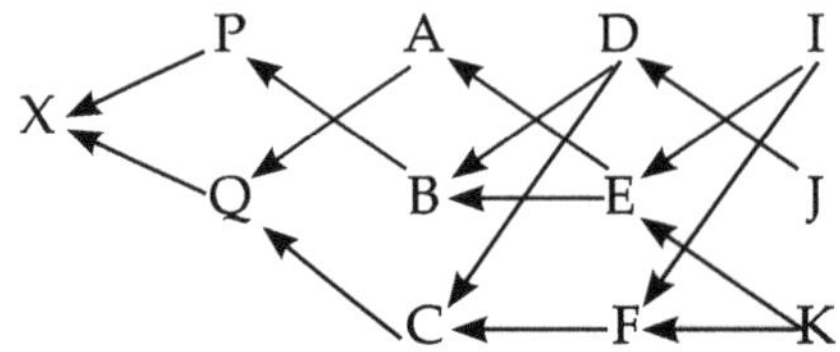


***Solution:***

$$F_{(1)} = F_A = F_B = F_C = 1/16\,(3 + 8 \times 0 + 4 \times 0 + 0) = 0.1875$$

$$F_{(2)} = F_P = F_Q = 1/16\,(3 + 8 \times 0.1875 + 4 \times 0 + 0) = 0.28125$$

$$F_{(3)} = F_X = 1/16\,(3 + 8 \times 0.1875 + 4 \times 0.28125 + 0) = 0.3515$$

## NUMERICAL SOLVED EXAMPLES

**Exercise 17.1:** Find out the inbreeding coefficient of the individual Z from the following ***pedigree No. 1*** by Wright's method.

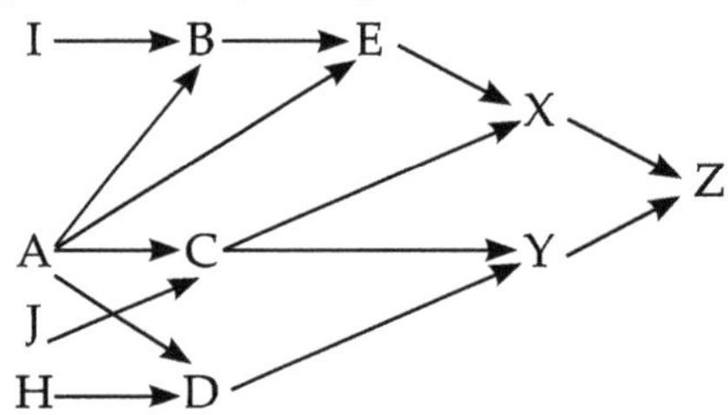


***Pedigree No. 1:***

**Solution:**

Calculations to compute $F_Z$, in the above pedigree, require the following three steps -

1. To find out the *common ancestors and the common inbred ancestors*: In the above pedigree A and C, are two common ancestors neither of which is inbred. The X and Y are inbred but their inbreeding coefficient will not be considered because they are the sire and dam of Z, neither of the two is common ancestor nor direct relatives.

2. *Tracing of the paths between sire and dam of inbred individual through common ancestors*: The first path connecting X and Y is through the common ancestor C and this is Y C X making n = 2. Further X and Y are connected through A by 5 paths viz. YDACX, YDAEX, YDABEX, YCAEX and YCABEX. It is better to underline the common ancestor. Neither of the common ancestor (C or A) is inbred.

3. *Contribution of each path*: This is estimated as $(½)^{n+1}$ or as $(½)^{n+1}(1 + F_A)$ if the common ancestor is inbred . The contribution of path is then added to find out the $F_Z$ as below:

| Paths | Generations | Contribution of each path = $(½)^{n+1}$ |
|---|---|---|
| YCX | 2 | $(½)^3 = 1/8$ |
| YDAEX | 4 | $(½)^5 = 1/32$ |
| YDABEX | 5 | $(½)^6 = 1/64$ |
| YCAEX | 4 | $(½)^5 = 1/32$ |
| YCABEX | 5 | $(½)^6 = 1/64$ |
| YDACX | 4 | $(½)^5 = 1/32$ |

$$F_Z = \frac{1}{8} + \frac{1}{32} + \frac{1}{64} + \frac{1}{32} + \frac{1}{64} + \frac{1}{32} = 0.25$$

**Example 17.2:** Compute $F_X$ from the following pedigree

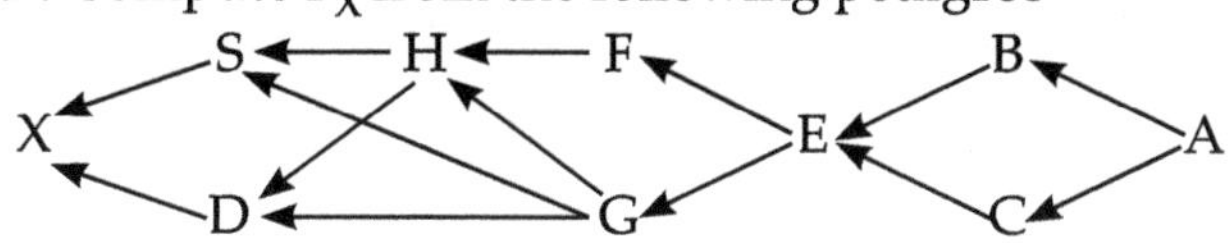


***Pedigree no.* 2**

**Solution:** There are three common ancestors (E, G, H) in the above ***pedigree no. 2***, out of which only two (E and H) are inbred.

(*i*) Common ancestor and common inbred ancestor: To compute $F_X$ requires the $F_E$ and $F_H$ because E and H are two common inbred ancestors whereas the common ancestor G is not inbred and hence $F_G$ is not required.

$F_{E'} = (½)^3 = 1/8 = 0.125$ for C AB path

$F_H{}' = (½)^3 (I + F_E)$ for G EF path

$= (½)^3(1+1/8) = 1/8(9/8) = 9/64 = 0.1406$

(*ii*) Tracing of paths and their contribution: Now, the $F_X$ will be computed as under:

| **Paths** *n* | **Generation** | **Contribution of each path** |
|---|---|---|
| DHS | 2 | $(½)^3$ (1+0.1406) = 0.1426 |
| DGS | 2 | $(½)^3$ (1/8) = 0.125 |
| DGHS | 3 | $(½)^4$(1/16) = 0.0626 |
| DHGS | 3 | $(½)^4$ (1/16) = 0.0625 |
| DGEFHS | 5 | $(½)^6$ (1+ 0.125) = 0.0176 |
| DHFEGS | 5 | $(½)^6$ (1 + 0.125) = 0.0176 |
| | | $F_X$ = 0.4278 |

**Note:** In this case, the $F_H$ in the paths DGHS and DHGS has not been considered because the individual H is not the common ancestor in either of the path but

H is the intermediate between parents (S and D) and the common ancestor (G). Secondly, the path DHFEGHS is not allowed because the individual H has come twice in this path.

**Example 17.3**: Prepare the variance-covariance chart (Table) using coancestry method from the pedigree No. 1 given above and compute the inbreeding coefficient of the individual Z.

**Solution:**

1. Arrangement of all the individuals in the pedigree in N x N table. All the individuals in the pedigree are arranged in the N x N table according to their birth (generation) where N is the number of individuals. Amongst the individuals of base generation, only the common ancestors are taken in N x N table. The parents of all the individuals in the table are also written.

2. Computation of the variance and covariance: The inbreeding coefficients of the animals in the base generation are assumed to be zero. Secondly, the genic covariance or coefficients of relationship of the animals of base generation are also assumed to be zero.

**Variance-Covariance Table**

| Sires - | A | A | A | A | C | C | X | |
|---|---|---|---|---|---|---|---|---|
| Dams - | I | J | H | B | E | D | Y | |
| Individuals | A | B | C | D | E | X | Y | Z |
| A | 1.0 | ½ | ½ | 1/3 | ¾ | 5/8 | ½ | 9/16 |
| B | | 1.0 | ¼ | ¼ | ¾ | ½ | ¼ | 3/8 |
| C | | | 1.0 | ¼ | 3/8 | 11/16 | 5/8 | 21/32 |
| D | | | | 1.0 | 3/8 | 5/16 | 5/8 | 15/32 |
| E | | | | | 5/4 | 13/16 | 3/8 | 19/32 |
| X | | | | | | 19/16 | ½ | 27/32 |
| Y | | | | | | | 9/8 | 13/16 |
| Z | | | | | | | | 5/4 |

(*i*) Genic variance of A and its genic cov. with other individuals

Var. A = 1 + ½ (genic cov. of A's parents)

= 1 + ½ (0) = 1.0

Since parents of A are unknown and hence unrelated

$\text{Cov}_{AB}$ = ½ (cov. of A with parents of B) -

= ½ ($\text{Cov}_{AA}$ + $\text{Cov}_{AI}$) = ½ (1.0 + 0) = ½

Since A & I are unrelated.

$\text{Cov}_{AC}$ = ½ ($\text{Cov}_{AA}$ + $\text{Cov}_{AJ}$) = ½ (1.0 + 0) = ½

$\text{Cov}_{AD}$ = ½ ($\text{Cov}_{AA}$ + $\text{Cov}_{AH}$) = ½ (1.0 + 0) = ½

$\text{Cov}_{AE}$ = ½ ($\text{Cov}_{AA}$ + $\text{Cov}_{AB}$) = ½ (1.0 + ½) = ¾

Since A & B are parent offspring

$Cov_{AX} = \frac{1}{2}(Cov_{AC} + Cov_{AE}) = \frac{1}{2}(\frac{1}{2} + \frac{3}{4}) = 5/8$

$Cov_{AY} = \frac{1}{2}(Cov_{AC} + Cov_{AD}) = \frac{1}{2}(\frac{1}{2} + \frac{1}{2}) = \frac{1}{2}$

$Cov_{AZ} = \frac{1}{2}(Cov_{AX} + Cov_{AY}) = \frac{1}{2}(5/8 + \frac{1}{2}) = 9/16$

(*ii*) Genic variance of B and its genic cov. with others:

Var. B = 1.0 + ½ (Cov. of its parents A & I) = 1.0

Since A & I are unrelated

$Cov_{BC} = \frac{1}{2}$ (Cov of B with both parents of C)

$= \frac{1}{2}(Cov_{BA} + Cov_{BJ}) = \frac{1}{2}(\frac{1}{2} + 0) = \frac{1}{4}$

Since B & J are unrelated

$Cov_{BD} = \frac{1}{2}(Cov_{BA} + Cov_{BH}) = \frac{1}{2}(\frac{1}{2} + 0) = \frac{1}{4}$

$Cov_{BE} = \frac{1}{2}(Cov_{BA} + Cov_{BB}) = \frac{1}{2}(\frac{1}{2} + 1.0) = \frac{3}{4}$

$Cov_{BX} = \frac{1}{2}(Cov_{BC} + Cov_{BE}) = \frac{1}{2}(\frac{1}{4} + \frac{3}{4}) = \frac{1}{2}$

$Cov_{BY} = \frac{1}{2}(Cov_{BC} + Cov_{BD}) = \frac{1}{2}(\frac{1}{4} + \frac{1}{4}) = \frac{1}{4}$

$Cov_{BZ} = \frac{1}{2}(Cov_{BX} + Cov_{BY}) = \frac{1}{2}(\frac{1}{2} + \frac{1}{4}) = 3/8$

(*iii*) Genic variance of C and its genic cov. with others

Var. C = 1.0 + ½ (Cov.of its parents A & I) = 1.0

Since A & I are unrelated

$Cov_{CD} = \frac{1}{2}(Cov_{CA} + Cov_{CH}) = \frac{1}{2}(\frac{1}{2} + 0) = \frac{1}{4}$

$Cov_{CE} = \frac{1}{2}(Cov_{CA} + Cov_{CB}) = \frac{1}{2}(\frac{1}{2} + \frac{1}{4}) = 3/8$

$Cov_{CX} = \frac{1}{2}(Cov_{CC} \div Cov_{CE}) = \frac{1}{4}(1.0 + 3/8) = 11/16.$

$Cov_{CY} = \frac{1}{2}(Cov_{CC} + Cov_{CD}) = \frac{1}{2}(1.0 + \frac{1}{4}) = 5/8$

$Cov_{CZ} = \frac{1}{2}(Cov_{CX} + Cov_{CY}) = \frac{1}{2}(11/16 + 5/8) = 21/32$

(*iv*) Genic variance of D and its genic cov. with others

Var. D = 1.0 + ½ (Cov. of its parents A & H)

$= 1.0 + \frac{1}{2}(Cov_{AH}) = 1.0$

$Cov_{DE} = \frac{1}{2}$ (Cov of D with parents of E)

$= \frac{1}{2}(Cov_{DA} + Cov_{DB}) = \frac{1}{2}(\frac{1}{2} + \frac{1}{4}) = 3/8$

$Cov_{DX} = \frac{1}{2}(Cov_{DC} + Cov_{DE}) = \frac{1}{2}(\frac{1}{4} + 3/8) = 5/16$

$Cov_{DY} = \frac{1}{2}(Cov_{DC} + Cov_{DD}) = \frac{1}{2}(\frac{1}{4} + 1.0) = 5/8$

$Cov_{DZ} = \frac{1}{2}(Cov_{DX} + Cov_{DY}) = \frac{1}{2}(5/16 + 5/8) = 15/32$

(*v*) Genic variance of E and its genic cov. with others

Var.E = 1.0 + ½ (genic cov. of A B) = 1.0 + ½ (½) = 5/4

$Cov_{EX} = \frac{1}{2}$ (Coy of E with parents of X)

$= \frac{1}{2}(Cov_{EC} + Cov_{ES}) = \frac{1}{2}(3/8 + 5/4) = 13/16$

$Cov_{EY} = \frac{1}{2}(Cov_{EC} + Cov_{ED}) = \frac{1}{2}(3/8 + 3/8) = 3/8$

$\text{Cov}_{EZ} = ½ (\text{Cov}_{EX} + \text{Cov}_{EY}) = ½ (13/16 + 3/8) = 19/32$

(*vi*) Genic variance of X and its genic cov. with others

Var. X = 1+ ½ (genic cov. of its parents C & E))

$= 1 + ½ (\text{Cov}_{CE}) = 1.0 + ½ (3/8) = 19/16$

$\text{Cov}_{XY} = ½ (\text{Cov}_{XC} + \text{Cov}_{XD}) = ½ (11/16 + 5/16) = ½$

$\text{Cov}_{XZ} = ½ (\text{Cov}_{XX} + \text{Cov}_{XY} = ½ (19/16 + ½) = 27/32$

(*vii*) Variance of Y and its genic cov. with Z

Var. Y = 1.0 + ½ (genic cov. of Y's parents)

$= 1.0 + ½ (\text{cov}_{CD}) = 1.0 + ½ (¼) = 9/8$

$\text{Cov}_{YZ} = ½ (\text{Co}_{XY} + \text{Cov}_{YY}) = ½ (½ + 9/8) = 13/16$-

(*viii*) Variance of Z

Var. Z = 1+ ½ (genic cov. of the parents of Z)

$= 1 + ½ (\text{cov}_{XY}) = 1 + ½ (½) = 1 + ¼ = 5/4$

**3. *Computation of the $F_Z$:***

$F_Z$ = ½ (genic cov. of the parents of Z)

$= ½ (\text{genic cov}_{XY})$

$= ½ (½) = ¼ = 0.25$

or $F_Z$ = Genic variance of Z - 1.0

$= \text{Cov}_{ZZ} - 1$

$= 5/4 - 1.0$

$= 1.25 - 1.0 = 0.25$

= fxy

**4. *Computation of relationship coefficient ($R_{XY}$)***

$R_{XY} = \text{Cov XY}/\sqrt{(1 + F_X)(1 + F_Y)}$

$= ½ /\sqrt{(1 + 19/16)(1 + 9/8)}$

$= ½ /\sqrt{(1.1875 \times 1.125)}$

$= 0.5/1.1558$

$= 0.4326$

***Example 17.4:*** Compute the coefficient of relationship between D and E from ***pedigree No. 1*** by path coefficient method.

***Solution:*** In computing the coefficient of relationship (by path coefficient method) the same steps and same rules are applied which are used to compute inbreeding coefficient by path coefficient method and finally putting the values of genic covariance and genic variance in the formula.

In this pedigree, the common ancestor of D and E is only one which is A. The A is not inbred and hence $F_A = 0$. The individual D is also not inbred and hence $F_D = 0$ whereas the second individual (E) is inbred. It is thus required first to compute the inbreeding coefficient of E ($F_E$).

The $F_E$ is computed as: $F_E = (½)^{1+1} = (½)^2 = ¼ = 0.25$ for AB path.

The genic covariance between D and E is computed by tracing the paths, finding the contribution of each path and then summing over all the paths:

$$(\text{Cov}_{DE}) = \Sigma (½)^n (1 + F_A)$$

| Common ancestor and paths | n | Contribution of each path |
|---|---|---|
| DAE | 2 | $(½)^2 = ¼$ |
| DABE | 3 | $(½)^3 = 1/8$ |
| | | Sum = ¼ + 1/8 = 3/8 = 0.375 |

$$R_{DE} = \Sigma (½)^n (1 + F_A) /\sqrt{[(1 + F_D)(1 + F_E)]}$$
$$= 0.375\,(1.0) /\sqrt{[(1.0)(1.25)]}$$
$$= 0.375/\sqrt{1.25}$$
$$= 0.375/1.118 = 0.3354$$

**17.5.** Compute $F_X$ from the following pedigree by path coefficient method, coancestry method and by preparing covariance table using coancestry method.

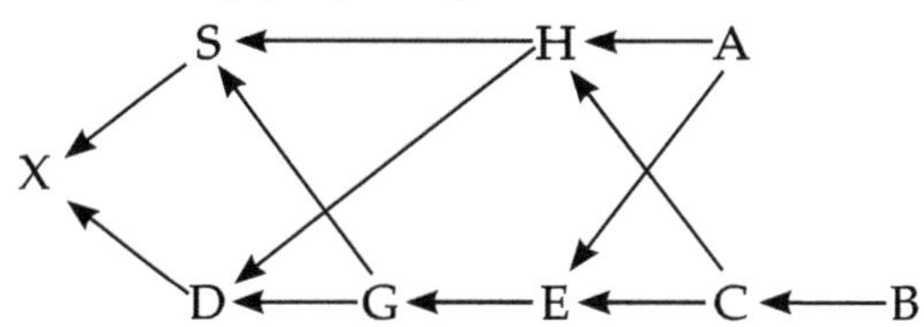


***Solution:***

**(*i*)** *Path coefficient method:*

| Paths | n | Contribution |
|---|---|---|
| S **H** D | 2 | $(½)^3 = 0.125$ |
| S **G** D | 2 | $(½)^3 = 0.125$ |
| S H **A** E G D | 5 | $(½)^6 = 0.0156$ |
| S G E **A** H D | 5 | $(½)^6 = 0.0156$ |
| S H **C** E G D | 5 | $(½)^6 = 0.0156$ |
| S G E **C** H D | 5 | $(½)^6 = 0.0156$ |
| | | $F_X$ = Sum = **0.3124** |

**(*ii*)** Coancestry method:

$$\mathbf{F_X = r_{SD}} = r_{GH.\,GH} = ¼\,(r_{GG} + 2r_{GH} + r_{HH})$$

$$r_{HH} = ½\,(1 + F_H) = ½$$

$r_{GH} = ½ (r_{EH})$
$r_{EH} = ¼ (r_{AA} + 2r_{AC} + r_{CC})$
$r_{AA} = ½ (1 + F_A) = ½$
$r_{AC} = 0$
$r_{CC} = ½ (1 + F_C) = ½$
$r_{EH} = ¼ (½ + 0 + ½)$
$r_{GH} = ½ ( ¼ ) = 1/8$
$r_{GG} = ½ (1 + F_G) = ½$

Thus, $r_{SD} = ¼ (½ + 2 \times 1/8 + ½)$
**= 0.3125**
**= $F_X$**

***(iii)*** Covariance table using coancestry method:

The following covariance table has been prepared as per procedure followed in example 17.3 by using the general formula for computing covariance of X with Y as:

$Cov_{XY}$ = ½ (Cov of X with sire of Y + Cov of X with dam of Y)
= ½ ($Cov_{XS}$ + $Cov_{XD}$) since Y = ½ (S + D)

Where, S = sire of Y
D = dam of Y

and $Cov_{XX} = 1 + F_X = 1 + ½$ (Cov of parents of X)

| Sires | - | - | - | A | A | - | G | G | S |
|---|---|---|---|---|---|---|---|---|---|
| Dams | - | B | - | C | C | E | H | H | D |
| Individuals | B | C | A | H | E | G | S | D | X |
| B | 1.0 | 0.5 | 0 | 0.25 | 0.25 | 0.125 | 0.1875 | 0.1875 | 0.1875 |
| C | | 1.0 | 0 | 0.50 | 0.50 | 0.250 | 0.375 | 0.375 | 0.375 |
| A | | | 1.0 | 0.5 | 0.50 | 0.25 | 0.375 | 0.375 | 0.375 |
| H | | | | 1.0 | 0.50 | 0.25 | 0.625 | 0.625 | 0.625 |
| E | | | | | 1.0 | 0.50 | 0.50 | 0.50 | 0.50 |
| G | | | | | | 1.0 | 0.625 | 0.625 | 0.625 |
| S | | | | | | | 1.125 | 0.625 | 0.875 |
| D | | | | | | | | 1.125 | 0.875 |
| X | | | | | | | | | 1.3125 |

The above table of covariance is usd to calculate the inbreeding coefficient of an individual ($F_X$) and the relationship coefficient between any two individuals as shown in example 17.3.

$$F_X = \text{Cov}_{XX} - 1 = 1.3125 = \mathbf{0.3125}$$

or $$= \tfrac{1}{2}\,\text{Cov}_{SD} = \tfrac{1}{2}\,(0.625) = 0.3125$$

## EXERCISES:

**17.1** (*a*) Find out the inbreeding coefficient of individual X produced from half sib mating with inbred sire (common ancestor).

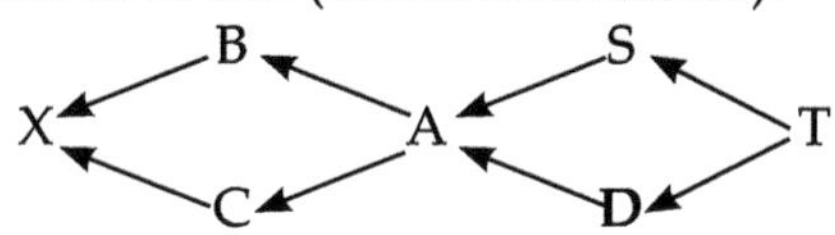


$F_X = 0.1406$

(*b*) Find out the inbreeding coefficient of individual Z produced from full sib mating with inbred sire.

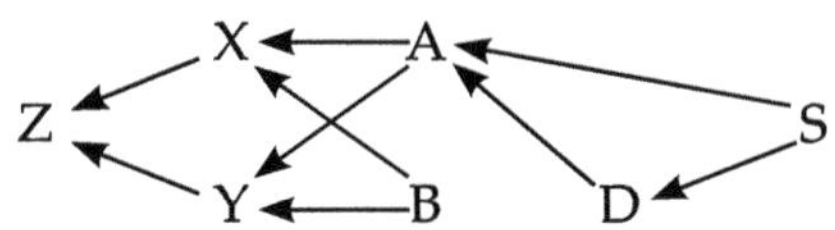


$F_Z = 0.28125$

**17.2.** Work out the inbreeding coefficient of an individual X produced from sire daughter mating with inbred sire of HS mating.

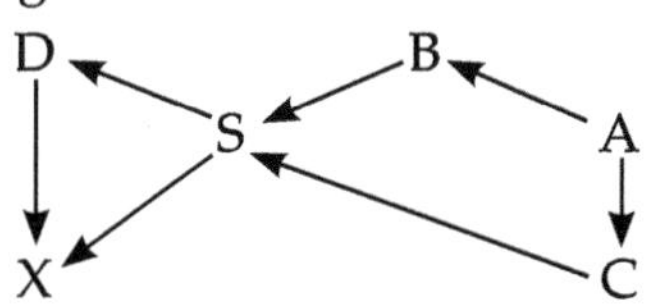


$F_X = 0.28125$

**17.3.** Compute the inbreeding coefficient of the individual X from the following pedigree.

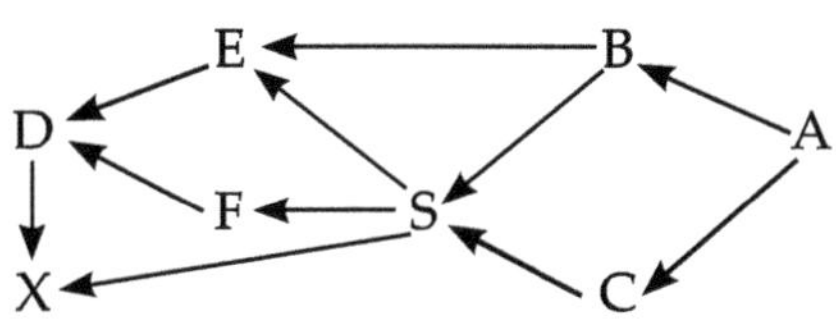


$F_X = 0.35193$

**17.4.** Work out the coefficient of relationship between two full sibs produced from sire daughter mating with inbred sire.

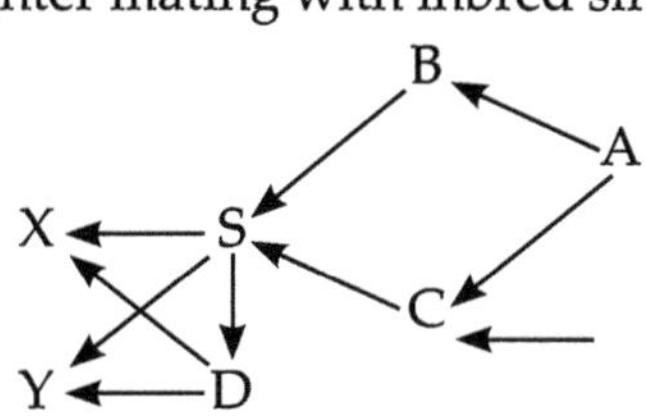


$R_{XY} = 0.6341$

**17.5.** Compute $F_X$ from the following pedigree

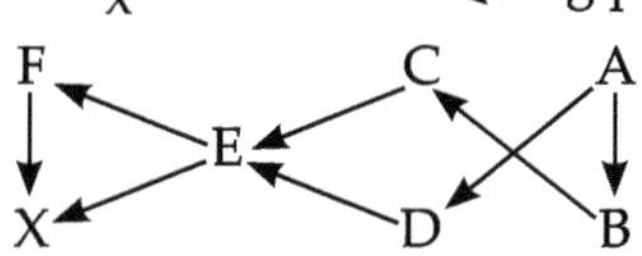


❑❑❑

# 18 Mating System II. Outbreeding

Based on genetic relationship, there is another mating system other than inbreeding, when the mated pairs are unlike in pedigree. This implies that the mated pairs have no common pedigree but they are genetically unrelated up to normal depth of pedigree and hence have the genetic effect opposite to inbreeding.

## 18.1 DEFINITION OF OUTBREEDING

The mating of unrelated individual is called as out breeding or genetic disassortative mating which is opposite or complementary to the inbreeding. The unrelated individuals may not necessarily belong to the same herd or different herds of the same breed (out crossing), but may also belong to different breeds (crossbreeding) or even to different species (species hybridization or species crossing).

## 18.2 FORMS OF OUT BREEDING

Out breeding can occur in several ways. There are mainly four different forms of out breeding which have been given as under.

### 18.2.1 Out Crossing

This is the mating system when the mating occurs among the unrelated individuals of the same breed. The progeny produced by outcrossing are called out-cross. The outcrossing system is to use the selected best available sires on the females of a herd. This is the most common breeding system practiced by breeders and has been responsible for most of the changes in livestock breeds. The outcrossing system is effective to bring marked changes in livestock characteristics for the reason that genetic variability exists within breed. The out crossing within a herd by use of selected sires is also called as ***selective breeding***.

The selection of sires and their use on the females of the same herd results in close herd and hence the replacements come from the same herd. An opinion is prevalent about the close herd that inbreeding increases rapidly in close herd. However, the increase in homozygosity within a herd or pure breed is reduced by use of selected sires and sometimes there is a selection of females too. The animals selected for breeding purposes are having less than average homozygosity because the performance of the selected animals is superior. This is because the inbreeding results in an increased homozygosity and tends to depress the performance. Therefore, the selected animals being superior in performance are expected to be less homozygous. Thus outcrossing with selection is capable for genetic change and improvement.

This system of selection and outcrossing is very effective for characters governed by additive effect of genes having high heritability. The outcrossing is practiced to exploit intra-herd genetic variability. This system is also effective that it does not allow the fixation of undesirable genes and hence brings improvement. However, the selection combined with outcrossing have an effect that the response to selection after 10-15 generation starts declining and in some cases after 20-30 generation of continuous selection the response is ceased. Thus a point of selection limit is reached. This point or level at which there is a failure to respond is often called as a ***plateau*** and the population is called ***plateaued population***. Thus, plateau is a phenomenon when selection is not effective. The effectiveness of selection determines the usefulness of outcrossing. However, in large farm animals hardly any economically important trait has reached the point when selection is not effective system for genetic improvement particularly for traits which are largely under the control of genes with additive effects. The outcrossing combined with selection is effective to offset the adverse effect of inbreeding.

**Top crossing:** This mating system is a form of outcrossing and is like grading up system. It is the mating of females to the last male in the top side of the pedigree. It is also used to refer the mating of purebred (inbred) males with unrelated females. The concept of top crossing can better be understood with the help of an example. Suppose a new strain has been developed by crossbreeding (like Karan-Fries) and further improvement is required in this newly developed strain. It is then advisable to introduce pure exotic inheritance in the herd of new strain developed earlier. Therefore, the mating of females of new strain (Karan-Fries) with purebred sire of Holstein Friesian breed will be known as "top crossing"

### 18.2.2 Cross Breeding

The mating of animals from different established breeds is called crossbreeding. The progeny produced is called crossbred. This system has major role in livestock improvement by developing new breeds of livestock. It is rightly said that purebreds of today were the crossbreds of yesterday.

***Purpose of crossbreeding***: The crossbreeding is practiced for the purposes mentioned below:

(*i*) *Commercial purposes*: The crossbreeding is done for commercial production of livestock for two reasons:

(*a*) *To take advantage of heterosis*: The two breeds differ genetically and hence the crossbreds are expected to show heterosis. This makes the crossbred more productive and better than either of the parental breeds. The crossbreeding has become quite common in pigs and poultry.

(*b*) *To combine good qualities of two or more breeds*: The distinct breeds have different good qualities. One breed may be superior for one or more traits than the other breed(s) while other breed(s) may be superior for some other desired traits. For example, exotic breeds of cattle from European origin are better for high milk production but have low disease resistance whereas the Zebu breeds of cattle are poor for milk production but have better disease resistance. The crossbreeding may combine these good qualities of the two species in one breed. This is known as the ***complementarity in crossbreeding***.

(*c*) The crossbreeding is effective for traits that cannot be improved by selection within a breed due to less genetic variability. New breeds of livestock can be evolved/ developed by crossbreeding.

(*ii*) *Genetics of characters*: The crossbreeding is also used to study the behavior of characters in hereditary transmission. The crossing of inbred line is particularly more useful in this respect.

***Types of crossbreeding***: The mating of individuals from different populations of strain, breed or species is known as crossing. The crossing of different breeds (cross breeding) can be practiced in different ways depending upon the number of breeds and the manner of their crossing. The major forms of crossbreeding are the regular crossing and composite crossing.

**1. Regular or systematic crossing:** This is making of the same cross on regular basis to take the advantage of heterosis and complementarity. The crossing exploits non-additive gene effects through heterosis and the additive gene effects through complementarity when two or more characters complement each other. The amount of heterosis depends on the environment, genetic variability between two populations involved in crosses and the non-additive gene effects. The heterosis may be parental (maternal or paternal) referring to the performance of animals as parents and the individual heterosis referring to non-parental performance of the individual.

There are two basic methods of regular crossing. These are known as specific crossing and rotational crossing.

**(A). Specific crossing**: Depending on the number of breeds used in crossing and the manner they are crossed, the specific crossing is grouped as:

**(*i*) Two breed crosses**: Two pure breeds are crossed together. This may be restricted with crossing of only the purebreds (purebred crosses) or may be extended to crossing the crossbreds (*inter se mating*) or with males of pure breed (*back crossing or crisscrossing*).

**(*a*) Pure breed crosses**: It is the mating of two purebreds together but their crossbred progeny are not used in breeding programme. This produces offspring which are 100% heterozygous showing 100% ***individual heterosis*** but no parental heterosis because the parents are not crossbreds.

**(*b*) Inter se mating**: This is the crossing of crossbred progeny having the same level of inheritance of two breeds like crossing of $F_1$ with $F_1$. This creates a number of genetic groups.

**(*c*) Back crossing**: It is the mating of crossbred animals to purebred animal of either breed used to produce progeny of first generation ($F_1$) like $F_1$ x $P_1$ or $F_1$ x $P_2$ where $P_1$ and $P_1$ are two purebreds used to produce $F_1$. This cross is made to exploit maternal or paternal heterosis. As the maternal heterosis is generally of more importance, the crossbred parents should be the females ($F_1$females). The back crossing between crossbred female and purebred males shows 100% ***maternal heterosis***. The back cross progey are, on an average, 50% less heterozygous (or say 50% homozygous and 50% heterozygous) than $F_1$ and hence they show only 50% of individual heterosis.

**(*d*) Criss crossing**: It is similar to back crossing except that both the parental breeds ($P_1$ and $P_1$) are used alternately for mating of crossbred progeny in each generation.

**(*ii*) Three breeds cross or triple crossing**: In this system of crossbreeding three breeds are used. The first cross animal ($F_1$) is mated to the animals of a third breed like (AB) female mated with males of third breed (C).

This results in full utilization of maternal heterosis because the dam is 100% heterozygous being crossbred as well as of individual heterosis because the progeny are also 100% heterozygous.

The first cross and three breed cross show full utilization of ***individual heterosis*** because they are both 100% heterozygous but the amount of heterosis may not be the same for the reason that different crosses are involved and the amount of heterosis is greater with some crosses than others.

**(*iii*) Four breed crosses or double two breed crosses:** This involves the crossing of crossbred females produced by crossing two breeds (A and B) with crossbred males produced from crossing another two breeds (C and D). Thus, mating is between (AB) x (CD). This is used to produce commercial progeny. This enables full exploitation of both maternal and paternal heterosis as well as individual heterosis for the reason that both parents are crossbreds.

**(B) Rotational Crossing:** It the mating of three breeds in a rotational manner. The males of three breeds are used in regular sequence (rotation) in successive generations on crossbred females of the previous generation.

The major disadvantage of rotational crossing is that it does not allow any exploitation of complementarity. Thus, it is good to use when the population show little or no complementarity but show heterosis. However, it produces less benefit from heterosis. The advantage of rotational crossing is that all female replacements are obtained from crossing programme itself and do not therefore have to be specifically bred in a separate population as they do in case of specific crossing.

The summary of fraction of different types of heterosis as indicated by average heterozygosity from different forms of crossbreeding is given as under:

| Type of crossing | Fraction of heterosis | | |
|---|---|---|---|
| | **Individual** | **Maternal** | **Paternal** |
| **Straight bred** | 0 | 0 | 0 |
| Two breed cross | 1 | 0 | 0 |
| **Back cross:** | | | |

*Contd...*

*Contd...*

| | | | |
|---|---|---|---|
| Cross bred female x purebred male | ½ | 1 | 0 |
| Purebred female x crossbred male | ½ | 0 | 1 |
| **Three breed crosses:** | | | |
| Cross bred female x pure male | 1 | 1 | 0 |
| Pure bred female x crossbred Male | 1 | 0 | 1 |
| Four breed cross | 1 | 1 | 1 |
| **Rotational crossing** | | | |
| 2 sire breeds | 2/3 | 2/3 | 0 |
| 3 sire breeds | 6/7 | 6/7 | 0 |

**2. Composite crossing to produce synthetic breed:** This is alternate to regular crossing. It is done by producing one or few crosses between two or more population to produce a single population having genes from each of the population. This single population of a mixture of various crossbred populations having various level of inheritance is called a synthetic or composite. This synthetic population is improved by selection within it. In the past, this crossing system with selection has resulted in the development of new breeds e.g. Santa Gertrudis, Jamaica Hope, AMZ, Australian Friesian Sahiwal, Norwegian Red and white and Belmout Red breeds of cattle which were developed by selection within a synthetic. Similarly synthetic populations of cattle to evolve new breeds have been produced by crossing of Bos Taurus breeds with heat tolerance and tick resistance of Bos indicus breeds.

The theoretical expectation about heterosis is that 1-1/n of the heterosis present in $F_1$ is retained in $F_2$ and in subsequent generations (Dickerson, 1973) where n is the number of different populations contributing equally to the synthetic.

***Advantages of Crossbreeding:*** It introduces new desirable genes in the herd. Secondly, it helps to evolve new breeds of livestock. A number of new breeds of different livestock species have already been developed. It also helps in the study of genetics of characters. Moreover, it exploits the hybrid vigour.

***Disadvantages of crossbreeding:*** The breeding merit of crossbreds is reduced due to heterozygosity. It is costly as it requires maintenance of two or more breeds.

### 18.2.3 Grading Up

In this mating system, the females of non-descript, scrub or native (deshi) type are mated with the sires of a pure breed. This is one type of crossbreeding. The crossbred females are back crossed to the pure bred sires to produce the progeny with 75% genes from the pure breed. The crossbred females in successive generations are mated with sires of pure breed. Thus, after six generations of crossing the graded females with the purebred sires produce the progeny which have 98.4% genes of the purebred. Thus, the upgrading system of mating changes the genetic composition without creating new genes by transmitting good quality genes of an improved breed.

Grading up is advantageous to produce purebreds after a few generations. Secondly, it is less expensive. Thirdly, this system helps to prove the genetic potentialities of the sire.

This system (grading up) has limitation. The most important is that the genotype-environmental interaction puts a limit to use this system. This is because the purebred sire which gives good results in one environment may not give similar results in other environment. Mostly, the sires of a pure breed are brought from another environment.

### 18.2.4 Species Hybridization

The mating is between the animals of two species. This system is the widest (extreme) possible type of out-breeding. The following are the example of species hybridization:

Species hybrid obtained Species crossed

1. Mule: Male ass (Jack) x females horse (Mare)
2. Hinny: Female ass (Jennet ) x Male horse (Stallion)
3. Zebroid: Zebra-horse hybrid
4. Asbra: Ass x zebra (in Africa)
5. Pien niu Cattle - yak cross in Tibet
6. Cattalo: American buffalo bull(B.bison)x B.taurus
7. Jatsa ($F_1$ male): Mithun x cow
8. Jatsamin ($F_1$ females): Mithun x cow
9. Jechha ($F_1$ males): Mithun x siri cow
10. Jessam ($F_1$ females): Mithun x siri cow

The species hybridization generally results in sterile progeny. The mules and Hinny are sterile. The crossing of Bison with B.taurus cow produces sterile males but the females are fertile. The cattalo was developed by backcrossing the females to bison and cattle.

## 18.3 EFFECTS OF OUTBREEDING

The out breeding is opposite to inbreeding and hence its effects on the genetic structure of population as well as on population performance are also opposite to inbreeding.

### 18.3.1 Genetic effects

The out breeding tends to increase heterozygosity because the mated pairs are genetically unrelated (having different alleles). The important and peculiar characteristic of out breeding is that the maximum heterozygosity is attained in first generation outcrosses and goes on decreasing in subsequent generations by random mating of outbred. The decrease in heterozygosity in subsequent generation is due to the segregation of genes which consequently increases the homozygosity. The two parents are homozygous (AA and aa) and their $F_1$ progenies are all heterozygous (Aa) while the $F_2$ progenies are produced in a ratio of 1AA: 2 Aa: 1aa. Further the crossbred of $F_1$ generation are likely to be uniform in traits related to physical fitness particularly when the parents are homozygous for different alleles of a particular pair. However, the out-bred cannot breed true because of heterozygosity and hence the selection is less effective among out-bred.

### 18.3.2 Phenotypic effect (Change of mean on crossing)

#### *Heterosis*

The outbreds are heterozygotes in which the effects of undesirable recessive genes are hidden by the effect of favourable genes (dominant). Therefore, the outbreds are more vigourus. The word heterosis was coined by Shull (1914) to describe the increased vigour of outbreds/crossbreds relative to their parents. One of the purposes of out breeding system of mating is to take the advantage of heterosis.

**Definition of heterosis:** The **heterosis** is defined as the amount by which the mean of $F_1$ generation, produced by crossing two breeds, exceeds to its better parent. Thus, the heterosis (H) is indicated as: $H = F_1 > P_1 > P_2$.

*Positive and negative heterosis*: The better parent is generally taken with greater mean and denoted by the letter $P_1$ where the poor parent is designated by the letter $P_2$. However, the better parent may also have low mean depending on the character under study like age at first calving, the cow with low AFC is a better cow and in this case $P_2$ having low AFC will be a better cow. When $P_1$ is a better parent (milk yield) the $F_1$ should have higher milk yield than its $P_1$ parent ($F_1 > P_1$). This is positive heterosis. On the contrary, when the $P_2$ is the better parent (AFC) the $F_1$ should have lower mean than its lower parent ($P_2$) and thus for heterosis to occur $F_1 > P_2$. This is negative heterosis. The positive heterosis is often called as **hybrid vigour**. In some cases, the crossbred progeny exceed even the better parent.

In animal breeding, the heterosis is taken as the superiority of $F_1$ progeny relative to the average of their parents (mid parent value). Thus, the heterosis is the amount by which the $F_1$ population mean exceeds to the mid-parent value. Therefore, the heterosis (H) is taken as the superiority of outbreds over the mean of their parents and hence:

$$H = F_1 > (P_1 + P_2) / 2$$

The heterosis is the measure of the effect of out breeding.

**Estimation of heterosis:** The amount of heterosis is estimated by comparing the mean value of purebred and crossbred animals as follows:

***Heterosis*** = mean of $F_1$ progeny – mean of parental breeds

This can also be expressed in terms of percentage as:

$$\% \text{ heterosis} = \frac{\text{mean of } P_1 \text{ peogeny} - \text{mean of parent breeds}}{\text{mean of parent breed}} \times 100$$

The heterosis as a deviation from mid-parent value can be estimated in the following three ways-

(*i*) *When parental information is known:* The heterosis in this case is estimated as a difference between the mean of a particular crossbred generation and parental mean generation ($P_1$ and $P_2$) as under:

$$H_{F1} = F_1 \text{ mean} - \text{mid-parent value}$$
$$H_{F2} = F_2 \text{ mean} - \text{mid-parent value} = \tfrac{1}{2} H_{F1}$$
$$H_{B1} = B_1 \text{ mean} - \tfrac{1}{4}(3\bar{P}_1 + \bar{P}_2)$$
$$H_{B2} = B_2 \text{ mean} - \tfrac{1}{4}(\bar{P}_1 + 3\bar{P}_2)$$

(*ii*) *When performance of purebred parents is not known:* In large livestock cross breeding programme, the performance of exotic breed is not available particularly under Indian conditions. In such situation, two procedures are used-

(*a*) ***Statistical approach:*** The heterosis is estimated taking the means of different grades of crossbred population as dependent variable and means of sire's breed, dam's breed and level of heterozygosity are as independent variables.

(*b*) ***Biometrical approach*** to estimate heterosis is based on fitting of genetic model in terms of gene effects to the means of crossbred generations. Nitter (1978) estimated the heterosis as twice the difference of $F_1$ and $F_2$ means as:

$$H_{F1} = 2\,(F_1 \text{ mean} - F_2 \text{ mean})$$
$$= 2\,F_1 \text{ mean} - (B_1 \text{ mean} + B_2 \text{ mean}) \text{ since}$$
$$\bar{F}_2 = \tfrac{1}{2}(B_1 \text{ mean} + B_2 \text{ mean})$$

(*iii*) *Gene frequency method*: The amount of heterosis (H) in terms of gene frequencies and gene effect is as under:

$$H_{F1} = dy^2$$

Where, d = non-additive gene effect and

y = genetic diversity of mated parents ($y = p_1 - p_2 = q_2 - q_1$)

= Difference in gene frequency).

$y^2 = 2pqF$.

The amount of heterosis increases with the degree of genetic differences between the two mated population and is limited by the barrier of interspecific sterility.

## 18.4 RESULTS OF CROSSBREEDING

In India, the crossbreeding has been done in different species by mating of females of native breeds with sires of exotic breeds. The important findings on crossbreeding results on different species have been given here-

### 18.4.1 (a) Crossbreeding Zebu cattle with Exotic breeds in India

The following crossbreeding projects on cattle in India were started. As a result some crossbred genotypes/strains/breeds have been evolved as given below -

**(*i*)** The crossbreeding of dairy cattle in India was started by European tea planters and Christian Missionaries in North Bengal, Assam and Nilgiri hills. The exotic animals were imported.

**(*ii*) Taylor:** This breed of cattle was evolved in 1875 around Patna, Bihar using Shorthorn and Channel Island bulls from UK on local cows. This breed has extincted due to less number of animals, their backcrossing to local animals and using no selection in crossbreds. The breed is almost extinct.

**(*iii*)** *Military Dairy Farms:* The first Military Dairy Farm (MDF) was started at Allahabad in 1886 with the purpose to supply milk to British troops in India. The crossbreeding was started in 1889 by importing and using Friesian, Jersey, Ayrshire and Shorthorn sires initially but in 1922 Ayrshire bulls were stopped using, whereas HF were used extensively. The crossbred cows were maintained at different MDF's. The criss crossing (backward and forward crossing) with exotic and indigenous breeds to maintain the exotic inheritance around 62.5 % (5/8) had been the breeding policy prior to 1952 and later on back crossing with Sahiwal bulls was started at MDF's, producing the nimals of wide range from zero to 100 %. The Friesian crosses performed better than Ayrshire and Shorthorn (Kartha, 1934). In 1952, in view of conservation of indigenous breeds like Sahiwal, Red Sindhi, Tharparkar, Gir and Hariana breeds, the crossbreeding was stopped. But again in 1958 through a resolution made by reorganization committee, the crossbreeding was restarted with HF at all Military Farms in 1960 by importing bulls from Holland. The Red Dane bulls were used for few years at MDF, Bangalore but were replaced by HF bulls.

**(*iv*) Jersind:** At Allahabad Agricultural Institute, Allahabad the selective breeding was used to improve Sahiwal, Gir, Hariana and Kankrej breeds from 1910 to 1924, and in 1923 Red Sindhi breed was also included. The crossbreeding was started using HF, Brown Swiss, Guernsy and Jersey sires on zebu cows of Sahiwal, Red Sindhi, Gir, Hariana and Kankrej during 1924 – 34. The crossbreeding was restricted to Red Sindhi with Jersey since 1934 having 3/8 to 5/8 Jersey inheritance and a strain **Jersind** was developed (1953). The cows produced about 1600 kg milk in first lactation.

**(*v*)** The crossbreeding was started in 1910 at NDRI, Bangalore. The Hariana cows were crossed to Ayrshire initially (1910). Subsequently other crosses were made like Ayrshire x Sahiwal (1913), Ayrshire x Red Sindhi (1917) and HF x Red Sindhi (1938). Other breeds like Tharparker and Jersey were also involved.

**(*vi*)** The crossbreeding was started in 1919 at Govt. Livestock Farm, Hosur under erstwhile Madras Govt. with the objective to evolve a new crossbred cattle by crossing Red Sindhi and Sahiwal with Ayrshire bulls and interbreeding among $F_{1's}$. The crossbred cows had better production performance over Red Sindhi. The crossbreeding of cattle using exotic sires was also carried out in Nilgiri area of

Madras State but there was no success due to the use of inferior and graded bulls without any selection.

***(vii)*** The ICAR during 1955 sponsored crossbreeding project in hilly region and heavy rainfall areas at 12 locations of 10 states at Palampur, Darjeeling, Shillong, Imphal, Cherapunji, Ranchi, Vishakhapatnam, Hyderabad, Coorge, Ootacmund, Nuyattnakara, and Chakalakudi. This involved Jersey x Red Sindhi crosses and then using the crossbred bulls on local cattle. The Jersey x Sindhi crossbred bulls were brought from AAI, Allahabad for crossbreeding in H.P. The cattle breeding farm was established at Katuala in Mandi district (HP) by importing Jersey bulls from Australia. In addition, some Dexter cattle were imported from UK in 1956 for use in crossbreeding by HP Govt.

***(viii)* Jerthar:** The crossbreeding in rural areas of Bangalore and Mysore was started in 1955-56 with Jersey bulls by the then Mysore Govt. The project was started at NDRI, Bangalore in 1958 to cross Tharparkar cows with Jersey bulls of American and Australian strains. Interbreeding of first generation progeny was adopted. The exotic blood of 50 % was maintained among the crossbred animals and a new breed '**Jerthar**' was developed. The cows produced about 2100 kg milk in a lactation.

**(*ix*)** Indo – Danish project of crossbreeding was started at Hessarghata, Karnataka (1961), Ind0 – German Project at Mandi, H.P. and Almora (U.P.) in 1963.

**(*x*) Sunandini:** Indo-Swiss crossbreeding project was started in 1963 in Kerala using Brown Swiss and Jersey semen on local non descript and Red Sindhi x non descript graded cattle. A new breed of cattle **'Sunandini'** was evolved by Kerala Livestock Development Board (1964) from crosses of local cows with BS bulls. This strain has 50 % exotic inheritance of HF, BS and Jersey. The animals under field conditions produced about 1400 kg milk in a lactation.

**(*xi*)** In 1961, it was decided by A.H. wing of Board of Agr. & A.H. to have the crossbreeding project in areas of suitable feeding and management conditions to maintain crossbred cattle and that only one exotic breed should be used in one area. In the same year the Central Council of Gosamvardhan also arrived at the same conclusion to have crossbreeding.

**(*xii*)** The crossbreeding was initiated at Harringhata (W.B.) and NDRI, Karnal. The crossbreeding at Harringhata included HF, Brown Swiss and Jersey for crossing Hariana cows to produce $F_1$, $F_2$, ¾ and 5/8 in reciprocal crosses. This was the largest project financed by ICAR, UNDP, and USAID under PL 480. This project was merged as a unit of AICRP of ICAR and subsequently the project was undertaken by the State A.H. Department without financial support of ICAR.

**(*xiii*) Karan Swiss:** The crossbreeding work was initiated in 1962 at NDRI, Karnal involving Brown Swiss as exotic breed with Red Sindhi and Sahiwal as Zebu breeds. Various crossbred groups were produced which did not show significant evidence of non additive gene action for growth, milk yield and AFC. The Brown Swiss inheritance is around 50 %. The various crossbred groups were merged and selective breeding was practiced. The creation of synthetic population following selection resulted into a new breed in 1980 named as **'Karan Swiss'.** The average AFC is about 32 months, FLMY of about 2600 kg with 4.2 – 4.4 % fat.

***(xiv) Vrindavani:*** The ICAR started in 1966 at IVRI, Izatnagar and PAU, Hissar a coordinated research project *"Behaviour Patterns of Zebu Crossbreds"* on Hariana Cattle to cross them with HF, BS and Jersey bulls. This project was lateron merged with AICRP on cattle in 1968 at 3 more centers viz. lVRl, HAU, JNKVV, Jabalpur; M.P.A.U.,Rahuri and A.P.A.U.,Lam (A.P.). In addition to Hariana breed, Gir was included at JNKVV & MPAU and Ongole at Lam. The exotic breeds used were HF, BS, and Jersey.

Later on during 1973, Haringhata Farm (West Bengal) was also included where the crossbreeding was already going on since 1957-58. This project showed superiority of HF followed by BS and Jersey. The project arrived at a conclusion that crosses involving different indigenous breeds did not differ and the half breds were better over higher crosses and hence HF x native crosses at 50 % exotic level of exotic inheritance should be used to evolve a new breed with further improvement by selection. The IVRI with continued *inre se* mating had developd ***Vrindavani*** strain of cattle (1968).

***(xv) Karan Fries:*** Another crossbreeding project was also started at NDRI Karnal in 1971 which involved HF, BS and Jersey semen on Tharparkar cows. A number of crossbred groups were produced and it was observed that two and three breed crosses did not show significant manifestation of heterosis for growth and milk yield. All the crossbred groups forming the synthetic population were merged in 1980 and the selective breeding was practiced in this population. This resulted into a new breed of cattle named as **'Karan Fries'** which has around 50 % Friesian inheritance. The animals are producing about 3700 kg milk per lactation with 3.8 to 4.0 percent fat. The AFC is around 30 – 32 months.

**(xvi) *Frieswal:*** The ICAR also started financing Military Dairy Farms since 1985 to evolve national crossbred cattle involving HF and Sahiwal. The breed ***'Frieswal'*** has been evolved by stabilizing at 5/8 HF level.

***Impact of crossbreeding***: The crossbreeding had the following impacts:-

- Crossbred cattle population was increased.
- There had been an increase in milk production and hence an increase in per capita of milk availability.
- An increase in growth rate and reproductive efficiency in terms of age at puberty, AFC, calving interval.

**Conclusions:** The following conclusions can be drawn on the basis of the results of crossbreeding dairy cattle:-

- Significant increase was observed in milk yield and reproductive efficiency of crossbreds.
- HF crosses and Jersey crosses proved superior over other exotic breeds for plane and high areas, respectively.
- Regarding level of exotic inheritance, the optimum was found between 50 to 62.5% for better growth, production and reproduction.

- There was a decline in milk yield of $F_2$ over $F_1$ due to use of poor quality crossbred bulls.
- The crossbreds gave 30-60% higher milk yield under challenged feeding than controls given ration as per NRC requirement. This suggested to re-examine the feed requirement for crossbred cattle so as the production potential can be fully exploited.

**Constraints of crossbreeding cattle in India:** There had been some of the following constraints of crossbreeding:

- Higher initial investment and maintenance cost.
- Non-availability of proven exotic bull semen.
- Less availability of quality green fodder.
- Higher disease incidence among crossbreds and lack of optimum veterinary care.
- Non-acclimatization of crossbreds to widely varying climatic conditions in India.
- A myth that crossbred male calves are not fit for draft purpose
- High culling rates in crossbred bulls (upto 40-70%)
- No well defined breeding policy for F I'S under field conditions.
- Problem of proper milk disposal of crossbreds.

**Realization:** The crossbreeding experiences had made the animal breeders to make the following realizations:-

- The indigenous recognized breeds should be maintained and not involved in crossbreeding with exotic breed.
- There should be balanced feeding of crossbred cattle.
- There is a need of proper health management to control mastitis and other infectious diseases.
- Realistic approach should be followed for culling of inferior/diseased animals.

### 18.4.1 (b) Crossbreeding Zebu cattle abroad

Indian cattle called as Zebu cattle in Europe and South America but Brahma in United States have been used in crossbreeding programme for development of new breeds abroad. Brahma cattle were developed from two zebu bulls in 1854 at Louisiana. The Brahma cattle developed in South America have been used for crossbreeding with European breeds. A number of beef breeds have been developed in the United States using Brahma crosses with European breeds.

The new breeds developed in U.S.

| New Breeds | Breeding policy adopted |
|---|---|
| 1. Santa Gertrudis: | 3/8 Brahma and 5/8 Shorthorn (southwest Texas) |
| 2. Brangus: | Brahma x Angus |
| 3. Braford: | Brahma x Hereford |
| 4. Charbray: | Charolais x Brahma |
| 5. Beefmaster: | 15 % Brahma x Hereford x Shorthorn |
| The breeds developed by crossing of exotic cattle among themselves: | |
| 6. Guernsey: | Brown & White Cattle of Brittany x Brindle cattle of Normandy |
| 7. Ayrshire: | Cross of many strains of cattle. |
| The new breeds of cattle developed in other countries are as under - | |
| 8. Australian Milking Zebu: | 20-40 % Zebu (Sahi. & RS) and rest Jersey |
| 9. Australian Friesian Sahiwal: | Sahiwal sires x HF cows |
| 10. Mambi: | ¾ HF x ¼ Zebu, developed in Cuba |
| 11. Siboney: | 5/8 HF x 3/8 Zebu, developed in Cuba |
| 12. Pitanqueiras: | 5/8 Red Poll x 3/8 Zebu, developbed in Brazil |
| 13. Jamaica Hope: | 80 % Jersy, 5 % Holstein, 15 % Sahiwal (Jamaica) |

### 18.4.2 Crossbreeding buffaloes

The genetic diversity among two types of buffaloes (river and swamp) exists to the extent of different number of diploid chromosomes. The riverine buffaloes have 50 chromosomes whereas the swamp buffaloes have 48 chromosomes. However, crossbreeding of two types of buffaloes produce fertile hybrids ($F_J$) with diploid number of chromosomes equal to 49.

The crossbreeding between riverine breeds of buffaloes has also been carried out. As a result new breeds of buffaloes have been evolved -.

***(i)* Mehsana**: Murrah x Surti buffaloes (Gujarat)

***(ii)* Godawari**: Murrah x Local buffaloes (AP)

(***iii***) **Kujang** was also evolved by using Murrah in Orissa.

***(iv)*** Murrah x Surti crossbreds at NDRI were produced but no breed was evolved. The crossbreds had shown increase in milk production and reproduction traits, the milk yield per day of calving interval was higher in crossbreds (3.4 kg) than Murrah (3.2 kg). The performance of these crossbreds was very close to Murrah.

**Crossbreeding Buffaloes Abroad:** The crossbreeding of buffalo has been done to improve milk and meat potential of swamp buffaloes with river buffaloes in other countries like China, Vietnam, Malaysia, Thailand, Indonesia, Brazil and Philippines.

In Bulgaria, the Murrah x Bulgarian crossbreds were found close to Murrah for body weight at different ages whereas the milk yield and fat in crossbreds were close to mid parent value.

### 18.4.3 Crossbreeding sheep

The crossbreeding of Bikaneri (Magra) ewes with Australian Merino rams was started at GLF, Hissar and a new breed of sheep ***'Hissardale'*** was developed. It has 75% exotic inheritance.

Similarly, ***'Kashmir Merino'*** breed was developed by crossing Gaddi, Bhakarwal and Poonchi ewes with Merino and Rambouillet rams. This has 50 to 75% exotic inheritance.

As a result of crossbreeding at CSWRI Avikanagar, two new synthetics (chokla and Nali synthetics) for fine wool have been developed. These new synthetics produce more wool of fine quality having desirable medulation percentage and fiber diameter. The crossbreeding of indigenous breeds (Gaddi at Tal in HP, Patanwadi at Dantiwada in Gujrat and Nilgiri at Sandynallah in TN) with Soviet Merino and Rambouillet showed improvement for wool production and quality together with an improvement in body weight. Three diverse native breeds (Chokla, Malpura and Jaisalmeri) were crossed with Rambouillet at CSWRI Avikanagar. The results of crossbreeding have shown that any exotic fine wool breed can be used in crosses with native breeds, for improvement in wool quality and quantity, with 50 to 75% exotic inheritance depending on the climatic and nutritional conditions.

Almost 60% of the total export of wool and woolens from India is obtained through the export of carpets. The Indian carpets are cheaper compared to that made in Iran, Afghanistan and China. This may be due to low quality of wool used or due to low labour charges in India. The important carpet wool breeds in India are NaIi, Chokla, Bikaneri (Magra), Marwari, Pugal, Jaisalmeri, Malpura, Sonadi, Deccani, Patanwadi and Bellary. But of these, Chokla, Nali and Patanwadi produce finer wool.

The indigenous **carpet wool breeds** have been crossed with exotic fine wool, mutton and dual purpose breeds. Malpura breed was crossed with Rambouillet, Suffolk and Dorset; Sonadi with Suffolk; and Muzaffarnagri, Coimbatore, Nellore and Bellary with Corriedale. The results showed improvement in wool quantity and quality (in terms of low medullation percentage and low fiber diameter).

Among the native sheep breeds, the North Indian Carpet wool breeds are heavier than South Indian mutton breeds. The native breeds viz. Malpura, Sonadi, Muzaffarnagri, Mandya and Nellore were crossed with Dorset and Suffolk whereas Coimbatore, Nellore and Bellary were crossed with Corriedale. The results showed better combining ability of Suffolk with Sonadi and Muzaffarnagri. Three new synthetic mutton strains were developed by crossing of Malpura and Sonadi with Dorset and Suffolk at CSWRI Avikanagar, crossing of Nellore and Mandya breed with Dorset and Suffolk by stabilizing 50% exotic inheritance. These new strains were named as Avimaans, Mandya synthetic and Neelore synthetic, respectively.

In India Malpura breed of sheep is used by some people of Rajasthan (Khatik) to produce **lamb pelts** on small scale. In 1975, 200 ewes and 50 rams of Karakul were imported in India from USSR. These were distributed to CSWRI and Khumbathang in Kargil district (J&K). The Karakuls adapted well to hot arid conditions of Rajasthan (Bikaner) as well as in the cold arid region. The native breeds of India

(Chokla, Nali, Malpura and Sonadi) were crossed with Karakul at Avikanagar. The Malpura and Sonadi crosses performed better than Nali and Chokla crosses for pelt quality. Further, Malpura lambs produced better pelts than Sonadi but the results of karakul crossbreds were reverse (Sonadi Karakul halfbred produced better pelts than Malpura Karakul crosses). The results indicated that crossing of coarse carpet Indian breeds with Karakul had good potential for pelt production.

The detais of the new breeds/strains developed in India are as:

| Breeds developed | Place | Breeds used | | Exotic inheritance |
|---|---|---|---|---|
| | | Native | Exotic | |
| *Fine Wool:* | | | | |
| 1. Bharat Merino | (CSWRI, Avikanagar) | Chokla & Nali | Rambouillet & Merino | 75 |
| 2. Avivastra | (SCWRI, Avikanagar) | Chokla & Nali | - do - | 50 |
| 3. Nilgiri synthetic | (TNVASU, Sandynallah) | Nilgiri | Merino | 62.2-75 |
| 4. Patawar synthetic | (GAU, Gujarat) | Patanwari | Rambouillet & Merino | 50 |
| 5. Kashmir Merino | (J & K) | Bhakarwal, Gaddi & Poonchi | Merino & Rambouillet | 50-75 |
| 6. Hissardale | (GLF, Hissar) | Bikaneri | Merino | 75 |
| *Mutton:* | | | | |
| 7. Avimans | (CSWRI Avikanagar) | Malpura & Sonadi | Dorset & Suffolk | 50 |
| 8. Nellore synthetic | (APAU, Palamner) | Nellore & Mandya | - do - | |
| 9. Deccani synthetic | (MPKV, Rahuri, MS) | Deccani | - | |
| *Carpet Wool:* | | | | |
| 10. Avikalin | (CSWRI, Avikanagar) | Malpura | Rambouillet | 50 |
| 11. - | RAU, Bikaner | Magra | - | - |
| *Pelt Production:* | | | | |
| 12. Indian Karakul | (CSWRI, sub station Bikaner) | Malpura, Sonadi, Marwari | Karakul | 75 |

The ***exotic breeds of sheep evolved abroad*** are as under –

1. Columbia: Rambouillet x Lincoln rams
2. Panama: Lincoln x Rambouillet rams
3. Targhee Rambouillet x Long wool breed
4. Romlet: Romney Marsh rams x Rambouillet ewes
5. Suffolk: Dark face South Down rams x Norfolk strain
6. Corriedale: Merino ewes x Lincoln ram
7. Karakul: Crossing of two local breeds.

### 18.4.4 Crossbreeding goats

The crossbreeding of goats has been made between native breeds as well as with exotic breeds. The bucks of Jamunapari and Beetal breeds were used to mate the Bengal does. These two crosses showed an improvement of 76 and 90%, respectively in the milk production over purebred Bengal. This improvement was due to large genetic differences between the breeds crossed. But the reciprocal crosses i.e. when high yielding dams of Jamnapari were crossed with low yielding bucks (Behgal), did not show heterosis. The crossing of Sirohi with Beetal showed imperceptible (very slight) improvement because the genetic divergence between the two breeds was less. There was considerable increase in milk production when Black Bengal was crossed with Barbari over the Black Bengal.

For meat components three large size indigenous breeds (Jamunapari, Barbari and Black Bengal) were crossed and 3 breed crosses were produced to study the meat parameters at B.R. College, Agra.

The crossbreeding of Indian breeds of goats with exotic breeds was done for milk (Alpine and Anglo-Nubian), for meat (Boer and Toggenberg) and for fibre (Angora) to assess the production potential of the crosses for various products during pre and post independence period. However, this programme was discontinued after short term experiment for certain reasons of high mortality and mobidity.

During fifties, the state Governments imported few Angora goats for use in M.S., HP and UP. The flocks at Pipalkoti (UP) could not adapt whereas the flocks in MS and HP performed better. Three-fourth and seven-eighth crosses of Angora x Deccani produced 0.183 and 0.454 kg mohair respectively. Whereas purebred Angora, and half, JI. and 7/8 Angora x Gaddi crosses in HP produced 0.475, 0.105, 0.192 and 0.369 kg mohair respectively.

Later on, AICRP on goats by ICAR was initiated for genetic improvement of goat breeds at CSWRI, Avikanagar and then shifted to CIRG Makhdoom, Mathura. The exotic breeds included Alpine, Saanen for milk and Angora for Mohair for production of crossbreds. The different crosses made were - Half and three-fourth Saanen x Beetal, and half and three-fourth Alpine x Beetal at NDRI Karnal; half and three-fourth Alpine x Malabari, and half and three-fourth Saanen x Malabari at KAU, Mannuthy (Kerala). The Beetal crosses were found superior than Malabari crosses. Secondly, the Saanen crosses were superior to Alpine crosses. Milk yield of crosses increased up to 75 % exotic inheritance. However, the higher crosses of Malabari did not prove better than half breds. The $F_2$ and $F_3$ generations had low milk production than $F_1$.

The crossbreeding with Saanen of local Deccani goats at Pune and with Alpine of Sirohi at Ajemer was started.

Under AICRP on goats at Rahuri (MS), the half bred Angora x Sangamneri could not produce Mohair but V. crosses produced finer fibres whereas 7/8 crosses produced 1.1 kg mohair per clip. White Marwari goats were crossed with Angora at CSWRI, Bikaner. The results obtained were similar to those at Rahuri. Indian Mohair strains have been evolved at Rahuri and Bikaner.

### 18.4.5 Crossbreeding results of poultry

Desi fowl were crossed with White Leghorn and RIR at Izatnagar and Nagpur. This increased the egg production up to 186 and 170 eggs for the two graded lines with egg weight of 60 gm for both lines at lzatnagar. Similarly there was an improvement in egg production and egg weight at Nagpur.

The cross of WL and RIR with desi fowls at Hisar showed heterosis for body weight at all ages, egg production and feed consumption. Similar performance of WL and RIR crosses was observed at Udaipur. The egg production in two lines increased by about 38 eggs over purebred desi but in reciprocal crosses using desi males over WL and RIR females decreased egg production.

The European community sanctioned an International Collaborative Project in 1992 on evaluation of Poultry Genetic Resources. Four native chicken breeds (Aseel, Kadaknath, Naked Neck and Frizzle Fowl) along with their $F_I$ and $F_2$ crosses with Dahlem Red were evaluated for their growth, egg production and quality, feed efficiency, genetic distance and disease resistance. In all 29 genetic groups comprising purebreds, $F_l$ and $F_2$ were produced and evaluated. The crossbred males of all the genetic groups were heavier than purebreds at all ages and there was heterosis in body weight. The heterosis was observed up to $15^{th}$ week of age. The sexual maturity of various $F_1$ and F 2 crossbreds ranged from 180 to 212 days against the range of 194 days (Kadaknath) to 218 days (Dahlem Red) of purebreds. The egg production of crossbred $F_1$ showed significant increase but slightly lowers for $F_2$ than $F_1$.

A number of crossbred genotypes have been evolved at different centres by the ICAR institutes and SAUs -

| Crossbreds | Breed cosses (Male x Female) | | Place | Purpose |
|---|---|---|---|---|
| 1. Vanraja | Cornish Synthetic popn. | | ICAR, Hyderabad | Dual |
| 2. Grama priya | Synthetic White Leghorn | | -do- | -do- |
| 3. Cari gold | RIR (Selective breeding) | | CARI, Izatnagar | Egg |
| 4. Kalinga brown | RIR (Selective breeding) | | CPBF, Bhubneswar | Egg |
| 5. Giriraj | - | - | UAS, Dharwar | Dual |
| 6. Nandanam | RIR popn. | - | TNVASU, Chennai | - |
| 7. Krishna-J | Synthetic | - | JNKVV, Jabalpur | Egg |
| 8. Yamuna | Khadaknath New Hampshire | | - | - |
| 9. Gram Laxmi | Australorp White Leghorn | | KAU, Kerala | Dual |
| 10. Cari Nirbhik | Aseel cross | - | CARI | 190 eggs |
| 11. Cari Shyam | Khadaknath cross | - | CARI | 210 eggs |
| 12. Upcari | Frizzle cross | - | CARI | 220 eggs |
| 13. Hitkari | Naked neck cross | - | CARI | 220 eggs |

## NUMERICAL EXAMPLES

**Example 18.1** Calculate the heterosis in $F_1$ and $F_2$ progenies considering the milk yield in 300 days of Sahiwal and Holstein Friesian cows as 1600 and 5600 hg respectively and their $F_1$ crossbreds as 3900 kg.

**Solution:** Heterosis $F_1 = \overline{M}_{F1} - \overline{M}_P = 3900 - ½\,(5600 + 1600)$

$= 3900 - 3600$

$= 300$ kg

$$\% \text{ heterosis in } F_1 = \frac{300 \times 100}{3600} = 8.3$$

Heterosis ($F_2$) = 1/2 heterosis in

$F_1 = \overline{M}_{F2} - \overline{M}_P$

$= 1/2 \times 300 = 150$ kg

**Example 18.2** In a poultry farm, the average egg production per year in breed A was 150 and in breed B was 180. When two breeds were crossed reciprocally, the average egg production was increased to 200 eggs per year. Calculate the percentage of heterosis.

**Solution:** Heterosis $F_1 = \overline{M}_{F1} - \overline{M}_P = 200 - ½\,(150 + 180)$

$= 200 - 165$

$= 35$ eggs

% heterosis = (35 x 100)/ 165 = 21.2

❑❑❑

# 19

# Poultry Breeding

Poultry occupy an important place in agricultural economy of India particularly by providing employment to urban and peri-urban community, contributing in agricultural GDP, and to supply highly nutrition protein (about 11-12 % ) at a relatively cheaper price besides poultry manure, feathers and other byproducts. The progress in poultry farming for both egg and meat has excelled the farming of other livestock species during the ending of last century in developing countries including India.

A number of poultry farms have been established involving important pure breeds for genetic improvement in these breeds through selection and testing for their combining ability to produce economically efficient layer and commercial broiler chicks. As a result the production levels of both the layers and broiler chickens have attained the objective due to genetic approach and improvement in the environmental inputs for exploitation of the new genotypes evolved. There has been appreciable increase in the production level of layers as well as of broiler chickens. The egg production has increased at the rate of over 6% per annum during the past 20 years. There have been developed the genetically improved layers with an average egg yield of about 300 eggs per year which is about thrice than that of the indigenous layer. The increase in egg production has been to the extent that the average egg yield in India has surpassed the world average as well as the average of egg yield of the developing countries. However, the production remained lower than the average of developed countries and there exists the potentiality of further improvement in the productivity. Therefore, there is still a tremendous scope for further improvement in the productivity of the existing high yielding stocks as well as local breeds. Thus there is a need of using the modern breeding techniques and improved feeding and management practice in poultry industry.

## 19.1 PLACE OF POULTRY IN ANIMAL KINGDOM

There are a number of species of Poultry and all these belong to the phylum chordate (with backbone), sub-phylum vertebrata (spinal cord), class aves (feathered, warm blooded, 4 chamber heart) and sub-class neornithes (no teeth). The neornithes are put in two orders, namely *galli form* (birds with short wings, legs) and *ansariform* (aquatic birds).

The turkey, Japanese quail, and pheasant belong to the family *meleagridiae* while fowl and pea fowl belong to the family *phasianidae* of the order galliform. The order ansariform (aquatic birds) has the family antidae which includes duck, goose and Muscovy duke.

## 19.2 CHROMOSOME NUMBERS

The genes which are the basic units of inheritance are located on chromosomes. All the species of poultry are diploids having chromosomes in pairs. The chromosome numbers of different species of poultry are given as under –

| Poultry Species | Local name | Zoological names | Chromosome No. |
|---|---|---|---|
| ***Order Galliform:*** | | | |
| Chickens | Murgi | *Gallus gallus* (domesticus) | 78 |
| Japanese quail | Bater | *Coturnix coturnix japonica* | 78 |
| Guinea fowl | Titri/Tatiri | *Numida meleagris* | 80 |
| Pea fowl | | *Pavo cristatus* | 80 |
| Turkey | | *Meleagris gallopavo* | 82 |
| Pheasant | | *Phasianus colchicus* | 82 |
| Domestic pigeon | Kabootar | *Columbia livia domestica* | 82 |
| **Order *Ansariforms (Aquatic birds):*** | | | |
| Ducks | Batakh | *Anus platyrhnchos* | 80 |
| Goose | | *Anser anser* | 80 |
| Muscovy duke | | *Cairina maschata* | 80 |
| ***Big birds:*** | | | |
| Emu | Large Austr. bird | *Dromaius novaehollandiae* | 82 |
| Ostrich | Shuturmurg | *Struthio camelus* | — |

There are two types of chromosomes regarding their size in different avian species. These are of very large size known as macro-chromosomes which have definite recognizable features and second types are of small size known as micro-chromosomes. The cytologically distinct macro-chromosomes are 8 pairs including sex chromosome pair (Z and W) and the rest 30 pairs are of small size (micro—chromosomes) which are cytologically indistinguishable.

The sex chromosomes in poultry are designated as Z and W instead of X and Y of the mammals and Drosophila. The male sex in avian species is homo-gametic (ZZ) and the female sex is hetero-gametic (ZW), unlike mammals. Some genes are also present on sex chromosomes like mammalian sex chromosomes. These genes present on sex chromosomes are called sex-linked genes and the characters controlled by sex-linked genes are called as the sex-linked characters, like in case of mammals and Drosophila. The inheritance of the sex-linked characters is just opposite to those of sex-linked characters of mammals. Therefore, a hen transmits the sex-linked character to her sons but not to her daughters whereas the cock can transmit the sex-linked characters to his sons as well as to the daughters.

This information on transmission of sex-linked characters is very useful for auto-sexing in day old chicks. The rapid feathering chicks, produced from mating of homozygous males for rapid feathering to hens for slow feathering, will be all females but slow feathering will be males. Not only this information on inheritance pattern of sex-linked characters is used for qualitative traits but is also used for quantitative traits. This can make use by making the reciprocal crosses. In case of sex-linked inheritance, the mean of crossbred progeny produced from crossing of two strains or lines will be close to sire line than dam line.

## 19.3 MEASUREMENT OF QUANTITATIVE CHARACTERS

The description of the genetic structure of population is character specific *viz.* different for qualitative and quantitative characters. The economic characters in poultry are little different to that in livestock species. It is thus essential to know about the way these characters are measured.

The characters of interest influencing the economy of poultry farming are mainly the egg and meat production. The different avian species are reared for these two purposes. Both egg and meat production and the related characters viz. body weight including body size and conformation, the age at which the egg production start, the interruption in egg production and life span of birds are all polygenic traits. The economic characters and the procedure of their measurement are given below –

### 19.3.1 Body Growth Traits (Mostly broiler traits)

*Body weight and related characters:* Body weight is an important character in all the species irrespective of the purpose they are reared. However, an optimum body weight has got its own importance. An optimum body weight is required for production of eggs of optimum size, though small body weight is preferred in layers whereas higher body weight is desirable in broilers. This is because small size birds are more efficient in egg production due to less feed consumption. The traits related to body weight are generally considered as broiler traits.

*(i) Body weight*: The birds are weighed of at about 6 – 7 weeks of age in case of broiler.

*(ii) Feed efficiency*: It is measured as a ratio of amount of feed consumed to the total gain in body weight.

*(iii) Feed consumption (Daily) per bird in a specified period*: It is taken as the total feed consumed to the number of days (period) multiplied by number of birds.

*(iv) Dressed weight*: It is measured as the carcass weight including viscera.

*(v) Dressing percentage*: It is the taken as a proportion of dressed weight multiplied by 100 to the pre-slaughter weight.

*(vi) Shank length*: It is the distance between hock and metatarsal joint (cm).

*(vii) Shank diameter*: It is measured in cm. by vernier calliper

*(viii) Keel Length*: It is the length of keen bone in cm.

*(ix) Breast angle*: It is taken as spread of breast bone in degrees ($^0$)

*(x) Body size:* It depends more on the bone size than on flesh. The fleshing of a bird is more a function of better feeding and management whereas the bone size has a genetic component to a greater extent and can therefore be improved by selection.

*(xi) Body conformation:* It is decided by body proportions of bone structure and flesh and it is of more significance in broilers. In poultry, the ratio of body weight: shank length is more important.

*(xii) Growth rate of body:* It is more important in broiler chickens. Faster growth lowers the cost of raising the broiler to market age. The growth rate can be improved through selection because it is highly heritable character. The gain in body weight decides the *feed efficiency* which is expressed as the ratio of feed consumed to body weight gain. In poultry the feed efficiency is also taken as the amount of feed consumed to produce one dozen of eggs.

### 19.3.2 Traits of laying birds

*(i) Age at sexual maturity:* It is an important character, taken as the age at which a pullet lays its first egg. It is counted from date of hath to the date of lay her first egg. Early maturity is preferred because the early maturing pullets produce more number of eggs during lifetime than late maturing. The sexual maturity is a trait governed by polygenes some of which are present on sex chromosomes. This character has moderate heritability and positively correlated with egg weight. Therefore, early maturing birds lay eggs of smaller size and hence too early maturity is not desirable. A good layer starts laying the eggs around 5 months of age.

*(ii) Feed efficiency for egg mass:* Recorded as a ratio of amount of feed consumed to the total egg mass produced.

*(iii) Feed efficiency per dozen eggs produced:* It is measured as the amount of feed consumed to the dozen eggs produced.

*(iv) Intensity and interruption in egg production*: The egg production is not a continuous process in poultry species but it is interrupted for some time and expressed in different ways. The factors determining the egg laying intensity of hens are sexual maturity, rate of laying, broodiness, persistency and pause. The moulting, broodiness and some environmental factors like stress, management and disease incidence induce pause (interruption or cessation of laying egg).

***(a) Rate of laying eggs or intensity of egg production***: The intensity of production is judged by clutch size. The clutch is the egg laying period. The number of eggs laid by a hen in a standard time interval is the rate of egg production or ***intensity of laying*** the eggs. The rate of lay is also estimated as percentage of eggs laid divided by the number of days in lay after age at first egg to a fixed date (See Example 19.1). This intensity may also be expressed as the percentage of eggs laid in a time interval or in terms of clutch size. The clutch size varies greatly among birds of the same breed as well as among clutches of the same bird. The interval between clutches is short with longer clutch in good layers. The character clutch is moderately heritable.

The egg laying ability of a hen is determined by keeping record of egg production. This can be done by keeping the birds in separate cages or *trap nesting* them. The trap nest is a cage made of metal with provision of slope towards rear side so as the egg may roll backward. There should be enough litter material to avoid egg breakage. It should have a single hinged door to open inside of cage so that the birds may not open the door to escape from cage. For every 2 or 3 layers it requires one trap nest. The trap nesting is advised after every 1.5 to 2 hours. The trap nesting is required not only to know the egg production of individual hen, but it also helps to determine the breeding worth of sire and dam for egg production and various egg quality parameters.

***(b) Persistency:*** It is length of laying eggs. The continuous laying of eggs up to 72 weeks is a good ***persistency.*** But the poor layers start cessation of laying eggs earlier due to moulting. It is better for egg production that laying period should be longer before moulting period starts. There is partial or complete cessation of egg production when moulting occurs in laying hens. Thus, the persistency of laying hens is taken in terms of moulting. The moulting is shedding of old feathers and their replacement by new feathers. The moulting should take place after about 72 weeks of age. The layers which start moulting in the production cycle before 72 weeks also stop laying the eggs and hence called low persistent. Good layers moult late. The poor layers take long time for moulting. The selection for improvement is done on the basis of annual egg production rather than on part egg production, though these are genetically correlated. The persistency of egg production and annual egg production are highly correlated.

***(c) Broodiness:*** The broodiness is the characteristic of a desi hen to sit on eggs for hatching. The broody hen takes the eggs under her chest, spread out her wings and sit silently on them. The birds show broodiness during which they stop laying the eggs and show the tendency for nesting. The broodiness is significantly negatively correlated with egg production (- 0.56) and highly heritable and selection of sire whose dam or sisters showed little or no broodiness is effective. The Asiatic breeds are more broody than Mediterranean breeds of poultry. There is variation in broodiness in heavy breeds like Australorp, RIR and White Rock. More broodiness is observed among interbreed crosses compared to the parental breeds. The complementary gene action is involved in broodiness and there is likelihood that the genes governing broodiness may be located on sex chromosomes. However, continuous selection has eliminated the broodiness in elite layers.

***(d) Pause:*** The interval between clutches may exceed for seven or more days. It is called as ***pause*** when there is interval of more than 7 days between two clutches of eggs. The period of broodiness of cessation of egg production is also the pause. The pause has the genetic component but some environmental factors like stress, management and disease incidence also induce pause. The pause may be due to digestive disturbances, cold, respiratory problems, parasites. The pause if occurs in winter it is called ***winter pause***. There is winter pause in temperate climate and also some pause during summer and rainy season due to hot and humid climate.

*(v) Measures of Egg production:* This is most important economic character. The number of eggs produced is affected by age starting from around 20 weeks of age,

reaching peak production in about 6 – 8 weeks and declining after first laying cycle which ended at about 72 weeks of age. The peak production reached in each laying cycle after the same period of laying the first egg (6 – 8 weeks).

The number of egg production is calculated for the flock as well for the individual bird. The egg production for flock is measured on the following bases viz.

(*a*) *Monthly average* – This is estimated as the total number of eggs laid during the month divided by the average number of birds for the month which is the average of the number of surviving hens on the first and last day of month. This is also called as survivor's average.

(*b*) *Hen day average* –It is calculated as the average egg production "on hen-day basis". It is obtained as the number of eggs laid during the month divided by the average number of hen days available during the month (hen day average).

*Hen day average* = No. of eggs laid / no. of hen days

The average number of birds during the month is estimated as the average number of hens per month which is obtained by dividing the actual number of hen days with number of days in that month. This is taken into account the mortality of birds by taking into account the number of days the hen performed before death.

(*c*) *Hen – housed average* – This is estimated as the total number of eggs produced during a given period by the number of bird in the flock at the beginning of the period (i.e. number of hens housed in the laying pen).

*Hen housed average* = No. of eggs laid / no. of birds housed

This is usually obtained on yearly basis. Thus total number of eggs produced during the year is divided by the number of birds in the flock in the beginning of the year. This does not take into account the mortality.

The egg production for individual bird is estimated in terms of the number of eggs laid up to a fixed age. Second measurement of egg production of individual bird is the rate of lay which is calculated as –

Rate of lay = (no.of eggs laid up to a fixed age/ no. of days in lay) x 100.

### 19.3.3 Egg Quality Traits

There are external appearance and internal egg quality traits.

***(A) External quality characters:*** The external quality of eggs is taken in terms of the external appearance of the egg from its size (weight), shape, shell colour and texture.

(*i*) *Egg weight (size):* The egg weight is also important trait and directly related to egg size because higher egg size means higher egg weight. The egg weight has moderate to high heritability depending on the population. The egg weight is negatively genetically correlated with laying rate as the birds with higher rate of laying produce eggs with low weight. However, this negative genetic correlation between two traits (rate of lay and egg weight) can be broken and both the rate and egg weight can be improved through application of breeding techniques which shows that pleiotropy is not the cause of genetic correlation. The age, nutrition, temperature and health status influence the egg weight. The temperature influences the egg weight through feed consumption which is adversely affected at high ambient temperature. Likewise, deep litter system has adverse effect than cage system of housing. The egg weight is higher of the first egg laid as well as after moulting.

The eggs are graded according to their size. The standard size of egg is about 56 gm. The eggs weighing between 50 – 60, 45 – 50, and 38 – 44 gm, are graded as large, medium and small size egg, respectively. The eggs weighing more than 60 – 79 gm are extra large or jumbo egg and more than 80 gm are called giant eggs whereas those weighing less than 35 gm are called dwarf or pevee egg.

(*ii*) *Shape of egg:* The shape of egg is normally oval and round. The egg should not be too round or too elongated. The spindle shape eggs are considered abnormal.

(*iii*) *Shape index:* It is expressed as a ratio of breadth of egg to the length of egg. This can also be expressed in percentage.

(*iv*) *Egg colour* or *shell colour:* It is a breed characteristic and is not related to quality in terms of nutritive value. However, most people prefer white eggs. The white, brown and blue eggs are observed. The meditarranean breeds (Leghorn, Minorcas, and Anconas) have white colour eggs whereas American and English breeds have brown shelled eggs and Araucana breed has blue coloured eggs.

***(B) Candling Test of egg:*** The candling test of egg is done in a dark room by passing the light from a lamp through the egg to the observer. This helps to know the internal quality like shell, air cell, egg yolk, and white.

The candling test detects easily the eggs with thin, porous or cracked shell. The air cell is an important character. It is usually at the large end of the egg and increased by evaporation of moisture from the egg. The air cell is fixed in position without bubbling or moving about. The bubbling of air cell gives the indication of staleness and weak shell membrane.

The egg yolk is not clearly seen by candling except as a faint shadow. The yolk is present near the center of the egg. The yolk in good quality egg does not move freely in the white albumin because the white is firm and thick (viscous). The eggs with thick albumin and high proportion of yolk are preferred. The yolk in poor quality egg moves freely and casts a darker shadow when it comes close to the shell. The germ development is not visible in good quality eggs. The egg quality is poor with greater germ development and the egg is not considered fit for consumption when its development reaching the stage at which blood begins to be seen.

***(C) Internal egg quality characters:*** Some characteristics of the egg can only be seen after the eggs are opened. These are the odour, flavor, quality of albumin (Haugh unit), yolk colour, yolk index, shell thickness and the presence of blood and meat spots.

(*i*) *Shell thickness:* It is an indication of breaking strength of egg. The shell thickness is positively correlated with shell strength. The thick shelled eggs give better hatchability. The shell thickness is normally from 0.33 to 0.38 mm and it can be measured by screw gauze. The shell thickness is estimated as the average of the thickness of 4 pieces of shell taken one each from the two ends and two from circumference. The shell thickness depends on availability of calcium carbonate. The utilization of calcium depends on the Calcium: phosphorus ratio and vitamin $D_3$. The shell thickness is generally less in summer.

*(ii) Yolk colour:* A disc having 15 colours ranging from light yellow to orange is available. The yolk colour is matched with the colour of disc. The normal yolk colour is yellow. The normal pH of yolk is 6.6 (acidic) and of albumin 7.2 (alkaline).

*(iii) Yolk Index:* It is expressed as a ratio of height of yolk (mm) to the average width of yolk (mm). The height of yolk should not be less than 1/3 of its width (diameter). Better quality yolk should not flatten and also should not rupture.

***(iv)** Haugh Unit:* This is a measure of egg white or albumin quality. The Haugh Units increases with the increase in height of albumin and egg weight. It is measured as:

$$H.U. = 100 \log (H + 7.57 - 1.7 W^{37})$$

Where, H = Height of albumin (mm), measured by spherometer,

W = Weight of egg (gm)

The values of $1.7 W^{37}$ are taken from table prepared according to the egg weight varying from 40 to 75 gm.

There is Haugh meter also and it gives direct reading in haugh units according to the weight of the egg.

*(v) Blood and meat spots:* The eggs with blood spots (generally on yolk) and meat spots (in albumin) are not preferred. These are recorded and expressed as percentage of total eggs under study.

### 19.3.4 Fertility and Hatchability traits

These are expressed in percentage and affected more by environmental factors like age, season, nutritional status of birds, management and diseases etc. and have low heritability estimates. To get good fertility and hatchability the poultry ration should have proper amount of minerals and vitamins.

The fertility is measured in percentage as:

*Fertility* (%) = (No. of fertilized eggs x 100)/ No. of eggs set.

The hatchability is expressed in the percentage on two bases as:

(*i*) *Hatchability* % = (No. of chicks hatched x 100) / No. of eggs set

(*ii*) *Hatchability* % = (No. of chicks hatched x 100)/ No. of fertile eggs set

### 19.3.5 Survivability

It is calculate in percentage at a particular age. The incidence of diseases in poultry not only affects the viability but also other characters. The susceptibility/ resistance of disease is a polygenic trait. The viability is more affected by environment factors and hence has low heritability estimate.

## 19.4 METHODS OF MATING IN POULTRY

Before discussing the mating system it is better to know about mating. The mating is essentially required to produce fertilized eggs for multiplication of flock. The fertilized eggs are incubated for hatching.

The important point in mating of poultry is ratio of male to female. This depends on a number of factors viz. breed, body weight, age, season, health and physical condition, and sexual activity of the male. Generally 10 – 20 hens are allotted to each male. More number of hens may be allotted in case of light breeds, in spring and winter season, and in case of young male. Secondly, summer mating should be avoided because of poor fertility. The males of light breeds (Leghorns) can cover more number of females than males of heavy breeds (RIR, White Plymouth Rock) because the males of light breeds are sexually more active. Likewise the cockerels can cover more number of females than older males.

The ***mating types*** or **methods of mating** in poultry may be natural or artificial. The natural mating is allowed in any of the following four ways -

(1) *Pen mating:* In pen mating a group of hens are allowed to mate with a cock in a pen which is of the size of 8 x 6 ft. The cock is allowed to live freely with about 10 hens in the breeding pen. The pen mating is used for pedigree hatching. The pen mating has the disadvantage to flock mating that the fertility is low in pen mating because the male mate may prefer to mate more frequently with certain females than others (preferential mating) for certain reason like plumage colour.

*(2) Flock mating:* It is also called mass mating. Large numbers of hens are kept with many cocks in a single pen. More number of females can be put to a cock in this system (male to female ratio). This ratio may be 1:12 to 15 females in case of light breeds and 10 to 12 in case of heavy breeds. The pedigree records can not be kept because the ggs can not be marked for their parentage. The fertility is higher and hence preferred for commercial purpose.

(3) *Stud mating*: The hens are allowed into male's pen one by one till mating is complete. In this system the male is kept in a pen or coop (2 x 3 ft size) separately and the female birds are brought to this male one by one for mating. The female bird is replaced with another female aftr mating with the first female is accomplished. It is thus labourious. Howevr, it has the advantage to get mate many females with a male of high genetic merit so as to obtain many offspring. The fertility in this system is higher even to flock mating.

(4) *Shift mating:* The males are shifted in the breeding pens. It is a method of family breeding where the B.V. of more number of males is required to be tested to identify superior males. In this mthod of shifting the males, a female can be mated to several males. This makes easy to evaluate the breeding worth of female. Howevr, this shifting method has a problem of proper parentage of the progeny due to the reason that fertility is maintained for about 2 weeks after shifting the cock. It has the advantage that large number of males can be tested in a limited space.

***Artificial Insemination***: The artificial mating is the deposition of the cock semen artificially in the genital tract of hen and called as *artificial insemination (A. I.).*

*Semen collection and A.I.:* The semen is collected by massaging gently the soft side of abdomen between gizzard and pelvic bones. The message causes enlargement and swelling of copulatory (ejaculatory) organ and the semen is collected. The semen should be collected between 2 to 4 PM once in two days. The volume of cock semen in one ejaculate is 0.4 ml and it may be used to inseminate 3 hens. To inseminate the hen its both the legs are gasped, cloaca is inverted and 0.1 ml semen is deposited in vagina with a syringe about 3 – 4 cm deep inside the genital tract.

## 19.5 PEDIGREE RECORD KEEPING

The identification records of individual and its pedigree (its parents and ancestors) as well the performance records are required to keep pedigree records. This is known in poultry as ***pedigree hatching.*** The record keeping starts from mating of hen to hatching of chick. Maintenance of records is essential to trace the pedigree (ancestry), to know the performance of individuals, and to evolve a new strain with desirable characters.

It is a difficult task to keep pedigree records in poultry. It needs to identify the eggs and hatching chicks for their parentage. This is difficult due to the fact that many pullets are kept together in a single pen and all of them lay eggs of the same colour and size to some extent and futher these eggs are kept in the incubator in mass for incubating/ hatching. Therfore, the pedigree hatching needs some steps with precautions at evry step. These steps are as -

(*i*) *Identification of breeding males and females:* This is done by using leg bands or wing badges of aluminium or plastic,

(*ii*) *Single sire mating (Pen mating):* The eggs laid are identified for the cock number,

(*iii*) *Identification of eggs:* This can be done by housing the pullets in individual laying pens (cages) or by use of trap nest.

(*iv*) *Marking of eggs:* This can be done with a fairly blunt and soft pencil which must give a water proof record. The eggs are marked for sire and dam number. Such marked eggs are called pedigreed eggs.

(*v*) *Handling of pedigreed eggs:* The eggs are sorted out daily pen wise and hen wise within a pen. The eggs are set for hatching weekly on the same day of the week.

(*vi*) *Hatching of eggs:* The eggs laid by a hen in a week are set in a setting tray in pen wise sequence with a record of sequence in the hatching register. The eggs are checked for fertility by candling on $4^{th}$ day of incubation and again candled on $18^{th}$ day of incubation to check and remove the dead and infertile eggs.

(*vii*) *Identification of chick at hatching:* The hatched out chicks are put identification mark by putting wing bands of aluminium in the right wing.

## 19.6 SELECTION BASED ON PHENOTYPIC CHARACTERS

The breeding of any species involves ***selection*** of genetically best individuals together with culling of the inferior stock and ***mating system*** followed (inbreeding and out breeding) for genetic improvement of important economic characters.

The selection and culling are two but same faces of allowing the individual to reproduce or not depending on the merit of the individuals. The selection is done based on phenotype and based on breeding value of the animal.

The selection may be *intra-population* either for single or multi-traits or the election may be on the basis of performance of a cross produced by crossing two or more populations, known as *inter-population selection*. The inter-population selection based on cross performance is used to develop elite crosses (high productive strains). The intra-population selection in pure lines is done before inter-population selection.

The use of control population is important in poultry breeding to measure the environmental effects. The random bred control population or repeat mating control is maintained to make comparison.

The selection based on the phenotypic characteristics in poultry is very important aspect of selection. The following points need to consider in phenotypic selection of birds for egg production (Pullets) and for meat production (Broilers).

### 19.6.1 Selection for egg production

The important factors for egg production are the size, colour and number of eggs. The pullets should be selected at about 20 weeks of age when they start laying.

*Egg production:* A normal laying bird lays about 100 – 200 eggs in a year and 6 – 15 eggs in first 40 days; good layer lays between 200 -250 eggs per year and 16 – 25 eggs in first 40 days; and excellent layers lay more than 250 eggs a year and more than 25 eggs in first 40 days are excellent layers. The birds laying less than 100 eggs in a year and less than 6 eggs in first 40 days should be culled.

The pullet should be quick maturing (20 weeks to lay first egg), with good confirmation, show alertness, lack of broodiness and capable to lay more number of eggs in a given time. The sexual maturity, broodiness, persistency, rate of egg laying and pause are the factors affecting the egg laying intensity. Therefore, these points should be considered in selecting the hen for egg production.

The birds having good family record of external characters, without any pause (not more than 7 – 8 days) and produced continuously 3 or more eggs or 50 eggs in 60 days period with egg production intensity of more than 85 %, with rapid feathering and late moulting of about 365 days from first egg are desirable characters in selection for egg production. The eggs produced should be of optimum size (55 – 60 gm) with better internal quality of eggs.

*Score card of external parts:* The *score card* procedure on the basis of external features is helpful in selection in case the other records are not available. The desirable external characteristics of hen to lay large number of eggs are flat broad back; short well apart legs; well proportionate body with consistent skeletal development with its size; relatively short, curved and strong beak; short and stocky neck; bright and prominent eyes; well developed ear lobes; and active, alert and docile temperament.

The good layers should have normal, red and fully expanded comb and wattles; oval, large and wet vent; the distance between pubic bones of more than 2 fingers; distance between keel bones of more than 4 fingers (6.6 cm); flexible pin bone; soft abdomen; wide and smooth shanks;

*Pigmentation:* The past record of egg production can be estimated from the pigmentation of ear lobes, beak, shanks and vent. The yellow pigment xanthophylls of feed is utilized for colouring the yok during laying period and so the body losses its reserve supply gradually from body parts (vent, earlobe and eye rings, beak, and shanks) which become bleach in approximate period as mentioned above under culling criteria.

### 19.6.2 Selection for broiler production

The important points to consider in selection for broiler production are the growth rate, feed efficiency, body conformation, viability, skin colour and feathering.

The *growth rate* is the most important trait and it is easy to improve the growth rate for its high heritability. This needs to select the birds on the basis of body weight at 8 weeks of age. A broiler should have 1.2 kg body weight at 6 weeks of age. The *feed efficiency* of 1 kg weight gain after 2 kg feed is ideal.

Regarding *body confirmation,* the broiler should have long, broad and fleshed breast with broad and well fleshed back, and with heavy and moderately long shanks.

The *skin colours* are qualitative characters with white colour as autosomal dominant and dark colour as recessive. However, yellow colour is influenced by feed stuffs.

The *early feathering* is a desirable character both in broilers and layers. The early feathering avoids the problem of pin feathers at processing.

## 19.7 CULLING CRITERIA

The culling is the removal from the flock of the unproductive birds and with any defect. The purpose of culling is done to increase the performance of the flock and hence to increase the profit.

*Reasons of culling:* The birds are culled for a number of reasons which are considered important parameters for culling. These are poor growth, poor feathering, broody hens, late maturity, low feed conversion efficiency, low egg production and defective body parts.

The phenotypic appearance of some body parts of the birds helps in differentiating between good layers and poor layers (laying capacity). The birds should be culled having distance of < 3 fingers between keel bone and pelvic bones and < 2 fingers between pelvic bones; hard and thick pelvic bones; body with thick and rough skin, fleshy body with fat deposition; body with small, round and thick bones; shrunken, dry and pale wattles and comb; small, dry and round vent; round shanks with rough scales and covered with flesh; small, shrunken and pale eyes; weak, downward and retracted head.

The *growing birds* are culled with poor body shape, long legs, untrue to type, droopy feathers, poor feathering up to 8 weeks of age and looking dull. The body shape (well proportion body) is important in broilers.

The birds with yellow ear lobes, eye rings, beak, shank, and vent should be culled. The yellow pigmentation is due to carotenoid pigment xanthophyll from ration (yellow maize). The laying birds lose the pigment for utilization in yolk formation resulting in bleached appearance of exposed parts. The vent loses the pigment earlier (1 – 2 weeks of laying eggs) followed by eyes and ear lobes (2 – 4 weeks), beak (6 – 8 weeks) and lastly the shank (12 – 20 weeks). The yellow pigment reappears in the same order of disappearance but twice fast when the birds stop laying the egg. Similarly the birds with dull and sleepy eyes, small and pale comb, bright plumage should also be culled.

The birds are culled on egg production basis. The time of moulting gives an indication of laying capacity. The moulting period is late in good layers and some time they continue laying the egg during moulting. The culling of poor layers increase the profit of poultry farming by saving the cost of feed, labour and maintenance, and increase in average egg production.

**Flock Replacements:** It is necessary for economic production to maintain the flock size which is reduced as a result of death and culling/selling the poor producers. It is desirable that about one half of the flock is replaced each year by new chicks. Thus the flock is renewed every two years. It is better to multiply the flock by incubating the fertilized eggs (hatching) at the farm itself rather than purchasing the new stock.

## 19.8 HATCHING

The hatching out chicks from fertile eggs is called ***incubation.*** The *incubation period* in chicken is of 21 days. It is variable from 17 to 42 days in different species of poultry viz. Japanese quail (17 days); Pigeon (18 days); parrot (19 – 25 days); pheasant (24 days); duck, turkey and pea fowl (28 days); guinea fowl (27 days); geese (31 days); swan (35 – 40 days); and ostrich (42 days).

The incubation or hatching may be natural or artificial. The natural incubation is the setting of eggs under a broody hen while the artificial incubation is through incubator (machine) by providing ideal temperature, ventilation and humidity to fertile eggs set in an incubator.

The *natural incubation* is done by a broody hen. Broodiness is a natural instinct of hens to sit on eggs till hatching out of chicks and to take care of hatched chicks. The broodiness is prominent in Desi hens. A broody hen show some indications of broodiness like that a broody hen does not lay eggs, she becomes dull and inactive, prefer dark and warm corner, makes clucking noise if disturbed, ruffles her feathers and keep sitting when approached. She collects the eggs with her beak, takes them under her chest, spread out her wings and sits on eggs silently. A broody hen covers 6 – 8 eggs for hatching. The natural incubation is cheaper, suited for small scale, requiring no turning of eggs and care of chicks because the broody hen does itself the turning of eggs and care of hatched chicks.

The *artificial incubation* is done on large scale incubation of eggs in a machine called incubator. The temperature and moisture of incubator are important consideration. The temperature should be 102, 102, and $103^0$ F during $1^{st}$, $2^{nd}$ and $3^{rd}$ week of incubation, respectively. Low temperature of incubator increases the time of incubation and number of weak and abnormal chicks (crippled) as well as death rate. The RH should be maintained 58, 60 and 70 % during $1^{st}$, $2^{nd}$ and $3^{rd}$ week of incubation, respectively. The optimum humidity is maintained either by humidifying device or by keeping open trays of water at the bottom below the egg trays or by sprinkling warm water on eggs after every turning. The turning of eggs is also important. The eggs are turned 3 - 4 times daily. This is done to prevent sticking of embryo to one side of egg because sticking reduces the hatchability. However, eggs are not disturbed after 18 $^{th}$ day of incubation.

It is better to keep the eggs in rows by keeping the narrow end of eggs down for proper respiration of embryo and easy turning of eggs. Always fresh eggs are set for incubation because the hatching percentage is reduced upto 5 days of staorage of eggs and the eggs older than a week fail to hatch.

## 19.9 BASIS OF SELECTION - GENETIC SELECTION

The ***selection criteria*** are same as used for other farm animals viz. individual, pedigree, family, progeny selection, and combined selection. The *selection methods* for single and multi trait selection are also the same. In case of poultry, the family selection, combined selection, and selection for combining ability need special consideration.

### 19.9.1 Family Selection

The term family means those individuals which have genetic relationship to each other due to their common ancestor (descent). The family selection in poultry is also taken in another way viz. as a unit of selection. The whole of the family is selected or rejected as a unit of selection based on family mean as a deviation from population mean, disregarding to the individual's performance except its contribution to the family mean. The family averages are worked out and the superior families as per requirement (desired numbers) are selected. The efficiency of family selection is increased with increase in family size. This is because the environmental deviations of larger size family are cancelled out and the family mean is close to the breeding value. The family selection is advantageous for traits of low heritability.

Family selection is practiced for selecting pullets for egg production because the performance of pullets (egg production) is also included in estimating the family mean. But the cocks are evaluated for egg production based on the performance of their sisters (sibs) and hence sib selection is practiced. The sib selection is therefore practiced for evaluation of males for egg production.

There is hierarchical structure of poultry population. This is because several sires (*s*) are each mated to a number of dams ***(d)*** and several progenies ***(n)*** are produced from each mating. There are dam families (full sib family) and sire families (full and half sib family). The family selection in poultry has been discussed by Osborne (1957) by using the sire and dam family averages.

***(i) Dam family (FS family) selection:*** The response from selection of female on the basis of dam family performance in a family of size ***n*** can be estimated as:

$$\Delta G_{DF(Female)} = 0.5\, i_F\, \boldsymbol{\sigma}_P\, h^2 \frac{[(n+1)}{\sqrt{2n\{2+(n-1)\,h^2\}]}}$$

The males are selected on the bases of their full sister's performance and the expected response ($\Delta G_{FS(Male)}$) may be estimated as:

$$\Delta G_{FS(Male)} = 0.5\, i_M\, \boldsymbol{\sigma}_P\, h^2 \frac{[(n\,r)}{\sqrt{n[1+(n-1)\,t]}}$$

Where, r = 0.5 which is genetic relationship between full sibs,

$t = 0.5\, h^2$, the correlation between the phenotypic values of FS for the trait.

***(ii) Sire family (HS and FS family) selection:*** A number of dams ***(d)*** are mated to a sire the number progenies ***(n)*** are available, making the family size dn. The expected genetic response to sire family selection for females is estimated as –

$$\Delta G_{SF\,(Female)} = 0.5\, i_F\, \boldsymbol{\sigma}_P\, h^2 \frac{[(n+nd+2)}{\sqrt{4nd\,\{4+(n+nd-2)\,h^2\}]}}$$

The males are selected on the basis of the performance of their half sisters (HS selection) and the expected response from selection of pullets is estimated as -

$$\Delta G_{HS(Male)} = 0.5\, i_M\, \boldsymbol{\sigma}_P\, h^2 \frac{[(n\,r)]}{\sqrt{n[1+(n-1)\,t]}}$$

Where, r = 0.25 which is genetic relationship between half sibs,

$t = 0.25\, h^2$, the correlation between the phenotypic values of HS for trait.

It is important to note that to get total response in progeny generation the expected response to selection for both sexes is added.

The size of sire family is quite large than dam family and hence sire family selection is easy. The pullets of a sire may be kept in a single pen for recording egg production without requiring individual trapnesting. On the other hand, the progenies of different dams need housing separately for individual trapnesting. The half sib size is more than full sibs and hence HS family may give more information than FS, despite less genetically related. Moreover, the FS performance is biased upward for sharing common environment particularly during early life.

### 19.9.2 Combined Selection

The information from at least two sources are considered in combined selection. These may be individual's own performance and the family average. This is done to get high genetic gain compared to individual selection or family selection alone. This is particularly true for characters of low heritability like egg production. The combined selection is done by giving proper weight to the individual records and to the averages of sire and dam families. Osborne (1957) selection index for increase in egg production is discussed here.

**OSBORNE INDEX:** Osborne (1957) developed a selection index based on combining information of records of the individual and its family members (sire family and dam family average) by giving proper weights in poultry for improving egg production. This is known as Osborne index. The indices were developed for two sexes separately as:

$$\text{Selection of Pullets } (I_F) = (P - \overline{P}) + W_2 (F_D - \overline{P}) + W_3 (F_S - \overline{P})$$

$$\text{Selection of Cockerels } (I_M) = W_2 (F_D - \overline{P}) + W_3 (F_S - \overline{P})$$

Where, $P$ = individual's own records

$\overline{P}$ = flock mean

$F_D$ and $F_S$ = dam and sire families average, respectively

$$W_2 = \frac{2n(1-h^2)}{4+(n-2)h^2}$$

$$W_3 = \frac{4\bar{n}d(1-h^2)(2-h^2)}{[4+(\bar{n}-2)h^2][4+(\bar{n}+\bar{n}d-2)h^2]}$$

where $n$ = no. of pullets per dam

$\bar{n}$ = average number of pullets per dam

$d$ = number of dams mated to a cock (sire)

$h^2$ = heritability of the character.

The expected response from *dam family selection* from selection of pullets and selection of cockerels by FS selection as well as the expected response from *sire family selection* (HS and FS family) for selection of females (pullets) and from selection of males (Cockerels) by HS selection based on Osborne index is obtained as outlined above under family selction.

The expected response from *combined selection* for Osborne index combining individual, dam family and sire family averages is estimated as –

$$\Delta G_{CS} = 0.5\,(i_M + i_F)\,\boldsymbol{\sigma}_P h\left[\left\{\frac{i_F}{(i_M+i_F)}\right\} R_1 + \left\{\frac{i_F}{(i_M+i_F)}\right\} R_2\right]$$

Where, $R_1$ and $R_2$ are multiple correlation coefficients between index and genotype of trait

$$R_1 = h\sqrt{(A+B+C)}$$

$$A = \frac{(n-1)}{2n\,(2-h^2)}$$

$$B = \frac{[(d-1)\,(n+2)^2]}{[4\,n\,d\{4+(n-2)h^2\}]}$$

$$C = \frac{(s-1)\,[n\,(1+d)+2)]^2}{[4\,s\,n\,d\{4+n+n\,d-2)h^2\}]}$$

$$R_2 = \frac{h}{2}\sqrt{(D+E)}$$

$$D = \frac{n}{d}\left[\frac{(d-1)}{\{4+(n-2)h^2\}}\right]$$

$$E = \frac{(s-1)\,(d+1)^2}{[s\,\{4+(n+n\,d-2)\,h^2\}}$$

Where, i = standardized selection differential

n = average number of progeny pr dam

d = average number of dams per sire

s = number of sires

**ABPLANALP INDEX:** Another index of combined selection was proposed by Abplanalp (1974) for selection of pullets and cockerels as:

$$\text{Selection of Pullets } (I_F) = b_1\,(P - P_{DF}) + b_2\,(P_{DF} - P_{SF}) + b_3\,(P_{SF} - \bar{P})$$

$$\text{Selection of Cockerels } (I_M) = b_4\,(P_{DF} - P_{SF}) + b_5\,(P_{SF} - \bar{P})$$

Where, P = individual record

$P_{DF}$ and $P_{SF}$ = dam and sire family averages

$\bar{P}$ = flock mean

$b_1$'s = weights given to different selection criteria as:

$$b_1 = \frac{G_1}{P_1} \text{ where, } G_1 = \left[\frac{n-1}{n}\right] 0.5\boldsymbol{\sigma}^2_G;\ \boldsymbol{\sigma}^2_G = 4\,\boldsymbol{\sigma}^2_S = 4\,\boldsymbol{\sigma}^2_D$$

$$P_1 = \left[\frac{n-1}{n}\right] \sigma^2_W \left(\frac{n-1}{n}\right)$$

$$b_2 = \frac{G_2}{P_2} \text{ where, } G_2 = \left[\frac{d-1}{d}\right] 0.5\,\frac{\boldsymbol{\sigma}^2_G}{n} + \boldsymbol{\sigma}^2_D$$

$$P_2 = \left[\frac{d-1}{d}\right] \left[\frac{\sigma^2_w}{n} + \sigma^2_D\right]$$

$$b_3 = \frac{G_3}{P_3} \text{ where, } G_3 = \left[\frac{(s-1)}{s}\right] \left[\frac{0.5\sigma^2_G}{nd} + \frac{\sigma^2_D}{d} + \sigma^2_S\right]$$

$$P_3 = \left[\frac{(s-1)}{s}\right] \left[\frac{\sigma^2_W}{nd} + \frac{\sigma^2_D}{d} + \sigma^2_S\right]$$

$$b_4 = \frac{G_4}{P_4} \text{ where, } G_4 = \left[\frac{d-1}{d}\right] \sigma^2_D$$

$$P_4 = P_2 = \left[\frac{d-1}{d}\right] \left[\frac{\sigma^2_W}{n} + \sigma^2_D\right]$$

$$b_5 = \frac{G_5}{P_5} \text{ where, } G_5 = \left[\frac{(s-1)}{s}\right] \left[\frac{\sigma^2_D}{d} + \sigma^2_S\right]$$

$$P_5 = P_3 = \left[\frac{(s-1)}{s}\right]\left[\frac{\sigma^2_W}{nd} + \frac{\sigma^2_D}{d} + \sigma^2_S\right]$$

n = no. of progeny per dam

d = no. of dams per sire

s = no. of sires

$$\text{Expected response} = \Delta G = \frac{i}{\sigma 1}(b_1G_1 + b_2G_2 + b_3G_3)$$

$$\sigma_1 = \sqrt{}(b_1^{\,2} P_1 + b^2_2 P_2 + b^2_3 P_3)$$

= Standard deviation of index

i = intensity of selection

Expected response in correlated trait ($\Delta G^{12}$):

$$\Delta G^{12} = \frac{i}{\sigma 1}(b_1G_1^{\,12} + b_2G_2^{\,12} + b_3G_3^{\,12})$$

Where, $G_1^{\,12} = \dfrac{G_1G^{12}}{\sigma^2_G}$

$$G_2^{\,12} = \frac{G_2G^{12}}{\sigma^2_G}$$

$$G_3^{\,12} = \frac{G_3G^{12}}{\sigma^2_G}$$

$G^{12}$ = genotypic covariance between two traits 1 and 2.

### 19.9.3 Selection For Combining Ability

**Modern breeding systems in poultry:**

In poultry, the non-additive genetic variation is exploited using the *recurrent selection* and *reciprocal recurrent selection*. The two or more population are crossed and hence inter population selection based on cross performance is done. The RS and RRS have become the *modern breeding systems in poultry* to improve performance of the crossbreds. The details of the procedures of RS ans RRS have been given in chapter 15.

## 19.10 MATING SYSTEMS

**Mating systems** in poultry are also of two type viz. *inbreeding and out-breeding,* as in case of large farm animals.

The mating of related individuals is called *inbreeding*. The inbreeding may be close breeding, line breeding and strain breeding, as defined earlier in the chapter on inbreeding. The out breeding is opposite to inbreeding. The different types of crossbreeding have been given earlier in chapter on crossbreeding. The same are applicable for poultry.

The crosses in crossbreeding may be single cross involving cross of 2 lines/ breeds, 3-way cross involving crossing of 3 lines/breeds and like wise 4-way cross involving 4 lines/breeds and multiple crosses involving the crossing of more lines/ breeds.

The multiple crosses in poultry breeding are useful to identify that which lines nick well. When all the lines/breeds are crossed in all possible combinations, it is called as diallel crossing. The diallel mating system is used to estimate GCA and SCA among crosses to detect best cross combination. A number of experimental designs are used for this. These designs are diallel analysis, partial diallel, triallel analysis, quadriallele (double cross hybrid or 4 breed crosses), line x tester analysis, and triple test cross analysis. The other types of designs are North Carolina Design 1, 2 and 3. The NCD 1 is nested design involving the mating of each sire to a different set of dams and n progenies (FS) from each mating are tested with r replicates of a randomized block experiment, the NCD 2 is a factorial design when sires and dams are mated in all possible combinations producing equal number of FS in each mating, and NCD 3 is a design in which each male is backcrossed to both parents.

## NUMERICAL EXAMPLES

**19.1** Calculate the rate of laying for the 4 hens whose egg production to 280 days has been as 99, 95, 85, and 80 during the net laying period of 128, 125, 122, and 116 days, respectively.

**Solution:** Rate of lay = (No. of eggs laid/no. of laying days)

Hen no. 1 = (99/128) x 100 = 77.3

Hen no. 2 = (95/125) x 100 = 76.0

Hen no. 3 = (85/122) x 100 = 69.6

Hen no. 4 = (80/116) x 100 = 68.9

**19.2.** There were 1500 hens at a poultry farm on September 1, 2010. The number of hens and the eggs laid along with the number of hens died and culled on different days during this month were recorded. It was found that 50, 75, 100 and 80 hens were either died or culled after day $7^{th}$, $14^{th}$, $20^{th}$ and $25^{th}$. Total eggs produced during the month were recorded as 30,000.

Estimate the (i) average egg production per bird on monthly basis, (ii) hen day basis, and (iii) percentage of egg production.

***Solution:*** Total number of hen days = 30 x 1500 = 4500

Loss of hen days:

(*i*) 30 -7 = 23 (hen days) x 50 (no. of hens lost) = 1150

(*ii*) 30 - 14 = 16 x 75 = 1200

(*iii*) 30 - 20 =10 x 100 =1000

(*iv*) 30 - 25 = 5 x 80 = 400

Total loss = 3700

Actual number of hen days = 45000 – 3700 = 41300

No. of hens on September 30 = No. of birds at the beginning
– Total no. of birds lost
=1500 – (50 + 75 + 100 + 80)
= 1500 –305 = 1195

Average number of birds per month = (No. of hens at beginning + No. of birds at the end) / 2
= (1500 + 1195)/ 2 = 2695/ 2 = 1348

(*i*) Average egg production per bird on monthly basis
= Total eggs produced/ no. of bird/month
= 30,000/1348 = 22.2 Answer

Average number of birds during month = Actual no. of hen days/ no. of days in a month (on hen day basis)
= 41300/30 = 1377

(*ii*) Average egg production per bird = Total eggs produced/ Av. no. of birds (on hen day basis)
= 30,000/1377 = 21.7 or 22 Answer

(*iii*) Percent egg production = Total eggs produced/ Actual no. of hen days
= (30000/41300) x 100 = 72.63 % Answer

**19.3** In a poultry farm 2249 pullets were kept in laying pens and daily egg production was recorded through trapnest for a period from July 1, 2011 up to October 31, 1911. The egg production of this flock and the number of birds in different months have been given as under –

| Days in different months | No. of birds | Total hen days | Total no. of eggs laid |
|---|---|---|---|
| 31 (July) | 2249 | 69193 | 36780 |
| 31 (August) | 2212 | 66112 | 46273 |
| 30 (Sept) | 2187 | 67503 | 49045 |
| 31 (Oct) | 2162 | 56552 | 44140 |
| **(123) Total** | **8810** | **259360** | **176238** |

Calculate the different measures of egg production for the flock.

**Solution:**

(*i*) Average hen days $= \frac{\text{Total hen days}}{\text{Total days}}$

$= \frac{259360}{123} = 2108.6$

(*ii*) Hen day per bird $= \frac{\text{Total eggs laid}}{\text{Av. hen days}}$

$= \frac{176238}{2108.6} = 83.5$

(*iii*) Hen housed /bird = Total eggs laid/no. of hens housed

$= \frac{176238}{2249} = 78.4$

(*iv*) Hen housed percent $= \frac{(\text{no. of eggs laid x 100})}{(\text{no. of hens housed x total number of days})}$

$= \frac{(176238 \times 100)}{(2249 \times 123)}$

$= 63.7$

(*v*) Hen day production (%) = (no. of eggs × 100)/ total hen days

$= \frac{176238 \times 100}{259360}$

$= 67.95$

Now, the ***egg production average per bird per month*** for any month may be calculated on different basis as -

*Monthly average* egg production (For July month):

Average no. of birds during the month

$= \frac{(\text{No. of bird in the beginning } + \text{no. of birds at the end of month})}{2}$

$= \frac{(2249 + 2212)}{2} = 2230.5$

Monthly average egg production / bird / month:

$= \frac{\text{Total eggs laid}}{\text{average no. of birds during month}}$

$= \frac{36780}{2230.5} = 16.5$ eggs/hen / month

(*vii*) *Hen-day average* egg production (For July month):

Actual no. of hens during month

$$= \frac{\text{Actual no. of hen-days}}{\text{no. of days in the month}}$$

$$= \frac{69193}{31} = 2232.0$$

$$\text{Hen-day average /bird/ month} = \frac{\text{Total eggs laid during the month}}{\text{Actual no. of hens during the month}}$$

$$= \frac{36780}{2232} = 16.5 \text{ eggs/bird/month}$$

*Hen-housed average* egg production/bird/month (For July month):

$$= \frac{\text{Total eggs laid during the month}}{\text{no. of birds housed in the begining}}$$

$$= \frac{36780}{2249} = 16.35 \text{ eggs/bird/month}$$

It is thus concluded that each bird in this flock produced about 17 eggs during July month. Likewise, the egg production per bird per month can be calculated for other months from the given data.

**19.4.** Estimate the expected response for egg production based on dam family, sire family and combined selection using Osborne index combining individual, dam family and sire family averages for egg production recorded from 1260 progenies produced by mating of 6 dams each with 35 sires and each mating produced 6 pullets. Assume heritability of egg production as 0.2 and standard deviation as 16 eggs.

(*i*) Estimate the expected response from ***dam family selection*** for selection of pullets and for selection of cockerels by FS selection. A total of 40 dam families from amongst 240 dam families were selected to produce 240 pullets for breeding to produce next generation whereas 40 cockerels from 20 dam families were selected out of 240 families.

(*ii*) Estimate the expected response from ***sire family selection*** (HS and FS family) taking 36 as family size. Consider that 7 sire families out of 40 sire families were selected to get 240 pullets to produce next generation and 40 cockerels from 4 sire families were selected out of total 40 sire families.

(*iii*) Estimate the expected response from ***combined selection*** using Osborne index combining individual, dam family and sire family averages when 210 pullets were selected from 1260 pullets and 35 cockerels were selected from 420 cockerels.

**Solution:**

(*i*) Expected response from dam family selection:

The first step is to estimate the intensity of selection:

| Sex | Proportion selected | Intensity (*i*) |
|---|---|---|
| Females | $\frac{40}{240} = 0.166$ | 1.49 |

| | | |
|---|---|---|
| Males | $\frac{20}{240} = 0.0833$ | 1.38 |

Expected Response from selection for pullets:

$$\Delta G_{DF(Female)} = 0.5\ i_F\ \boldsymbol{\sigma}_P\ h^2 \left[\frac{n+1}{\sqrt{2n}\left[2+(n-1)h^2\right]}\right]$$

$$= 0.5 \times 1.49 \times 16 \times 0.2 \quad \left[\frac{6+1}{2\times 6\left[2+(6-1)0.2\right]}\right]$$

$$= 2.384\ [7/\ 6] = 2.384 \times 1.166 = 2.78 \text{ eggs}$$

Expected response from selection of cockerels by FS selection:

$$\Delta G_{FS(Male)} = 0.5\ i_M\ \boldsymbol{\sigma}_P\ h^2 \left[\frac{nr}{\sqrt{n}\ \sqrt{1+(n-1)\ t}}\right]$$

$$= 0.5 \times 1.38 \times 16 \times 0.2 \left[\frac{6\times 0.5}{56\left[1+(6-1)\ 0.1\right]}\right]$$

(since, $r = 0.5$ and $t = 0.5\ h^2$)

$$= 2.208\ [\ 3/\ 3] = 2.21 \text{ eggs}$$

Total response for both sexes = 2.78 + 2.21 = 4.99 eggs

***(ii) Expected response from sire family selection*:**

The first step is to estimate the intensity of selection:

| Sex | Proportion selected | Intensity (i) |
|---|---|---|
| Females | $\frac{7}{40} = 0.175$ | 1.47 |
| Males | $\frac{4}{40} = 0.010$ | 1.75 |

Expected response from selection of pullets:

$$\Delta G_{SF(Female)} = \frac{0.5\ i_F\ \boldsymbol{\sigma}_P\ h^2\ [(n + n\ d + 2)}{\sqrt{4\ n\ d\{4 + (n + n\ d - 2)\ h^2\}]}}$$

$$= \frac{[0.5 \times 1.47 \times 16 \times 0.2][6 + (6 \times 6) + 2]}{\sqrt{4 \times 6 \times 6\ [4 + (6 + 34) \times 0.2]}}$$

$$= \frac{2.352 \times 44}{\sqrt{144\ (12)}} = \frac{103.488}{41.5692} = 2.49 \text{ eggs}$$

Expected response from selection of cockerels by HS selection:

$$\Delta G_{HS(Male)} = \frac{0.5\ i_M\ \boldsymbol{\sigma}_P\ h^2\ [(n\ r)}{\sqrt{n\{1 + (n-1)\ t\}]}}$$

$$= \frac{0.5 \times 1.75 \times 16 \times 0.2\ [\ 36 \times 0.25}{\sqrt{36\{1 + (36-1)\ 0.1\}]}}$$

$$= \frac{2.8 \times 9}{12.7279} = 1.98 \text{ eggs}$$

Total response for both sexes = 2.49 + 1.98 = 4.47 eggs

***(iii) Expected response from combined selection:***

The first step is to estimate the intensity of selection:

**Sex** Proportion selected Intensity (i)

$$\text{Females } \frac{210}{1260} = 0.1666 \ 1.50$$

$$\text{Males } \frac{35}{420} = 0.0833 \ 1.38$$

$$\Delta G_{CS} = 0.5\,(i_M + i_F)\,\boldsymbol{\sigma}_P h \left[\left\{\frac{i_F}{i_M+i_F}\right\} R_1 + \left\{\frac{i_F}{i_M+i_F}\right\} R_2\right]$$

Where, $R_1$ and $R_2$ are multiple correlation coefficients between index and genotype of trait

$$R_1 = h\sqrt{(A + B + C)}$$

$$A = \frac{(n-1)}{2n\,(2-h^2)}$$

$$B = \frac{[(d-1)\,(n+2)^2]}{[4\,n\,d\{4+(n-2)h^2\}]}$$

$$C = \frac{(s-1)\,[n\,(1+d)+2)]^2}{[4\,s\,n\,d\{4+n+n\,d-2)h^2\}]}$$

$$R_2 = \frac{h}{2}\sqrt{(D+E)}$$

$$D = \frac{n}{d}\left[\frac{d-1}{4+(n-2)h^2}\right]$$

$$E = \frac{(s-1)\,(d+1)^2}{[s\,\{4+(n+n\,d-2)\,h^2\}}$$

i = standardized selection differential ($i_F$ = 1.50 and $i_M$ = 1.38)

n = average number of progeny per dam = 6

d = average number of dams per sire = 6

s = number of sires = 35

[The students may find the solution of $\Delta G_{CS}$ by working out the values and putting the values in above formula]

**19.5** In a flock of poultry, the average egg production was 100 eggs with heritability of egg production of 0.2. Find out the Osborne's index value of a pullet and a cockerel with the following observations of egg production –

| Pullets | I | $F_D$ | $F_S$ | n | d | Cockerels |
|---|---|---|---|---|---|---|
| A | 120 | 110 | 105 | 7 | 8 | X |

Where, n = no. of pullets per dam,

d = no. of dams mated to a sire

$F_D$ = averages of dam

$F_S$ = average of sire family

I = Individual' record

**Solution:** The Osborne index values are very simple and easy to to work out compared to other methods. These are as under for males and females -

$$I_F = (I - \bar{P}) + W_2 (F_D - \bar{P}) + W_3 (F_S - \bar{P})$$

$$I_M = W_2 (F_D - \bar{P}) + W_3 (F_S - \bar{P})$$

The first step is to work out the values of weighting factors ($W_2$ and $W_3$) as under -

$$W_2 = \frac{2n(1-h^2)}{4+(n-2)h^2}$$

$$= \frac{[2 \times 7 (1 - 0.2)]}{4 + (7 - 2)\, 0.2}$$

$$= \frac{11.2}{5} = 2.24$$

$$W_3 = \frac{4\bar{n}d(1-h^2)(2-h^2)}{[4+(\bar{n}-2)h^2][4+(\bar{n}-\bar{n}d-2)h^2]}$$

$$= \frac{4 \times 7 \times 8\,(1 - 0.2)\,(2 - 0.2)}{[4 + (7 - 2)\, 0.2][4 + \{7 + 7 \times 8 - 2\}\, 0.2]}$$

$$= \frac{322.56}{(5)\,(16.2)} = \frac{322.56}{81} = 3.98$$

***Indexing pullets***: The individual's performance, dam family and sire family averages are used to find out the index for females (Pullets) as -

$$I_F = (120 - 100) + 2.24\,(110 - 100) + 3.98\,(105 - 100)$$

$$= 20 + 22.4 + 19.9 = 62.3$$

It is customary to work out the indices for all the pullets whose dam family and sire family averages are higher than flock average. This is done to rank them for selection of required number of pullets with superior merit.

***Indexing cockerels***: The dam family and sire family averages are used for indexing cockerels. Thus,

$$I_M = 2.24\,(110 - 100) + 3.98\,(105 - 100)$$

$$= 22.4 + 19.9 = \mathbf{42.3}$$

In case of selection of cockerels, the index values are estimated only for those of superior 5 to 6 sire families as well as best dam families of the best sires.

**19.6** The egg production data were recorded from a flock of 1440 birds sired by 30 sires each being mated to 8 dams producing 6 progeny per dam. The $\sigma^2_G$ was estimated to be 35.0 and $\sigma^2_W$ as 115.0. The $\sigma^2_S = \sigma^2_D = 0.25\sigma^2_G$. Construct Abplanalp index for selection of pullets and cockerels.

**Solution:**

First the $b_i$ values will be estimated to construct combined index for selection of pullets and cockerels from the following formulae as -

$$b_1 = \frac{G_1}{P_1} \text{ where, } G_1 = \left[\frac{n-1}{n}\right] 0.5\sigma^2_G;$$

$$P_1 = \left[\frac{n-1}{n}\right] \sigma^2_W \left(\frac{n-1}{n}\right)$$

$$b_2 = \frac{G_2}{P_2} \text{ where, } G_2 = \left[\frac{(d-1)}{d}\right]\frac{0.52\sigma^2_G}{n} + \sigma^2 D$$

$$P_2 = \frac{d-1}{d}\left[\frac{\sigma^2_W}{n} + \sigma^2_D\right]$$

$$b_3 = \frac{G_3}{P_3} \text{ where, } G_3 = \left[\frac{(s-1)}{s}\right]\left[\frac{0.52\sigma^2_G}{nd} + \frac{\sigma^2_G}{nd} + \frac{\sigma^2_D}{d} + \sigma^2_S\right]$$

$$P_3 = \left[\frac{(s-1)}{s}\right]\left[\frac{\sigma^2_W}{nd} + \frac{\sigma^2_D}{d} + \sigma^2_S\right]$$

$$b_4 = \frac{G_4}{P_4} \text{ where, } G_4 = \left[\frac{d-1}{d}\right] \sigma^2 D$$

$$P_4 = P_2 = \left[\frac{(d-1)}{d}\right]\left[\frac{\sigma^2_W}{n} + \sigma^2_D\right]$$

$$b_5 = \frac{G_5}{P_5} \text{ where, } G_5 = \left[\frac{(s-1)}{s}\right]\left[\frac{\sigma^2_W}{nd} + \frac{\sigma^2_D}{d} + \sigma^2_S\right]$$

$$P_5 = P_3 = \left[\frac{(s-1)}{s}\right]\left[\frac{\sigma^2_W}{nd} + \frac{\sigma^2_D}{d} + \sigma^2_S\right]$$

n = no. of progeny per dam = 6

d = no. of dams per sire = 8

s = no. of sires 30

Substituting the required values in different formulae, the $P_i$ and $G_i$ values as well as the $b_i$ values estimated are as under -

| | | |
|---|---|---|
| $G_1 = 14.58$ | $P_1 = 95.83$ | $b_1 = G_1/P_1 = 0.1521$ |
| $G_2 = 10.208$ | $P_2 = 24.42$ | $b_2 = G_2/P_2 = 0.418$ |
| $G_3 = 9.868$ | $P_3 = 12.23$ | $b_3 = G_3/P_3 = 0.8069$ |
| $G_4 = 7.6563$ | $P_4 = 24.42$ | $b_4 = G_4/P_4 = 0.3185$ |
| $G_5 = 9.4613$ | $P_5 = 12.23$ | $b_5 = G_5/P_5 = 0.7736$ |

$$\text{Selection of Pullets } (I_F) = b_1 (P - P_{DF}) + b_2 (P_{DF} - P_{SF}) + b_3 (P_{SF} - \bar{P})$$
$$= 0.1521(P - P_{DF}) + 0.418 (P_{DF} - P_{SF}) + 0.8069 (P_{SF} - \bar{P})$$

$$\text{Selection of Cockerels } (I_M) = b_4 (P_{DF} - P_{SF}) + b_5 (P_{SF} - \bar{P})$$
$$= 0.3185 (P_{DF} - P_{SF}) + 0.7736 (P_{SF} - \bar{P})$$

**19.7.** From the information given and estimated in Q number 19.6 above, work out the expected response in egg production and correlated response in egg weight, taking the genetic covariance between two traits as – 5.6.

**Solution:**

(*i*) ***Expected response*** (No. of eggs, ΔG):

(*a*) Female selection $(\Delta G_F) = \dfrac{i}{\sigma_1}(b_1G_1 + b_2G_2 + b_3G_3)$

$$\sigma_{I(F)} = \sqrt{(b_1^2 P_1 + b^2_2 P_2 + b^2_3 P_3)}$$
$$= \sqrt{[(0.1521)^2 \times 95.83 + (0.418)^2 \times 10.208 + (0.8069)^2 \times 12.23]}$$
$$= \sqrt{14.4464} = 3.8$$

$$\Delta G_F = \frac{i}{\sigma_1}(b_1G_1 + b_2G_2 + b_3G_3)$$
$$= (1.49/3.8)\ (0.1521 \times 14.58 + 0.418 \times 10.208 + 0.8069 \times 9.868)$$
$$= 0.3921\ (2.2176 + 4.2669 + 7.9624)$$
$$= 0.3921 \times 14.4469 = 5.66 \text{ eggs}$$

(*b*) Male selection $(\Delta G_M) = \dfrac{i}{\sigma_1}(b_4G_4 + b_5G_5)$

$$\sigma_{I(M)} = \sqrt{(b_4^2 P_5 + b^2_5 P_5)}$$
$$= \sqrt{[(0.3185)^2 \times 24.42 + (0.7736)^2 \times 12.23]}$$
$$= \sqrt{2.4772 + 7.3191} = 3.1299$$

$$\Delta G_M = \frac{i}{\sigma_1}(b_4G_4 + b_5G_5)$$
$$= (1.38/3.1299)\ (0.3185 \times 7.6563 + 0.7736 \times 9.4613)$$
$$= 0.4409 \times (2.4385 + 7.3192)$$

$= 4.30$ eggs

Average expected response $= (\Delta G_F + \Delta G_M) / 2$

$= (5.66 + 4.3)/2 = 4.98$ eggs

(*ii*) ***Correlated response*** in egg weight ($\Delta G^{12}$):

(*a*) Through selection of pullets:

$$\Delta G^{12} = \frac{i}{\sigma_1} (b_1G_1^{12} + b_2G_2^{12} + b_3G_3^{12})$$

Where, $G_1^{12} = G_1\, G^{12} / \sigma^2_G = (14.58 \times - 5.6)/35 = - 2.3328$

$G_2^{12} = G_2\, G^{12} / \sigma^2_G = (10.208 \times -5.6)35 = - 1.6332$

$G_3^{12} = G_3\, G^{12} / \sigma^2_G = (9.868 \times - 5.6)/35 = -1.5788$

$G^{12}$ = genotypic covariance between two traits 1 and 2.

$= - 5.6$

$$\Delta G^{12} = \frac{i}{\sigma_1} (b_1G_1^{12} + b_2G_2^{12} + b_3G_3^{12})$$

$= (1.49/3.8)\ (0.1521 \times -2.3328 + 0.418 \times - 1.6332 + 0.8069 \times - 1.5788)$

$= 0.3921\ (- 0.3548 -0.6826 - 1.2739)$

$= - 0.91$ gm

(*b*) Through selection of cockerels:

$$\Delta G^{12} = \frac{i}{\sigma_1} (b_4G_4^{12} + b_5G_5^{12})$$

Where, $G_4^{12} = G_4\, G^{12} / \sigma^2_G = (7.6563 \times - 5.6)/35 = - 1.225$

$G_5^{12} = G_5\, G^{12} / \sigma^2_G = (9.4613 \times - 5.6)/35 = - 1.5138$

$$\Delta G^{12} = \frac{i}{\sigma_1} (b_4G_4^{12} + b_5G_5^{12})$$

$= (1.38/3.1299)\ (0.3185 \times - 1.225 + 0.7736 \times -1.5138)$

$= 0.441 \times (-0.39 - 1.171)$

$= - 0.69$ g

Average correlated response $= (\Delta G_F + \Delta G_M) / 2$

$= (- 0.91 - 0.69)/2 = - 0.80$ g

❑❑❑

# 20 Sustainable Animal Production

*The* ***system*** *is an act or practice to accomplish a task following certain rule or set of rules, a* ***sustainable production system*** *is one that gives better output with economic use of inputs without their exhaustion for future use, and hence the* ***sustainable animal production*** *involves the proper use of inputs in terms of breeding techniques to develop superior genotype of animals and giving them the sustainable environment of feeding, care and management to keep them healthy and productive through certain interventions based on new techniques (Tomar, S.S.).*

## 20.1 MILK PRODUCTION IN INDIA

India is a vast reservoir of animal genetic resources with large numbers of breeds of various animal species serving various purposes mainly under low input animal production system. The livestock rearing has emerged as a fast growth sector of Indian agriculture and has become a business which was a way of life in the past. In India, the milk production was 53.64 % from buffaloes, 40.96 % from cows (21.03 % by indigenous cows and 19.92 % by crossbreds) and rest 5.40 % from goats during 2006. The contribution of crossbred cows had been significant by producing 19.34 MT of milk which was 48.65 % of the total cow milk produced in India. It also needs to mention here that most of the bovine milk (94.3 %) is produced by only 14 states in the country.

The contribution of livestock and fisheries sector to the agricultural sector in India is 31.5 % of the value of the output, against the livestock contribution of about 40 % in the world agriculture. These sectors had contributed about 7.35 % to the total GDP in 2001 – 02. The milk had been the primary contributor to agricultural GDP and was higher than paddy, wheat and sugarcane.

The regular expenses of landless labourers and small/ marginal farmers of India are met from sale of raw milk in the local market produced by cattle and buffaloes and this support their livelihood during off season of earnings from cash crop. They keep 1 – 2 milch animals under mix farming system and these constitute about 70 % of the milch animals. Not only this but animal rearing by tribals and nomads is the only source of their earning and livelihood. Therefore, in view of livestock contribution, this livestock sector has a potential and a great role in the upliftment of rural poor.

At present India has the honour of number one position in the world for milk production (106 MT during 2010) which had increased six fold from 17 MT during 1951. This has increased the per capita availability to about 259 gm/day which is still short to recommended 280gm/day/head (ICMR, 1989). However, this is mainly attributed to large number of dairy animals, though the productivity of Indian dairy animals is low compared to the temperate breeds.

The milk productivity of animals per lactation per animal in India had been around 900 kg and scored its $35^{th}$ rank out of 44 top ranking countries in the world. This indicated scope for increase of milk productivity of our dairy animals. Not only this low productivity but the reproductive performance of our native animals (AFC, CI and other reproductive traits) is also poor. The low production and reproduction have been attributed to a number of reasons viz. low genetic potential, shortage of green fodder and feeds (concentrates), and poor management.

## 20.2 CONSTRAINTS OF LOW PRODUCTIVITY

The survey under National Agricultural Technology Project (NATP) has shown that there are following constraints impeding (obstructing) the improved productivity of animals under traditional systems of animal keeping for breeding, feeding, management and health care of animals by the farmers.

- Non availability of superior germ plasm and the breeding of village animals by locally available bulls of low genetic worth.
- Lack of awareness of farmers due their illiteracy for timely heat detection and insemination, thereby increasing the calving interval.
- Lack of infrastructure in the village for A.I., supply of semen of high quality bulls and trained A.I. inseminators, and the treatment facilities.
- No performance recording of animals by the farmers because they do not find its immediate use and also because of their illiteracy.
- Inadequate supply of green fodder and feeding of poor quality roughage and crop residues with low nutritive value.
- Inadequate feeding of concentrate and lack of mineral supplementation.
- Adoption of unscientific herd management practices.
- Lack or insufficient marketing facilities for milk & milk products as well as also of the animals.

Therefore, there should be a wholestic/ multi-faceted approach to increase the productive and reproductive performance of native animals of different species to make livestock production sustainable by adopting the sustainable animal production system. A strategic planning in different agro-climatic zones of the county is required for enhancing overall productivity of livestock by exploiting judiciously and utilizing sustainably the available resources. The sustainable milk production requires the sustainable breeding strategies not only through genetic and breeding interventions but also adopting the nutrition and management interventions for different animal production systems.

***Challenges facing the dairy sector:*** These include the followings -

- Small herd size with poor production potential;
- Lack of superior quality bulls together with A.I. service network;
- A gap between requirement and availability of feeds and fodder for livestock and poultry (shortage of dry fodder, green fodder and concentrate of 10, 35 and 33 %, respectively). This is due to gradually declining in grazing land (fallow land, pasture and common land), limited allotment of only 4 – 5 % of the total cultivated land under fodder production for the reason of growing of cash crops, demand of feed and fodder ingredients by industry, non-availability of quality seeds of fodder crops, burning of straw in the field and wastage of green fodder during flush season;
- inadequate veterinary services for vaccination, disease diagnosis, surveillance and forecasting mechanisms leading to disease outbreak with high morbidity;
- Lack of interest of youths in livestock and poultry farming;
- Inadequate allocation of budget for livestock sector in policy framework;
- Lack of organized market.

## 20.3 ANIMAL PRODUCTION SYSTEMS

The output of any system is the result of some input and there is no output without input. There exists relationship between input and output of a system. The various input factors (biological, climatic, economic, social, cultural and political) combine and determine the output (production) of a system. When animals are involved in the production system, the output is some animal product like milk, meat, wool, draft, etc and such a production system is called as the animal production system (APS).

The category of animal production systems may be purpose based (animal product) viz. dairy, meat, wool, draft production besides skin and hide production. The APS in India are also classed depending up on the level of input factor, mainly the feeds and fodder ranging from very low input to very much intensive which impose limit on the output (animal product), survival and reproduction of livestock. Therefore, the APS have been classed as extensive (zero input or low input), semi-intensive (medium input) and intensive (high input) animal production system. Another similar classification of APS also on input base is done in to grassland base pasture and range management also called as pastoralism or nomadic APS, landless production system, mixed farming, and intensive farming.

In view of the increasing human and livestock population, the socio-economic sustainability is under stress and becoming increasingly a non-profitable proposition economically. The animal production systems particularly the crop-livestock integrated (mix farming) system is the determinant of the socio-economic sustainability.

***(i)* Extensive Animal Production System:** The main feature of this system is zero input or low input use and low production. Hence there are two variants of extensive system of animal production viz. zero input and low input APS. In this production system, the zero or low input factor impose restriction on output and hence results in low output with low survival and low reproduction of livestock. This system is forced to be adopted throughout the country in different agro-climatic zones by nomads as well as by landless and small/marginal farmers who have poor resources.

The *pastoral system* is mainly adapted in the desert part of India especially in Rajasthan and Gujarat for small ruminants (Sheep and goat) wherein the nomads/ owners with their belongings move with their animals (cattle, buffaloes, sheep and goats) to the area of availability of water and fodder. This system also prevails in hills for seasonal migration of sheep and goats.

The *landless livestock farming* is mainly adapted for poultry and pig farming. Landless labourers are also involved in cattle and buffalo farming wherein they feed their animals on scraped grasses /eatable weeds from cropping area, waste/ fallow lands, roads/ railway side lands. This system has contributed an increase in supply of milk, meat and eggs per person. This system has been assisted by small dairies by giving to the landless labourers the female calves for rearing to age at first calving on profit share basis.

The *extensive or zero input production system* is based on indigenous knowledge prevalent in remote areas, arid and semi-arid regions of the country like Banni area (Kachchh), Orissa and North-East region, adapted by pastoralists in different parts of the country. This system is well integrated with the local feed / fodder resources, plants where the animals develop capacity to survive under shortage of feed / fodders, prevalent diseases, drought and other adverse climatic conditions. The animals are well adapted to ecology of the area, use of crop by-products and have low requirement of feed and fodder, shelter and housing, health care and managed by family labour especially by family woman.

The herd size under this APS is also limited by keeping 1 – 2 cows/ buffaloes that are either non-descript or belong to locally adapted breed. The animals are mainly kept for milk production and draft power. The milk produced is mostly consumed in the family and the surplus milk which is less than 50 % of the milk produced is marketed through milk vendors.

There may be cited some examples for unique extensive production system with zero input for buffalo production in India. The zero input system is based on indigenous knowledge adopting seasonal and spatial grazing system that are complementary to natural feed / fodder resources and symbiotic with agro-ecosystems.

- ***Banni buffaloes,*** the only source of livelihood for maldharies (traditional animal breeders) in Kachchh, are reared under locally adapted extensive production system wherein the buffaloes graze during night in Banni grassland.

- ***Chilika buffaloes*** are reared with almost zero input by the people in the villages on the bank of salt lake "Chilika" in Orissa. These animals enter in knee deep water of the lake during evening, graze overnight on the weeds inside the lake, and return to the owners by the morning.
- ***Manda buffaloes*** are maintained by tribals' in remote areas across mountain terrain in Koraput district of Orissa, and used for draught in lower hills.
- ***Manipuri*** buffaloes, found in the hills as well as valley regions of Manipur state, are reared under extensive (zero input) production system. They are let loose in the forest for grazing and stay in forest even at night in the hilly region.

**(*ii*) Semi-intensive (Medium input) APS:** The medium levels of input and management are provided under this system. The restricted levels of input factors limit the output, survival and reproduction to some extent. This Crop-livestock integrated system is dominated by medium farmers having better resources in comparison to poor landless/ marginal farmers who adapt zero input or low input system. The farmers keep a mix herd of local cattle/ buffaloes, crossbreds and also buffaloes of superior quality (good dairy breed).

The semi-intensive / low input or *mixed farming (crop – livestock) system* is prevalent in states still having large uncultivated land in arid, semi-arid, hilly highlands, and sub-humid regions. This system is characterized by combining grazing with stall-feeding of planted forage and crop residues. Under semi-intensive production system, animals are allowed grazing according to the season and supplemented with concentrate, mineral mixture, green and dry fodder as per the productivity (milk yield) of animals, lactation stage of cows/buffaloes, milk marketing facilities and economic status of the owners. In this system lactating and pregnant buffaloes are provided concentrates, green and dry fodders, whereas dry, growing stock and male calves are least cared.

The animals are maintained in small numbers under highlands mix farming system in Himalayan region particularly in H.P. and J & K. The dairy production by small scale farmers is widespread in Uttarakhand Hills, and animals are kept grazing during day time but they are fed crop residues and tree leaves in the evening/night hours. This system of animal production is adopted in most of the states in India, marketing of raw milk is done and more than 50 % of the milk produced is marketed. The advantages of semi-intensive / mixed crop-livestock production remain a minor contributor to dairying in India due to non availability of infrastructure of marketing facilities. However, this system of mixed farming is an additional source of food and income to the farmers during off crop season and also covers the risk of crop failure. The animals contribute to the draft power and manure which enrich the soil.

***(iii) Intensive (High input) production system:*** This is also known as stall feeding production system. The animals are kept and managed with high input to have high output, survival and reproduction of animals. The cows, crossbreds and buffaloes kept under this intensive system are of high genetic merit. The owners use high inputs and technology locally developed or imported one. High level of productivity of the animals and their genetic improvement is the objective. The animals are provided intensive care and management of breeding, feeding, housing and health care. This system is adapted at organized livestock farms of State/Central Government, SAU's, ICAR institutes, progressive commercial farmers and some NGOs. Large herd comprising more than 50 adult heads are maintained. All the milk produced is sold either as liquid or as milk products.

The milk production system in the country is changing very fast. This intensive production system is generally followed in out skirt or within urban / metro cities near fluid milk consumption centers. The livestock farms in urban peripheries have the advantage of better access to markets and the decrease in unit costs of the support services such as input supply, animal health services and milk marketing with increase in production.

The variant of this intensive APS is predominantly found in and around large cities, such as Delhi, Mumbai, Calcutta, Bangalore, Ahmedabad, and Surat where the buffaloes are usually kept for milk production and not for breeding purposes. After completing lactation majority of dry buffaloes are slaughtered for meat production. This system in urban and peri-urban areas purchased selected buffaloes of high milk production potentials from the breeding tract and after extracting last drop of milk and majority of them disposed in abattoir. The owners purchase all the feed inputs for feeding the animals from local market and hence the non-productive animals are immediately disposed off. Buffaloes under intensive production system are maintained under stall feeding and fed satisfactory diet comprising mixture of chaffed straw or stoves, concentrate mixture of locally available cakes, flours of grains, dal chunnies and bran etc through mixing with chaffed green fodders during feeding time. The compounded feeds and mineral mixtures are also provided. Recent trends have also been observed in rural area from Haryana, Gujarat, Maharashtra and Andhra Pradesh that the progressive farmers or having large land holdings maintain the buffaloes and crossbred cows under intensive production system. Some farmers also maintain genetically superior buffalo germplasm and engaged in buffalo marketing. They purchase superior germplasm from the breeding tract and provide better feeding and management conditions to attain optimum productivity and sold off as and when get good price.

The intensive production system is also adapted by poultry farmers with large poultry farms in the area where market facilities for inputs and products are available.

## 20.4 SUSTAINABLE SYSTEM

Now it is essential to understand about the sustainability of the animal production system. The FAO has defined the sustainable development as the management and conservation of the natural resource base and orientation of the technological and institutional change in such a way as to ensure the attainment and continued satisfaction of human/animals needs for present and future generations. In simple words, it is the management and conservation of the natural resource bases (land, water, plant and animal) using the resource conservation technology (RCT's). The RCT's are defined as any practice that improve the efficiency of the use of natural resources including water, air, fuels, inputs and people.

The RCT's must be successful in sustaining any improvement programme. A system is regarded as sustainable if the biophysical and socio-economical objectives of the system are met. The sustainable mean to keep alive. Thus, the system should survive to be sustainable. The sustainability thus means the proper use of resources so as these are not exhausted but are left for use by the generations to come. Therefore, the sustainable development, besides conserving the natural resources, should be technically sound, environmentally non degradable, economically viable, and socially acceptable. It is therefore essential that performance of the system be continuously monitored for productivity and natural resources to insure and improve sustainability.

The manpower plays an important role in making the system sustainable. The manpower involves not only the farmers but also any agency (Govt., NGO) responsible for conducting research, training, extension activities for the benefit of the farmers, policy making and implementation, market regulation, finance and creating infrastructure. The policy makers should understand the problems of the farmers and find the solution like technological development which should be cost effective, easy to use and acceptable by the farming community.

The livestock is an integral part of agriculture particularly in India and contribute significantly to the economic stability not only of the farmers but of the farming systems. Therefore, there should be proper planning for sustainable animal production and breeding.

## 20.5 DETERMINANTS OF SUSTANABLE ANIMAL PRODUCTION

The sustainable animal production system (APS) should be such as to take maximum advantage of the existing resources with optimum design and management without any damage to the environment to maintain or improve farm profitability and conserve all resources for future as well. The sustainable animal production should meet the four requirements of economic viability, technical feasibility, social acceptability and resource conservation ability.

The sustainable APS is determined by the followings:

- Critical evaluation, documentation and conservation of indigenous livestock breeds;
- Breeding strategies to produce superior germplasm under different animal production systems (using selective breeding through progeny testing for sire evaluation in organized herd as well as field progeny testing, upgrading, crossbreeding) with assisted reproductive and genomic technologies used for genetic improvement of livestock;
- Feed and fodder resources along with supplementation of minerals, diet supplementation (urea, molasses) and by-pass protein for improvement of feed utilization, application of biotechniques related to nutrition to increase animal productivity;

- Proper management of animals by providing optimum shelter and protection against adverse climate;
- To ensure proper animal health care;
- To execute the development programme effectively by planning and development agencies through motivating the farmers to use newly developed technologies, and
- To adopt the multi-purpose farming systems integrating livestock to maintain soil fertility.

## 20.6 WAYS OF ACHIEVING SUSTAINABLE ANIMAL PRODUCTION

In view of the gaps and shortcomings and the determinants of sustainable animal production listed above, it is imperative to think about the sustainable animal production otherwise the livestock production may not sustain in long run due to limited natural resources. The solution of future growth lies in the improvement of technology and service delivery systems to increase the productivity, processing and marketing by use of multifaceted approach including biotechnology in animal reproduction, genomic selection as a part of the breeding strategies for genetic improvement, feed and nutrition technology, optimum management practices particularly of shelter management to protect the animals from adverse climate, and disease control technologies.

**20.6.1.** ***Sustainable Animal Breeding Strategies:*** The development of breeding strategies results in permanent genetic improvement of animal population. The animal breeding strategies are formulated for identification and multiplication of superior germ plasm.

The *identification of superior germ plasm* involves critical evaluation, documentation and conservation of animal genetic resources (AnGR) of the indigenous livestock breeds, and formulation of breeding strategies. This requires the estimation of breeding value of animals (evaluation) for selection of females and males by running an effective progeny testing programme. Secondly, the identification of superior germ plasm requires the use of genome analysis techniques. The genome analysis is done by using of DNA markers (recombinant DNA technology), DNA finger printing known as DNA identification analysis and marker assisted selection.

The *multiplication of superior germ plasm* requires the area specific different breeding strategies for different breeds and species of livestock and poultry.

The use of assisted *reproductive biotecniques* which include A.I., frozen semen technology, embryo transfer technology (ETT) including MOET, *in vitro* produced (IVP) embryos, embryo cryopreservation and embryo splitting, and secondly, the *biotechniques in animal breeding* for genome analysis and manipulation (transgenesis and cloning) are further helpful in multiplication.

**20.6.2. Sustainable Environment**: The genetic improvement is not enough to harvest the maximum possible production and reproduction of animals. This is because the best possible genotype developed through breeding strategies does not exhibit to its full potentiality unless it gets proper environment. The proper and sustainable environment includes the feeding of balanced ration, proper management of animals by providing adequate housing for protection of animals from adverse climatic conditions and health care of animals.

The *improvement of farmer's animals* is very essential to make the animal production sustainable. This is because the animals kept by landless and small/ marginal farmers constitute 70 % of milch animals in the country. The improvement in farmer's herd requires the facilities to supply the superior germ plasm (semen of high genetic merit), insemination of animals at farmer's door step and performance recording of animals for genetic improvement. Secondly, the interventions related to feeding of balanced ration, housing for proper management of animals to protect them against adverse climatic factors, and timely health care are essential aspects of sustainable animal production.

❑❑❑

# 21 Documentation and Conservation of Germplasm

The nature has created variation everywhere on the earth in living and non-living things. The earth is diversified in every respect regarding topography (hills, plains, rivers, lake, and sea), vegetation (all field crops, all horticultural crops), all species of domestic livestock and poultry, varied agro-climate (temperate, sub-temperate, arid, sub-arid, humid and hot humid) and other environmental aspects. These variations have created the uniqueness and hence the beauty and have made possible the identification, every object being different from others. Not only this, but the variations are of ecological (ecological preference of different species), economic as well as of biological or evolutionary significance. India is no more an exception in varied diversity in every respect. India is essentially a sub-continent along with Pakistan, Bangladesh, Nepal, and Bhutan comprising South Asia.

## 21.1 BIODIVERSITY

Life is greatly diversified into various forms and there is no dearth of organic diversity in the world. There are innumerable forms of organisms on this earth. There are about 100000 species of animals listed in literature. The diversity exist to the extent that no two individuals are alike how closely genetically related they are except clones and identical twins. This is called the biodiversity. The biodiversity exist in domestic and pet animals as well as among wild life. The biodiversity in domestic animals is taken as the genetic diversity in the form of different breeds within species. The genetic diversity has resulted due to the process of evolution over thousands or millions of years during wild and domestic stages, and the efforts made by man to meet the market demand, in the present day context.

Since beginning man has tried to exploit many species of plants and animals. This leads to domestication of some species of animals having desirable characteristics for mankind. However, man has domesticated a mere handful of about 40 species useful for his own use like milk, meat, eggs, fiber, hides/skin, fuel, manure and draught power, etc. The various breeds and distinct animal types developed through selection and breeding practices and quest for development of need based animal types in different agro-climatic conditions have acquired specific morphological and physiological characters grouped as adaptation. The farm animals of India have the unique features of adaptation to adverse climatic factors, diseases and parasites of tropics, survival on inadequate quantities of feeds, fodder and water. The zebu cattle have the advantages over the Taurus with respect to greater resistance to heat, less susceptibility to ticks and gastrointestinal parasites, and a lower metabolic rate and nutrient requirements.

## 21.2 ANIMAL GENETIC RESOURCES (AnGR)

The AnGR includes all types of species and breeds that exist and man has domesticated. Masson, I.L. (1988) has published a book "A World Dictionary of Livestock Breeds, Types and Varieties". It is most comprehensive book on domestic AnGR.

According to world watch list, FAO, 1995, the numbers of breeds of domesticated livestock species are given in table 8.2.

## 21.3 SPECIAL FEATURES OF AnGR OF INDIA

The India has is a gold mine and vast reservoir of the germplasm of domestic animals and Poultry. This includes the various species viz. cattle, buffalo, sheep, goat, pig, yak, mithun, camels, equines (horses and ponies, mules, donkeys), poultry and ducks. Further there is a large biodiversity within each of these species having well recognized breeds viz. 30 breeds of cattle, 15 of buffaloes, 42 of sheep, 20 of goats, 6 of camels, 6 of horses, 18 of poultry in addition to other useful animals like mithun, yak, donkeys, pigs, etc.

According to FAO data bank (1998), India had 15.9 percent of world cattle population and 56.1 percent buffalo population. The sheep and goat had 5.3 and 17.4 percent of world population of these two species, respectively. The share of Indian pigs, horses, camel and chicken had been 1.7, 1.6, 7.7 and 2.5 percent of world population of these respective species.

The population density of animals is high but production potential is low compared to exotic breeds, except buffaloes. The stocking rate per sq. km had been found during 1982 to be: 58.5 cattle, 21.2 buffaloes, 14.8 sheep, 28.9 goat, 0.29 equines, 0.3 camel, 3.0 pigs, and 63.2 poultry with overall livestock density as 127.64 animals.

About 70 - 80% of the total indigenous animals are non-descript and hence remained undocumented.

All the breeds of various livestock species have not been properly documented and evaluated for their utility.

The Indian breeds of livestock species have better adaptability to cope with harsh climatic conditions and to certain diseases and parasites, have better efficiency to utilize coarse feed and fodder.

The crossbreeding programme of Indian cattle, sheep, goat and poultry for genetic improvement had resulted in marked decline in availability of pure breeds.

The bulls of various recognized breeds of cattle of superior genetic merit (progeny tasted bulls) are not available to the extent required.

The better female germplasm particularly the buffaloes are sold and taken to metropolitan cities for milk production but after completion of lactation they are slaughtered.

The farm livestock and poultry have their significant role in agriculture; providing various livestock products (milk, meat, skin, wool, dung, eggs) and draught power; national economy by contributing to the gross domestic product (GDP); employment to rural poor, landless and marginal farmers as well through animal based industry besides that some animal species provide companionship and entertainment.

## 21.4 MAIN FEATURES OF CATTLE AND BUFFALO POPULATION IN INDIA

About 80% of the total cattle population in India is non-descript. Further the large majority of descript cattle belong to draught and dual purpose breeds.

The cattle population in India showed an increasing trend from 155.3 millions in 1951 to 204.58 million in 1992. Further, it was observed that cattle population was 199.66 m in 1987 and thereafter remained oscillating around 200 m up to 2007 (Indian Livestock Census, 2007), except that it was 185.18 m during 2003 which was about 14 percent of the world cattle population.

The distribution of different classes of cattle (million) during 1992, 2003 and 2007 were –

| Years | Category | Total | Crossbreds | Indigenous | | |
|---|---|---|---|---|---|---|
| | | | | Total | Descript | Non-descript |
| 1992 | All cattle | 204.5 | 15.21 | 189.37 | 28.41 | 160.96 |
| | Adult female | 64.36 | 6.49 | 57.87 | 8.68 | 49.19 |
| 2003 | All cattle | 185.18 | 24.68 | 160.49 | | |
| | Adult female | 64.51 | | | | |
| 2007 | All cattle | 199.07 | 33.06 | 166.01 | | |

Livestock census was conducted breed wise for the first time in India in 2007. It was found that the crossbred cattle were 313.36 lakhs and exotic cattle were 17.32 lakhs, totaling 33.06 m. Among the *crossbred cattle* of 313.36 lakh, the Jersey crossbred cattle were predominant (202.79 lakh) followed by H.F. crosses (77.95 lakh) and very few numbers of other crosses (32.61 lakh) including Red Dane and Brown Swiss crosses as a result of their dilution by breeding with other crosses. Among the *exotic cattle population* of 17.32 lakh, Jersey cattle comprised 10.37 lakh, H.F 5.57 lakh and others 1.38 lakh.

There were 18032 indigenous cattle bulls and 431 crossbred cattle bulls in the country in 1992.

India has excellent quality of draught cattle which significantly contributed as draught animal power. Besides it, there are some milch type and dual purpose breeds which also provide milk and good quality bullocks.

The FAO had estimated 61 breeds of cattle in India but there are about 26-30 well defined breeds according to Indian literature.

The population density of cattle is very high (58.5 cattle per sq. km) which is increasing constantly. Due to increase in population number, the overall milk production has increased in the country but the milk production per animal is decreasing.

In India, human: livestock population is declining. The cattle: human ratio has declined from 430 per thousand in 1961 to 278 per thousand in 1981 and 165 per thousand in 2011. The reason has been an increasing pressure of human population against declining or almost oscillating the cattle population around 200 million during this period.

The indigenous cattle are, able to perform under harsh climate and stressful environment, less susceptible to disease, able to consume and capable to convert low quality coarse and dry fodder more efficiently into animal products (milk, meat, eggs, wool, etc) and thereby can be raised on zero input system in terms of feeds/fodder and health care.

The crossbreeding for breed improvement programme has resulted into the reduction of high producing animals of indigenous breeds. Moreover, indigenous cattle are also being replaced by milch breeds of buffaloes.

The cattle of India vary in size and body conformation perhaps due to their origin, adaptation to different agro-climate conditions and the use which they have been put to. The cattle breeds of India have been divided into **milch breeds** which produce good quantity of milk but not produce good quality bullocks, **dual purpose breeds** which produce fairly medium amount of milk and produce bullocks of good draught powers, and the **draught breeds** providing only a little amount of milk but their bullocks are good for draft purposes.

The ***buffalo population*** in India also showed an increasing trend from 43.4 m in 1951 to 81.21 m in 1992 (only 43.81 m breedable), 89.91 m in 1997and 96.90 m in 2003 (56.8 % of world buffalo population). Asia had 165.45 m buffaloes in 2003 which was 97.1 percent of world buffalo population (170.458 m).

The world buffalo population is estimated to be approximately 177.247 million spread in some 42 countries of which 171 million (97%) are in Asia, while approximately 5.38 million (3%) are in rest of the world (FAO 2008). India possesses 111.3 million buffaloes (56.7 percent of world's buffaloes), has 13 recognized breeds plus about 17 lesser known defined population groups and world's best milch breeds: Murrah, Mehsan, Jaffrabadi, Banni and Nili-Ravi. During the last 25 years (1982 to 2007) population of buffaloes increased about @ 2.0 percent per year.

Buffaloes are now reared in almost all the states. Buffaloes outnumber cattle in U.P., A.P., Rajasthan, Gujarat, Punjab and Haryana which account for two-third of the country's total buffaloes as against 30% of the total cattle of India in these states. Interestingly, crossbred comprise over half of the total cattle population in the buffalo dominated states, Punjab and Haryana.

Buffalo has been rightly christened as a black gold mine of India. They provide milk, meat, draft power, hides and skins and other valuable by-products. Dairy farming is the single largest contributor to India GDP. The export of animal products indicates the importance of Buffalo in export earnings from animal products. The total export earnings in 2012-13 was Rs. 20,130.90 crore, in which, Rs. 17,400.60 crore was from buffalo meat export and around Rs. 700 crore from buffalo dairy products.

## 21.5 UNIQUE FEATURES IN INDIGENOUS BREEDS OF LIVESTOCK

Some of the breeds of different livestock species found in India have some special and unique features as indicated below:

| Species and Breeds | Special/ unique features |
|---|---|
| **Cattle:** | |
| Sahiwal, Red Sindhi, Gir, Rathi | Milch type, heat and drought tolerance |
| Hariana, Ongole | Dual purpose, heat and drought tolerance |
| Nagori | Excellent drought type for hot climate |
| Vachur | Miniature cattle |
| Punganur | Draft cattle |
| Umblicherry | Excellent for wet ploughing |
| Siri | Dual purpose, high altitude breed |
| **Buffalo breeds:** | |
| Bhadawari | High fat content (10 – 12 %) |
| Jerangi | Small size (Deer buffalo) |
| **Sheep breeds:** | |
| Chagthangi | Fine wool Alpine sheep of high altitude |
| Garole | High fecundity, survive under saline condition |
| Magra | Lustrous carpet wool |
| Marwari, Jaisalmeri, Deccani and Chokla | Hardy and capable of travelling long distances |
| Mechari | High quality skin |
| Chokla, Patanwari | Best carpet quality wool |
| Mandya, Muzaffarnagri | High quality and palatable meat |
| **Goat Breeds:** | |
| Barbari | Short oestrus cycle, high yielder |
| Jamnapari | High yielder |
| Black Bengal | High ovulation rate and kidding percent |
| " & Surti | Tender low fat meat and delicate skin |
| Chegu, Changthangi | Low fiber diameter (8 – 10 m) |
| Andman | High tolerance of salt and sea water |
| **Poultry Breeds:** | |
| Kadaknath, Tellycherry | Meat of medicinal value |
| Naked neck | Ability for convection, improved adult fitness |
| Frizzle fowl | Ability for convection |

## 21.6 DOCUMENTATION (MANAGEMENT) OF AnGR

The approach for managing effectively the AnGR over time to better understand, use, develop and maintain them in order to meet the challenge should include the following aspects:

(*i*) Identification, characterization, evaluation, documentation and conservation

(*ii*) Training of manpower for successful completion of the above activities

(*iii*) To develop breed improvement programmes on national level for proper management of AnGR (Utilization)

The breed identification and characterization based on phenotypic characteristics was done in early thirties on the basis of limited and cursory observations. This resulted that many breeds having distinct attributes were not recognized. On the other hand, a number of varieties of a breed in different species have been reported over time e.g. Deccani, Patanwadi, and Nellore breeds of sheep. The varieties of Deccani are Lonad, Sangamneri, Solhapuri (Sangla), and Kolhapuri; of Patanwadi are the fine fleeced Patanwadi, Coarse fleeced (Kathiawadi) and meat type coarse fleeced (Charotari). Therefore, there is a need to identify, characterize, evaluate and document the AnGR for monitoring the status of breeds. To accomplish these tasks require the livestock survey.

***Livestock Survey***: The livestock census is conducted species wise in most of the countries including India. This does not serve the purpose of conservation of AnGR. The breed wise livestock census data is essentially required to know the population and production trends so as to enlist the endangered breeds. This requires a well planned survey of AnGR for their identification, characterization, evaluation, documentation and conservation. The survey programme helps to provide information on breed descriptors, the status of a breed required for its conservation and use by the planners, scientists and development agencies. It also helps to provide information on unique characteristics and to enlist the elite animals of a breed.

The survey of a breed should ensure the recording of the following information:

- Geographical distribution and native environment.
- Demographic information (age and sex wise)
- Qualitative and quantitative characteristics of morphological, productive and reproductive traits of the breed as a whole and of unique animals, elite producers.
- Collection of blood samples for genetic evaluation (characterization) from 50 unrelated animals of a breed.

**Breed Descriptors:** This means the information on habitat, physical and morphological characteristics, performance and genetic architecture of breed. The habitat of a breed should include the history of breed and geographical distribution (home tract), soil, topography, climate, feeds and fodder availability. These may be recorded from district headquarters of the breeding tract of the breed. The morphological and physical characteristics include the body size, the size and shape of different physical traits like horns, ears, tails, body colour etc. The data to be recorded on performance include the information on growth in terms of body weight at different ages, production (milk, fleece, eggs, meat etc.) and reproduction traits. These can be collected either from the farmers or by field recording. The genetic architecture of a breed may be known from blood groups, biochemical polymorphism and DNA markers. To develop breed descriptors of a breed requires the survey in the breeding tract. The purpose of descriptor is to facilitate valid comparison and classification of breeds within a species.

**Identification:** The term identification with respect to conservation of AnGR consists of:

- Country and source of data viz. published scientific papers, case studies, annual reports of different agencies involved in AnGR, thesis, store of unpublished data etc.
- Species, breed or population and location within country where breed is found.
- Animals of a breed with unique characteristics or unique genetic potential.
- Breeds and animals that may be at risk and declining in number.

**Characterization:** It is the description of a breed in terms of breed descriptors. This includes the distillation of all available knowledge (published and unpublished) which contributes to the reliable prediction of genetic performance in a defined environment.

**Evaluation:** It is contemporary comparison of performance records of two or more breeds under the same environment. Therefore, evaluation should provide characterization information for two or more breeds.

**Documentation:** It is simply the collection (to compare in detail) of existing data.

**Methodology of characterization:** Depending on the level of knowledge, the time and means available, the different methods of breed characterization can be used.

A net work project with NBAGR, Karnal, India as the nodal agency for genetic characterization and conservation of indigenous breeds of livestock and poultry came into operation in later half of VIII five years plan. This project aimed to develop breed descriptors, to characterize the breeds for phenotypic traits (qualitative and quantitative) and molecular genetic characterization, and finally to conserve the germplasm of elite/unique animals and birds. The technical programme for survey and characterization of AnGR under Network Project is as under:

To conduct survey in 3 districts by one supervisor and 4 enumerators in each district to determine the demographical and geographical distributions of breeds and to record information on morphological and physical characteristics, performance and utility traits together with feeding and management practices for animals of different age groups *viz.* 1-3 months, 3-12 months, yearlings, milking and dry animals, and the breeding males.

Three different levels of description varying from general to the specific can be used. These are as given below:

Macro: Phenotypic characterization, conservation

Meta: Phenotypic characterization

Micro: Phenotypic and genetic characterization

The details are as under:

***(i) Macro level***: This consists of on-farm surveys based on stratified sampling procedure to collect information about herds and animals. The basic information are collected about the location, village identification of the herd and the animal details like type or breed, sex, age and some simple measurements (body height etc.). The information may also be collected by direct questioning the farmer about the farming system and herd production

***(ii) Meta level:*** It is based on a monitoring of on-farm husbandry practices. This is an individual recording system for performance of identified animals so as to allow comparing breeds with respect to the mean and the phenotypic variance of performance traits. To accomplish the task, a stratified random sampling procedure is most appropriate for sampling of farms and animals within the farm. The number of farms and individuals within farm to be sampled (selected) depends on the .variability among farms, among animals within farm and the breeding structure within farm particularly the number of sires used per year. The number of farms should not be less than 40 and that too using two or more sires whereas the number of animals per breed should not be below 400. The information should be recorded on the sampled animals on all the performance traits viz. physical traits (linear measurements, body score), performance traits (milk yield, fleece yield, body weight, fertility, viability), health records (abortion, blood and faecal sample, tick infestation), and environmental conditions like location, climate, feed resources (feed type and availability), management (shelter/housing, watering frequency, herding practices), herd structure (animal movement), breeding structure (number of males and females, castration, use of A.1. and age distribution of animals).

***(iii) Micro level:*** It is complementary to macro and meta studies for comparing and testing animal populations. The micro level is based on selected methods that can be used to gain knowledge of the genome of individual animals within a population. It is considered a detailed study measuring the performance of animals of selected population under controlled and standardized conditions for a precise phenotypic and genetic evaluation of a population. The sampling should follow a stratified random sampling procedure assuring as broad a genetic sample from as many different farms as possible and not more than one sire per farm. A total of 200 breed-able females and 10 males per breed are required for estimating breed differences whereas for estimating genetic parameters require at least 20 sires each mated to 20 dams. However these numbers can be built up over time.

## 21.7 ANIMAL DATA BANK

The data base or data bank is collection of data on any aspect of life in well organized, meaningful and easily accessible way. A database file is for storage and maintaining the records.

The data bank for animal genetic resources has a key role to play in listing and describing the breeds, in documenting the erosion of these resources and in highlighting the breeds that are likely to disappear in near future unless specific action is taken.

The first data bank was initiated by European Association of Animal production (EAAP) in 1980's organizing the survey of EAAP member countries and established a data bank at the Institute of AB&G in Hannover, Germany to store information on European breeds. The EAAP also tried later on in 1990 to recover data from non-EAAP countries but could not success due to poor response. In 1991 the FAO was assigned the work to establish a data bank in Rome to cover all the non-European countries and that EAAP in Hannover would keep data bank solely on Europe. The data bank in Rome is called the FAO Global Data Bank for domestic livestock. The types of information stored in data bank are as under:-

***General information***: Species, geographical location of breed, local name, synonyms and international name of the breed after Mason (1988), name of organization provided the information.

***Origin and development of breed***: Origin of breed, wild or primitive, population size, risk status, use of A.1. and storage of semen.

***Breed description***: Coat colour, horn description, hair or wool type, adult weight and size, genetic characters (marker genes, chromosomal aberrations)

***Use and qualities of breed***: Main use, special quality like disease resistance, adaptability to harsh climate

***Management condition:*** Type of management, housing period, feeding. ***Performance trait information***: Various types

***Additional information***: Estimates of genetic distance to other breeds, analysis of genetic material of the breed, conservation programme in operation for the breed.

The sources of information for data bank are the Dictionary of Mason (1988), scientific literature, and FAO/UNDP Global Breed Survey.

The Global Data Bank first dealt with only seven species viz., cattle, buffalo, sheep, goat, pig, horses, and ass. The documentation on camel and poultry breeds started later on.

The FAO Global Data Bank had 2047 entries of which 86 (5%) are wild or feral, 502(25%) are descended as a variety, sub variety or strain of a breed and 164 (8%) are found in more than one country. Out of 2047, there are 564 cattle, 68 buffaloes, 600 sheep, 259 goat, 245 pigs, 247 horses and 64 asses.

The FAO had developed Domestic Animal Diversity Information System (DAD-IS) for management of animal genetic resources. It is a multi language, multi faceted, computerized information and communication system. This provides support for teaching, research, formulating policies, early warning for conservation of DAD. It now contains information on about 5300 breeds of 35 species of domestic animals of 180 countries. It is used to monitor breeds, their population and status, contains performance and production attributes of breeds.

## 21.8 CONSERVATION VS PRESERVATION

The different indigenous (Indian) breeds of different species of farm animals and poultry are getting diluted and facing degeneration. The *dilution of a breed* is in terms of purity of breed. There is a decline in the number of purebred animals conforming to the model attributes of the breed. Some of the breeds of farm animals of India are under risk of their extinction. Therefore, there is a need for conservation of different breeds. The technical aspects of conservation of animal breeds were considered jointly in 1980 by FAO and UNDP and laid down definite procedures after analyzing the problems for establishing gene and data bank. The efforts in this direction in India were started with the establishment of National Bureau of Animal Genetic Resources, Karnal under the control of ICAR.

The word conservation is closely related to the preservation. The preservation covers the continued maintenance of genetic variability and required when a breed reaches an endangered level of population or near to extinct. The preservation and multiplication may also be required in case of individuals having unique traits or exceptional genetic merit.

The conservation includes the preservation along with up-gradation (improvement) of the genetic potential and management of a breed for use in future. Thus, the conservation covers both continued maintenance of genetic variability, improvement and sustainable utilization by exploiting the genetic variability. The conservation can also be defined as the management of the biosphere of human use for benefits in present time together with maintaining its potential to meet the future needs.

## 21.9 REASONS OF THREAT TO GENETIC DIVERSITY

The different indigenous breeds of different species of farm animals and poultry in India are essentially the result of evolutionary processes. They are integral part of agriculture. However, they are now getting diluted and facing degeneration. There is a decline in the availability of purebred animals conforming to the model attributes of the breed. Some of the breeds of farm animals of India are under constant risk of their extinction.

The endangered breeds or the breeds which need attention for their conservation

| Species | Breeds at risk |
|---|---|
| Cattle | Red Sindhi, Sahiwal, Tharparkar, Vachur, Panganur, Mewati, Bargur, Kankatha, Kharigarh, Siri, Krishi valley |
| Buffaloes | Bhadawari, Nili-Ravi, Toda |
| Sheep | Malpura, Chokla, Jaisalmeri, Magra, Lohi, Pugal, Munjal, Gurez, Bhakarwal, Karnah, Kashmir valley, Poonchi, Hisardale, Nilgiri, Muzaffarnagri, Bonpala, Changthangi |
| Goat | Barbari, Beetal, Jamnapari, Surti, Jhakrana, Chegu, Changthangi |
| Camel | Double humped, Jaisalmeri, Sindhi |
| Horse | Bhutia, Manipuri, Spiti, Zanskari |
| Poultry | All the breeds |

The following reasons can be assigned to affect the domestic animal diversity and dilution and declining trend of population of breeds-

**1. Breeding systems and breed dilution:** Two breeding systems *viz.* upgrading and crossbreeding had resulted in dilution and threat of breeds.

*(i) Upgrading:* There had been the unrestricted interbreeding among different breeds particularly in rural areas. The native animals/ breeds are crossed with improver breed for better economic returns. This has resulted into the mixture of various breeds. Bhadawari buffaloes have been upgraded with Murrah. The sheep breeds like Malpura and Chokla showed declining trend possibly because of introduction of Marwari inheritance through migratory flocks. There is also intermixing of Jaisalmeri and Chokla sheep. The reason is that recommended breeding policies are neither mandatory to follow nor these are backed by the support service.

*(ii) Crossbreeding:* The efforts have been made to improve the productivity of indigenous breeds of livestock and poultry by introducing exotic germplasm of superior genetic merit through crossbreeding. The crossbreeding has been widely used as a major force for achieving rapid genetic improvement than selection within breed. The use of A.I. with frozen semen and ETT have further made easy and fast to introduce elite germplasm all over the globe. However, the crossbreeding has adverse effect like overall reduction of genetic variability in a species and also to put at risk to the low producing but well adapted breeds. Thus, the crossbreeding has resulted in the dilution of breed and declining the number of purebred animals. There is thus threatening to the virtual replacement and elimination of many native breeds, viz. Sahiwal and Red Sindhi cattle, almost all native breeds of sheep of J & K state and Beetal goats have been crossed with exotic breeds. Therefore, there is no planned strategy for conservation of indigenous breeds.

2. *Geographical reorganization:* The breeding tract and organized farms of some breeds of cattle (Sahiwal, Red Sindhi), buffalo (Kundi, Nili-Ravi), sheep and goats have gone to Pakistan.

3. *Purpose based farming system:* The different breeds are used to serve different purposes of the owner viz. milk and draft in case of dairy animals; meat, milk, wool/fiber in case of sheep and goats; eggs and meat for poultry, etc. The breeds that do not serve the purpose are neglected. They are either not kept or used for crossbreeding. The Nilgiri, a fine wool breed of sheep of Tamil Nadu is at risk because wool is not required in South India, also there is no market of wool whereas the mutton is in great demand in the area. Toda buffaloes of T.N. are getting endangered due to neglect. Some breeds like Vachur and Panganur cattle have lost for the reason that they did not get recognition at proper time.

4. *Economic viability:* The economically useful breed is automatically conserved whereas the breeds with poor performance and declining economic returns to farmers are losing their existence e.g. Mewati, Kankatha, Kherigarh, Bachur cattle.

5. *Modern agricultural practices:* There had been change in agricultural production system after 1970 viz. mechanization and a trend of growing cash crops. The large scale mechanization has reduced the requirement of draft power. The draft animals have lost their utility and became uneconomically viable because of their low milk production ability and intensification of mechanized agriculture.

The agriculture production system has been completely changed due to increase of population pressure, high density and availability of low inputs. As a result more land has come under irrigation for crop production. Thus there has been reduction of land holding and common grazing area. This has caused the reduction in herd size also. It is not possible for small farmers to keep a bull for breed improvement with poor resources and hence the breeding of his animals has become dependent on the bull available in or around village, irrespective of the breed and genetic merit of the bull. This has caused the genetic dilution and reduced performance of the progeny produced.

Muzaffarnagari sheep in U.P. and Munjal sheep in Punjab are decreasing in numbers due to non-availability of grazing area. The waste land, grazing area and forests are declining due to their conversion in to irrigated cropping area. Likewise mass degradation of forest area has shown continuous decrease of Mithun population which is a browser of broad leaved trees/shrubs. All these has resulted in reducing the requirement of draft power, genetic dilution and reduced performance of progeny.

6. *Other factors:* Increase in livestock population is also causing deterioration due to inadequate inputs of feeding, health care, etc.

Some sheep breeds like Changthangi, Tibetan and Bonpala are at risk due to smaller flocks scattered in large area of fragile ecology. The flood and avalanches (snow storms) have also caused endangerment of some breeds.

## 21.10 ENDANGERED LEVEL OF A BREED

The number of breeding females of a breed is the main determining factor to decide the endangered status. The estimates of minimum population size for different species to declare the endangered status have been given by different workers. However, it depends on the species, local circumstances like breed management system, rate of crossbreeding, rate of decline and utility of breed. The mortality rates, growth rates, sexual maturity, fertility rates in different species are different and hence the endangered status differs for different species.

According to FAO, the survival risk of the breed should be studied when breed population size reaches 5000 breed-able females and total population of 10, 000. The population size in thousands for consideration of endangered status of a breed under Indian conditions for different species have been suggested as under by NBAGR, Karnal (ICAR).

| Species / status | > 10000 Normal | 5000 – 10000 Insecure | 1000 – 5000 Vulnerable | 100 – 10000 Endangered | < 100 Critical |
|---|---|---|---|---|---|
| Cattle | 25 | 15 – 20 | 5 – 15 | 2 – 5 | < 2 |
| Buffalo & goats | 30 | 20 – 30 | 10 – 20 | 5 – 10 | < 5 |
| Sheep | 50 | 30 - 50 | 15 – 30 | 8 – 18 | < 8 |
| Camel & Horses | 20 | 15 – 20 | 5 – 15 | 2 – 5 | < 2 |
| Pigs | 10 | 5 – 10 | 1 – 5 | 0.5 – 1.0 | < 0.5 |

## 21.11 NEEDS OF CONSERVATION

The native breeds need to be conserved for the following purposes-

***(i) Genetic insurance:*** The native breeds have been developed over thousands of years as a result of evolutionary processes, have better adaptability to harsh climate, tropical diseases, heat tolerance with low management inputs in terms of health care, low quality feeds and fodder being capable to convert them more efficiently into animal products (milk, meat, wool, eggs). They are integral part of agriculture and have genes or gene combinations which are associated with adaptability and producing ability. Thus, loss of such germplasm means loss for specific adaptive traits which may be introduced in highly productive germplasm through new techniques of biotechnology. It is not known what might be needed in future and hence conservation of native breeds is required as a genetic security.

***(ii) Scientific study:*** The conservation of native germplasm provides useful research material in genetics, biochemistry, physiology, morphology and anatomy, and to understand the process of evolution and domestication.

***(iii) Economic potential in particular environment:*** The crossbred animals fail to exploit their genetic potential under Indian conditions of low input and harsh climate, more sensitive to tropical diseases and their production level goes down beyond $F_1$,s. Therefore, the conservation of native breeds is required to produce $F_1$,s for future use.

***(iv) Environmental considerations:*** Every organism has its own role to play within an ecosystem and loss of germplasm has adverse effect on the ecosystem which is hazardous to the existence of mankind. The domesticated breeds are an integral part of our ecosystem.

***(v) Cultural and ethical requirements:*** A breed if not economical, it should be preserved for cultural and public interest of historical importance, being part of natural heritage, culture and ecosystem. It is also required on ethical and moral grounds.

***(vi) Energy source:*** India has a number of breeds of draft cattle used to perform a number of agricultural operations. Non-renewable source of energy used in mechanized agriculture may be exhausted sooner or later. Under such condition, the animal draft power will be required and hence these breeds should not be allowed to extinct.

## 21.12 PRINCIPLES OF CONSERVATION

The conservation of AnGR should be based on the following basic principles

***1. Population size:*** The stock should be maintained with optimum population size above the level of risk.

***2. Characteristics of the stock:*** The stock for preservation/conservation should be of the following types

* Select pure form,

* Having special traits,

* Select diverse stock

***3. Environmental conditions:***

* ***Maintain and conserve the locally*** adapted breeds and that too in the same location.

* Maintain live animals providing the similar feeding, management and environmental conditions under which they had been traditionally kept

***4. Breeding methods:*** The genetic merit and diversity should be maintained using appropriate breeding programme.

## 21.13 METHODS OF CONSERVATION

The conservation of AnGR can be done *in situ* as well as *ex situ,* for which two methods of conservation are called as the *in situ* conservation and *ex situ* conservation.

### 1. *In situ* conservation

This involves the maintenance of live population of animals in their adaptive environment (original and natural conditions) and animal population continues to evolve and be developed for more sustainable use. There are two aspects of *in situ* conservation viz., active and passive. The active *in situ* conservation is equivalent to breed development through animal breeding programme whereas the passive *in situ* conservation is the maintenance of live animal populations within their environments for the breeds which are at risk of loss.

This method is best for conserving a breed. The sufficient genetic variability in the breed population can be maintained if the sample size is fairly large. The proper breeding plan for genetic improvement can effectively work to make the breed economically viable. Moreover, the breed can gradually adapt to the changing environment over time. However, the limiting factor for its implementation is the high cost involved in maintaining large herds /flocks.

The *in situ* conservation can be done in a better and scientific way at the organized farms under the control of Govt. to ensure purity of breed. This method is also feasible in field conditions i.e. in herds maintained by farmers and NGO like BAIF, AFPRO, Goshalas. The conservation can be done in field conditions by forming breed societies.

***Institutional herds***: Some important indigenous breeds are kept by ICAR institutes, SAU's animal breeding farms, central and state Govt. livestock farms. However, the population size is too small for improvement through selection. The conservation of breeds on organized farms together with improvement in their productivity can be done by the followings:

- Effective population size
- MOET
- Nuclear herd both ONBS and CNBS
- Improving the management and environment

The organized farms have small number of breedable populations. It is, therefore, very important to know the minimum population size of a breed for *in situ* conservation so as to avoid the risk of inbreeding and genetic drift which are increased with the decrease in population size. The organized farms having a nucleus herd of 100 breedable females' at least can run the selective breeding programme for Improvement and *in situ* conservation.

The details of MOET, ONBS etc. have been given elsewhere in this book (Chapter 30)

***Farmers herds***: The individual farmer maintains small herds/flocks. Therefore, they can not keep the male germplasm of good quality. This forces them to mate their females with the unselected bulls of low genetic merit available in and around the village. This result in production of non-descript population. Secondly, no data recording system is followed in farmer's herds/flocks. These problems of maintaining the purebred population of indigenous breeds and data recording can be solved:

- By forming breed societies for development, improvement and conservation of indigenous livestock breeds. The breed society may be constituted having Distt. Collector, Deputy Director AH/Distt. Livestock Officer/VAS, Block Pramukh/ Pradhan, Farmers, Animal Breeders etc.
- By covering the area of breed under Herd Registration Scheme for data recording and

- By giving incentives to the farmers in different ways viz. an amount for keeping purebred animal, provision of semen and A.I. facilities and support price of products from purebreds etc.

The following strategies to increase fodder resources may be helpful for *in situ* conservation of indigenous breeds:

- To maintain the common grassland and other feed/fodder resources for grazing of animals.
- To reduce the grazing pressure on depleted vegetation and feeding resources by improving migrating grazing system during lean period (summer) in a way to move the animals to the forests, riverside wasteland and harvested agricultural fields and return to native place in rainy season.
- Re-seeding of wastelands/grasslands and controlled grazing
- Controlled grazing in forests because livestock and grassland ecology are complimentary to each other for the overall maintenance of sustainable ecological balance.
- Planting the high yielding species of plants which provide optimal lopping in common grasslands and pastures, so as the livestock can be supplemented with tree lopping when grazing resources are depleted in lean period. This is known as silvipasture and multi tier vegetation.

Besides above strategies to increase the fodder resources, the following programmes will be helpful for *in situ* conservation:

- Improving economic efficiency of indigenous breeds through use of superior germplasm by selective breeding, better health coverage and providing balanced ration
- To create public awareness about the local breeds for their suitability and special attributes under nature conditions.

## 2. *Ex situ* conservation

The *in situ* conservation may not be feasible for the breeds which are economically not viable and hence they may be lost due to economic pressure. The *ex situ* conservation can be done to handle this situation. The *ex situ* conservation can be *in vivo* and *in vitro.*

***In vivo* method**: It is *ex situ* conservation of live animals in small numbers at a place away from the main breeding tract of the breed viz. organized herd of research station, central/ state government animal farms, bull mother farms, zoo and breeding park. This type of conservation has the limitation in terms of population size to avoid the adverse effect of inbreeding. It is not possible to keep the large herds/ stlocks of breeds which are economically not viable. Inbreeding in small population is unavoidable and hence the genetic defects appear. Therefore, the breeding population should be maintained in such a manner that the inbreeding is minimum and the performance of the population showed improvement over the years so as the breed becomes self sustaining. The effective population size can be maintained on organized farms or with farmers in the native breeding tract or under livestock breed safari/parks on the pattern of wild life.

The live animal reservoir/breed safari/park may help to preserve the natural habitat keeping all species of plants, animals and other organisms. It will be a good system to conserve the precious germplasm of breeds of domestic animals as a self sustaining unit. This may also attract the tourist industry as in developed countries. These may serve as the amusement parks if managed properly keeping diverse population of distinct breeds and may thus provide opportunity for urban people to know about the animals.

***In vitro* method:** Storage of living cells for long period of time. It is done by

(*i*) deep freezing of sperms (semen) and oocytes,

(*ii*) deep freezing of embryos, and

(*iii*) storage of DNA.

The *in vitro* conservation has the following basic objectives

- Regeneration of endangered breed
- New breed development
- Supporting the *in vivo* populations
- Research for determining the effect of single major gene
- DNA studies and genome mapping

***(i) Cryopreservation of sperms and oocytes***: The semen (sperms) of all the species is suitable for deep freezing. The technique of freezing, storing and thawing of semen is simple and a routine one. Likewise, the cryogenic storage of oocytes as a portion of ovaries from slaughtered animals can be done for most of the species of farm animals, except cattle. The conservation of haploid genome as semen, but not the oocytes, has great limitation of the requirement, for IVF of the animals of opposite sex or both types of cells (sperms and oocytes) of the same breed.

***(ii) Cryopreservation of embryos***: It has been found successful with varying degree in case of cows, buffaloes, sheep, goats and horses. The embryo being a diploid zygote contains all the genetic information and hence it is an excellent tool of conservation. But it is relatively expensive to obtain embryos. Secondly, the cryo-preservation techniques need further refinement for economical use. It requires 8 to 12 embryos to be frozen to get one live animal. This can be used for conservation of at least endangered breeds. About 300 frozen embryos are required for conservation of a breed assuming the pregnancy rate of 40% and survival rate of 50%.

***(iii) Storage of DNA:*** The genetic material can be preserved as the cryogenic storage of NDA. This method avoids the spreading of disease during its transportation. The storage of un catalogued DNA is already possible but the genome maps of different species are not yet available for want of which, the storage of DNA could not become the normal method of preservation.

***(iv) Recent/New approaches***

***Embryonic stem cells***: These are obtained from culture of inner cell mass of a young blastocyst. These are known as totipotent (undifferentiated) cells which are cultured *in vitro* for a very long period. These cells can be multiplied indefinitely and can be obtained relatively easily from cultured young embryos or cultured cells {primordial germ cells) and are routinely kept frozen before use. The cells have potential to develop into several embryos i.e. the single fertilized zygote gives rise to many embryos capable to frozen for a very long time.

***Somatic cells cloning***: The live animal can be produced from stored somatic cells. The cloning produces series of exact replica/copy of the concerned animals from which the somatic cells have been used. Two techniques (Roslin and Honolulu) which are in use at present enucleate the egg cell to eliminate majority of the genetic information. The Honolulu method is quicker than Roslin as well as it does not require the *in vitro* culture of donor cells. In Roslin technique, the donor cell is forced to remain in dormant cell stage and cultured *in vitro* to produce multiple copies of same nucleus. The enucleated recipient cells and the nucleated donor cell are fused by electric pulse to develop an embryo. The survived embryo is incubated in sheep's oviduct and finally placed in the uterus of surrogate mother ewe. In Honolulu method, the unfertilized mouse egg cells are used as the recipient of donor nucleus. The egg cell, after acceptance in new nucleus, is grown in a chemical culture. The developing embryo is transplanted in surrogate mother.

## 21.14 UTILIZATION OF INDIGENOUS AnGR

See section 22.1 (Breed Improvement).

## 21.15 STRATEGIES FOR CONSERVATION

The conservation of livestock genetic resources in the country should have the broad based objectives with following lines of action:

**1. National survey programme**

- To determine the demographical and geographical distribution of breeds in their home tracts
- To characterize the different breeds of livestock species with respect to their morphological, productive and reproductive characteristics, management practices
- To identify superior germplasm and rare variants within the specific breed.
- A national data bank should be established to keep complete information on all AnGR of the country
- To identify the breeds/types which are declining and heading toward endangerment or extinction
- Planning improvement and conservation programme

**2. *In situ* conservation programme**

- Herd registration programme should be started for all the breeds with active participation of the farmers for which they need incentive and technical support
- Farmers will not come forward to conserve such breeds which are not economical to them. Such breeds should be kept at Govt. farms
- The breed societies should be formulated. The breed societies and voluntary agencies should be given financial and scientific support by the Central and State Govts. for conservation of native breeds.
- The capacity of Govt. animal farms should be increased with all types of infrastructural facilities to keep the indigenous breeds of livestock for their *in situ* conservation
- Different breeds should be adopted by different organization viz. Central and State Govts Farms, SAU's, ICAR institutes, A.H. Deptt. etc. located in home tract of the breed. They should be responsible for the characterization and further improvement and conservation of the breed. One farm for each breed in the breeding tract should be established.
- Inter-institutional associated herd programme should be developed for genetic evaluation and conservation.
- Goshalas having purebred animals should be supported by providing semen from bull mother station free of cost.
- The crossbreeding should be restricted to non-descript population and low productive animals of the breed.

**3. *Ex situ* conservation programme**

- India has good infrastructure for semen collection, storage and deep freezing. Wherever such facilities are available, it should be made mandatory for them to keep reserve the semen doses in addition to their requirement. The semen and embryo banks should be established.
- Training programmes should be organized.

## 21.16 AGENCIES FOR IMPROVEMENT AND CONSERVATION PROGRAMMES

There are a number of departments, institutions and agencies involved in different aspects of AnGR improvement programmes like research, education/ training/ manpower development, animal health services, counseling and guidance to farmers etc. These are as under:

1. Department of A.H. and Dairying, Govt. of India.
2. I.C.A.R.
3. SAU's and Veterinary colleges
4. State Animal Husbandry Department
5. Non-Govt. Organizations (NGO) which are mainly on dairy animals

(*i*) *Gaushalas*: There are about 1000 gausalas engaged in cattle development in India. Some of them are:

* Shree Gaushala Trust, Bhiwani (Haryana) maintains about 1000 crossbred animals of Jersey and Friesian crosses.

* Shree Gaushala Society, Panipat (Haryana) maintains over 200 cattle of HF, Jersey, Hariana and Sahiwal

* Amritsar Pinjrapol Gaushala maintains a large herd of Sahiwal breed (600 animals)

* Kasturbagram Krishi Gaushala, MP has about 150 purebred G ir cattle.

* Lok Bharti Gaushala, Gujarat has Gir and its crosses with HF and Jersey Gorakshan Pinjarapol Sanstha, Ahmednagar (MS) maintain Gir and its crosses of HF and Jersey

* Bombay Pinjarapol has about 1300 purebred Gir cattle.

* Sabarmati Ashram Gaushala, Gujarat, produces bulls and supplies the liquid and frozen semen as well as provides training in A.1.

* Nasik Panchvati Pinjarapol has also the facilities of A.1. work.

* Bombay Gow Rakshak Mandli has devoted to research and development of dairy cattle

(*ii*) *Dairy cooperatives*: The cooperative has played significant role in dairy development programmes by providing to their members the inputs viz.,

A.I. and Veterinary aids etc. The cooperative sector in the year 1987 had 8 bull mother farms and 11 frozen semen bull stations. Village cooperatives have the involvement of lay inseminators.

***(iii) Voluntary organizations***: These are as under:-

- Bhartiya Agro-Industries Foundation (BAIF), Urlikanchan (MS) is involved in crossbreeding of cattle in rural areas under the sponsorship of State Govt., Sugar Cooperatives, Milk Unions, industries and other funding agencies. It has covered about 8000 villages through 400 cattle breeding centers of 6 states in the country. The BAIF has the infrastructural facilities of Bull Mother Farm, Frozen Semen Lab, Research Lab, Demonstration Farms and Training Institute.
- Action for Food Production (AF PRO), Aligarh (UP) is involved in development of good breed of draught animal with improved milk production. It has its crossbreeding work at farm and in the nearby villages.
- Raymond's Embryo Research Centre (MP) is involved in MOET technology
- Janaki Devi Bajaj Trust, Maharastra, has also started cattle crossbreeding work in tribal belts of Pune district.

***(iv) Private companies***: These are the commercial breeding companies and the produce processing companies which are dealing with poultry, sheep and goat keeping.

**6. Other Departments**

* National Dairy Development Board – Operation Flood
* Deptt. of Biotechnology
* Deptt. of Environment

## 21.17 ICAR-NATIONAL BUREAU OF ANIMAL GENETIC RESOURCES, KARNAL, HARYANA/ 1984

*Historical Background and Growth*: The I.C.A.R. established with the objective to conserve the indigenous AnGR and for sustainable utilization and livelihood, the twin institutes of National Bureau of Animal Genetic Resources (NBAGR) and National Institute of Animal Genetics (NIAG) on 21$^{st}$ September, 1984 at SRS of NDRI, Bangalore. Both these institutes were shifted to Karnal campus of NDRI building in 1985 and later on in 1994 were shifted in the campus at Makrampur block of N.D.R.I., Karnal. These were merged in 1995 in to a single unit as NBAGR. This NBAGR is at present functioning as a nodal organization in the country with the aim of identification, evaluation, characterization, documentation, conservation and sustainable utilization of livestock and poultry genetic resources of the country.

The institute has developed ***data base information system*** on AnGR of India to provide information on various breeds of livestock and poultry, ***database on genetic characterization of AnGR*** to store data on relevant published literature, Statistical methods and software for data analysis, and the computerized information system for management of gene bank; ***methodology for phenotypic characterization of AnGR*** for conducting survey for characterization of farm AnGR, mechanism for registration of livestock and poultry genetic resources, models for conservation of various livestock species with the aim of breed improvement and enhanced livelihood security, to revive breeds from critical stage and sustainable utilization of breed; ***parentage verification protocols in different species*** (Cattle, buffalo and camel).

The institute is also providing services to livestock stakeholders like parentage verification, semen of endangered breeds of different livestock species, DNA sequencing, skill development training to stakeholders and consultancy, planning and policy making for characterization, conservation and utilization of AnGR.

❑❑❑

# 22 Animal Breeding Strategies

The various animal husbandry programmes for livestock and poultry improvement have been launched by Govt. of India through different research and development organizations. As a result, the suitable *animal breeding techniques* of selection and multiplication of superior germ plasm (sire evaluation, field progeny testing, ONBS, crossbreeding) have been developed besides development of infrastructure facilities for network of A.I., animal health coverage, improved management practices and milk marketing facilities (Cooperative dairies engaged in milk production, collection, processing and marketing) which have encouraged the farmers of the country. As a result, India has come up on top position in milk production in the world. In 1950-51, the country's milk production was only 17.0 million tones which had increased to 7 folds of producing 121.8 MT of milk in 2010-11. After 1970, the annual growth rate of milk production of the country had been around 4 – 5 %. The improvement has also been made in wool production (27.0 million kg in 1951 to 45.6 million kg in 1995) and egg production (2881 million eggs in 1961 to 36600 million in 1998). The annual per capita egg has increased from 7 in 1961 to 36 in 1998.

The indigenous species are well adapted to local conditions. The zebu cattle, sheep and goat breeds of India have acquired, by natural selection, certain adaptation traits viz. high degree of heat tolerance, resistance to certain disease and parasites and ability to survive on low feed and fodder resources. However, based on population and production statistics, there is a wide gap between requirement and per capita availability of milk, meat, eggs and wool in the country. This is due to the reason that zebu cattle have low potential for milk production, the indigenous sheep have also low level of performance for fleece production, its quality and mutton production and also likewise is the poor production potential of indigenous goats, pigs and poultry. Thus, there is a scope for further increasing the production ability of our animal population.

The poor production potential for milk, meat, egg and wool of indigenous breeds of animals in India has been for the following reasons –

- Poor genetic potential and improper breeding policies
- Poor environment which includes inadequate nutrition, lack oh health coverage and poor husbandry/management practices.

## 22.1 BREED IMPROVEMENT

The utilization of indigenous AnGR in a better way for their conservation is most relevant point to consider. In view of poor genetic potential of indigenous livestock and poultry, it has been the challenge to the animal breeders to improve the production potential to a desired level without sacrificing the adaptation traits. The scientific breeding of animals based on formulation of *breeding plan* is required to change the genetic specifications of animals in accordance to human needs and hence to make the animals more useful to the man. The change in genetic constitution of population is thus the pre-requisite for the improvement and development of a new genotype (breed). The *selection and mating systems* are two important components of breeding plan and the tools of animal breeder to work for bringing genetic improvement of a breed as well as to evolve a new breed to meet the human need.

The *population size* also plays its role mainly through selection. This is because the selection is more effective in a large population due to the reason of ample scope of selection in large population by culling the low producers, and secondly the large population is free from the effect of random genetic drift. Thus, there is low chance of being lost the best gene and the best genetic material and also there is no chance of inbreeding in large population.

Further, it is true that no mating system without selection is effective in changing the genetic structure of population. The genetic improvement through selection is possible by selecting the genetically superior animals present either in the herd (within herd improvement) or by introducing from outside the genetically superior animals (introduction or migration). The possibility within herd improvement through selection depends on the followings:

- Performance level of the herd,
- Variability in selection objective, and
- Population size.

However, both selection and mating systems require the existence of genetic variability to bring genetic change. The combination of selection and mating system to change the genetic structure is called the breeding plan. ***The formulation of breeding plan*** involve the followings –

- To decide the type of production required (*selection objective*) viz. milk, meat, egg, wool etc. This depends on the choice of owner and the market demand,
- The choice of breed as a foundation stock. This depends on the availability, utility and adaptability of the breed.
- To make a decision whether genetic improvement is possible within herd / flock or introduction of superior genes from outside is required. Basically, the improvement can be done in two ways *viz.* selection within population and introduction of superior germplasm.

The decision on introduction / migration or within herd improvement rest on the following factors –

- Performance level of the herd in comparison to other herd of the same breed or of other breed. This requires the estimation of average values of economic trait (selection objective),
- Genetic variability present in the herd / flock. This requires the estimation of heritability and genetic correlation among traits,
- Comparative genetic value and adaptation of the introductions (G-E interaction),
- Heterotic effect.

The introduction of new genes (migration) is recommended when performance of a herd is lower than other herd(s) and the genetic variability within herd is very low.

### 22.1.1 Selection

Now the important consideration is the *utilization of indigenous genetic resources* which implies the use of *breeding methods for optimum production* or to make genetic improvement to the desired level. This requires the formulation of proper *breeding plan* on scientific basis which involves selection and mating systems.

The genetic improvement through selection (response to selection, $\Delta G$) per year depends on certain factors discussed already in chapter 24. It is required to give due consideration to these factors. In addition to these factors which affect response to selection, the following points should also be considered.

The selection criteria for males and females are little different. The culling of low producing dams, selection of replacement young females born to high yielding dams and breeding them (replacements) by proven sires have been recommended. Therefore, genetic improvement can be made by selection of outstanding dams of future sire and by selection of males that will produce future cows and bulls.

### 1. Female Culling/Selection

It is important that the surplus and uneconomical female animals are culled. The culling results in considerable improvement in milk production, also increases the availability of limited feeds and fodders to the rest high producing females, and thereby reduces the cost of milk production. It is essential that to attain high and economic levels of milk production in dairy animals, wool production in sheep and egg production in poultry and their further improvement. The culling of females should be done considering the following plan-

(*i*) The females are initially culled for any off breed character, development and genetic defect, low growth rate and late maturity, and any breeding problem.

(*ii*) The adult female animals are culled for a number of reasons viz. reproductive problems (anoestrus condition, repeat breeding), utero-vaginal disorders (prolapse of uterus, retention of placenta, metritis), udder problems (mastitis, teat block), other diseases, and finally for low milk production (voluntary culling).

(*iii*) In case of female selection, the individual selection is advocated rather than based on the performance of its relatives. The dairy cows and buffaloes are selected for increase in milk production mainly, though it is also desirable to select for improvement of other characters viz. low age at first calving, shorter service period, normal lactation length, and hence proper calving interval, etc. Therefore, the individual selection of female may be for single trait or for multi traits. The multi trait selection has 3 methods viz. tandem selection, I.C.L. method, and selection index method (See next chapter).

The final criteria of culling the females is based either on the basis of the anticipated performance of the female at subsequent age which is known as *expected producing ability (EPA)* or on the basis of genetic value or breeding worth.

It is of interest to cull the female animals for their low productivity in future during their life time. This requires ranking the females on the basis of their E.P.A. The simple method to estimate the E.P.A. of dairy cows was given by Lush (1945) who called it as the most probable producing ability (MPPA) as given else where in this book. This increases the accuracy of selection.

The E.P.A. of dairy animals is estimated from the following formula –

$$\text{E.P.A.} = \mu + \left\{\frac{nr}{[1+(n-1)r]}\right\} (x - \mu)$$

Where, n = number of records

R = repeatability of milk yield

To maximize the intensity of selection based on milk production, it is essential that to minimize the culling based on the basis other than milk production. It is possible to provide better and balanced feeding, proper management for housing and reproduction and health management.

*Phenotypic selection:* As explained in chapter 9.

Basic requirements: The population under selection should have its adaptability to the environmental conditions under which selection is to be carried out.

The initial performance level of the population under selection should also be considered. Moreover, the vital characteristics of foundation population (fertility, abortion, sex ratio, mortality rate, involuntary culling etc) and genetic parameters of important traits should be known.

Uniform environmental conditions regarding feeding and management should be provided to the population under selection or the records should be adjusted for non-genetic factors.

Culling of young females: The young females from inferior dams should be culled.

Culling of adult females: The adult females are culled for a number of reasons viz. reproductive problems (anoestrus condition, repeat breeding), utero-vaginal disorders (prolapse of vagina/uterus, retention of placenta, metritis), udder problems (mastitis, teat block, blood in milk), other diseases and also for low production (voluntary culling).

*Grouping of adult females*: All the adult females at the farm are listed with their phenotypic values of economic traits and graded as elite (superior in most of the traits), superior (meeting standards equal to elite but inferior in one or more traits), good and poor. Adult females should be selected as per standard set for the traits and the number of young replacement(s) available to maintain herd strength.

*B.V. estimation*: The final selection of female is done on the basis of expected B.V of trait under selection. In case of females of farm animals, the individual selection is advocated rather than based on performance of relatives.

The expected breeding value of a cow is estimated as:

$$\text{E.B.V.} = h^2 (X_C - X) \text{ for single record}$$

$$= \left\{\frac{h^2 n}{[1+(n-1)r]}\right\} (X_C - X) \text{ for more records}$$

The accuracy of selection will be as:

$$\text{Accuracy} = h \text{ for single record}$$

$$= \sqrt{\frac{n\,h^2}{1+(n-1)r}} \text{ for more records}$$

The *multi trait selection* is more effective instead of single trait selection. The index selection is more effective than other methods of multi trait selection.

The genetic improvement can be speed up by using embryo transfer technology (ETT) than normal way.

## 2. Male Selection

The individual selection of males in farm animals is not possible. Therefore, the males are selected on the basis of the performance of relatives.

***Male selection is more important*** to bring genetic improvement than female selection, though most of the economic characters in farm animals are expressed in females (sex-limited). The importance of male selection is because of the followings-

- Dams leave limited number of progeny whereas sires leave more number of progeny. The use of A.I. and frozen semen technology has further extended the extensive use of a sire in many herds.
- The sire is said "more than half of the herd". The genetic improvement through sire path is more than 50 percent. Only few sires are needed for breeding a herd for which selection is more intense among males than females. This makes the selection differential larger in case of males.
- Sire selection is equally important in grading up and crossbreeding programmes, besides within herd improvement.

The sire selection can be completed in the following steps –

***(i)* Preliminary Selection of males:** It is essential to examine the males for breed characteristics, physical defect, body growth and health, maturity age, testicles, libido, semen production and quality as well as freezability of semen.

***(ii)* Selection based on relative's performance:** The individual selection for males is not possible for the reason of sex limited nature of economic characters. Therefore, selection based on relative's performance is done as under -

***(a)* Selection of males based on pedigree:** The expected predicted difference (E.P.D,) of male calves/kids/lambs based on dam's performance is estimated as:

$$\textbf{E.P.D.} = \left\{\frac{0.5h^2n}{[1+n-1]r}\right\}(D - X)$$

Where, $h^2$ = Heritability of trait

n = number of records of dam

r = repeatability of trait

D = Average of dam's records

X = Herd average

The accuracy of male selection based on dam's record will be as:

$$\text{Accuracy} = 0.5\sqrt{\frac{nh^2}{1+(n-1)r}}$$

The E.P.D. based on the record of dam and paternal grand dam will be:

$$\text{E.P.D.} = \frac{\{0.5\,h^2\,n_1}{[1+(n_1-1)\,r]\}}(D - X) + \left\{\frac{0.25h^2n_2}{[1+(n_2-1)r]}\right\}(G - X)$$

Where, $n_1$ and $n_2$ = number of records of dam and paternal grand dam

G = Average of grand dam's record

***(b)* Progeny Selection:** The final selection of sire for further use in the herd is based on its progeny performance. In view of small population size, the P.T. should be done in more than one herd. The E.B.V. of sires can be estimated by any method of sire indices (Chapter 26).

The *test mating* should be carried out for testing the bulls carriers of recessive genes so as such bulls are not used in PT programme. A sire can be proved free from recessive gene based on five normal progeny produced from mating of sire with homozygous recessive females.

The *effective use of PT* requires accurate progeny testing of bulls, intensive selection among PT bulls and extensive use of selected bulls. This requires large population and A.I. facilities.

It is better to run the associated or farmer's herd PT programme in view of small population in organized herds/flocks and the nucleus herd should be raised to provide the males of high genetic merit for breeding.

*Mating Plan/System:* A set of 10 bulls should be used in breeding the selected females, out of these 10 bulls, 2 should be tested and 8 should be under testing in each cycle of one or two calving interval, depending on herd size. The female should be mated with male as per the nature of genetic variability viz. additive or non-additive genetic. In case of high heritability, the breeding plan should be the individual selection with *selective breeding* (mating best with the best) and the information on pedigree, family and progeny tests should be taken as supplement to increase accuracy of selection or for sex linked traits or for traits expressed in later life or after death of animal. When the genetic variability of non-additive type is important, the selection alone is not important but selection for *combining ability* (inbreeding followed by selection or crossing the lines) should be practiced.

### 22.1.2 Introduction of Superior Germplasm

The non-descript animals can be improved by *upgrading*. The genetic improvement of well known breeds can be done by *selective breeding* (within herd improvement as discussed above) or by introducing the genes of the same breed from another herd of high genetic merit (*Out crossing*) or of other breed (*crossbreeding*). The crossbreeding in the present day context is the migration / introduction of genes of genetically superior exotic breeds in indigenous breed. This exploits the breed differences and heterosis for the reason that there are large genetic differences between two types of breeds (Sub-species). Thus, the genetic superiority for milk production, fleece production and quality can be combined with good qualities (adaptive traits) of indigenous breeds through crossbreeding (complementarity in crossbreeding).

***Development of synthetic population:*** It combines the inheritance of two or more breeds. The synthetic population can be produced by *inter-se* mating after having the desired level of exotic inheritance. The population is established by *inter se* mating of first crossbred generation ($F_1$), if equal proportion of two breeds is optimum. When higher proportion of exotic inheritance is desirable, the *inter se* mating is started after making backcrossing to exotic breed (forward cross) for one or two generations. But the *inter se* mating declines in performance in next generation because of the decrease in heterozygosity. The decline is more in a population produced by *inter se* mating of $F_1$ but the decline can be avoided by *inter se* mating of population with higher level of exotic inheritance. A synthetic population is stable in the first generation produced by *inter se* mating after which the genetic structure will remain constant in the absence of selection. The heterosis is exploited in individual, maternal and paternal traits in synthetic breed programme. Secondly, the population is self contained and needs no outside genetic material.

The amount of heterozygosity depends on number of breeds forming a synthetic population and the difference in gene frequency between breeds. The proportion of maximum heterozygosity retained in inter se *population is:*

$$H = 1 - \sum p_i^2 \text{ (Dickerson, 1973)}$$

Where, $p_i$ is the proportional contribution from i $^{th}$ breed

or, $$H = 1 - \frac{1}{n} = \frac{(n-1)}{n} \text{ for n breeds each contributing equally (1/n)}$$

$H = 2pq$ in case of two breed synthetic population

Where, p and q are the proportion of the two breeds.

The heterozygosity is maximum with equal proportion of breeds($p = q$).

However, the genetic distance among temperate breeds as well as among tropical breeds is less. Therefore, there is slight increase in heterozygosity by including another temperate breed in crossbreeding programme. This is the reason that little is gained by using more than one exotic breed and one native breed. The breeds to be crossed must be genetically different to exploit heterosis.

### 22.1.3 Nucleus Breeding Schemes (NBS)

The genetic improvement in productivity per animal in the shortest possible time with nominal cost is the main aim of the animal breeder. The conventional breeding programmes viz. selection and mating systems have made significant contribution in the genetic improvement of livestock in developed countries by large scale field progeny testing schemes but the rate of genetic improvement had been low in developing countries like India. The main reasons have been the non-availability of sires of high genetic merit in required numbers, poor spread of A.I. due to lack of infrastructure, small size of farmer's herd, high cost of data recording and the selling of animals before time in field whereas in organized herds the main reasons had been the less intense selection particularly for female side, small population size and putting more emphasis on non-productive traits, etc.

The ***Nucleus Breeding Scheme,*** considering the above situations, can be used for changing the breed structure for increasing the overall genetic merit of the breed. The NBS is of two types: Closed NBS and Open NBS, depending on the direction of gene flow. A nucleus herd is created which is used entirely for production of males for breeding in the population. A breed is structured in such a way that all the animals are not allowed to make contribution for genetic improvement but very few are given this opportunity. In India, a breed of livestock is kept at organized/institute farms known as organized herd and kept by farmers known as commercial herd/flock or village herd/flock which are very small to the extent of 1-2 animals per farmer but collectively the village herds constitute more population of animals of a breed. However, the genetic improvement can be made in organized herd for obvious reasons.

An organized herd under NBS is divided in two groups: Nucleus herd and test herd (multiplier). The nucleus herd is constituted of elite females of high genetic merit and is of a size of about 10-15 % top ranking females of the total herd. The aim is to maximize the genetic gain in nucleus herd to pass on to the test herd and village herds. The nucleus herd breeds its own male and female replacements and occasionally introduces a sire or dam from another nucleus herd. The test herd takes the males and sometimes the females from the nucleus herd to produce sufficient breeding stock to meet the demand of commercial herds. Thus the genetic gain achieved in nucleus herd is passed on from nucleus herd to multiplier and then to village herds.

***(i) Closed Nucleus Breeding Scheme (CNBS)***: There is one way gene flow only with the direction from top to down herds i.e. Nucleus to Multiplier to commercial/ village herds. Thus, the improvement is made in nucleus herd and passed on to multiplier and then to the commercial herds and the only source of cumulative genetic progress in village herd is that occurred in the nucleus herd. Therefore the improvement in nucleus herd is essential otherwise no improvement will occur in other herds because the improvement in nucleus herd is transmitted to other herds.

The time taken in transfer of genetic progress from one herd (nucleus) to the next, known as improvement lag, can be reduced by adopting any of two practices viz. transferring of males and females of nucleus herd directly to the commercial herds and keeping the males and females in the lower herd for short time before replacing them with younger stock.

As there is no gene flow to the nucleus, it is so called as the Closed Nucleus Breeding Scheme which is mainly used in pigs and poultry to avoid the risk of introducing diseases in the nucleus flock.

***(ii) Open Nucleus Breeding Scheme (ONBS):*** In this scheme, the gene flow is both ways viz. downward from nucleus to other lower herd (multiplier) and upward from lower herd to upper herd (nucleus) by introduction of superior animals from other herds. Therefore, the superior animals from commercial herd are introduced into nucleus herd. This reduces the rate of inbreeding in the nucleus herd and increases the genetic progress because the superior animals are also available with farmers. This scheme is mostly used in cattle, buffalo, and sheep.

The ONBS can run in an organized (pedigreed) herd at institutional (Govt.) farm or by forming breed societies. The ONBS is run by a group of breeders or breed societies and hence require a close cooperation between breeders. They cooperate in forming and running of ONBS for getting breeding bulls in turn from nucleus herd. This is called as *cooperative breeding scheme.*

The progeny generation of nucleus herd is reared, recorded and the males are evaluated on the basis of the performance of their sibs (paternal half sibs) and their own performance. The males with high genetic merit for trait under selection can be used in the base population (multiplier) for genetic improvement through natural service or A.I. The ONBS can be operated with and without E.T.T.

***ONBS using MOET or MOET Nucleus scheme***: When MOET (Multiple Ovulation and Embryo Transfer) is used in NBS, it is called the MOET Nucleus Scheme. The MOET is a composite technology which involves a number of processes viz. super ovulation, estrus synchronization among recipients, A.I. of donor, embryo recovery from donor, and finally embryo transfer in the recipient females. Generally fresh embryos are transferred but these can also be cryopreserved for further use in case the recipients are not available on the day of embryo flushing or for the purpose of preservation of gene pool.

The MOET augments the reproductive potential of donor females, which are of high genetic merit and are given hormonal treatment. This results in shedding of more number of ova (eggs) which are fertilized by A.I. The embryos so formed are recovered from uterus of donor by flushing and transferred to the recipient cows of low genetic merit. In this way, more than one progeny (about 8) from donor cows per year can be obtained. About four good quality transferable embryos can be obtained in one flush and 4 flush can be taken in one donor cow. The conception rate of transferred embryos in recipient (the successful rate of implantation) is about 50%.

*Common features of MOET- Nucleus scheme*: An elite herd (or a few elite herds) of males and females is set up and the selection is made at early age based on family information (sibling test from the information of full and half sisters).

The generation interval is considerably reduced for the reason of sibling test of selection and not using the performance of the daughters. Thus the accuracy of selection is less under sibling test but this in accuracy is out weighted by the reduced generation interval.

The selection and testing is done within elite or nucleus herd (s) under controlled environment. By controlling environment, $h^2$ should increase and so improves the accuracy of B.V. estimation. This makes possible the greater degree of control on intensity of selection, generation interval and rate of inbreeding. This is the *major advantage of MOET nucleus breeding scheme.*

The rate of twinning can be increased by transferring two embryos. There are then increased chances of freemartin which can be reduced by embryo sexing. Thus more number of progeny per donor can be obtained without increasing the herd size of recipients.

The problem of getting offspring from high yielding cows that are infertile due to disease or injury may be overcome using E.T. For example, it may allow Brucella positive donors to have Brucella negative offspring.

The females carrying harmful recessive genes causing diseases can be detected by progeny testing of females with E.T.

***Types of MOET nucleus scheme:*** The MOET nucleus schemes are of 3 types on the criteria and age of selection. These are as under:

*(i) Juvenile MOET*: In this scheme the selection of bulls and cows is done at an early age before first breeding. The selection criteria are the ancestor's performance only. The generation interval in this scheme is less.

*(ii) Adult MOET scheme*: In this scheme the selection of males and females is done on different criteria viz. males are selected on the basis of ancestor's performance (dam, sibs of sire and dam) plus their full and half sisters records whereas the females are selection on the basis of ancestor's performance plus their own sibs and their own performance. The generation interval in this scheme is longer than juvenile MOET scheme. The bulls are kept for a longer time for collecting the records of their sibs and hence the cost of keeping bull is high.

*(iii) Mixed or hybrid MOET*: In this scheme, the progeny testing of males required for nucleus replacement is carried out in nucleus herd. The mixed MOET can be juvenile mixed MOET or adult mixed MOET.

### 22.1.4 Other Methods of Genetic Improvement

The followings are some of the methods of genetic improvement of farm animals:

- Exploitation of non – additive genetic variation
- Marker Assisted Selection
- Decanalization (Unmasking of genetic variation by environmental shock treatment).

## 22.2 BREEDING STRATEGIES FOR OPTIMUM PRODUCTION

The livestock improvement programmes for utilization of indigenous species have been discussed in chapters 3 - 6.

### Constraints in formulating uniform breeding policy

.It is not feasible to have a *uniform breeding policy* for different breeds of either species to bring genetic improvement in view of the followings

- Large number of breeds,
- Different utility of different breeds (milch, draft and dual purpose dairy animal; milch, meat, coarse & fine wool and pelt production sheep; milch, meat and mohair/pashmina production goats; egg, meat production poultry).
- Larger proportion of population (70 – 80 %) being non-descript with their low performance potential compared to well recognized breeds,
- Different agro-climatic regions, and different animal production systems,
- Different socio-economic level of farmers (small, marginal and landless), etc.

The broad framework of National Breeding Policy of cattle and buffalo was postulated as early as in 1962, on the basis of which states have to evolve their own breeding policy; of course in tune with the overall guidelines prescribed by Govt. of India and was reviewed and revised in accordance with the changing needs and availability of newer technologies.

The cattle and buffalo population and diversity in terms of breeds and the genetic variation in production and adaptive traits, breeding objectives (traits or combination of traits) and production system, determine the breeding policy. Moreover, the genetic improvement programme should be based on perception, wishes and needs of farmers, which would enhance acceptability and facilitate easy implementation of policies. The requirements for sustainable breeding strategies in a given agro-climatic region, are the details of population size of breed along with farmers socio-economic status, agriculture and cattle/buffalo production systems, available feed and fodder resources, cow/buffalo production potential, utilization pattern, breeding organizations, infrastructure and developmental facilities. The region specific cattle/buffalo breeding strategies can be planned for conservation and genetic improvement under different production systems.

The relevance of breeding programme for genetic improvement in the prevailing scenario, objectives / goals of the programme and breeding strategies determine the success or failure of the breeding programme. Moreover, the breeding programme without considering the nutritional or disease aspects of the area where the population lives may not be advantageous. The availability of sufficient infrastructure to support the programme like A I and veterinary services, procurement of milk and supply of other inputs at the doorsteps of the cow/buffalo keepers are important for genetic improvement in farmer's herd.

### 22.2.1 Breeding Zebu Cattle

The cattle breeding programmes are currently being run by different agencies involving the crossbreeding of Indian cattle with exotic breeds particularly HF and Jersey, and in some states BS and Red Dane had also been used in the past. This has resulted in to a large crossbred population during last two decades and replaced the low producing non-descript population. This crossbred population need to be further improved by selection based on Progeny Test. This crossbreeding programme for genetic improvement needs to be supported by providing sustainable environment in terms of increasing the feed and fodder resources, management practices and health coverage. However, the indiscriminate crossbreeding programme has also resulted in to the endangerment of indigenous breeds.

The infrastructure facilities for A.I. including liquid nitrogen plants have been created. There are 55 semen production units out of which 42 are in Govt. sector and rest 13 are with NDDB, NGOs like BAIF, Pune and private agencies having 2472 bulls producing 39 million semen doses. These stations are maintaining about 40-50 bulls at each centre. There are about 65871 A.I. centres covering 34 million inseminations per annum and each A.I. centre may cover about 800 breedable females. These A.I. centres are capable to cover 40 % of the breedable cattle population (around 40 million) for which about 150 million doses of frozen semen from genetically superior bulls are required. To meet this requirement need about 30, 000 bulls of different breeds and crossbreds. To cover rest 60 % population also requires huge number of selected breeding bulls assuming 2 bulls per hundred cows.

Similarly, the selected crossbred bulls are required to cover the crossbred cattle population of the country. The selection among crossbred population is essential requirement of genetic improvement and establishment of a new breed from crossbred population. This is due to the break-down of genetic superiority (heterotic gene combination) as a result of gene segregation in $F_2$ generation. The break-down of gene linkage and G-E interaction also play important role to cause a decline. The task of producing selected crossbred bulls can better be assigned to Military Dairy Farms having large number of crossbred females of HF x Sahiwal stock (about 24 000 breedable females).

The followings are the recommended breeding policies for cattle improvement:

1. ***Selective breeding within zebu breeds***: Only 15 – 25 % of the cows belong to the well recognized breeds. The indigenous breeds have been developed as per agro-climatic and ecological regions of the country on need based of the area, able

to thrive on the stressful conditions of the area (limited feeds and fodder supply). The State and Central Govt. has established the livestock farms to maintain and improve these breeds. However, there had been lack of proper breeding policies. This requires the setting up more animal farms in the breeding tract of each native recognized breed with all the facilities to identify and multiply the superior germ plasm. The selective breeding policy for improvement of native breeds should be area specific as –

Punjab, Haryan, Western U.P. for Hariana breed,

Rajasthan for Tharparkar and Nagori breeds,

Gujrat for Gir and Kankrej breeds,

Karnataka for Deoni, Hallikar and Amritmahal breeds,

Tamil Nadu for Kangayam breed, and

Andhra Pradesh for Ongole breed.

The milch and dual purpose breeds need to be improved genetically by selective breeding through PT programme, preferably in their home tract in associated herds. The selective breeding is expected to yield about 1 % genetic gain per year in organized herds whereas the genetic improvement in farmers herd is expected to be higher around 10 %. The indigenous breeds of cattle need to be improved by selective breeding in their home tract. The ONBS with or without MOET is a better technique. The cattle breeds developed by selective breeding in India are: *Gangatiri breed* of U.P. developed from Hariana breed, *Rathi breed* of Rajasthan developed from a mixture of Sahiwal, red Sindhi and Tharparkar, and *Krishna valley* of Karnataka, etc.

**2. *Grading of non-descript cattle with improver breeds:*** Large majority of cattle in India (75 - 80 %) is non-descript with low milk production capacity of about 200 kg per lactation. The grading up of these non-descript low producing animals with bulls of improver breeds (high level of milk production like Sahiwal, Tharparkar, Gir, Red Sindhi, Ongole and Kankrej) is required in the areas of poor resources of feeds and fodders, the farmers do not like crossbreeding with exotic breeds, and where the facilities for processing and marketing of milk are not available. The bulls to be used for grading up should be the progeny of superior dams having lactation yield of more than 2000 kg for those belonging to milch breeds and around 1500 kg for dual purpose breeds. This grading up programme at national level requires the setting up of bull mother farms of the native improver breed of the area and A.I. facilities at farmer's door and the demonstration programmes.

The different improver breeds of different states should be used for grading up the farmers animals with bulls of known improver breed of the area viz.

Sahiwal and Tharparkar bulls to be used on local cattle of Haryana, Pb, UP, MP and WB;

Tharparkar bulls for Kerala, Tripura, AP, Bihar, U.P., Haryana, Pb, and Rajasthan;

Red Sindhi bulls to be used on local cattle of HP, J & K, UP, Bihar, Orissa,

Kernataka, Kerala, Assam and Arunachal Pradesh;

Gir and Sahiwal bulls for Gujarat, Rajasthan, MP and Maharastra;

Hariana bulls for local cattle in almost all the states except south India;

Ongole bulls for MP and AP local cattle;

Kankrej bulls for local cattle of Gujarat;

Kangayam bulls for Kerala local cattle;

Hallikar bulls for TN and AP local cattle;

The upgrading programme is expected to make an improvement in milk yield from 200 to about 500 to 800 kg in first generation and in subsequent generation an improvement of more than 5 % annually is expected.

**3. *Crossbreeding (grading up) of non-descript cattle with exotic cattle*:** The crossbreeding of local indigenous cattle with sires of exotic breeds is recommended in the areas near metropolitan cities and industrial cities/towns where market facilities for milk are available and the good resources of feeds and fodders are also available. In general, HF bulls should be used in irrigated plains (Delhi, Haryana, Punjab, UP, Gujarat and Karnataka) and Jersey in hills and coastal areas (Assam, Arunachal, Bihar, Goa, HP, Haryana, Punjab, J & K, MP, MS, Orissa, Rajasthan, UP, WB, Manipur, Southern states. The level of exotic inheritance should be restricted from 50 – 75 %. Brown Swiss is being used in Kerala. A number of crossbred strains of cattle (*Frieswal, Karan-Fries,* etc) have been developed in India as well as in other countries of the world like Ayrshire, *Jamaica Hope, Guernsey, Santa Gertrudis, AMZ, Beefmaster, Brangus.* The crossbreeding reduces the age at first calving from 40 to 50 months to 30 to 35 months, reduces the dry period and improve the milk yield by 6 – 7 times. However, the exotic level of inheritance should be kept below 75 %. Such a crossbreeding programme to run in the country requires the frozen semen by setting the bull mother farms and the regional centres to supply frozen semen, and A.I. facilities under village conditions

**4. *Conservation of Indigenous breeds of cattle:*** There are some of Indian breeds of cattle which are threatened to extinction. These are Burgar, Kankatha, Kherigarh, Krishna velley, Mewati, Ponwar, Panganur, and Vachur. The programmes for the conservation should be taken up as a part of breeding policy. The ICAR is supporting the conservation programme for Panganur and Vachur breeds.

### 22.2.2 Breeding of Buffaloes

The national breeding policy for buffalo is selective breeding of the major buffalo breeds for milk, grading-up of the non-descript and low producing breeds of buffaloes with Murrah bulls. Most of the states have well defined breeding policy, with the main features being in consonance with the national policy. The defined / lesser known buffalo population found in different states selective breeding is the appropriate system looking to the feed and fodder availability in various agro-climatic zones. Also Murrah graded population in various part of the country should be improved through selective breeding. Central Institute for Research on Buffaloes recommended the buffalo breeding policy for enhancing productivity of well defined buffalo breeds by selective breeding. It has been advocated to use genetically superior bulls of defined breeds for the improvement of non-descript buffaloes through grading-up in the country.

Two breeding strategies are recommended for improvement of Indian buffaloes-

1. ***Grading up of non – descript buffaloes:*** The non-descript buffaloes should be upgraded with superior bulls of Murrah, Nili-Ravi, Mehsana and Surti breeds. This will increase the milk production by 2-3 times from about 500 kg milk of non-descript buffaloes to about 1200 kg of first generation upgrade buffaloes. The continuous upgrading of upgraded buffaloes in 4-5 generations will replace the non-descript low producing buffaloes by comparatively high producing upgraded buffaloes which may be very nearer in conformation to the well defined breeds. A breed of buffalo, *Godawari* of A.P. was developed by crossing non-descript buffaloes with Murrah bulls. This requires the development of infra-structure facilities to cover farmer's herd.

Murrah breed has been recommended for upgrading the non-descript buffaloes of the states of UP, Bihar, WB, Haryana, Rajasthan, MS, Karnataka, AP, and TN. The Surti is the breed of choice for Karnataka, Kerala, Gujarat and Rajasthan. The bulls of Murrah and Nili breeds have been recommended for Punjab.

The brief outline of breeding policy for genetic improvement through grading up of non-descript buffaloes of different regions in India are as under:

| Sr. No. | Area/region | Agro-climatic conditions | Breed to be used for grading-up |
|---|---|---|---|
| 1 | North India | Where sufficient feed and green fodder available | Murrah |
| 2 | Western Region | Well irrigated area | Murrah Breed |
| | | Semi-arid region | Mehsana Breed |
| | | Arid region (limited feed and fodder available) | Banni Breed |
| 3 | North – East region | Semi-arid and Humid climatic conditions | Murrah Breed |
| 4 | South India | Semi-arid and Humid climatic conditions | Murrah Breed |
| | | Arid climatic conditions | Mehsana and Banni Breeds |

2. ***Selective breeding for well defined breeds***: The buffaloes in India produce about 57 % of the total milk. The milk yield of buffaloes under farm condition varies generally from 1600 – 2000 kg in one lactation of about 300 days whereas under field condition is lesser than this, averaging about half of this quantity. The selective breeding is recommended for further improvement of recognized breeds of buffaloes. The selective breeding under PT programme for buffaloes on large scale needs to strengthen the existing organized farms of well known breeds like Murrah, Nili Ravi, Mehsana, Surti, Bhadawari, Jaffarabadi and, Nagpuri. Some more farms should be developed for production of breeding bulls. The infra-structure facilities need to be developed for covering the farmer's herd like bull

mother farms to raise high pedigreed caves, facilities for testing, freezing of semen, storage of frozen semen, A.I. facilities at farmer's door, technical manpower for A.I., pregnancy diagnosis, health care and milk recording.

***Murrah brred*** is preferred as a breed of choice in the states of Haryana, Punjab, Western UP and Delhi. ***Surti breed*** is preferred in the states of Gujarat, Rajasthan and Karnataka, Kerala. ***Mehsana and Jaffarabadi*** are the breeds of Gujarat whereas Nagpuri and Pandharpuri are the breeds of choice for Maharastra. The ***Nili Ravi*** has good population in Punjab whereas Bhadawari are present in UP. It will also be better to run PT programme in associating different herds as well as farmer's herds to overcome the constraint of small population size. The genetic improvement of about 1.0 to 1.5 % annually is expected through selective breeding in organized herds but more than this improvement is expected in farmer's herd depending on the quality of the sires used for breeding. The buffalo breed, *Nili-Ravi*, was developed through selective breeding of Murrah buffaloes in Punjab.

**3. *Crossbreeding of Indian breeds of buffaloes*** has resulted in development of new breeds viz. *Mehsana* was developed by crossbreeding Murrah and Surti breeds; *Godawari* was developed by crossing of non-descript buffaloes with Murrah in Krishna Valley region (Andhra Pradesh). The production performance of Godawari buffaloes has been reported at par with Murrah under network project on buffalo. *Kujang* was evolved by using Murrah in Orissa. Similarly, the non-descript population of buffaloes may be crossed with superior bulls of improver breeds.

**4. *Salvage of best germ plasm of buffaloes:*** It is a very much burning issue that no efforts are being made to save the superior germplasm of buffaloes that are taken to metropolitan cities regularly to meet the milk requirement and to earn the money and then to slaughter these superior buffaloes after they complete their lactation period. Not only this loss of adult superior buffaloes but their male and female calves are also not well cared and left to die to save the milk as well as due to their poor management and high incidence of diseases. To save these buffaloes by returning back to their breeding tract should also be a part of their breeding policy at Govt. level. The alternate option to save such good germ plasm is that the ovaries of slaughtered animals should be utilized through oocyte culture and in-vitro fertilization with transfer of such embryos to low producing buffaloes so as to produce genetically superior progeny.

### 22.2.3 Breeding Sheep

The fleece weight and quality traits have medium to high heritability estimates and correlated with body weight. Thus, the animals can be selected on the basis of live weight. The selective breeding has made some improvement in average fiber diameter and medullation percentage in Deccani breed at Pune, Bikaneri and Lohi breeds at GLF, Hisar.

The non-descript population should be either upgraded with recognized breeds of the locality or may be crossed with fine wool exotic breeds (Rambouillet, Merino) and mutton breeds of exotic origin (Suffolk, Dorset).

*Fine Wool:* The crossbreeding breeding of native breeds of hilly region, Nali and Chokla of Rajasthan, Patanwadi of Gujarat and Nilgiri of TN with Rambouillet and Merino rams has been recommended for fine wool breeding keeping 50-75 % exotic inheritance. Some strains of sheep have been developed through crossbreeding programmes of native breeds of sheep with rams of exotic breeds viz. *Kashmir Merino* (cross of Gaddi, Bhakarwal and Pooncchi ewes with Merino and Rambouillet rams), *Bharat Merino* and *Avivastra* (Cross of Chokla and Nali ewes with Merino and Rambouillet rams), *Nilgiri synthetic and Patanwari synthetic* (cross of ewes of respective native breeds with Rambouillet rams) and *Hissardale* (Cross of Bikaneri ewes with Merino rams).

*Carpet wool:* Selective breeding for improving carpet wool in breeds like Marwari, Jaisalmeri, Magra and Pugal should be considered in development programme incorporating six monthly body weight, fleece weight and the North Indian carpet wool breeds (Nali, Magra). The inferior carpet wool breeds (Malpura, Sonadi, Jaisalmeri, Muzaffarnagri) may be improved for better carpet wool quality by crossing them with exotic fine wool or mutton breeds stabilizing the exotic inheritance around 50 % exotic inheritance. Similarly, the crossbreeding of other native breeds like Deccani, Bellary, Shahabadi and Chhotanagpuri with exotic breeds can improve the carpet wool quality. A sheep strain named as *Avikalin* has been developed by crossbreeding Malpura ewes with Rambouillet rams at CSWRI, Avikanagar for producing ideal carpet wool. *Munjal breed* of sheep was developed by crossbreeding of Nali and Lohi breeds in Punjab.

*Mutton:* The mutton production characters have medium to high heritability estimates and hence respond to selection but with low rate. Therefore, the *upgrading* of non-descript population and crossbreeding of inferior native breeds with superior native breeds or with exotic breeds can bring rapid improvement. The native breeds viz. Malpura, Sonadi, Muzaffarnagri, Mandya and Nellore should be crossed with Dorset and Suffolk. A crossbred strain of sheep for meet, *Avimans,* has been developed at CSWRI, Avikanagar by crossing Malpura and Sonadi ewes with rams of Dorset and Suffok breeds.

*Pelt production:* The crossbreeding of coarse carpet wool Indian breeds with Karakul has good potential for pelt production. The native breeds like Malpura and Sonadi can be used for crossbreeding. *Indian Karakul,* a crossbred strain has been developed at sub-station of CSWRI, Bikaner by crossing Marwari, Malpura and Sonadi ewes with rams of Karakul exotic breed. The *Karakul breed* of sheep was itself evolved by crossing two local breeds of the area Karakul (Black lake), Bokhara (Central Asia).

### 22.2.4 Breeding Goats

The breeding of goat for improvement should be on the following lines:

The non-descript females should be upgraded by mating with bucks of high genetic merit of recognized breeds for improvement of milk and meat.

The selective breeding within recognized breed is recommended, involving farmer's flock by selecting best males from breeding tract of each breed of milch and meat type.

Crossbreeding for meat with Boer breed of South Africa has been undertaken in Maharastra in farmer's flock with encouraging results. This can be started in Rajasthan also because the Boer breed belongs to a relatively dry region. The crosses of local and Sirohi goats with Boer have shown good growth. Crossbreeding is also recommended for pashmina fiber production in Changthangi breed in J&K state with exotic breeds from Russia (Orenburg which is best quality Pashmina breed).

Conservation of indigenous breeds for milk (Jamnapari), meat (Black Bengal) and fiber (Chegu, Changthangi): The threatened breeds like Surti, Beetal, Jamnapari should be conserved.

It is also emphasized here that successful implementation of the long term programme also needs the coordination between different agencies viz. State A.H. Deptt.; NBAGR, Karnal; CIRG, Makdoom, Mathura (UP).

### 22.2.5 Poultry Breeding Practices

Some strains of poultry have been developed in India by selective breeding (*Cari Gold* and *Kalinga Brown evolved* from selective breeding of RIR) for egg production, and by crossbreeding for egg production *(Hitkari, Upkari, Krishna)* and for dual purpose of eggs and meat production (*Vanraja, Grama priya, Giriraj).*

The ***modern poultry breeding practices*** being recommended and practiced in poultry for egg and meat production have been given as under -

The poultry is reared for egg and meat production. A large proportion of total cost is the cost of feed. Thus the efficiency of feed conversion to egg and meat is important in profitability. The breeding of specialized type of chickens is required for egg and meat production.

The breeding systems used for commercial production of hybrid chicks are –

(*i*) Intra and inter-population selection and then crossing of improved strains/ lines so as to produce a commercial line/strain,

(*ii*) Producing of inbred lines with 0.50 F produced by 3 generations of FS mating or 6 generations of HS mating and the crossing of such lines that have better combining ability.

**(1)** ***Breeding for egg production:*** The chicken for egg production require to breed them for high egg production of optimum size with better egg quality, low body size with less feed consumption and hence less feed cost per egg, and low mortality.

The *egg production traits* in poultry (egg number, size and weight) have moderate heritability estimates and hence respond to selection. Therefore, selection for these traits directly is recommended. However, an increase in egg number results in the production of smaller eggs of lighter weight.

The exploitation of non-additive genetic variation is also important through crossing different lines/breeds in all possible combinations. The recurrent selection and reciprocal recurrent selection should be used. On the basis of combining ability (GCA and SCA), the commercial hybrids are produced.

The size, colour and number of eggs per year are important for egg production. Now a-days, the high yielding egg type chickens are produced by three or four-way crosses. However, the pure line improvement is required to produce high yielding chicks before using the combination of cross-line and pure-line performance. This gives better results than selection either on pure-line or cross-line performance. This breeding system is practiced keeping in view the low heritability of egg production and livability which may not be improved on the basis of pure-line performance alone.

The poultry breeders have developed the strain or line-crosses of White Leghorn of egg-type chickens. The breeds used for crossing are the females of RIR, New Hampshire and Australorp with White Leghorn males. The selection of males and females for improvement in egg production with compensation of egg weight using ICL is done based on Osborne index for combined selection.

Direct selection for egg quality is advisable. The superior parents should be identified through performance testing and then by progeny testing. The best parents can be used in family selection and exploited by crossing.

***(2) Breeding for broiler production:*** The growth rate is of great concern for meat birds (***Broilers***). The fast growing birds converts its feed most efficiently. The meat birds must have a large breast with fleshy thighs and short legs. The feed consumed should be directly converted to edible meat and the fat deposit inside the body cavity should be reduced. The traits of interest in broilers are the body weight, carcass weight, proportion of breast meat to total carcass, shank length, feed consumed per unit of dressed carcass and disease resistance.

The chicks for meat production (broilers) are bred for high body weight at juvenile age or marketing age, high feed efficiency and low mortality. There is a need to develop parent stock with high production and reproduction to achieve either goal of high egg production or more meat.

The *meat production traits* have fairly high heritability estimates and hence respond to selection by performance testing. There should be selection against fat deposition inside the body cavity. The egg production in broiler is also important at least to multiply the highly selected birds. To meet this objective, a sire of meat line is crossed with a female of egg laying line to produce the market hybrid.

The broiler traits are highly heritable and also exists negative genetic correlation between production and reproduction traits. This is against the production of better broilers. Therefore, the inbreeding and crossing is not so important for production of broilers but the development of specialized sire and dam lines has great significance in broiler production because the two lines have different contribution for the reason that lesser males are required than females.

The ***dam lines*** are selected for early growth and reproductive traits viz. age at maturity, egg production traits (egg number, egg size, and weight), egg quality, and hatchability. Whereas the ***male lines*** are selected for early growth, conformation, feed efficiency, fertility, and carcass traits. The breeding plan for commercial broiler production is the intense pure line selection followed by crossing the lines keeping in mind to avoid inbreeding. The crossing of these specialized sire and dam lines utilizes the additive and heterotic effects of gene combination in the progeny and reduces the cost of production of day old chicks. Thus, better performance of crossbreds in traits of interest is expected than purebreds. Moreover, the crossbreds are less susceptible to environmental factors than their parental lines and hence result in uniform performance. Therefore, crossbreeding is advocated for broiler breeding.

The breeds used in crossbreeding for production of commercial broilers are New Hampshire, White/red Cornish and White Rock. This should involve the male lines of Cornish strain or the synthetic population with variable amount of Cornish, Hampshire and Rock inheritance and the female line of either closed flock of Hampshire and Rock strains or their strain crosses.

*Specialized Breeding:* It is also advocated to introduce some major genes to develop the breeding stock with better adaptability in tropical climate like India. For this purpose the major genes like naked neck (loss of neck feathers and secondary feathers) and frizzle feathers (curling and reduced feathering) which have their effect of improving the convection ability have been successfully used. This practice is useful for developing both egg layers and broilers.

*Random sample testing:* The performance testing of the new strains developed for either purpose is essential. The testing can be done at several locations through random sample test for both egg layers as well as for broilers. The random sample tests are conducted at Central Poultry Breeding Farms located at Gurgaon, Bhubaneshwar, Bangalore and Bombay under Government of India.

### 22.2.6 Pig Breeding

- Maintenance of sufficient breeding stock in government and institutional farms
- Establishment of breeder pig farms, establishment of nucleus breeding farm and application of modern reproductive technologies for faster dissemination
- Upgrading and cross-breeding programme for non-descript pigs ensuring sustainability
- *In situ* conservation, documentation and registry to create indigenous pigs as breed
- Import of frozen semen, Popularization of artificial insemination (AI) with liquid and frozen semen in pig
- Distribution of piglets to the farmers through the breeder farms for formation of pig village and Pig group formation with 10-20 farmers per group
- Subsidies to traditional pig farmers to promote indigenous pig rearing
- Training programmes to the pig farmers for modern technology transfer,
- Promotion of integrated farming system including piggery

❑❑❑

# 23 Biotechniques for Animal Breeding

The present knowledge of genetic mechanism to transmit the hereditary characteristics from generation to generation has arrived from the studies at the following three levels -

- ***Phenotypic level*** (patterns of transmission of phenotypes to next generation),
- ***Cellular level*** (process of cell division and fertilization), and
- ***Molecular level*** (process of genes expression to produce their effects through chemical reactions).

Mendel (1822-1884) made unique contribution to the knowledge at phenotypic level by demonstrating that hereditary factors are discrete entities without being admixing in the hybrid, occur in pairs in each individual and one factor of each pair comes from each parent following the law of segregation.

Sutton and Boveri (1902-04) explained the ***cytological basis of heredity*** (Mendelian inheritance at cellular level). The cytological basis states that the chromosomes occur in pairs in somatic cells, carry the hereditary factors (genes) which also occur in pairs, divide exactly and are distributed equally in daughter cells due to their segregation during cell division (meiosis), one member of each pair of chromosome is paternal in origin while the other member is maternal in the fertilized ovum which is the result of fertilization (union of egg and sperm) and give rise to the new individual. In other words, it is the ***physical basis of inheritance*** which includes the genetic material, its particulate nature and distribution in chromosomes, chromosome theory of inheritance, and its transmission from parents to offspring by the process of cell division (meiosis) and fertilization.

The process of transmitting the genetic material by the parents to the offspring for maintaining the genetic continuity from one generation to the next is called as the ***bridge of inheritance.***

Watson and Crick model of DNA structure helped to understand the ***chemical basis of heredity*** (at molecular level) by which the genes express themselves. The chemical basis of DNA structure enunciates that:

- Genes carry genetic information (in the form of the order of purine and pyrimidine bases in the DNA molecule),

- Genes must duplicate itself with great precision (double strand DNA structure and the separation of the two strands of the helix),
- Genes may change in their chemical composition through change in one or more of the bases to create new arrangement of sequence of bases in DNA molecule (mutation), and
- Genes translate the genetic information to develop the characteristics of the individual.

In short, the chemical structure of DNA (nucleosome concept) and its function (replication and protein synthesis) form the chemical basis of inheritance.

The traditional or conventional animal breeding programmes to bring genetic improvement in animal production since domestication involved selection and mating systems based on performance recording and crossbreeding, form the traditional biotechnology. During the last three decades, newer technological breakthrough in the field of reproduction, genetics and related aspects (nutrition, health care) have opened the new area of animal improvement. These new techniques using biological systems or their substances at molecular (gene) level are the areas of modern biotechnology.

## 23.1 BIOTECHNOLOGY

The convention on Biological Diversity defines ***biotechnology*** as any technique or a combination of techniques that uses biological systems, living organisms or derivatives thereof, to make or modify a product or process for specific use. Many biological techniques have been used by man since long for manufacturing of many products like vinegar, beer, wine, bread. The biotechnology comprises the techniques of modern biology used to bring changes in livestock production, processing and utilization of products and health care (vaccines, drugs, diagnostic probes, etc).

Biotechnology has the promise to manipulate the biological systems for the benefit of mankind using gene-based techniques:

- More recent biotechniques in the field of *genetics* like QTL and MAS, gene transfer and nuclear transfer for genetic manipulation of animals have demonstrated the potential for genome analysis and manipulation of genome.
- The newer *reproductive biotechniques* have improved the reproductive performance of farm animals in addition to A.I. and embryo transfer.
- There are opportunities for improving the *nutritive value* of low quality animal feeds by chemical and biological treatment, and genetic manipulation of rumen microbes.
- There is also wide scope of biotechnology in improving the *animal health* through vaccines and genetically engineered immune-diagnostic and immune-prophylactic agents.

## 23.2 QTL MAPPING DESIGNS

The QTL are generally mapped by two methods based on crosses between lines differing in mean values, and based on the segregating populations. The crossing of inbred strains or lines showing genetic differences for the trait of interest is most powerful to map QTL. This method is efficient in fixing the alternate allele at both the QTL and the marker loci due to maximum linkage disequilibrium between the loci in the $F_1$. All the alleles should be in association (coupling phase). The homozygosity and QTL in association (coupling phase) is the result of divergent artificial selection of parental populations for trait of interest and subsequent inbreeding. It is possible to locate QTLs when the alleles are in dispersion but to a lesser extent than when the alleles are in association.

The $F_1$ generation produced from crossing of parental inbred population is back crossed to either one or both the parental lines or inter se crossing is done to produce $F_2$ generation, Both the backcrosses and $F_2$ generation are segregating. The $F_2$ can be used to produce recombinant inbred lines if the species can tolerate inbreeding whereas the cross analysis can be used in the species which have long generation interval and do not tolerate inbreeding, provided they are available and are likely to be fixed for alternate QTL alleles. In random bred population, the linkage analysis is least efficient to detect QTLs because the linked loci are likely to be in linkage equilibrium.

The procedure of QTL-marker loci linkage analysis involves the scoring of individuals for their genotype at marker locus and their phenotype for the quantitative trait. The significant difference in mean phenotype of quantitative trait among marker genotypes indicates the association of QTL with marker. The number of QTL by this procedure (QTL-marker linkage analysis) is always underestimates of the number of loci. This is because two QTLs closely linked together may appear as one if they are in coupling phase (association) or they may not be detected at all in repulsion phase (dispersion).

The experimental crosses have been used as mapping design in case of pigs and poultry but not in case of cattle due to non-availability of inbred lines, low reproduction rate, long generation interval and high cost. In cattle, the daughter design and grand- daughter design are used as mapping designs.

***(i) Daughter design:*** The grouping of progeny is made according to the marker allele received from a heterozygous parent. The offspring of parent heterozygous for marker locus and a linked QTL, receives linked QTL allele along with a particular marker allele, in the absence of recombination. The progeny of two groups with respect to the presence of alternative alleles at the linked QTL will differ in their mean value for quantitative trait. The presence of a linked QTL near the marker is indicated if the two progeny groups (receiving alternative marker alleles from their common parent) differ significantly. The model is as under –

$$Y_{ijk} = S_i + A_{ij} + e_{ijk}$$

Where, $Y_{ijk}$ = trait value of the $k^{th}$ daughter of i $^{th}$ sire having marker allele j,
$S_i$ = effect of i $^{th}$ sire,
$A_{ij}$ = effect of of j $^{th}$ allele of i $^{th}$ sire,
$e_{ijk}$ = random residual error

This is the marker QTL association based on the production records of the daughters of a single sire and hence called the daughter design to map the QTL. This design has the limitation to establish the marker QTL association that it requires large number of daughter groups. This design works only for a short time because the marker genotype gets fixed in few generations of selection. Moreover, in a segregating population, the sire although heterozygous for a marker, will be homozygous for the linked QTL and hence the daughters of two groups based on marker will not differ for quantitative trait. Thus it will not give accurate results. Therefore, only a fraction of the segregating QTL present in the population will be revealed based on the progeny of a single sire. In view of this it requires to pool the data for a number of sires and use the t-test or ANOVA but it has its limitation because the linkage relationship between marker and QTL differs among individuals.

***(ii) Grand-daughter design:*** Weller et al (1990) suggested an alternate to daughter design which uses the grandsire heterozygous for a marker and a linked QTL. The sons of the grand sire are genotyped based on specific marker alleles received from their sire (grandsire) and divided in two subgroups. The genetic merit of sons (based on performance of their daughters which are the grand daughters of the heterozygous grand sire) of two subgroups (sons of grandsire) is estimated for quantitative traits. In this design, only the sons are genotyped. Thus, sons instead of daughters of a heterozygous elite sire are genotyped for markers and divided in two groups based on marker allele received from grand sire. The son's marker genotypes and their predicted genetic merit are used as a measure of the association between marker and QTL. The model to progeny test each son to estimate his value for quantitative trait is as under –

$$Y_{ijkl} = G_i + A_{ij} + S_{ijk} + e_{ijkl}$$

Where, $Y_{ijkl}$ = observation of trait of l $^{th}$ grand-daughter, daughter of k $^{th}$ sire, son of grandsire i receiving j $^{th}$ marker allele
$G_i$ = effect of i $^{th}$ grand sire
$A_{ij}$ = effect of j $^{th}$ allele of i $^{th}$ grandsire
$S_{ijk}$ = effect of k $^{th}$ son with j $^{th}$ marker genotype of i $^{th}$ grandsire
$e_{ijkl}$ = random residual error

The marker-QTL analysis has some requirements viz. genetic linkage analysis, gene mapping, association of marker loci to linked QTL, and use of correlation in genetic improvement of economic traits. The number of QTL that can be used in selection programme with their large effect on a trait is likely to be small and their number with moderate effect is limited by the total genetic variation. On the contrary, it is difficult to identify the QTL with small effect, their estimation will be less accurate, and hence their individual contribution in selection response will be low. The selection increases the frequency of loci with large desirable effect. Any gene with large or moderate effect is largely exploited after 2-5 generations of selection but the exploitation may take 5-10 generations if the initial frequency of favourable allele is low.

***QTL status:*** With a view of better selection of farm animals to increase their productivity and health, there are consortium of genomics and bioinformatics working on whole genome assembly status of cattle, buffalo, sheep, goat, pig, poultry, camel, horse and rabbit. The QTL status (total number of QTL for number of traits) for different species of farm animals reported so far is as under –

| | Cattle | Sheep | Pig | Chicken |
|---|---|---|---|---|
| No. of QTL | 5207 | 639 | 6346 | 2736 |
| No. of traits | 378 | 184 | 577 | 257 |

## 23.3 APPLICATIONS OF MOLECULAR MARKERS

The polymorphisms observed at the DNA sequence level have been playing a major role in human and livestock genetics for the following aspects.

(I) Short range or immediate applications
1. Parentage determination
2. Individual identity
3. Genetic distance estimation and evolution
4. Determination of twin zygosity and freemartinism
5. Identification of disease carrier (heterozygous) animals
6. Sex determination

(II) Long range applications
7. Gene mapping
8. Breeding programme (Marker assisted selection, Genome selection)
9. DNA fingerprinting
10. Transgenic breeding

## 23.4 MOLECULAR MARKERS IN BREEDING PROGRAMMES

The molecular markers have enormous application in selecting the best animal as parent of next generation. The molecular markers lay emphasis on identification of quantitative trait loci (QTL) in following different ways –

### (1) Marker Assisted Introgression (MAI)

The introduction of a favourable allele from the donor breed into recipient breed based on phenotype is termed as the introgression. In animal breeding it is done by making a cross of recipient breed with donor breed and back crossing the heterozygous ($F_1$) progeny with donor breed. The incorporation of the most desirable alleles into a group of selected individuals/strain/breed is less common in animals than in plants. The identification of animals for favourable allele based on the use of molecular markers is called the marker assisted introgression (MAI). The introgression may be extended from gene level to QTL when more number of molecular markers with optimum positions to QTL is used. The MAI requires large population size in case of polygenic trait involving more number of QTL. However, the long generation interval, low reproductive rate, and higher rearing cost limit the use of MAI in livestock breeding compared to that used in plant species.

The introgression can be applied in animals for major genes. The examples are the Booroola gene for increased fertility in sheep, and genes for trypanosome tolerance for N'Dama cattle. The MAI will increase the rate and accuracy compared to conventional selection based on phenotype.

## (2) Marker Assisted Selection (MAS)

The selection procedure for a trait using genetic marker as a criterion of selection is called marker assisted selection (MAS). The MAS is a form of indirect selection of superior breeding stock. The molecular markers have been found to be linked with the genes controlling the economic traits (grain quality, resistance to disease and pests) in many crop species and in animals.

The gene marker technology as MAS in animal breeding was used by the breeders in the beginning of 1990s. This was used to remove deleterious gene (such as halothane in pigs or fish taint in brown-shelled chicken eggs) or to select favourable conditions based on some marker information (such as the polled condition in cattle).

The MAS depends on identifying associations between genetic markers and linked QTL (quantitative trait loci) with useful effects. The association between the markers and the QTL depends on the distance between the markers and target traits.

***A genetic marker is a DNA segment associated with a trait*** and segregates in a predictable pattern. Thus, the genetic marker facilitates the tagging of individual gene or small chromosomal segment affecting the trait. The linkage analysis helps in screening the genome for genes with a large effect on trait. The marker and phenotypic data on an individual can provide more information than phenotypic data alone. The MAS is done in two steps of identification of marker loci liked to QTL of economic character and the use of linkage association between QTL and a marker locus. The close is the linkage between marker and QTL more is the success of MAS.

***Advantages of MAS***

*(i)* The MAS utilizes the complete genetic variation due to additive genetic value, dominance effect and epistatic variance. Therefore, the MAS has the advantage that it gives more rapid and accurate prediction of phenotypic value.

(*ii*) The MAS is ***more efficient*** for traits of low heritability because heritability of marker is 1 without genotyping error and the genetic variance is more explained by the marker than economic trait.

(*iii*) The MAS is ***faster*** as it requires lesser time than progeny testing. It can be used in early age for selecting young sires before putting them in PT programme, and hence reduces the generation interval. Thus the MAS in this way of selecting sires at younger age increases the rate of genetic gain (15 - 30 % additional genetic gain) by reduction in generation interval, higher selection intensity and increasing accuracy of selection.

(*iv*) MAS do not need to maintain more number of animals and that too for longer time. Thus, it is ***cheaper*** when the cost of genotyping is low than phenotyping.

(*v*) The MAS has the prospects of selection for traits which may not be improved by conventional methods viz. resistance to disease etc.

(*vi*) MAS can be used for sex limited traits and for the traits expressed in old age or after death of animal (carcass traits, body composition), and survival rate. It *is* ***easy*** for difficulty to record the phenotype. Marker genotype can be determined from early collected samples of hairs, blood, etc taken at the time of birth of animal and hence genotyping is possible before phenotyping for the trait.

***Limitations of MAS***

(*i*) The MAS has the limitation that it focuses only on those regions that influence the trait of interest but unaccounting rest of the genome and hence the genetic variation. Thus, MAS explains only a limited part of genetic differences beteen individuals.

(*ii*) The application of MAS is difficult work. It requires prior knowledge of the association of marker and QTL affecting the trait of interest as well as the location of QTL. It must therefore be implemented within families.

(*iii*) It needs large number of offspring for linkage analysis.

(*iv*) It requires family specific information and many individuals from a family.

Examples of QTLs-marker linkage in cattle:

Some workers have analyzed the QTL-marker linkage for different traits in cattle. Some of the examples to cite are:

- Association of beta-lactoglobulin allele with higher fat content;
- Significant effect of kappa casein and beta casein on milk protein percentage;
- Linkage between microsatellite markers with loci controlling yields of milk, fat and protein;
- Association of genetic disorder (weaver disease – progressive degenerative myoencephalopathy) in cattle with increase in milk production.
- Association of three growth hormone (B, C and D) with decrease in birth weight and shoulder width of calves at birth;

## (3) Genomic Selection (GS)

The GS is a form of MAS proposed by Meuwissen *et al* (2001). As a result of breakthrough in DNA sequencing and genome technologies the selection strategies have been shifted from single gene or QTL selection to whole genome selection. The selection of individual based on the molecular breeding value assessed through evaluating complete genome (including all the genetic markers located throughout the genome) rather than one or few markers is called genome selection. The genetic markers covering the whole genome are used so that all QTL are in linkage disequilibrium with at least one marker.

Goddard and Hays (2007) mentioned that the underlying principle of GS is the exploitation of the linkage disequilibrium of QTL with one or more genetic marker(s). Therefore, the DNA markers covering the complete genome are used so that all the genetic variation is explained by the markers. In other words, the decision of selecting an individual based on genomic estimated breeding value (GEBV) using molecular markers is the genomic selection. The GS exploit the population wide linkage disequilibrium between QTL and genetic markers spread over the genome. The success of GS depends mainly on utilizing the next generation sequencing approaches and the associated bioinformatics tools to identify the SNPs (Single Nucleotide Polymorphism).

The GS in dairy cattle has already been applied by a number of countries like USA, Canada, Australia, New Zealand, Denmark, Netherland, Germany, Ireland and Norway. However, the GS remains limited as a technology in most of the countries and it will take years to come up a shift to this technology for use in animal breeding.

***Implementation of Genomic Selection***: The genomic selection can be used to estimate the genetic merit of animals by using the genome wide SNP panels as markers and statistical methods that explain the effects of large numbers of SNPs simultaneously. The large panels of SNP markers are available and can be used in genomic selection to increase the accuracy of prediction of selection for younger animals. The BV can be predicted with high accuracy using genetic markers alone. However, its validation is required in a population which is different from reference population. The reference population is the one in which the marker effects are estimated. Thus, the GS uses reference population (called as training set) and a validation set different from reference population.

The GS is implemented in three stages of discovery dataset (information of reference population or training set), validation of prediction equation (validation set), and selection. The reference population contains the phenotypic information of population collected over a range of environment, molecular marker scores (based on reference population), and pedigree information (kinship). The genomic BVs of new genotypes are predicted based only on the marker effect. This method of estimating BV is based on marker genotype instead of the QTL genotypes.

A large number of SNP are assayed on a moderate number of animals from discovery dataset which have information on phenotypes of the relevant traits.

A prediction equation using markers as inputs to predict BV is derived from this data set of performance records. This equation is validated in a validation sample of animals having records of the traits and genotypes for the markers to be used commercially. Thus, the prediction equation is tested for its accuracy on the independent sample of validation set. The validation set contains the selection candidates which are selected based on marker effects estimated in reference population. Now the selection candidates are genotyped for the markers and the prediction equation estimated in the discovery data is used to calculate GEBV. The selection candidates are genotyped but not evaluated based on phenotype.

**Advantages of Genomic Selection**

- The GS based on SNP increases the accuracy of selection,
- It reduces the generation interval because it can be used to select the dairy sires before the PT results are known.
- The selection intensity can be increased by screening hundreds of samples.
- The GS can double the rate of genetic gain. The genetic progress is faster with genomic selection in view of increase in accuracy and intensity of selection and reduction of the generation interval.
- Shaffer (2006) has shown through a simulation study involving economic aspects that GS breeding strategy in place of traditional PT scheme could save up to 92 % of breeding cost mainly due to reduction in generation interval and increase of selection accuracy for bull dams.
- The genomic selection is possible without information on the phenotype (character under selection) if there exists and established the association between chromosome segments and phenotype.
- The GS can be easily applied for traits with low heritability and the genetic change is slow, traits difficult to measure and the traits expensive to measure.
- The GS takes into accounts the complete genome, giving greatest emphasis on the regions with large effect and hence it also accounts for the ambiguous genetic variation in rest of the genome. Thus, the GS is a better option for the reasons that the economic traits in livestock are polygenic that are distributed throughout the genome mostly with small effect. The GS is in contrast to the MAS. The MAS is based only on the genome regions influencing the economic trait under selection without taking into account the rest of the genome and its genetic variation.

## 23.5 APPLICATION OF MOLECULAR GENETICS

There have been developed the following ***molecular techniques*** for use in molecular genetics –

- Discovery of restriction endonucleses
- Reconbinant DNA technology including amplification and cloning of genes/ DNA segments,

- Polymerase chain reaction (PCR) technique to amplify DNA sequences,
- Construction and screening of DNA libraries,
- Electrophoresis, and autoradiography,
- Hybridization techniques (Southern blot, Northern blot, and Western blot) for molecular analysis of DNA, RNA, and proteins,
- Molecular analysis of genes and chromosomes for physical maps of DNA molecules based on restriction enzyme cleavage sites and nucleotide sequences of genes and chromosomes,
- Microarrays and gene chips technology, and
- Gene sequencing techniques, etc.

***Applications*** of the above techniques of molecular genetics are as under -

- DNA profiling (Molecular markers & their application),
- Human gene therapy,
- Diagnosis of inherited human diseases,
- Production of eukaryotic proteins in bacteria (Human growth hormone),
- Production of transgenic animals,
- Animal cloning, and
- Stem cell technology, etc.

❑❑❑

# 24 Biotechniques in Animal Reproduction

A number of activities and techniques are involved in reproduction of farm animals. All these techniques can be dealt as conventional (traditional) methods and the advance biotechniques to accomplish the different activities related to animal reproduction.

## 24.1 REPRODUCTIVE CHARACTERS OF DIFFERENT SPECIES

The reproductive characters include the age at sexual maturity, age at first calving/lambing/kidding/farrowing, post-partum breeding interval, service period, gestation period, calving/lambing/kidding/farrowing interval, heat cycle, duration of heat, ovulation time, time to breed the females in heat, etc.

### 24.1.1 Cattle

The heat/estrus/*oestrus* is the intense sexual urge of sexual excitement shown by female mammals. The female animals show some *signs of heat* viz. mounting on other animals, slight rise in rectal temperature, frequent micturition, drop in milk yield, decrease in feed intake, bellowing, mucous discharge from vulva and restlessness. The estrus is shown for some period (heat period) during which the female shows some outward signs of heat but others do not show the signs, called as silent heat or shy-breeder in which the heat can be detected by use of teaser bulls.

The *age at puberty* is 2 to 2.5 years. The *duration of heat period* is 12 – 20 hours. The *heat cycle* (recurrence of heat) in cows is of 19 – 23 days (21 days). The female reproductive cell or egg is released from the ovary within 12 hours after end of heat (*Ovulation time*). It takes from 4 – 9 hours for the sperms to travel from the vagina to the oviducts where the fertilization normally takes place. The *optimum time to breed* cows is mid heat to the end of heat.

The cells of fertilized eggs multiply rapidly after the egg enters the uterus. The body organs and parts start forming after about 10 days when the outer layer of the fertilized egg breaks up. The heart-beat of embryo begins 22 – 24 days after conception. The embryo floats unattached in the uterus fluid for about a month after which the embryo membranes unite with the buttons (caruncles) on the inside of the uterus. The embryo is then called fetus and starts getting nutrition through blood from the capillaries in the buttons of the uterus.

The *gestation period* is 280 -285 days in cows. The *calving interval* is a function of service period and gestation period and depends more on the length of service period which has wide variability depending on a number of factors that are more of environmental in nature. The *service period* is the function of *post-partum breeding interval* (PPBI - time interval from calving to exhibition of next heat) and the number of services required for conception (S/C). The PPBI also varied widely. However, with proper care and feeding most of the females come into heat within 100 days of calving.

### 24.1.2 Buffaloes

The age at puberty is 2.5 years. The *heat cycle* in buffaloes is of 19 – 23 days (21 days). The *duration of heat period* is 10 – 16 hours. The female reproductive cell or egg is released from the ovary within 7 hours after onset of heat (*Ovulation time).* It takes from 4 – 9 hours for the sperms to travel from the vagina to the oviducts where the fertilization normally takes place. The *optimum time to breed* buffaloes is mid heat to the end of heat. The *gestation period* is 310 days in buffaloes. The PPBI varied widely. However, with proper care and feeding most of the females come into heat within 100 days of calving.

### 24.1.3 Mithun

The sexual maturity is attained at 2.5 to 3 years, gestation period of 272 days, service period of 60 – 180 days. Its productive life is 12 years and total life span is about 20 years.

### 24.1.4 Yak

The age at puberty to bred first time is about 2 years. The breeding season is generally July to August.

### 24.1.5 Sheep

The sheep attain sexual maturity at the age of about 2 years. The sheep are seasonally polyestrus which implies that ewes have many heat cycles during a breeding season. There are mainly two *mating seasons* in Indian sheep *viz.* summer (March - April), autumn (Sept – October). The lambs from summer mating grow well. The *estrus cycle* is of 17 (14 – 20) days. The *estrus duration* is of about 30 hours. The service to the ewes in heat should be given during end of heat period as the ovulation occurs about 24 to 30 hours after the onset of heat. The *gestation length* is 150 days. However, generally there is only one lamb crop a year in sheep.

### 24.1.6 Goats

The *sexual maturity* is very early, even before one year of age, generally between 9-10 months. However, the goats should not be bred before 18 months of age so as they kid first time at the age of 2 years (*age at first kidding*). The incidence of estrus is high from May to October (*breeding season).* The *duration of estrus* is 24 to 48 hours but in some breeds it may be of shorter duration like Beetal goats. The *ovulation time* is in the end of heat and the goats should be mated between 10 – 15 hours after showing heat. The *gestation period* is of about 150 days. The goat may kid twice in a year or thrice in two years. The goats have a special feature that a pregnant doe accept the buck. The *life span* of a goat is about 12 years.

### 24.1.7 Rabbits

The male and female rabbits mature at about the same age of about 5 – 6 months whereas the *breeding age* in larger meat producing breed is little delayed. The *gestation period* is about 31 days. The female rabbit does not have *oestrus cycle* and can stay in heat for several days. There is no regular period of heat and the *ovulation* occurs only after mating. The acceptance or refusal of service indicates whether the female is in heat or not. A doe accept mating and ovulate when its vulva become red in colour which happens in 90 % cases. The *ovulation* occurs 10 – 12 hours after mating, and induced by coitus stimuli. The buck after successful mating makes a typical sound or cry and fall down to one side of the doe. The gestation period is 30-32 days. A doe can be rebred and become pregnant after 7 days of kindling (act of parturition in rabbits). *Average kindling per year* is about 4 producing 25 kits per doe.

The kids are born hairless with eyes closed and they open their eyes after 10-14 days. Weaning is done after one month of age. The sex of young rabbit can be determined after one week of age. The active growth period in rabbit is 3 – 10 weeks of age and they gain 30-35 gm daily. The sex of the rabbit can be determined after 7 days of age.

### 24.1.8 Camel

The cow-camel shows signs of sexuality nearly at 2 years of age but it is not allowed to serve before 4 – 5 years. The *adult age (puberty)* is about 5 – 6 years among males whereas in females it is 4- 6 years. The peak *breeding season* (*rutting period*) is from November to March. The *estrus duration* is of 3 – 4 days. The sexual act takes about 15-25 minutes. The male produces *gurgling sound* and *froths from the mouth* whereas she camel goes on *bleating* during the time of copulation. The *gestation period* is of about 375 (365- 405) days. *Oestrus occurs* after about one year of parturition and hence the females breed only once in two years. The *calving interval* is of about 650-700 days. The female (cow camel) can bred up to 20 – 30 years of age. The *lactation period* is of more than a year. The sexual instinct of the male is suppressed for more part of the year. They are in *rut* commonly known as *mush* for a few months only and generally become mush after the age of 6 years. The camel may live up to 40 years of age.

### 24.1.9 Equines

This includes horses, donkeys and mules.

***(a) Horse:*** The horses have lower reproductive rate than other animals. The *age of puberty* in a filly (mare) is 15 to 18 months but it is bred at later age of about 3 years. The race and show mares are bred after the end of their career in race and show. The stallions attain maturity at 4 – 5 years of age. The best *breeding season* for horses is spring. The mares after foaling (*Post-partum heat*) come in heat (foal-heat) within 7 – 10 days. The *estrus cycle* is of 21 days with *duration of heat* varying from 4 – 7 days. The optimum *time to breed* a mare is in late heat on 7$^{th}$ day of heat. There is single birth with seldom twins. The average *gestation period* is 340 days but it is about 1 – 3 days longer for stallion foals than fillies.

**(b) *Donkeys:*** The donkeys bred first time at about 3 years of age (*age at puberty*). There is *no breeding season* and hence can bred throughout the year. The *gestation period* is about 160 days, *estrus cycle* is of 21 – 28 days, heat persists for 3 – 7 days and optimum *time to breed* is on the 3 $^{rd}$ day of heat. The females can be bred 6 – 8 weeks after parturition (*PPBI*). The milk composition of its milk is almost similar to mare's milk. The life span is about 17 years.

### 24.1.10 Pigs

Gilts are generally bred at about 8-9 month of age, farrow at about 13 months of age (*age at first farrowing*). The *estrus cycle* is of 21 days. The *ovulation* occurs late in the estrus, about 48 hours of the onset of estrus and the ovulation rate usually ranged between 10 – 20 ova. The fertilization occurs 6-10 hours after service. The *gestation period* is about 114 days. The normal litter size is 8 – 10 piglets.

## 24.2 CONVENTIONAL PRACTICES OF ANIMAL REPRODUCTION

The following conventional/traditional practices with some interventions are followed for breeding the female farm animals to optimize their reproductive efficiency:

**1. Heat detection in females:** The female animals (cows) in the beginning of heat (estrus) up to first 8 hrs show some behavioural symptoms like isolation from other animals, decrease in feed intake, reduction in milk, restlessness with frequent urination, bellowing and stretching movements of lumbosacral region, licking other animals, slight increase in body temperature, clear and thick mucus secretions from vulva, swelling of vulval lips and vaginal mucosa becomes slightly red.

The mid heat (standing heat) period starts after 8 hrs of onset of heat lasting for about 10 hrs during which the animal should be inseminated. During this mid heat, the animal becomes excited, show interest in other animals allowing them to mount on her as well as mount her-self on others, pronounced redness and swelling of vulva, vulva secretions become very thin, watery with streaks hanging from vulva down to the ground with less consistency of mucus. The behavioural symptoms start disappearing after 18 hours period of heat (late heat). The heat symptoms are less intense in buffaloes and that too for a shorter time for which the buffaloes are known for silent heat.

**2. Time of Insemination**: Timely insemination of female animals is very important aspect of animal reproduction to make the breeding programme successful. The ovulation occurs after 28 hours from onset of estrus allowing a capacitation time of 6 – 10 hrs prior to the time of ovulation and descend of ovum. The animals should be inseminated after about 12 hrs of onset of estrus. The animals observing in heat during morning should be inseminated in the evening and those observed in heat during evening should be inseminated in the next morning.

The animals must be watched closely between days 16 – 25 of the previous heat. Such animals which were not inseminated in last heat and do not show heat, should be got examined by a veterinarian.

**3. Method of insemination**: The natural mating should be strictly avoided because the bull available around the village for natural service is not of superior genetic merit and hence the A.I. should be done with frozen semen of good quality from bulls of high genetic merit.

**4. Pregnancy diagnosis:** The pregnancy diagnosis at the earliest is essential to improve the reproductive efficiency by detecting the repeat breeders and their treatment. Early pregnancy diagnosis can be done by traditional and biological/ biochemical methods. The traditional methods include visual observation, abdominal palpation and palpation of female genetalia through rectum, service records and return to heat. The rectal palpation is widely used in large animals. This method requires the removal of faeces/dung by hand from rectum, holding the cervix and uterine horns gently, palpation and feeling for the asymmetry of uterine horns and balloon like appearance of horns due to pregnancy. The traditional methods are not reliable and require expertise with their further limitation of not detecting the pregnancy before 45 days and also in small animals like goat and sheep.

Some biochemical methods are now being used which can detect the pregnancy by 21 days and also in small animals. The pregnancy diagnosis is done by level of progesterone hormone in milk or blood on 20 -24 days post breeding. The milk sample or blood sample is taken on 20 -24 days and tested for immunological assay of progesterone. The level of more than 7.3 ng/ml progesterone in milk and 1ng/ml in blood plasma of cows, buffaloes and goats is an indication of pregnancy.

**5. Post-partum reproductive management:** An optimum calving interval is essential which depends on post-partum breeding interval (PPBI). The animals exhibiting heat within 60 days post-partum normally are better but in some cases the heat is delayed due to certain reason primarily the poor management which results in loss in body weight due to inadequate feeding. The heat is also delayed due to parturition difficulties, retention of placenta and its manual removal or infection in genitalia due to any reason which may cause metritis, endometritis, pyometra. In some animals there is delayed involution of uterus and persistent corpus luteum which causes anoestrus. Therefore, the animals require proper care and management. Normally the females should be bred after 45 – 60 days post-partum for optimum breeding efficiency so as the calving interval is not too much long which adversely affect the lifetime calf and milk production and the overall profitability.

To optimize the calving interval the animals should exhibit heat within 45 days post-partum and it require proper management and feeding of animals so that the animal may not loose body weight. Care should also be taken that there is no genital infection. Thus, there is a need to check the genitalia for complete involution of uterus, examine the ovaries for the presence of corpus luteum activity and against genitalia infection. The uterus should be douched with 0.5 % sterile Lugols solution 10 – 20 ml (intra-uterine) and treat the animals against infection.

**6. Treatment of sub-fertile animals:** The reproductive performance is no less important than milk production in dairy animals and is the most important to influence the economics of dairy farming. The most important determinants of reproductive efficiency are the age at which the puberty is attained (age at puberty), the production starts (AFC), service period, calving interval which is a function of the PPBI and the number of services required for conception. The heifers become a liability to dairyman and dairy industry if do no attain puberty at proper age, do not conceive at proper time and do not have regular calving per year. The most important physio-pathological/reproductive problems of dairy animals are the delayed puberty and maturity, poor growth rate, low body weight gain, puberal anestrus, silent heat (anestrum), post-partum anoestrum and repeat breeding. These reproductive problems make the animal sub-fertile and lower the reproductive status of dairy animals.

The reproductive efficiency in terms of above indices can be increased by feeding balanced ration, providing optimum management for body weight gain and growth, and sexual health. In adult animals the reproductive efficiency can be increased by reducing the number of sub-fertile animals by providing better post-partum reproductive management. It is most essential that 65 – 75 % of the animals in the herd are conceived with first service with an average of 1.5 to 1.7 services per conception, should not have any reproductive problem, resulting a calving interval of 13 to 14 months.

In view of the above reproductive problems which result in sub-fertility, the following measures should be used:

(*i*) Examination of reproductive organs and their message per rectum for those showing delayed puberty/maturity as well as three applications of Lugol paint on cervix at 10 days interval,

(*ii*) Supplementation of mineral mixture to weak animals @ 25 gm daily for 15 – 30 days and vitamin A to those with smooth and inactive ovaries,

(*iii*) Induction of oestrus by some expert by the application of folligon (PMS) 1000-1500 IU intra muscular, lutalyse (PG) 25 mg intra muscular in anoestrus animals with functional ovaries and 5 mg in case of post-partum anoestus, and to induce ovarian activity/ovulation Gn RH @ 200 μg dose per animal,

(*iv*) Periodic deworming and ectoparasitic control and other health measure to make the heifers healthy,

(*v*) Timely heat detection and insemination.

## 24.3 REPRODUCTIVE BIOTECHNOLOGIES

The biotechniques in reproduction have made significant contribution in increasing the reproductive efficiency, global transport and multiplication of genetic material, and conservation of unique genetic material for use in future. These techniques are A.I., E.T., embryo production and preservation, semen and embryo sexing, embryo splitting and cloning.

**(1) *Artificial insemination (AI):*** The AI has played a valuable role to bring genetic change in animal populations. The specific advantages of A.I. are:

(*a*) Dissemination of superior germ plasm by widespread use of outstanding males. The AI and cryopreservation has enabled the use of a single bull in several herds at different places of different countries and continents for breeding about one lakh cows in a year.

(*b*) Progeny testing under different environments (different herds at different places) so as to increase the rate and efficiency of selection.

(*c*) Introduction of new genetic material (semen) by importing semen instead of live animals.

(*d*) Frozen semen can be used in future even after the death of sire.

(*e*) Reduction in spreading the sexually transmitted diseases viz. Brucellosis, Vibriosis, Trichomoniasis, Infectious Bovine Rhinotracheitis, Blue Tongue.

**(2) *Frozen semen technology*:** It is also a valuable technique whereby the semen is stored for longer time without loss of viability for use in A.I. This may promote conservation of endangered genotype, though it requires cryopreservation facilities.

**(3) *Embryo transfer technology (ETT):*** The ETT has proved a powerful technology for genetic improvement of farm animals. This involves the collection of fertilized egg called embryo from donor female and transfer to the uterus of recipient female which foster the transferred embryo. It is a composite technology involving selection of donor and recipients, estrus synchronization, super ovulation, A.I., embryo recovery and evaluation, cryopreservation, and embryo transfer and sometime micro manipulation of embryos. These stages in embryo transfer have their importance in successful transfer of embryo. This technology improves the fertility of females and brings faster genetic change.

The technique of ET is as old as it was first time done in 1890 in rabbit by Walter Heape. This technique was then used in goat (Warwick and Berry, 1932), rat (Nichols, 1933), sheep (Warwick etal, 1933) and mouse (Fekete and Little, 1942). However, in bovines, the transfer of embryo was reported in 1949 by Umbaugh but first bovine calf was reported by Willet etal (1951). The first successful E.T. in human was done by Steptoe and Edwards (1978).

SOET and MOET: The ET can be done through single ovulation (SOET) or through multiple ovulations (MOET). Single naturally ovulate egg or embryo from elite females is harvested and transferred to another recipient female in case of SOET. The donor cows come in heat again after about 21 days of oestrus cycle and ovulates the egg to become pregnant. Thus, a superior embryo can be harvested every month from elite female without any hormonal treatment and hence with minimum input. In case of MOET, more numbers of embryos (Multiple embryos) from elite female are harvested. The MOET can be implemented by the procedure given here.

(*i*) *Selection of donors and recipients*: The donors are selected either on the basis of genetic superiority (milk production etc) or for conservation of rare germplasm. The recipient is selected on the basis of its reproductive organs capable of receiving, nourishing it to full term and delivering a healthy calf with ease at calving.

(*ii*) *Estrus Synchronization*: It is used to increase reproductive efficiency, to breed a large number of animals in short time, to schedule the calving at required time suited for better milk production in season of sufficient supply of green fodder. The synchronization can be done by reducing the length of luteal phase of oestrus cycle with single injection of 25 mg prostaglandin $F_2 \alpha$ lutalyse during luteal phase (day 8 to 12) of oestrus cycle and or its analogues like carboprost tramethamine, or extending the length with exogenous progesterone through intra vaginal sponges. Injection of 2 doses of $PGF_2 \alpha$ at 11 days interval may be given at any stage of oestrus cycle.

*(iii) Super ovulation*: The donors selected are those which are genetically superior whereas the recipients are those having low production. The donors are super ovulated by injecting gonadotrophins mainly FSH (follicle stimulating hormone) between days 8 and 12 of oestrus cycle. The gonadotrophins stimulate growth of multiple follicles and hence induce multiple ovulations. FSH is injected for 4-5 days daily twice. The super ovulation can be induced by pregnant mare's serum gonadotrophin.

*(iv) A.I. of donors*: The donors are inseminated during super oestrus with semen from fertile and high genetic merit bulls.

(*v*) *Embryo recovery/collection*: The embryos are recovered 5 to 7 days after oestrus. The uterine horns of the super ovulated donors are flushed using phosphate buffer saline solution containing antibiotics and serum using a 2way Rusch catheter. The outflow is passed through an embryo filter.

(*vi*) *Embryo evaluation:* This is also important step of ETT. The embryo is spherical composed of cells (blastomers – 16 cell stage) surrounded by a gelatin-like shell. The cell numbers increase geometrically (1-2-4-8-16-etc) during embryonic development. The 32-cell stage is called morula. Good quality embryos are ether transferred or cryopreserved if recipients are not available.

*(vii) Embryo cryopreservation:* The embryos are preserved by cooling at $5^0$ C or in liquid nitrogen.

(viii) *Embryo transfer*: The embryos are transferred to the recipients. This is done either surgically or non-surgically similar to A.I. The successful pregnancy (implantation) rate is about 50%.

***Advantages of E.T.:*** The ETT has following advantages:

- Rapid multiplication of rare genotype,
- Low cost in transporting embryos instead of living animals which also require quarantine,
- Rapid improvement of livestock to produce elite herd,
- Twinning rate can be increased,
- Prenatal sex determination,
- Control on sexually transmitted diseases,

* Animal conservation programme and embryo bank.

***(4) In vitro Fertilized (IVF) Embryos:*** The embryos can also be produced *in vitro* (laboratory). This involves the *in vitro* maturation (IVM) of oocytes, followed by *in vitro* fertilization (IVF) with the use of *in vitro* capacitated spermatozoa. The newly formed zygotes are cultured in suitable media to develop up to the transferable stage. This is called the *in vitro* produced embryos (IVP). The *in vitro* production (IVP) of embryos technology requires the following steps:

(*a*) Collection of immature oocytes - Following a local anesthesia, an ultra sound probe is inserted into the vagina to puncture the follicle > 2 mm diameter using a transvaginal curvilinear transducer. This is known as ovum pick-up (OPU)

(*b*) Maturation of oocytes (IVM) - in culture medium preferably TCM 199

(*c*) Sperm capacitation - by use of Heparin

(*d*) *in vitro* Fertilization (IVF) - in fertilization medium

(*e*) Culture of embryos - In culture media

(*f*) Embryo sexing, freezing and transfer to the uterus

The IVP technology has the several advantages over the conventional ET. Two techniques can be compared as under:

| OPU + IVF = IVP embryos | Embryo transfer (ET) |
|---|---|
| Donors need no hormonal stimulation | Required |
| IVP embryos can be obtained from | Nearly 30% animals do not produce viable embryos |
| Prepuberal donors can be used | Not |
| Infertile animals can be used for OPU | Not |
| No information on reproductive status of donor is required | Require |
| Pregnant donors up to third month of pregnancy can be used for OPU | Only non-pregnant donors are used |
| Aspiration of follicles can be done twice a week over a long period | Flushing can be repeated after 2 months |
| Once flushing gave about 14 transferable embryos or 6 pregnancies | Only 5 transferable embryos with 3 pregnancies |
| Oocytes of same donor can be fertilized individually with semen of any bull | Only one sire is used on a donor at one time |

The IVP embryos technology thus allows the repeated collection of oocytes for bulk production of high quality embryos from livestock of high economic value or unique genotype or endangered genotype so as to propagate such genotype at a much faster rate.

**(5) *Embryo cryopreservation***: Like freezing of semen, the freezing of embryos is possible and is a practice in ET programme. It is useful as a conservation programme for endangered or unique genotype so as to propagate such genotype at a faster rate.

***(6) Sexing of sperms or embryos:*** It is most desirable technology of reproduction for want of more number of female offspring in dairy farming. In cattle, the X – chromosome bearing sperms contain about 3.8 % more DNA than Y – chromosome bearing sperm. The sexing of semen and embryos can be done to produce the progeny of desired sex. The sex of the progeny can be determined with 90 % accuracy. The semen sexing (separation of X and Y chromosome bearing sperm) can be done with the help of a cell sorter/ cell flow cytometer. The PCR technique is used to amplify Y – chromosome specific DNA sequences for sexing of embryos.

***(7) Embryo splitting:*** The embryo splitting is a beneficial technique in which the embryo is divided into halves to have two embryos or into one quarter to get 4 embryos or sometimes to get 16 embryos from single embryo. The aim of embryo splitting is to obtain greater number of offspring from a given donor. The halving or quartering of embryos increases the intensities of selection for females nearly to that of males. The identical twins can be thus produced which will facilitate sire testing, reduction in insemination cost, testing of maternal traits, to study the genotype-environment interactions and to have information on carcass traits of other twin in meat animals by sacrificing one twin. However, the survival rate of split embryos is less.

## 24.4 E.T. WORK IN INDIA

The E.T.T. has been adopted successfully in cattle by many institutions in India. APAU Tirupati had produced first E.T. calf (cattle) in 1986 in the country. The National Institute of Immunology (NII) also had ET calf (cattle) in January 1987 and Sabarmati Ashram Gaushala, Bidaj, in December 1987 produced first buffalo calf through E.T.

Govt. of India initiated in 1987 a National Service and Technology Project on E.T.T. to be operated through Deptt. of Biotechnology with NDDB as the lead implementing agency at Sabarmati Ashram Gaushala, Bidaj as the main E.T. laboratory. This project had also four collaborating centers/agencies viz. IVRI Izatnagar, NDRI kamal, NIl New Delhi and Central Frozen Semen Production and Training Institute Hessarghatta. It had also four required ET centers viz.

(*i*) Sree Nasik Panchwati Pinjarapole, Nasik, with donor animals of pure Gir, HF and crossbreds of HF and Jersey;

(*ii*) Buffalo Breeding Centre, Nekorikallu (Guntur, AP) exclusively to work on buffaloes;

(*iii*) Animal Breeding centre, Salon (Rai Bareili, UP) to produce HF and its crossbreds;

(*iv*) Central Frozen Semen Production and Training Institute, Hessarghatha (Karnataka) to work on Hallikar, Amritmahal and HF crossbreds.

❑❑❑

# 25 Biotechniques In Animal Nutrition

The only factor responsible for the continuous functional behavior of an animal is the regular supply of food. The food gives energy to the body to carry out the vital processes such as digestion, metabolism, respiration, blood circulation, etc. The food is also responsible for the building and repairing of various body tissues and ultimately growth of the young ones and also for the work and production of an animal. The food is made up of some constituents known as nutrients because these are essentially required for the nourishment and sustenance of living bodies. The nutrients are supplied to the living bodies through food and hence called the food nutrients. All the food nutrients are present in the soil. The plants are capable to take up these nutrients from the soil and convert them in the presence of solar energy in vegetative matter and seeds which are the food materials with stored energy for consumption by animals and humans. Thus, the plants build up and conserve the food nutrients (constituents of food) during their growth and maturity and are the primary food sources for animals. The animals by eating these plants utilize the food nutrients and convert/transform them (stored energy of plant materials) through various metabolic processes in to animal protein and fat.

## 25.1 MAJOR AND MICRO NUTRIENTS

The food nutrients either provide energy or aid in the release of energy from other nutrients. The nutrients depending on their quantities present in the body and required by the living body are known as major and minor (micro) nutrients.

**(1) *Major nutrients*** are carbohydrates, proteins and lipids. These nutrients provide energy to the body and form part of tissue structure. The water is also essentially required for structural and metabolic functions and playing its role as a solvent and a medium for biological reactions.

***(i) Carbohydrates:*** The carbohydrates are main source of energy providing nutrients to the body system through their metabolism. The surplus carbohydrates are converted and stored by the animal body in the form of fat. The carbohydrates are classified as:

*Mono saccharides* or *simple sugars:* These are classed depending on the number of carbon atoms present in the basic units of single molecule like glyceraldehydes (triose); erythrose, threose, erythrulose (tetrose); xylose, ribose, arabinose, xylulose, ribulose (pentose), glucose, galactose, mannose, fructose, sorbose (hexoses).

*Disaccharides:* These contain two molecules of monosaccharides as basic units like maltose (2 glucose units), sucrose (fructose + glucose), and lactose (glucose + galactose);

*Polysaccharides:* These are polymers of simple sugars like starch, cellulose, dextrin, glycogen, insulin, pentogens made up of pentose units (xylan), hemicelluloses (pentogens + hexosans), and lignin (polymer of some complex alcoholic compounds).

***(ii) Proteins:*** These are polymers of amino acids. Proteins differ in their amino acid make up. The proteins are the major part of the protoplasmic structure of plant and animal tissues. The proteins are essential for growth and for repairing of the wear and tear of tissues. They are also responsible for the formation of muscular tissues, skin, hair, wool, feathers, hoofs, horns and blood cells. The feeding of excess protein serves as source of energy in the body.

The protein requirement of the body is taken in terms of amino acid requirement. Some amino acids are synthesized by the body but those that are not synthesized at all or in required quantity are called essential amino acids. The amino acid requirements of ruminants are met by microbial protein synthesis in the rumen, though young pre-ruminant calves require all the essential amino acids through diet. The proteins are classified as:

*Simple proteins:* These yield amino acids on hydrolysis like albumins, globulins, glutelins, prolamine, protamines, histones and globins which are soluble, and keratins, collagens and elastins which are insoluble;

*Conjugated proteins:* These contain non-protein group (prosthetic group) like lipoproteins, chromoproteins, phosphoproteins, glycoproteins and nucleoproteins; and

*Derived proteins:* These include various products of protein degradation like proteans, metaproteins, proteoses, peptones and peptides.

(iii) ***Lipids:*** These are the esters of fatty acids which have high levels of energy because they have predominance of hydrocarbons in their structure. Fatty acids are the essential components of lipids. The lipids are classified as:

*Simple lipids*, viz. fats and oils (which are the esters of fatty acids with glycerol), and the waxes (which are esters of fatty acids other than glycerol);

*Compound lipids* (containing additional groups along with fatty acid esters) viz. phospholipids (lecithin, caphalin), sphingolipids, cerebrosides, gangliosides; and

*Derived lipids,* obtained on hydrolysis of lipids. These include fatty acids, alcohols, pigments, sterols etc. The sterols are solid alcohols of high mol. wt. like cholesterol which is synthesized in liver.

The fats are concentrated form of energy and provide 2.25 times more heat and energy compared to carbohydrates.

**(2) *Micro-nutrients:*** The vitamins and minerals are the micro-nutrients required in minute amounts. These are equally important and required for animal body as the major nutrients are required.

(*i*) ***Vitamins:*** These are organic compounds required in assisting the metabolic processes and formation of tissue structure, though they do not give energy and also not the part of tissue structure. The vitamins are classed as:

Fat soluble vitamins: Vitamin A, $D_2$, $D_3$, E, K

Water soluble vitamins: Vitamin $B_1$, $B_2$, $B_6$, $B_{12}$, C (ascorbic acid), Niacin, Biotin, Folic acid, Pantothenic acid, Choline

(*ii*) ***Mineral elements:*** These are also required by animal body and these can be classed as major and minor elements.

*(a) Major or macro elements* are required in relatively large amount and used mostly in the synthesis of structural tissues. The major ones are bone forming elements (calcium, phosphorus, magnesium), interrelated minerals of body fluids involved in water and electrolyte balance (sodium, potassium, chlorine) and component of protein, thiamine and biotin (sulphur).

*(b) Micro elements* are required in trace amounts. These usually function as activators or as a component part of enzyme system and have functional role to play in animal physiology to regulate body processes. Among these minor elements, some are blood forming elements (iron, copper, and cobalt) and others are iodine, manganese, zinc, molybdenum and selenium. In addition to these, some other elements like chromium, fluorine, nickel and vanadium may be important. Proper mineral balance is required even for regular heart beating.

## 25.2 FEED RESOURCES

The roughages and concentrates constitute livestock feeds. The roughages may be in green or in dry form, called as green roughages and dry roughages, respectively. The roughages have coarse texture and high level of crude fibre for which called as roughages. The concentrates provides nutrients and categorized as energy or protein concentrates. The different feed resources in the country vary in different agro-climatic zones. At national level, the feed resources are classed as greens, crop residues and concentrates.

**(*i*) *Green roughages:*** These are the cultivated fodder crops (leguminous and non-leguminous), perennial and seasonal grasses, forages from grasslands (grazing/forest area comprising fallow lands, grazing area of permanent pastures, cultivable wastelands and edible weeds) and fodder leaves from trees.

India has a total geographical area of about 328.8 million ha with a net cultivable of 193 m ha (46.5 %), out of which net irrigated land cover of only 60 m ha (53.5%) and only 4.8 % of the total cultivable land is used for growing feed and fodder production. Due to the limited use of land (4.5 % of total land) for fodder growing, the green fodder availability to animals is inadequate.

The *green fodder crops* are harvested before flowering and seed setting so as these are remained as succulent, palatable and with low lignin content. The non-leguminous fodder crops are sorghum, bajra, ragi (millet crop), oats, barley, maize, green mustard whereas berseem, lucerne, pea, cowpea, beans, pulses, lentils are leguminous fodder crops which are rich in protein and minerals like calcium and phosphorus.

The *grasses* includes Napier grass, Guinea grass, Para grass, Rhodes grass, Sudan grass, Spear grass, Dub grass, and Anjan grass. The clovers are the cultivable leguminous grasses comprising white clover, red clover, crimson clover and alike clover and belong to Trifolium species including Berseem. Lucerne is also most important grass giving several cuttings.

The permanent *pastures and grasslands* are so called waste lands in the country (India) accounts for only 3.5 % of the total geographical area. The grasslands have several kinds of species of grasses including weeds grazed by the animals

*Tree fodders* are mainly used in hilly areas and arid zones. Different fodder trees are found either naturally or planted depending on the topography, type of soil and climatic conditions. The camels and goats also consume Neem leaves.

The *area under forest* is about 22.6 % of the total area in the country. The forests are important habitat of the tribal and ethnic population of the country for their livelihood by rearing livestock. However, these forests are no more available for free grazing for need of conserving the forest to protect the environmental deterioration and maintain ecological balance.

***(ii) Crop residues:*** These are the major feed resource (50 % or more) of fodder (dry fodder) to Indian cattle. The cereal crop residues are low cost feeds for ruminants supporting a major share and sustainability to the livestock farming system but may not maximize the output (milk). These are usually the cereal plants after grain removal and constitute to about 60 % of the total dry matter produced from green fodder, pasture and concentrates. These include coarse straw/fodder from coarse grains, sorghum and pearl millet straw, fine straw (wheat and paddy straws), straws from legumes and pulses, and sugarcane tops.

***(iii) Concentrates:*** These are feed ingredients which include the followings -

(*a*) Cereal grains (sorghum, oat, barley, maize, millets);

(*b*) Cereal byproducts like brans (wheat bran, rice bran and rice polish) and chunnies of pulses; and molasses,

(*c*) Oil cakes (mustard cake, GNC, cotton seeds/cake, soybean meal/cake).

The cereal grains are rich in starch to the extent of about 85 percent and about 10 % protein. The grains of coarse cereals (Jawar, bajra, maize and small millets) form major share (about 90 %) among the cereal grains used to feed livestock in our country and only 10 % constitute the fine cereals of wheat and rice obtained from godowns of FCI after they are spoiled being affected seriously by mycotoxins. The grains spared for livestock are mostly fed ony to two classes of animal viz. high producing dairy animals (Cows and buffaloes) and poultry, commercial layer and broiler chicken.

Further, the concentrates are agro-industrial byproducts like bran, chunnies and polices with little quantity of oil cakes or meal. The maize is mostly used for feeding to dairy animals, pigs and poultry whereas the horses are given the oats and barley. The rice polish is fed to poultry but the brans form the livestock feed. The pigs are also fed the potato, and sugarbeet.

The molasses is either not fed to the animals or it is not available because mostly it is used in large quantities for alcohol industry.

The oil cakes and meals are rich in protein. The mustard cake, cotton seed cake and GNC are fed to dairy cattle. The soybean meal is mostly preferred for poultry feed as a protein source. The cotton seed has gossypol which is anti-nutritional factor but destroyed by soaking or heat treatment. The oil cakes are in short supply due to their low production because the main emphasis is given on grain production for human consumption and secondly these oil seeds are exported to rich countries as a requirement to earn foreign exchange for importing goods required to meets other needs. This leaves no choice except to import the goods available in the country and acceptable to other countries. .

## 25.3 ENERGY REQUIREMENT

The animal body requires food nutrients to meet out the maintenance and production requirements. The process of digestion initiates as soon as the food enters the buccal cavity. The food is ingested and the digestive enzymes act on food in alimentary tract and associated secretary organs (salivary glands and numerous small glands). This results in to breakdown of nutrients in to simple units for their absorption. The undigested waste material is eliminated.

The animal body gets energy from nutrients, called free energy. The energy is utilized as well as released by two different types of reactions in the body regarding metabolism. These are the anabolic and catabolic reactions. The anabolic reactions are synthetic and utilize the energy (endergonic reactions) while the catabolic reactions are the breakdown processes (exergonic reactions) and release energy. The free energy released from exergonic reactions is trapped by certain intermediate high energy compounds like Adinosine triphosphate (ATP) which has the role during metabolic reactions and energy transformation in the body. The ATP is formed when energy is released and it is a measure of energy efficiency of the nutrients. There are similar high energy phosphate compounds like creatine phosphate present in skeletal muscles playing the role in transfer of high energy phosphate bonds and used as energy reserve in the muscular tissue.

The nutrient catabolism in the animal body occurs in three stages. The first stage is the break down of the major nutrients viz. carbohydrates, proteins and lipids in to their basic building stones of monosugars, amino acids and fatty acids, respectively and this occurs during digestive process in the alimentary canal. Subsequently, these fundamental units are further broken down in to much simpler common metabolites during preparatory stage of intermediary metabolism. The final stage is the breakdown of simple metabolites in to carbon dioxide, water and excretory products along with release of energy. The anabolic processes are performed resulting in reversal of the above three stages by forming the macromolecules, major nutrients and cellular constituents by utilizing the energy.

***Maintenance vs production requirements:***

The ***maintenance requirements*** of animals are of two types on the basis of the basic life processes and active factor. The *basic life processes* include respiration, circulation, maintenance of muscular tones and internal secretions, etc and need to continue life even when the animal does not receive any food. The active factors are related to carry some activities to remain active like the activities related to standing up, to lay down, eating and digesting food including its assimilation, etc (maintenance).

(*i*) The energy to carry out the *basic life processes* (respiration, circulation, etc) in the absence of food (fasting stage) comes from within the body. This energy is measured in a calorimeter or indirectly by indirect calorimetry as the energy released by a fasting animal. The fasting animal should be in a good health, post-absorptive state (so as no heat production is increased from the body), kept in a neutral environment of no stress of either heat or cold (to influence the minimum metabolic rate), and the animal is fully relaxed to have minimum muscular activity. The minimum fasting catabolism measured in this way is called *basal metabolic rate (BMR)*.

In addition to the energy, the maintenance of basic life processes also requires a minimum essential nitrogen catabolism and it is measured as minimum urinary excretion of nitrogen keeping the animal on nitrogen free, energy adequate diet. This is known as endogenous urinary nitrogen (EUN) expressed in mg per day.

(*ii*) The BMR plus the energy required to meet out the *active factors* (standing, laying, eating and digesting food) constitute the maintenance requirement of energy which keep the animal in energy equilibrium. The maintenance requirement for protein should also include the losses of nitrogen due to EUN and metabolic faecal nitrogen (MFN) plus the protein required for growth and renewal of epidermal tissues giving rise to nails, hairs, feathers (birds), etc. The active factor does not increase the protein requirement. The protein requirement for maintenance is estimated as the minimum protein required in maintaining nitrogen equilibrium, based on feeding trials.

The ***production requirements*** are needed to carry out some production functions like growth, reproduction, production (milk, meat, egg, wool, and draft work).

## 25.4 FEEDING REQUIREMENTS OF DAIRY ANIMALS

The nutritional requirements are met by roughages and concentrates on the basis of DCP and TDN.

*(i) Maintenance ration:* The daily maintenance requirement of DCP and TDN are 56 gm and 885 gm per 100 kg body weight, respectively. The production requirements of DCP and TDN in addition to maintenance requirements are

48 gm and 275 gm, respectively for the production of every I kg milk of 3 % fat. But if the fat percentage in milk is above 3 %, there should be an addition of 6 gm DCP and 50 gm TDN for every 1 % increase in fat percent.

In addition to maintenance and milk production requirements, a cow in the advanced stage of pregnancy (2 months before calving) requires an extra allowance of about 50 and 25 % of the maintenance DCP and TDN, respectively.

The nutritional requirement of working animals (Bullocks) depends upon the amount of work performed. For light work of 4-6 hours, the daily requirement of DCP and TDN are 115 gm and 1 kg, respectively per 100 kg body weight of the animal. The corresponding values for performing heavy work of 6-8 hours are 140 gm and 1.30 kg, respectively. The breeding bulls require 90 gm DCP and 900 gm TDN daily per 100 kg body weight.

The amount of dry matter (DM) is calculated @ 2 – 2.5 % of body weight of the cow. The bulls, bullocks and buffaloes need up to 3 % DM of body weight. About 2/3 quantity of total DM required is supplied from roughages and the rest (1/3) from concentrate.

To increase the productivity, it is essential to make provision of green fodder up to 70 % of total roughages. For the inclusion of 70 % green fodder of the total roughages, 50 % of the 2/3 quantity of total DM to be supplied from roughages should be supplied as green fodder and other 50 % as dry fodder. In other words, the total DM requirement should be met by dry fodder, green fodder and concentrate mixture in the ratio of 1:1:1, respectively.

The maintenance requirement is mainly met by roughages. A cow having body weight of 400 kg requires about 5 kg dry fodder, 15 kg green fodder and 1 kg concentrate for its maintenance requirement. The concentrate requirement for maintenance can be minimized to zero if the green fodder is berseem or lucerne.

*(ii) Production ration:* An average size dairy cow requires 1 kg concentrate mixture for every 2.5 kg milk production in addition to its maintenance requirement. But no concentrate is required for first 2.5 kg milk production if the cow is fed adequate quantity of leguminous fodder like berseem. The cow in advance stage of pregnancy should get an additional concentrate allowance of about 1 kg daily. The bullocks during heavy work and breeding bulls during the period of giving service require 2 kg concentrate mixture in addition to their maintenance requirement.

The concentrate mixture should contain 2 parts of grain, 1 part oil cake and 1 part bran or husk. Adequate quantity of common salt (1-2 ounces) should also be added in concentrate mixture. The feeding of mineral mixture (Super-mindif) @ 1 ounce daily with concentrate mixture to growing stock, and milch cows is also essential.

## 25.5 FEED AND FODDER REQUIREMENT

The existing dairy production systems are the crop-livestock integrated farming system, dairying by landless, nomadic system by the tribal and migratory communities, peri-urban system and intensive and scientific dairy system. The present system of production is extensive, crop production dominant, low input, crop residue and low productivity based but sustainable to landless, small and marginal farming community. The crop production draws maximum attention and resources because the livestock production is only an adjunct. The dairy production system needs extensive and unwieldy (bulky) infra-structural back up. The crop-livestock integrated production system may be suitable in rural situation but the intensive system of dairy farming needs green fodder and high level of concentrate feeding with judicious use of agro-byproducts such as oil meals and cereal byproducts (brans and husks) and also a need of feed industry to supply feed of about 120 million tones every year against the supply of mere 8 million tones. The intensive fodder cultivation to meet the annual green fodder requirement of over 3000 million tones fodder requires 20 % of the cultivable land of the country against 4.8 % land available. The annual production of coarse grains of about 30 million tones in the country needs to be diverted to the animal feed industry out of the old stocks that undergo storage deterioration and damage due to rats, insects and fungal contamination.

Besides the feed cost, the availability of green fodder and concentrates as per requirement and their quality are the main limiting factors of animal production. There is a gap between requirement and availability of livestock feeds in India. It has been estimated recently that the requirement of the different feed ingredients viz. green fodder, dry fodder and concentrates had been 648, 503 and 87 million metric tons (MMT) and their availability was being 491, 459 and 62 MMT, respectively. The estimates have indicated the feed deficiency of green fodder, dry fodder and concentrate by 23, 31 and 47 percent, respectively.

Not only the deficient quantity of these feed ingredients but the quality in terms of DCP and TDN of feeds is also poor. The requirement of DCP and TDN is 36 and 351 MMT but the availability is only 18 and 257 MMT amounting to a deficiency of 45 to 50 percent DCP and 35 to 40 percent TDN. The dry matter availability is deficient by 35 % from green fodder, 11 % from dry fodder and 28 % from concentrates.

In view of these, there is need of the day for substantial animal production to enhance the quantity as well as the quality of feeds and fodders. Attempts have been made in this direction and these are given here –

## 25.6 INCREASE IN FORAGE PRODUCTION

The available fodder can meet the demand of only 46.7 % of livestock. In view of the shortage of green fodder, there is need for growing nutritive green fodder to increase the production and reproduction of animals. The area under cultivable fodder crops is very less, about 4.5 % of total cultivable land. However, in some of the states like Punjab, Haryana, Western U.P., Gujarat and Rajasthan, the area under fodder production is about 10 % or even more. There is no scope for horizontal increase in this area due to a rider of population growth and increasing demand of food grains as well as cash crops. In order to increase the green fodder per unit area per unit time and to maintain regular supply of green fodder round the year, the efforts have been made in the following directions:

*(a) Production of quality seeds of fodder crops:* There are needed the programmes for production of fodder seeds of high yielding fodder varieties, drought resistant fodder crops and varieties of fodder crops of better nutritive value. The AICRP on fodder has developed some varieties of such fodder crops.

The Central Fodder Development Organization was founded in the IX plan by the Department of A.H. and Dairying to establish fodder calendar, introduction of fodder crops and production of seeds of fodder crops. Under the scheme, seven regional stations for forage production and demonstration at Kalyani (W.B.), Hissar (Haryana), Gandhinagar (Gujarat), Suratgarh (Rajasthan), Sahema (J & K.), Alamadhi (T.N.) and Hyderabad (A.P.) and a Central Fodder Seed Production Farm at Hessarghata (Bangalore) were established and assistance to state govt. for establishment of fodder bank to preserve fodder for use during drought and security period.

*(b) Crop rotations:* The AICRP on Forage crops has suggested suitable crop rotations for different agro-climatic zones to supply green fodder round the year.

*(c) Feeding of fodder tree leaves:* The leaves of most of the fodder trees have fairly high nutritive value in terms of DCP and TDN. The tree leaves and bushes also meet the calcium and vitamin A requirement of animals. Thus, different fodder trees and bushes can be grown depending on the soil type, topography and climatic conditions. Some of the important fodder species which can be grown in arid zones are: *Subabul (legume tree, leaves contain 16 % DCP and 70 % TDN on DM basis), Khejri (14 % DCP but high tannin), Ardu (19 % DCP), Vilayat Kikar (18 % DCP), Hedge Lucerne (22 % DCP), Augasthi (20 % DCP) and Shevri (24 % DCP).*

The fodder trees for hills are Arjun, Kachnar, Biul, Mulbery, Dudhilo, Nabaro, Gogun, and Rubinia.

However, the precaution should be taken not to feed the tree leaves more than 40-50 % of their total dry matter consumption of the animals and rest of dry matter may be met through grasses or straws supplemented with concentrate. This is because exclusively feeding the tree leaves to the animals result in symptoms of constipation. Moreover, many fodder trees have relatively high tannin content which reduce the protein availability to the animals

*(d) Fodder preservation for scarcity period:* The surplus fodder can be preserved in the form of silage and hay depending on the fodder crop and climatic conditions.

The silage is the preservation of green fodder in succulent form in air tight structures. The cereal fodder crops like *jowar, maize, oats* etc are more suited for silage making at flowering to milk stage.

The hay is a fodder crop harvested before it is totally ripe and dried. The fodder crops with thin stem and leafy, sufficiently nutritious and palatable after drying are good for making hay. Such crops are *Berseem, Lucerne, Oats, Anjan grass.*

*(e) Feeding of industrial byproducts*: The research conducted under AICRP on Agro-industry byproducts had pointed out that the Agro-industrial byproducts can be fed to growing and milch animals after their mixing in concentrates. The important byproducts are apple pomace, tomato pomace, babul pods and seeds,

cauliflower leaves, empty pea pods, groundnut hull, sal seed meal, mango seed kernel, watermelon seed cake, tapioca leaf meal and pulp, molasses, distillery waste, etc. These products are good source of DCP and TDN.

## 25.7 QUALITY OF FEEDS AND FODDERS

There is scarcity of green fodders and this scarcity is region specific and also varied among seasons. Under the green fodder scarcity conditions, the animals are fed only the coarse roughages (straws and other crop residues). The nutritive value of the crop residues/straws is very low because these are deficient in nitrogen, mineral and vitamin contents but have silica and lignin contents which results in their poor digestibility. The paddy straw is considered to be inferior compared to wheat straw and hence the farmers in Punjab and Haryana burnt the paddy straw in the field rather than feeding it to their animals.

The crop residues/ straws are of inferior quality feeds. The information available showed that wheat straw (on % DM basis) contains on an average, 0.25 to 1.0 % N, about 50 % cellulose, 30 % hemicellulose, 10 % lignin, with voluntary intake (Kg DM/100kg body wt) of about 1.0 and *in-vitro* digestibility of 20 to 55 %. These crop residues having low nitrogen but high fibre contents have low palatability, low intake by the animal and low digestibility. These attributes of crop residues (high fibre contents, low palatability and low digestibility) are attributed to the specific structure of their cell wall which has polysaccharides (cellulose, hemicellulose, and pectic polysaccharides), lignin, cutin and silica. There is lingo-cellulosic bond in cell wall of crop residues for which their energy is not liberated. High level of lignification as well as low content of nitrogen in straws is the barrier limiting the microbial fermentation/digestion of lignocellulonics. Thus, the key issue is to improve the quality and better use of crop residues for ruminants by cleaving this bondage to rumen microbial fermentation.

Moreover, the crop residues do not provide the required nutrients. Therefore, it needs supplementation of crop residues with green fodder and or concentrate as well to develop and use some nutritional technologies either by genetic manipulation and or by feed processing.

## 25.8 FEED PROCESSING TECHNOLOGY: INCREASE OF FEED QUALITY

Some efforts have been made in this direction by the nutritionists and some of the nutritional technologies (***feed processing technology***) have emerged to improve the nutritive value of crop residues and to provide better quality ration to the animals.

The various feed processing technologies are the physical processing (*total mix ration*) and chemical processing (*alkali treatment*) of feeds and fodders, use of *feed supplements* (feeding of mineral mixture and vitamins to increase feed value, and feeding of enzymes for microbial degradation of crop residues), mixed feed technology (total mix ration and complete feed block), nutrient bypass technologies (protein bypass and fat bypass technology), and genetic manipulation of rumen micro-organisms. These feed technology innovations are used with the aim to increase the palatability, voluntary intake and digestibility of low quality crop residues and to increase their feed value.

### 25.8.1 Physical processing

This includes chaffing (chopping), grinding and mixing, washing/ soaking and boiling or heat treatment.

The green fodders and crop residues are *chaffed* or chopped in chaff cutter to reduce their wastage and better consumption of chaffed fodder by the animal. The DM intake is improved by chaffing the fodder but it has no effect on digestibility. The chaffing practice is common in north India. The sugarcane tops which contain about % % Crude protein and about 50 % TDN is fed to the animals after its chopping.

The *grinding* of concentrate ingredients is a practice for the cattle, pig and poultry feeds for the purpose of thorough mixing in proper ratio, to supply in convenient packing, to prevent wastage and to ensure uniform consumption. The grinding decreases the particle size, increases surface area and bulk density for which the rumen microbial accessibility is increased and hence the feed intake is also increased. The grinding increases the intake of low quality more compared to high quality residues as well as by the small and young animals compared to older and large animals.

The grinding and *mixing* of roughages and concentrate ingredients as well as feed supplements are the feed processing technology. In northern India this practice is followed by the farmers in the form of *"sani making"* and it can be advocated in other parts of the country.

The *washing of straws* as such is practically not feasible because it is laborious and chances of spoiling if wet fodder is not fed immediately, though it softens the crop residue, removes dust and waste ingredients (oxalates, silica and excess potassium). The soaking of chaffed wheat straw (Bhusa) is preferred and it improves the palatability, feed intake and digestibility. The animal can chew easily the hard nodes of soaked straw otherwise left unconsumed. The other palatable feed ingredients (cakes and grain byproducts) can be mixed in the soaked wheat bhusa so as the animal can consume whole mass of mixed feed.

The *soaking and boiling* is condition specific. The soaking or heat treatment (boiling) of cottonseed is useful to destroy the *gossypol* in this feed, to destroy trypsin inhibitors in soybean and cyanogenic glycosides in linseed and tapioca. The soaking of concentrate also improves the palatability, avoid selective eating of ingredients by the animal and make better consumption.

### 25.8.2 Chemical Processing

During the last century, research has been conducted to improve the digestibility and nutritive value of crop residues by chemical treatments. The different chemicals used are alkali (sodium hydroxide, potassium hydroxide, calcium hydroxide and oxide, ammonium hydroxide, ammonia and urea). The crop residues are treated by these chemicals with 2 to 5 % solution with 40 – 60 % moisture level. After chemical treatment the crop residues are stored/ stacked for about one month under polythene cover.

The *alkali treatment* to crop residues alters the characteristic of specific structure of cell wall of crop residues (bondage between hemicelluloses and lignin) to saponify the ester bonds of carbohydrate with lignin. The chemical treatment is thus effective to cleave/ break the bond. As a result of alkali treatment, there is solubalization of hemicelluloses. The soluble cellulose and hemicellulose are more susceptible to microbial attack and highly digestible by micro-organisms of rumen. The increase in the digestion of cellulose and hemicellulose makes the structural carbohydrates in both lignified and unlignified plant tissues more digestible. This increases the palatability and digestibility of crop residues and hence the availability of energy to the animal is increased.

*(a) Sodium hydroxide treatment:* The NaOH treatment (3 - 5 % with moisture content of 20 - 30 % of DM) has some negative effects like load of chemicals on the kidney of animals to excrete the sodium ions in animal excreta, corrosive nature and cost of NaOH.

*(b) Ammonia treatment of straw:* The ammonia is a weak alkali. The straw may also be treated with 3 % liquid or anhydrous ammonia, by injecting it into straws in the stacks, and covering the stacks for about a month. This treatment increases the nutritive value of about 5 % DCP and 50 % TDN. It is more effective in a situation when the animals are not provided enough nitrogen to meet the nitrogen requirement of rumen micro-organisms for their proper growth and function. The ammoniation is effective to break the lingo-cellulose bond, to improve digestibility, to add NPN to treated residues and hence improve the feeding value of crop residues. However, the gaseous ammonia handling is dangerous.

*(c) Urea treatment of straws:* The urea is more preferred as a source of nitrogen in comparison to sodium hydroxide and ammonia in view of their nature. The wheat or paddy straw is treated with 4 % urea solution. The urea solution is prepared, mixed by spraying with the straws and kept for about a month under polythene cover. The urea treated straw is palatable and it can be fed with concentrate and mineral mixture even to the milch animals. The urea treatment is feasible, cheap, easy in availability and handling and more effective to make the straws more palatable. This increases the voluntary intake by about 85 % and digestibility by 40 to 50 %, increases the crude protein content from 3 to 8 % and TDN content from 40 to 55 % after one month of treatment, and gives better response to milk production and growth rate in young animals.

*(d) Urea-molasses-mineral mixture treatment of straw:* The ruminants have the unique ability to synthesize enough protein for maintenance through microbial action and hence the urea can be used as a NPN source. The straws being the poor source of nitrogen and minerals can be treated with 2 % urea, 15 % molasses and 0.5 to 1 % mineral mixture to increase the feeding value. The urea is dissolved in water and then added the molasses and mineral mixture. This mixture is sprayed on the straw.

*(e) Animal urine treatment of wheat straw:* The animal urine contains urea/ammonia and hence the straw can be treated with it and kept for month before feeding to animals. This urine treatment of wheat straw increases the DCP from 0 % to 1.3 – 2.9 % and the TDN from 40 % to 46 %, besides increasing the calcium and phosphorus balances. Thus, the nutritive value of wheat straw is improved.

### 25.8.3 Use of feed supplements

The feed supplements include the feeding of mineral mixture and vitamins to increase feed value and enzymes for microbial degradation of crop residues and to make *complete feed block.* This technology makes use of the physical and chemical treatment of the various feed ingredients and includes the *total mix rations, complete feed block technology and feeding of mineral mixture and feed supplements* to the animals. This technology has gained popularity to make complete feed or pellet feed.

*(a) Feeding of mineral mixture and vitamins:* The crop residues and other feeds like agro-industrial byproducts and non-conventional feeds are deficient in mineral contents. The mineral deficiency in soil has been caused by intensive crop production adapted by the farmers. This deficiency not only affect the growth of plants but also causes the mineral deficiency in the diet of animals which ultimately affect the reproductive efficiency of animals by increasing the incidence of repeat breeding in dairy animals. The mineral supplementation to the animals has improved their breeding and production efficiency as per data available. The chelated mineral technology has given better results for better absorbability and bio availability of minerals.

*(b) Feed supplements/additives (Exogenous fibrolytic enzymes):* The lingo-cellulose bond is partially broken down by enzymes produced by microorganisms. The lignin is not accessible to bacteria but it is degraded by fungi due to oxidative process. The utilization of crop residues by ruminants can thus be improved through enzymes. These are exogenous fibrolytic enzymes (EFE) and are mainly cellulose, hemicellulase, and xylanase which are mainly of fungal origin and available in powder or liquid form at cheaper rate. The use of enzyme additives improves the animal performance as a result of improvement in rumen fibre digestion which in turn makes an improvement in digestible energy intake. The enzyme solution is sprayed on the crop residues before feeding. These are thus called the exogenous enzymes. These exogenous enzymes increase the hydrolytic capacity of the rumen, improves the feed intake, palatability and nutritive value of crop residues and result in significant increase in growth of calves..

The favourable effect of live *yeast culture* (Saccharomyces cerevisiae and Aspergillus oryzae) on rumen environment in terms of reducing the concentration of granular starch causing buffering action has been observed. This increases the fibre digestion particularly under high concentrate feeding rather than in the animals fed on high fibre-low concentrate diet.

The *feed additives* like ionophores including monensin and lasalocid have also been reported to increase the propionate production and to reduce the methane production in rumen and thereby result in increase of feed conversion efficiency.

### 25.8.4 Mixed feed technology

Based on physical processing (grinding, soaking), chemical processing, use of feed additives and mixing of all the feed ingredients have developed the mix ration technology and complete feed block technology as given below:

***(i) Total mixed ration:*** This involves the grinding of feeds and fodder by reducing the particle size of various feed ingredients (roughages and concentrates) and mixing them in proper proportions. The combined feed can be enriched/ supplemented with minerals, vitamins and feed additives. Such a feed mixture becomes the balance ration as per nutritional requirement of the specific category of animals. The practice of mix ration feeding is locally known as ***"Sani Making"*** and is well practiced by the farmers of Northern India. The farmers have their own chaff cutters (hand or power driven) for chopping of fodders. This practice of ***"Sani"*** should be advocated and farmers should be educated about the beneficial uses of this practice in areas (eastern and southern part of country) where it is not practiced. This practice of grinding and mixing makes the feed a homogeneous mixture so as to avoid the selective eating of specific ingredients of the feed mixture by the animal. This practice of grinding and homogenous mixing increases the feed intake, reduces the wastage because it becomes palatable. Therefore, the *mix ration technology* makes better the feed conversion efficiency as well as better performance of the animals for their growth and production.

***(ii) Complete feed block technology:*** This technology has formulated the compound or complete feeds for livestock and poultry. There are some limitations of urea molasses mix supplementation in liquid form. Therefore, it is better to solidify the mix into the form of a block known as *urea molasses mineral blocks* (UMMB) which is cost effective and easily manageable. The technology of formation of UMMB has thus been developed to supplement the diet of ruminants fed on poor quality roughages.

The UMMB, in addition to NPN, also supplies some intact protein, minerals, vitamins and energy. The UMMB with wheat straw improves the intake of dry matter and crude protein, and also the digestibility of DM, CP, cellulose and hemicelluloses. The practice of feeding the animals UMMB *ad libitum* has been found to be better converter of wheat straw into biomass resulting fairly good growth of calves. The wheat straw supplemented with UMMB is sufficient not only to meet the maintenance requirement but it also improves the straw consumption and also increases the milk yield by about 10 % and general health of the animal.

The UMMB is the *complete feed block technology* and has many *advantages*. The straw can be supplemented (fortified) with other nutrients lacking in straws by adding concentrate ingredients (maize grain) and other feed additives like urea, molasses, minerals, vitamins, enzymes. The green fodder like Lucerne leaf meal or subabul leaf meal can also be added to make it a complete feed. The milk production can be increased by incorporating the concentrate like maize grain, bypass proteins (formaldehyde treated cakes), and bypass fat.

The other advantages of this complete feed technology are that its ingredients are easily available, the block can be easily prepared, and has long life on storage at a dry place. The stake volume of raw straw can be reduced five times by densification. The *complete feed block technology* make easy to transport the UMMB to long distance from surplus area to the scarcity area and to supply the dry fodder based diet in droughts and floods hit areas due to high density of UMMB while it is cumbersome to transport the raw straw. The UMMB improves the gradual intake (about 700g daily) by the animals and finally it is cheaper compared to oil cakes.

However, some *precautions* should be taken like that the UMMB should be hard enough and meant to serve as licks by the animals so as the urea nitrogen is released slowly and frequently to avoid ammonia toxicity. The UMMB should not have more than 10 % moisture and should be stored in a dry place. These should be offered in dry manger to avoid wetting. The moist UMMB become soft to facilitate swallowing. The degree of hardness determines the voluntary intake of the block. The soft blocks are gobbled (eaten up hurriedly with noise) but the hard ones may not be eaten at all. The UMMB through gradual licking supplies more amounts of protein and essential nutrients, and hence improves the nutrient availability.

The production of complete feeds has increased during recent years from about 2160 thousand tones in 1990 to 3330 thousand tones in 2011. The macro and micro nutrients are also added in complete feed. This process avoids the wastage, increasing the palatability and DM consumption as well as digestibility. It is also labour saving process and keeps the surrounding clean. The compound feed is being manufactured by the Compound Livestock Feed Manufacturer's Association (CLFMA) and milk federation of various states for feeding to dairy animals and poultry.

### 25.8.5 Bypass Nutrient Technology

The supplementation of feeds with oil cakes in the ruminant diet enhances the animal productivity but the oil cakes are costly. Moreover, the oil cakes contain some protein which is degradable in the rumen and utilized by rumen microbes as a source of nitrogen for their protein synthesis and thereby reducing the total nitrogen available for absorption in GI tract of ruminants. Besides this, there is also rumen hydrolysis and biohydrogenation of dietary fat and thereby decrease the energy density of the diets. Therefore, there is a need to modify the feeds and feeding conditions and to manipulate the digestive tract, especially the rumen so as to increase the nutrient supply to ruminants (bovines) and efficient utilization by them with the objective to increase the productivity and to reduce the cost of production of milk.

**(1) Bypass Protein Technology:** The physical and chemical treatments of oil cakes can protect the proteins from rumen degradation by allowing these proteins to escape/bypass the rumen to avoid their degradation in the rumen and get them digested in the lower tract. The major part of feed protein is degraded in the rumen and this part is called rumen degradable protein (RDP) and the lesser part of dietary protein that escapes rumen degradation to enter the lower tract for absorption mostly as amino acids is called as undegraded dietary protein (UDP) or bypass protein. Substantial amount of RDP is utilized by the rumen microbes as the nitrogen source for their protein synthesis and the rest is absorbed as ammonia being converted in to urea in liver, some part of which is recycled back to rumen and rest is excreted out through urine. Thus, there is wasteful ammonia production in rumen and thereby a decrease in total availability of amino acids in the intestine of ruminants. It is thus required to decrease this wasteful ammonia production and to increase the level of UDP in the diet so as to increase the total availability of amino acids in the intestine of animal. The increase in UDP of diet increases the flow of amino acids at the abomasums of the ruminants. Such a mechanism of increase in the level of bypass protein in the diet is called as bypass protein technology. Now there are two options available either to feed the bypass protein feeds to ruminants or to evolve the method of protein protection against degradation (bypass protein).

*Sources of bypass protein:* Regarding the sources of bypass protein, the studies have indicated that mustard cake and GNC are highly degradable, while linseed cake, soybean meal, deoiled rice bran and subabul leaf meal have medium degradability. But the cotton seed cake, coconut cake, maize grain, maize gluten meal, fish meal are the good source of bypass protein which means that these feeds have low protein degradability. Therefore, the highly degradable feeds like mustard cake, GNC, rape seed cake, sunflower seed cake and even the soybean meal require protection from being attacked by rumen proteolytic enzymes so as their degradation in rumen is reduced and their utilization is improved by the ruminants. But the feeds with low protein degradability need no protection.

The tree leaves contain 50 to 70 % of total nitrogen as bypass protein, but about 16 to 53 % of total nitrogen is un-digestible (called as ADI- nitrogen) in whole of the gastrointestinal tract. The presence of tannin in the tree leaves is responsible to bind the proteins irreversibly and capable of corroding the epithelial lining of GI tract. Thus it requires the detoxification of tannin content which may be achieved by some chemical or biological or biotechnological approach.

*Methods of protein protection:* Among the methods used to reduce protein degradability of highly degradable cakes, the heat treatment and formaldehyde treatments are effective.

(*i*) The heat treatment method has its limitation of higher cost and hence not cost effective besides that it may either under-protect or over-protect the proteins if not done properly. Thus, heat treatment method requires proper and fine tuning by giving optimum heat treatment of CNC and soybean cake at $150^0$ C for 2 hrs. The heat treatment of soybean cake at lower temperature for short time ($125^0$ C for 30 mint.) was not found effective to protect the proteins of soybean feed.

(*ii*) Regarding formaldehyde treatment, the optimum level of 1.2 gm/100 gm CP of the cake was found effective to treat the mustard and GN cake. This level of treatment substantially reduces the protein degradability and the treatment is cost effective due to low cost of formalin. The treatment requires only the formalin which is cheap.

The *course of action* of formaldehyde treatment is that it arrests the growth of mycotoxin forming fungus and hence there is no further increase in alfatoxin level during storage. The feeding of formaldehyde treated mustard cake protects glucosinolates in rumen from its degradation to thiocynate which is more severe toxic and goitrogenic. The formaldehyde treated mustard cake increases the thyroxin level in goats which thereby may be attributed to reduce the amount of thiocynate produced in rumen. Therefore, it has twin advantage by making the cake safe to the animals and also producing the safe milk for human consumption by such animals fed the formaldehyde feeds.

There are two way *additional supplies of nutrients* to mammary gland through bypass protein feeding. The first is that feeding of bypass protein lowers the ammonia production in rumen which means low synthesis of urea in liver and hence the energy required for urea synthesis is saved.

Secondly, the feeding of bypass protein results in excess supply of amino acids which can partly be used for synthesis of milk proteins in mammary gland and partly used for synthesis of glucose in the liver through gluconeogenesis. This will result in more supply of glucose to mammary gland, resulting in more synthesis of lactose and thereby increase in milk production as the lactose regulates the osmotic pressure of milk.

The experiments conducted on formaldehyde treatment have indicated significant increase in growth rate of calves and kids, with increased feed efficiency (feed requited per kg gain) at low feed cost per kg gain, and consequently the calves attain early maturity causing the increase in reproductive life due to start of reproduction at earlier age. The bypass protein feeding also significantly increased the milk production up to 8-16 %. The increase in reproductive efficiency of males has also been observed in terms of sexual behavior like libido score and seminal quality attributes of semen volume, mass activity and sperm concentration. The bypass protein feeding was also effective to reduce the number of services required per conception in goats.

The formaldehyde treatment of cakes is also better from animal health point of view, particularly to reduce the toxicity (alfatoxin level, thiocynate) and to reduce the cellular degradation of organs.

**(2) Bypass Fat Technology:** The animals require sufficient energy for milk production and this energy requirement is generally not met by the feeds and fodders available in India particularly for high milk producing animals. The dietary fat provided to the ruminants gets hydrolysis and biohydrogenation in the rumen. The energy density of animal diet can be increased by using the bypass fat technology without any adverse effect on fibre fermentation in the rumen. The dietary fat can be protected from ruminal hydrolysis and biohydrogenation through this technology without affecting the growth of cellulolytic organisms and the fibre fermentation in the rumen. This bypass fat technology provides the fat a protection (against ruminal hydrolysis and biohydrogenation) in converting fatty acids into calcium salt of fatty acids (CSFA) by fusion method or by double decomposition process. The use of this technology has resulted in a substantial improvement in milk production. The bypass fat manufactured in India by some firms is available.

### 25.8.6 Genetic Manipulation of Rumen Microbes

The efforts are being made for genetic manipulation of rumen organisms to obtain the genetically modified organisms which can degrade the lingo-cellulosic feeds. Some success has been achieved to clone the genes in E.coli for endoglucanase, xylanase, beta glucosidase and amylase. This further needs the biotechnological efforts in this direction.

## 25.9. ICAR-NATIONAL INSTITUTE OF ANIMAL NUTRITION AND PHYSIOLOGY, BANGALORE, 1995.

This ICAR institute has its mandate of conducting studies on physiological and nutritional aspects to biophysical translation of nutrients for productive functions, to understand the mechanism at cellular and molecular level for providing solutions of the related problems.

The institute has developed some technologies like area specific mineral mixture for livestock; specific mineral mixture for small ruminants; protection of prebiotics from agro-industrial byproducts; test to identify sub-fertile semen; detoxification of castor, neem and karanj cakes; district and state level database available of animals and feed and feed resources; database on enteric methane emission from ruminants.

The institute is providing services to stakeholders for the proximate analysis of feed and fodders (moisture, C.P., ether extract, crude fibre, ash, NDF, ADF and lignin), estimation of macro-minerals Ca, P, magnesium), micro minerals, heavy metals (lead arsenic cadmium, murcury), estimation of amino acids, alfatoxins, steroid hormones (progesterone, estradiol, testosterone), peptide hormones (FSH, LH, prolactin), estimation of urea, nitrate iodine, fluorine), microbial tests for microbial and fungal count, probiotic organism count and anaerobic count.

❑❑❑

# 26 Sustainable Animal Management

The success of an enterprise depends on judicious decision taken in utilizing properly the various inputs to get the maximum output. This aspect of proper and judicious decision is called management. Here the enterprise is animal rearing which involves the integrated application of the principles of breeding, feeding, housing and heath of animals. These are thus the essential aspects of livestock management. Improvement in productivity needs a good livestock management. There is a need to modify livestock management system with passage of time and need of the day in view of the sustainability of livestock production so as there is no harm to the environment or depletion of natural resources and inputs keeping in mind the social aspects too.

## 26.1 COMPONENTS OF LIVESTOCK MANAGEMENT

The general principles of livestock management deal with the essential aspects of breeding, weeding, feeding and heeding.

The ***breeding*** involves the selection of male and female animals of superior genetic merit of desirable characters viz. production and reproduction characters which are under the control of heredity and environment. The heredity gives the base to the animal to develop a particular character and the environment provides the opportunity to express the character.

The ***weeding*** is the culling of the uneconomic animals by way of controlling the reproduction to reduce their multiplication and unnecessary strain on inputs too.

The ***feeding*** to animals should be proper, adequate and quality to provide better environment for expression of genotype properly. The feeding aspect of animal rearing profitably has become a growing concern in view of the decrease in feed and fodders availability for the animals due to land use for purposes other than producing fodder crops. This needs proper planning of feeding practices based on easily available, cheap and low quality ingredients of feed.

The ***heeding*** means good management and general supervision of different activities related to general care of animals of different age groups and categories (care of new born calves/kids/lambs, etc.), housing, hygiene and sanitation, and to keep animals healthy and care of sick animals for their recovery.

The essential aspects of animal breeding, nutrition, management and health care have been given in separate chapters. This chapter has been devoted mainly to the care of animals of specific categories and the housing management.

## 26.2 CARE AND MANAGEMENT OF ANIMALS OF DIFFERENT CATEGORIES

The new born becomes the future productive livestock. The performance of newborn after their turning into adult livestock in terms of not only of their production and reproduction but also their health and survivability depends on their better care and management, in addition to that given during their productive period. It is therefore most essential to know about the care and management of animals at different ages and performing different functions viz. growth, pregnancy, parturition, and production of different livestock products.

### 26.2.1 Care and management of new born

First of all the mouth, nostrils eyes and years of new born should be cleaned with a clean cloth for the any adherence of mucus membrane, though in dairy animal the mother itself licks her calf dry. The artificial breathing should be given by pressing and relaxing alternately the chest wall, in case the calf is not breathing. The birth weight of new born should be recorded. Use the disinfectant on the naval cord, tie it 2.5 cm away from its body, cut the same and apply antiseptic for 2-3 days. Give assistance to the calf if it does not stand within 30 minutes. It is then essential to feed colostrum within first hour of birth and continue feeding up to 4 days thrice daily. This is essential to develop resistance against calf's disease as the colostrum has antibodies (gamma globulins) which are absorbed as such into blood stream due to permeability of the intestinal mucous membrane is high for absorption of large globulin molecules. The mineral and vitamins are present in large quantities in the colostrums. Moreover, it is laxative and helps to clear the meconium. In case of non availability of colostrums from the dam, the colostrums from other should be given or a mixture of 2eggs in 30 ml castor oil should be given. The enema with 1 teaspoon sodabicarb in 1 lit of warm water should be given if the meconium is not passed out.

After the period of colostrums feeding the calves should be given milk @ 10 % of body weight. The calf should be on milk feeding up to 3 months of age and then small amount of dry meal (linseed meal, oat meal, crushed maize, bran) should be fed. Mineral mixture and common salt may be added. The weaned calves should be trained for drinking milk from the bucket. Healthy growing calves can be given calf starter of high quality protein and low fiber content. At the age of 15 – 20 days, the calf can be given green fodder.

Disbudding to remove horns should be done after about 10 days of birth by applying caustic potash on the horn bud after cutting the hairs and cleaning the area with antiseptic. The hot iron or electric dehorner may also be used on horn button. Identification number should be given to each calf either by ear tags or tattooing and later on the banding is done which could be with hot iron or cold branding (dipping branding iron in liquid nitrogen). The deworming, grooming and dipping to control external parasites should be done periodically.

The proper record of all information should be kept in permanent register.

### 26.2.2 Care and management of young stock (heifers)

The heifers should be kept in a separate pen after weaning till first service and be shifted in another pen after first service. The heifers should be weighed to check the growth. Proper feeding is must for proper growth and timely sexual maturity. Heat detection should be started depending on the age and body weight of heifers. The heifer mostly show standing heat when the heifer stand and allow other animals to mount on her.

### 26.2.3 Care and management of cows during pregnancy and parturition

The pregnant cows should be provided proper feeding so that it gives birth to healthy calf, to ensure better production and show post-partum heat at proper time of about 40 days. The pregnant cows should not allow to walk long distance or to run fast or to fight other animals or frightened.

The pregnant heifer / cow should be allowed extra ration (concentrate mixture) of about 0.5 to 1.0 kg after 6 months of pregnancy for proper development of foetus. The steaming ration of about 1.5 kg daily should be given during last trimester or at least 2 months before calving and it should be increased to about 4 kg one week earlier to expected calving.

The cow in advanced stage should be separated in calving pen before about a week earlier to expected calving. The diet given should be light and laxative with supply of fresh drinking water every time. The cow in lactation should be dried up after 300 days of lactation or depending on the pregnancy period. This is done to recoup and prepare the cow for next lactation.

### 26.2.4 Care and management at calving

Let the cow calve normally without any disturbance. Care should be taken that cow does not eat placenta which is normally expelled within 4 to 8 hours of calving. The assistance of a professional should be taken in case of difficult calving and retention of placenta. The cow should remain in calving up to 3 – 4 days of calving and then shifted to milking shed.

### 26.2.5 Care and management at milking

Care should be taken to milk the cow with proper method of milking (full hand milking), to obtaining full amount of milk in a shortest period without causing any injury to the teats and udder. The teats and udder should be cleaned before and after milking by a clean cloth after dipping in antiseptic solution. Incomplete milking may leads to early drying of the cow and chances of infection. Every care should be taken for clean and safe milk production. The milking time should be fixed. The milking should be completed within 5 to 7 minutes after letting down the milk which requires ½ to 1 minute.

## 26.3 ANIMAL HOUSING

The environmental factors influence the production and reproduction of farm animals to a great extent. The environmental factors besides the feeds and fodders, care and management, and health control also include the climatic conditions under which the animal lives. The climatic conditions are very important to influence the production and reproductions of animals. Undoubtedly, the indigenous livestock are well adapted to heat stress, climatic change, having disease resistance, and also have ability to thrive on poor feeing regimen but have low productivity. The adaptation to hot climate occurs by reducing heat production and metabolic rate and feed intake. Greater productivity requires higher metabolic rate which is in conflict with adaptability to heat stress. Therefore, the genetic architecture of indigenous breeds for adaptability to adverse climatic conditions together with limited feed and fodder availability may not favour higher productivity. It is therefore essential to control the environmental stress with great care of animals and providing housing for productivity improvement.

It is therefore essential to reduce the excessive heat and cold stress so as to minimize/eliminate the losses due to environmental stresses. Provision of shelter (house) to the animals modifies the environment by reducing the adverse effect of heat and cold stress so as the animals become comfortable. This also helps to maintain/sustain the production, reproduction, and health of animals.

The animal housing is the essential pre-requisite for the animals of intensive animal production system rather than a luxury to provide shelter for protection from heat and cold stress and to provide sustainable environment for sustainable production and reproduction. The animal housing also make easy the integration of different activities of feeding, watering, cleaning, milking, handling, and removal of manure.

### 26.3.1 Points to consider in Housing infrastructure

Good houses are required for reduction in incidence of diseases, for more production of quality milk and for improving the reproduction. Therefore, in constructing the animal house, the following points should be kept in mind:

(*i*) The climatic conditions (temperature, rainfall) and soil of the area should be kept in mind. According to the climatic conditions, there are 5 regions of the country and need different types of animal houses to be constructed.

(*ii*) There is proper feeding of animals with convenience and labour efficiency, clean milk production, comfort and health of animal, economy in construction, use and maintenance, with facilities for proper supervision and disposal of waste products.

(*iii*) The proper ventilation is a major factor in house construction mainly where more numbers of animals are kept indoors under one roof. There should be proper ventilation system in animal house. The house without proper ventilation invites stagnant air to become gradually warmer and humid with accumulation of gases like ammonia causing heat stagnation and poor appetite and adaptability to the changed weather. These conditions cause likelihood of respiratory diseases and poor productivity of animals.

(*iv*) The urine and dung from the animal shed should be drained out to a drain outside the building.

(*v*) There should be made arrangement for free availability of water.

(*vi*) The animal houses should be properly disinfected to make free the animal shed from microbes that may cause diseases to the animals.

### 26.3.2 Types of housing

There two types of housing systems viz. close housing with conventional barn system and loose housing where in the animals are kept loose for most of the time except for milking time and administration of medicines to sick animals.

**I. Conventional barn system:** This is also called as tie barn or stanchion barn is particularly used for housing the dairy animals. In this system the animals are tied at neck by stanchion, neck chain or neck rope. The animals are tied for feeding, milking and while taking rest. The animals are less exposed to outside weather in this system. This system with ventilation and air movement is better for temperate climate.

**II. Loose housing system:** This is used to keep the animals loose in open paddock with a provision of shaded area with a common feeding manger and watering trough. This shaded area is used by the animals to take rest and retire during hot and cold hours. The milking of animals is done either in separate milking byre or after tying individual animal on the feeding manger providing the concentrate before milking. The loose housing system is generally good for most of the part of the country. This system is advantageous and preferred for keeping more numbers of animals at one place for a number of reasons like low cost of construction, easy and economic common feeding and water arrangements for all animals, easy heat detection through teaser bull, freedom of movement of animals and hence more comfort, accommodation of more number of animals by easy expansion of the area in a short time at low cost.

### 26.3.3 Animal houses for different agro-climatic zones

The type of houses required by the animals depends on the climate of the area. The different climatic zones in India and the type of houses to be required in different zones are as under-

*Temperate Himalayan region* which comprises the mountainous area of J.K, H.P., Punjab, Assam, W.B., and Uttrakhand where there is heavy rainfall and snowfall and frost during winter, there should be close structure of the house with wooden structure and sloppy roof.

*North-Eastern region* comprising plains of Assam, W.B., Bihar, Orissa, Eastern U.P., with rainfall of about 125 cm and rice straw as the staple feed for animals. The loose housing with sloppy roof should be constructed.

*Dry Northern region* comprises the plains of western U.P., Punjab, Haryana, Delhi, part of Gujarat and M.P., and Rajasthan. There is hot and dry air in this

region during day hours with low humidity and rainfall below 75 cm. The loose housing is better option in this region. The shades in open area and a ceiling below the roof is better option.

*Coastal region* comprises the states of M.S., Karnataka, Kerala, T.N., A.P. and Orissa with heavy rainfall of more than 500 cm and humid area. The animal houses should be pucca with sloppy roof.

*Southern region* of M.S., Karnataka, Kerala, T.N., A. P. and M.P. where there is moderate rainfall with humid atmosphere. The animal houses should be loose with thatch roof supported by wooden pillars.

### 26.3.4 Space requirement under loose housing system

The domestication has resulted to confinement and has restricted the movement of animals in closed confinement. This has lead to the behavioural stress on the animal in terms of the limited space provided in the house for free movement of the animal. The animal in the close confinement requires the following spaces –

* Space occupied by the animal body,
* Space required for changing the posture by the animal,
* Space required for changing the position by the animal in its free movement,
* Walking and resting space,
* Feeding and watering space, and
* Residual space in the corner of the building inaccessible to the animal,

The floor space and feeding space requirements for different category of dairy animals are as under –

| Category of animal | Floor space/animal (sq.ft.) | | Manger length per animal (inches) |
|---|---|---|---|
| | Covered area | Open area | |
| Calves (< 2 mth) | 9 – 12 | 18 – 20 | 16 - 20 |
| Calves (> 2 mth) | 15 - 18 | 30 - 40 | 16 - 20 |
| Heifers | 18 - 20 | 50 -60 | 18 - 24 |
| Cows | 20 -30 | 80 -100 | 24 - 28 |
| Buffaloes | 25 – 35 | 80 -100 | 28 – 30 |
| Pregnant cows | 100 – 120 | 180 – 200 | 24 - 30 |
| Bulls | 120 – 140 | 200 – 250 | 24 – 30 |

### 26.3.5 Cleaning and Sanitation

The cleaning and sanitations of animal houses including the floor, walls, mangers, and the feeds, fodder, water, bedding, utensils, etc. help to keep the animal healthy. It is also equally most important animal attendant should also be healthy. All the organic matter which is suspected to harbor microorganisms should be removed. The clean, dry and disinfection of sheds and its surroundings can prevent the spread of infectious diseases spread from physical contact, contaminated feed and water, and air. Animal sheds should be cleaned by washing and after that 1 to 2 % solution of phenyl should be sprayed on the floor every day.

The milking must be done under sanitary conditions for clean and hygienic milk production. The udder of the animal before milking should be washed preferably with clean sanitizing solution ($KMnO_4$ or chlorehexidine) and dry with a clean cloth. Every care should be taken for clean milk production.

The disease animals should be separated from healthy animals. The sick animals should be given proper treatment. The animals after returning from a fair or show should be dipped in a solution of disinfectant. The stray dogs should not be allowed to enter the cattle shed.

Here it is important to know about disinfectants. The lime, phenol or carbolic acid, formalin, chlorine or hypochlorites, alkalies and chloro-compounds are environmental disinfectants while iodine, idophores, alcohol, potassium permanganate, hydrogen peroxide are skin disinfectants. Fresh lime is used to disinfect the floors, walls and water troughs. Chlorine is cheap, powerful and convenient to use but corrosive which can be minimized by using hypochlorite. The phenol or carbolic acid (5 %) solution can be used for metals. The formalin (5%) solution in water can be used as fumigation.

## 26.4 ADVERSE EFFECT OF HEAT STRESS AND ITS AMELIORATION

The heat stress in hot and hot humid climate severely affects the animals. It is thus very essential to know the effect of heat stress on animal functions viz. physiological, metabolic, production and reproduction. Increase in ambient temperature and air moisture (humidity) beyond their upper level of tolerance ($25^0$ C and 50 % RH) cause heat stress and hence discomfort to the animal which is measured by temperature humidity index (THI). The THI is an index of thermal comfort and correlated with animal productivity. The optimum comfortable THI for livestock as well for man is between 65 to 72, beyond above from 72 to 78 it gives mild stress, and severe stress when it THI goes above 80. The change in livestock productivity is associated with a change in THI.

### 26.4.1 Adverse effect of heat stress

The heat stress leads to a change in biological functions of reducing the feed intake and nutrient utilization, reducing metabolic rate, and disturbing the energy and mineral balance, enzymatic reactions, hormonal secretions (adrenal-pineal-hypothalamic interaction) and blood metabolites. These changes are associated with reduction in production and reproduction of animals.

The milk production of dairy animals during summer months is markedly reduced due to heat stress and lesser availability of green fodder. Likewise, excessive cold during winter also causes reduction in milk production.

The heat stress during summer season resulting in high body heat and high uterine temperature not only increases the incidence of silent heat in cattle and buffaloes but also results in increased embryonic mortality due to poor embryo implantation. Not only the females are affected by heat stress but the males too also produce semen of poor quality and fertility. The climatic changes affect the production of not only of the milk but also leads to low production of meat, work and wool. The impact of climate change to heat stress is more through feed and fodder availability.

The increase thermal stress affect adversely not only the production and reproduction but also affect the health of the animal in terms of temperature related illness, morbidity and death. The heat waves for longer impair the cell-mediated immunity in young ones, high producers and other vulnerable livestock affecting all the physiological, biochemical and behavior patterns.

### 26.4.2 Amelioration of heat stress

In view of low production and reproduction due to stressful conditions there is a need for improvement in livestock production system particularly by providing the protection to animals from heat stress.

*(i) Animal Houses:* Provision of animal house has been mentioned above. Thermal relief to lower down the ambient temperature can be given to the animal by keeping them under shade of a tree.

*(ii) Heat reducing cooling systems:* The heat stress can also be lowered inside animal house by using heat reducing systems like providing fans, fog misting and sprinkling the water on animal body, particularly the high producing animals during summer months. The effect of these cooling systems can be observed from wet bulb temperature. The sprinkling of water on buffaloes or their water wallowing decreases their body temperature and respiration rate. This practice may be repeated twice or more time depending on peak hours of heat stress.

*(iii) Change in animal diet:* The adverse effect of heat stress on reduction of voluntary feed intake may be reduced through feeding strategies by changing the composition of the diet.

*Grain feeding:* The first approach is to reduce dietary fibre intake so that rumen may function properly. The amount of grain feeding should be increased.

*Dietary fat:* During summer or during negative energy balance, feeding of dietary fat to provide extra energy is effective and recommended. This is because there is less heat increment in the rumen on feeding dietary fat and provides energy with no thermal effect. The Indian livestock due to lower metabolic rate and low feed intake with better digestibility produce lesser methane.

*Protein feeding:* The quantity and quality of protein in animal diet should be improved to reduce heat stress because the cattle consume less feed due to heat stress and hence have a negative nitrogen balance.

*Mineral feeding*: The dietary levels of sodium, potassium helps to maintain water balance, ion balance and acid-base status during heat stress and hence their dietary level should be increased because they compete with potassium for intestinal absorption. This can be done by supplementing with sodium bicarbonate and magnesium oxide in the diet. This also helps to maintain pH of the rumen. The blood acid-base homeostasis and rumen pH can be maintained by feeding ammonium chloride or ammonium sulfate along with sodium bicarbonate.

*Feed additives*: The heat stress can also be reduced by feeding the feed additives which generally increase the heat dissipation and thereby decrease the internal body temperature. Addition of fungal culture in the diet reduces the body temperature and respiration rate.

*Niacin (nicotinic acid)* Niacin feeding induces vasodilation and hence transforms the body heat to the periphery.

*Antioxidants feeding*: Feeding of vitamin A and E, selenium and zinc reduce the heat stress, prevent mastitis, optimize feed intake and remove the effect of heat stress on milk production.

(*iv*) *Breeding strategy:* The proper *breeding strategies* may also be exploited to reduce the negative impact of heat stress. This is because many of the local breeds of livestock are adapted to different agro-climatic conditions under harsh living conditions of adverse climate (high temperature), drought and feed scarcity. This is the result of domestication of several thousand years under the challenges of climate and diseases. There is substantial genetic variance in heat tolerance among animals. Therefore, the inclusion of characters associated with heat tolerance in selection programme may be undertaken.

❑❑❑

# 27

# Animal Health Care

The animals have been the richest source of food, cloths, draft power, medicines, great companions of man, and so on. Therefore, the survival of people is dependent on survival of animals. The indigenous livestock have better genetic superiority for adaptation to withstand the stress of thermally adverse environment, better disease resistance, and better ability to thrive on low feeding regime. These qualities results in better survivability of indigenous livestock but they are poor producers. The indigenous livestock, besides having poor productivity due to poor genetic potential and poor availability of nutrients, under Indian hot climate adapt themselves by reducing the heat production and metabolic rate and low feed intake which in turn further reduce the productivity. However, the animal should be healthy to exploit its full genetic potential under good nutrition and with proper management for production and survival for a longer period.

## 27.1 HEALTH

The term ***health*** does not mean the normal existence without an evidence of disease but it is an indication about the condition of physiological functioning of the body parts and its various systems. Thus, the body parts of normal healthy living being should have normal structure and optimum physiological functioning according to age, sex and utility.

***Disease*** is defined as a condition of the body as a whole or of some parts or organ (system) of the body in which its functions are disturbed or deranged. Predisposition to disease is generally thought to be a hereditary phenomenon. As far as the resistance and susceptibility to a disease are concerned, these are quantitative traits being influenced by both heredity and environment because the ultimate deciding factor for their expression is the presence of a particular amount of toxin or pathogen at the point of threshold.

When a living being shows a positive reaction to a pathogen attack and recovers, it means that it has succeeded in building up of sufficient resistance by its ability to form antibodies to overcome the effect of the invasion. The ability of the host to combat disease through the formation of antibodies has genetic implications otherwise without the ability to form effective antibodies, any form of vaccination or immunization of the host would be a futile procedure.

The ***susceptibility to a disease*** is inter-dependent on:

- The inherited resistance of the host to disease,
- The inherent variation of pathogens giving rise to strains different in virulence,
- The environment which may materially influence the interaction of the host and the pathogen.

The cause variable for the prevalence of a disease is the joint effect of polygenes and of the environment. The cause variable could be the quantity of a substance from biochemical pathway or the rate of some biochemical reaction or the rate with which the development process progress and it is determined by the biology of the individual in the prevalence of a disease as a result of the action of polygenes and the environment. It is known that the prevalence of a disease is the result of environmental stimuli. No living being always suffers with a disease but it suffers under certain environmental conditions, e.g. pneumonia mostly prevailed during colder months, heavy smoking leads to coughing, sugar intake in excess amount for longer time cause diabetes, etc.

There are evidences that life span is determined by heredity and there also exists genetic variability for resistance to diseases (Hutt, 195). Whenever possible the diseases of livestock should be prevented by use of prophylactic measures like immunization and sanitation, better management practices keeping the animals in hygienic conditions and removal of stressors. The term hygiene indicates the living conditions and their effect on animal health, adopting safety measures to prevent the occurrence of disease and to keep the animal healthy. The hygiene simply deals with the sanitary care of animals. It also includes to know the causes of all preventable diseases and to find out the ways and means of removing or making ineffective those causes so as to increase the efficiency of animals by providing more favourable environment, particularly with respect to the water, air and housing accommodations that are well ventilated, well drained for removal of excreta (dung and urine) and properly disinfected.

## 27.2 SIGNS OF GOOD AND ILL-HEALTH OF ANIMALS

It is important for livestock owner to distinguish between healthy and sick animal so as the proper care and treatment should be given to the sick animal.

### 27.2.1 Signs of health

The animals show some signs of good health and ill-health. The *signs of good health* of animals are as under –

* The animal is active and alert in normal condition,

* The healthy animal has normal sleep having easy, normal and free posture without nervousness,

* The healthy animal takes feed and water regularly (better appetite - consuming normal feed and water),

* There is frequent mastication in ruminants during rest period,

* Other common signs of health are glossy hair coat; wet, bright, alert and clear eyes; wet muzzles, ears alert and moving frequently; tail movement is also quite frequent,

* The faeces are semisolid and yellowish urine in colour,

* There is no abnormal discharge from natural openings;

* There is no drop in milk yield or egg production of healthy animal/bird,

* The healthy animal has normal group behavior,

* The normal physiological parameters of healthy animals of different species with low variability are as under -

| Animal Species | Body temperature ($^0$F) | Pulse rate / mt | Respiration rate/ mt |
|---|---|---|---|
| Cattle | 101 – 102 | 40 – 56 | 14 - 22 |
| Buffalo | 99 – 102 | 40 – 45 | 16 - 18 |
| Sheep | 101.5 – 103.5 | 67 – 82 | 14 - 22 |
| Goat | 101.5 – 104.0 | 67 – 82 | 14 - 22 |
| Horse | 99.4 – 101.2 | 33 – 41 | 8 - 13 |
| Camel | 97 – 101 | 32 – 50 | 5 - 12 |
| Pig | 101 - 103.5 | 60 – 80 (heart beat) | 8 - 18 |

### 27.2.2 Signs of Diseased Condition (Ill Health)

The animals in *ill-health condition* besides specific disease signs, show the following some common and apparent signs –

* In ruminants, the first sign of ill-health is the cessation of rumination,

* There is discharge from the eyes, nostrils, and mouth,

* Loss of appetite to various degrees,

* The animal becomes restless and changes the postures frequently,

* The eyes of the animal becomes dull and dry, and pasty deposition of canthus, drooping ears, dry muzzle, dry and outstretched nostrils, dry and rough hair coat, salivation,

* Sick animal becomes dull and keeps its head downward,

* Hard (labored) and rapid breathing

* The change in physical appearance of dung viz. hard in case of impaction and constipation, semisolid to watery in diarrhea,

* There is rise in temperature, pulse rate and respiration rate than mentioned above for healthy animal,

* Reduction in milk yield in dairy lactating animals.

* Disease specific signs/symptoms

## 27.3 BENEFITS OF CONTROL MEASURES

The prophylactic measures and control measures to treat the sick animals are beneficial in the following ways:

- To check the spread of animal parasites and diseases, and to reduce the incidence of transmission of dreadful diseases through parasites,
- To reduce side effect and complication of diseases,
- To reduce the cost of treating the animals,
- To make more chances of the recovery and also reduce the recovery time of sick animals,
- To save the lives of animals, and thereby increase in their life span,
- To increase the growth rate of growing stock, thereby attaining early puberty/maturity, reduces the cost of raising up to maturity, start early production, increase the life span, increase in reproduction efficiency and hence increase the lifetime production
- To increase the productivity and economy of production of the animal,
- To make more profitable the animal farming and industry.

## 27.4 DISEASE RESISTANCE

The animals have inherent resistance to specific disease causing pathogens and parasites as well as a curative power to recover from many diseases and surgical conditions with or without treated/medicated against them.

*Resistance Mechanism*: There is the effective inherent immune mechanism in healthy animal's body to provide protection from pathogenic organisms present in the environment. This mechanism of innate resistance is effective against most of the pathogens and hence non-specific in nature.

The natural *defense mechanism* to afford effective protection against pathogens involve: intact skin and mucous membranes of the body, acidic pH of the gastro-intestinal tract, lacrimation, ciliated mucous membranes of the nasal passage and the sphincter muscles of natural orifices. The foreign body particles like bacteria are engulfed and destroyed by an effective defense mechanism provided by the reticulo-endothelial system (RES) of the animal body. Moreover, the body temperature and habitat of animal and the absence of specific microbes binding receptors (fimbriae or piles, the appendage of adhesion in bacteria and the adhesion by many bacteria is essentially required for production of disease in animal) may be helpful to give the effective inherent defense mechanism.

## 27.5 SYMBIOSIS AMONG ORGANISMS

Two living organisms may have some type of living association called as symbiosis which is generally the mutualism, commensalism and parasitism.

(*i*) *Mutualism* is the symbiosis which is highly interdependent between two partners in a way that none can survive without another and hence both the partners are benefited like rumen protozoa and ruminants. The bacteria live in animal tissues and benefit the host.

(*ii*) *Commensalism* is the association in such a way that one partner is benefited but the other is neither benefited nor harmed like various amoebae in caecum and colon of ruminants wherein no metabolic dependence exists between the partners. Commensalism is the state of living the bacteria in or on the host without causing disease, but may become pathogenic under certain circumstances called as potential pathogens.

(*iii*) *Parasitismic state* is harm causing state of bacteria to the host and called oblique pathogens. It is another type of symbiosis wherein one organism is benefited but another is harmed. The benefited organism is called ***parasite*** while the organism which harbour (lodge) the parasite and being harmed is called ***host***. The parasite lives metabolically dependent on or within another species (called host) for whole or part of its life, without giving any compensation to the host for the advantage taken. The parasite has more reproductive rate than the host as a result of which the host may be killed. The parasites may be *obligatory* (completely or wholly dependent on host) or *facultative* or *optional* (parasite which may live independent to the host if necessary).

## 27.6 DISEASE CAUSING AGENTS

In spite of the natural defense mechanism, the animals are affected by various parasites and diseases. There may be external parasites (ticks, mites which produce sarcoptic mange or cattle scab, lice, mosquitoes, flies, and parasitic fungus producing ringworm which is contagious disease) or internal parasites (gastro-intestinal worms, lung worms, liver fluke, blood parasites). The diseases are caused by bacteria, virus, fungi and protozoa and these diseases may be contagious.

**(1) *Bacteria:*** The bacteria are prokaryotes microscopic organisms examined under oil immersion (100 x); capable of independent existence, growth, multiplication and to carry other vital activities; found in three forms (spherical – cocci, rod shape –bacillus, spiral – spirochaetes and comma – vibro shaped or filamentous); widely present in soil, fresh and saline water, air, vegetation, animals and human beings; and capable of causing diseases in animals, fishes, insects and human beings.

The bacteria on entering in to the host body act as antigens due to bacterial proteins as well as lipopolysaccharides and stimulate the production of specific antibodies adapted to reacting with the bacterium and modifying its activities. Each species of animals has many specific antigens and produces antibodies corresponding to each antigen. This antigen – antibody reaction is the basis of protection of animals and humans against specific diseases caused by bacteria by using the sera or vaccines prepared from bacteria.

***Types of Bacteria:*** All the bacteria are not pathogenic. The bacteria are grouped in two type based on their nutritional requirements (autotrophic and heterotrophic).

(*i*) *Autotrophic bacteria* do not cause the disease in animals or man and hence these are not pathogenic.

(*ii*) *Heterotrophic* are disease causing bacteria. These bacteria are further of two types, namely saprophytes and parasites.

(*a*) *Saprophytes:* These do not cause any disease normally but become opportunistic to cause infection when the host has low resistance particularly under stress condition of nutritional deficiency, or sudden change of weather, or prolonged antibiotic or steroid treatment.

(*b*) *Parasites:* These are second type of heterotropic bacteria and live in animals in three states of *Symbiosis viz. Mutualism, Commensalism and parasitismic.*

***Mode of action of bacteria:*** The bacteria which are in parasitismic state produce diseases due to their two properties of *invasiveness* and *toxigenicity*.

The *invasiveness* is attributed mainly due to the presence of capsule on the cell surface of the bacteria. The capsule is a covering outside the cell wall of bacteria and this capsule protect the bacteria from normal defense mechanism of the host animal. The disease causing capacity to produce diseases by some pathogens like *pneumococcus* depends on invasiveness. The septicaemia in cattle is produced by *pasteurella multocida* by virtue of invasiveness. Such bacteria with invesiveness multiply at the site of infection either in host tissues (intra-cellular) like *Brucella abortus* and *Mycobacterium bovis* or outside the tissue (extra-cellular) like *Pasteurella, Klebsiella* and spread throughout the body of the host animal.

The *toxigenecity* is due to the production of toxins by bacteria. Some pathogens produce extra-cellular toxins which cause damage to the tissues and produce disease like *Clostridium tetani* responsible to produce tetanus through production of toxin multiplied at local site in the host. The extra-cellular toxin is produced outside the body in the contaminated food by *Clostridium botulinum* in case of food poisoning and the disease is caused by consuming the contaminated food.

Several bacteria have both the attributes of invasiveness and toxigenesity to produce the diseases like *Streptococcus pyogenes, Staphylococcus aureus and Clostridium perfringens* produce the toxins to make them enable to breach the tissue barriers. Another category of toxins produced by many bacteria is the endotoxins that are not released by the cells and differ from extra-cellular toxins, and also cause the disease like *Pasteurella multocida.*

***Bacterial Disease:*** The various bacterial diseases are Anthrax (Mata), Brucellosis, Black Quarter, Calf diphtheria, calf scours, Foot rot, John's disease, leptospirosis, listeriosis, mastitis, haemorrhagic septicaemia, tetanus, tuberculosis (T.B.), vibriosis, botulism, etc.

**(2) *Virus:*** The virus is a particle consisting of a nucleic acid core and a protein coat known as capsid. The capsid encloses the nucleic acid. The electron microscope is used to study the structural chemistry of virus. The mosaic disease of tobacco plant gave rise to the discovery of virus in 1892 by Beijernick and Lvanovski and later on the discovery of FMD in cattle caused by virus in 1898 by Loefler and Frosch. Now many viruses from plants, animals and humans have been recognized and they are having their involvement as etiological agents of diseases.

The virus spread directly or indirectly through vectors. The most important sources of spread are the attendants, infected cloths, feed and water troughs, insects or droplet infection.

***Viral diseases:*** The common viral diseases are Rinderpest or Mata, Foot and Mouth Disease, Bovine viral diarrhea, ephemeral fever (TDS); Pox diseases (cow pox, sheep pox, goat pox, swine pox, horse pox; fowl pox); rabies; Hog cholera; mark's disease and Ranikhet disease in poultry.

**(3) Fungi**: The fungi are unicellular (yeast) as well as multicellular (molds) chemo-heterotrophic eukaryotes. The yeasts reproduce by budding. The fungi reproduce by means of specialized cells called spores. The fungal diseases are important. The ringworm caused by fungus is zoonotic in nature creating problem to the people handling animals.

***Fungal diseases:*** The fungi cause some diseases like aspergillosis, blastomycosis, candidiasis, coccidiodomycosis, dermatophytosis, etc.

***(4) Parasites:*** The parasite is an organism (plant or animal) which lives on or within another organism of another species to get advantage without giving compensation in exchange.

The parasites may be classified in different ways depending on their habitat, degree of parasitism, duration spent with the host, etc:

***Habitat:*** (*Eectoparasites* which are found on the external surface of the host like ticks and mites, lice, flies (horseflies, stable flies, common house flies, warble flies, blow flies) and mosquitoes which are arthropods and classed as external parasites; or *endoparasites* which live with in the body of host and called as internal parasites like helminthic parasites which are the nematodes (round worms, ascarids), cestodes (tape worms *viz. taenia solium*), trematodes (liver flukes).

[The internal parasites may either reside inside the cells/tissues of the host and called as intracellular parasite like *Babasia, Theileria* or may live outside the cell in intercellular spaces like apeworms, nematodes (round worms), trypanosomes, or may inhabit the lumen of the organ called as *Coelozoic parasite* like G.I. nematodes, Fasciola spp. (Trematodes - liverflukes) in bile duct or may inhabit the tissues of the host like some *Tripanosomes*)];

***Degree of parasitism*** (*facultative* parasites which can exist independently but may become parasite if given the opportunity like larvae some flies causing myiasis, soil amoebae; or *obligate* parasites which can not live independent life but spend some part of its life as parasite like *liverfluke, blood protozoa*);

***Duration spent on/in the host*** (temporary or intermittent parasites which remain with the host for some time like *mosquito*, or permanent parasites which remain permanently within or on the host like *lice*).

***Mode of Harmful effects of parasites to the host:*** The parasites cause harm to the host in different following ways:

- Competition of nutrients with host sharing ingested food (cestodes like tape worms, gastro-intestinal nematodes like roundworms, trematodes like liverflukes),

- Destruction and feeding on host tissue by bite (ticks, mites) and sucking blood (ticks, mosquitoes, hookworms), lymph (midges) and exudates (lungworms),
- Producing toxins to cause tissue lyses like haemolysins, histolysins, anticoagulants (leech, hookworms), tick toxins (ticks),
- Insiting inflammation and its sequelae, hypersensitivity and allergic reactions,
- Causing mechanical obstruction and perforation in body organs/tissues like intestine (ascarids, hookworms, tapeworms), bronchi (lungworms), bile ducts (live flukes),
- Carrying other pathogens and transmitting parasitic, viral and bacterial diseases like malaria (mosquitoes), trypanosomiasis (horseflies), rickettsioses (ticks), swine influenza (lungworms), histomoniasis (caecal worms), filariasis (ticks, mites, insects) and black disease (liverfluke),
- Causing reproductive failure like abortion (trichomoniasis, trypanosomiasis) or causing hormonal imbalance resulting in to reproductive failure through anoestrus and delay in puberty,
- Reducing the appetite, feed utilization, growth and production of host due to creating panic (hypoderma, tabanus), terror (warble flies) and annoyance (ticks, lice, dipteran flies).

However, the parasite – host relationship may be successful if the balance is maintained between them on the principle of "live and let live". However, the ill effect and manifestation of disease in the host by parasitic infection depends on many factors like age, nutrition level and health status of host; strain and numbers of parasites attacking the host, genetic make up and immune status of the host, environmental stress on the host, site of infection in the host viz. a parasite invading in the central nervous system and artery cause more harm and quickly.

The parasites not only harm to the host animal but they are also capable of infecting and producing diseases in humans. There are parasitic zoonoses which have been recognized and may prove potential threat to humans.

***Parasitic diseases:*** Both the external and internal parasites have harmful effects to the host animals the ways mentioned above and also produce the diseases.

The *external parasites* (arthropods like ticks, lice, mites, mosquitoes and flies which are horse flies, stable flies, common houses fly, buffalo fly) harbor on external body of the host animals causing direct injury to them as well as cause skin diseases. Moreover, they also transmit fearful/ dreadful diseases caused by *protozoan parasites* (mentioned below).

**(5)** ***Protozosan diseases:*** The protozoa cause a number of diseases. The different parasitic diseases are *anaplasmosis* caused by anaplasma transmitted by ticks, biting flies and mosquitoes; *babasiasis* which is a group of diseases and caused by numerous species of babesia transmitted by blood sucking ticks; *coccidiosis* caused by coccidian; *theileriasis* caused by theileria transmitted by tick bite; *trypanomiasis* *(surra)* caused by trypanosome; *trichomoniasis* caused by trichomonads transmitted by coitus, hence a veneral disease and cause abortion and infertility in females.

**(6) *Non-infectious diseases*:** Some diseases are *non-infectious.* These are

(*i*) *Metabolic diseases:* Some metabolic diseases also occur due to disturbances in physiological functions as a result of underfeeding, malnutrition and deficiency of some nutrient (rickets, anemia, milk fever etc). ***Milk fever*** is caused by rapid fall in blood calcium and magnesium level in order to meet the demand for milk production after calving, ***Ketosis*** is caused by shortage or faulty utilization of sugar in the body (disturbance in carbohydrate metabolism), ***Tympany*** *or bloat* occurs due to large amount of gas formation in rumen as a result of overfeeding, ***Goitre*** caused by iodine deficiency

(*ii*) *Poisonous plants:* Some plants like Jawar and Sudan grass produce *prussic acid* before rainy season which is poisonous to animals. These initiate and upset the normal function of cells causing their degeneration and death.

*(iii)* Mechanical causes: There may be mechanical injuries like cuts (bruises, wounds), sprain, abscess (boil), fracture, sore teats, yoke gall and sore neck, horn cancer, and swelling of hardware due to barbed wire, sharp metal pieces which may either cause injury or may be swallowed by the animal.

(*iv*) Thermal causes: The thermal cause like heat stroke, sun burns and cold effect (chill blains) to the body results in rapid degeneration death of cells.

(*v*) Hereditary causes: Some are the hereditary diseases due to lethal genes. These cause defects in cells of the body.

Losses due to parasitic infestation and other diseases in farm animals cause a great loss to the animal owner and animal industry at personal and national level. There are huge losses to the national resources due to animal diseases through low productivity, cost of treatment and death of animals. Not only these loses, but some of the infectious animal diseases (brucellosis, tuberculosis, rabies, bovine spongiform encephalopathy) are serious zoonoses concerning to human health.

Therefore, prevention and control of parasites and diseases through better care and management and treatment of sick animals become most essential and important aspect of livestock farming. However, it is said that prevention is better than cure. The prevention from the attack of parasites and diseases is based on better diet, hygienic conditions and taking every care to keep the animals healthy and productive like protecting them from adverse climatic conditions, etc. It is therefore, advisable to take every care to keep the animals healthy to make the animal farming a viable, sustainable and economic enterprise.

## 27.7 CONTROL OF EXTERNAL PARASITES

The parasites are those organisms which live either on or within the body of another organism of other species for some part or whole of their life, and respectively called as external and internal parasites depending on whether the parasite survives on or inside the body of the host. Hardly any farm animal remains free from parasites during its life and these parasites cause huge losses of animals. It is therefore essential to control the external parasites by the preventive measures so as to avoid the damage and losses.

***Control measures of external parasites:*** The farm animals may be safeguarded from external parasites in some ways –

- ***Good management:*** Loose housing system and changing the grazing area in rotation, use of proper drainage system, ploughing the soil and lime application, burning the grasses in affected areas to destroy the parasites on the grasses
- Application of insecticides by spraying, dusting, dipping and logging on breeding places of parasites are some of the measures to be adopted, particularly in case of protozoan infections like *Trypnosomiasis, theileriosis, babasiosis*.

Some of the insecticides used as spray and dusting (mostly fortnightly) in animal shed as well as on animals are *malathion* or *texaphen* (0.5 to 1.0 % solution spray, and 4-5 % as dusting powder on animal body and animal shed), *DDT* (0.5 % solution spray, and 10.0 % as dusting powder on animal body and animal shed), *lindane* (0.25 % solution spray and 5.0 % as dusting powder).

***Precautions in control measures:*** The insecticides used in control measures are so toxic in nature that personal care of the treated animals is required to avoid the risk through licking of insecticide by the animal and the dusting powder should be used after adding in cow dung ash by rubbing the animal body (affected parts). Further care should be taken that the feeds/fodder and drinking water is not contaminated with insecticides, to keep proper concentration of the solution, to protect the face of the animal including mouth, nostrils, eyes, ears from spray and dusting. If the animal gets poisoning, warm and soapy water bath should be given to the animal to remove quickly the insecticide.

## 27.8 CONTROL OF INTERNAL PARASITES

The internal parasites harbor inside any part of the body (stomach, intestine, lungs, liver, and blood) of the host animal. Therefore, the internal parasites may be gastro-intestinal worms, lung worms, liver fluke, and blood parasites.

These parasites cause great loss by absorbing food material from the digestive tract of the host, sucking blood and tissue fluids, and producing poisonous materials in the animal system. As a result the production performance of the host animals goes down, also result in weakness and finally the host animal may die.

The internal parasites are transmitted from one animal to the other through feeds and water, during grazing in pasture or through some intermediate host as snails.

***Diagnosis of infestation:*** The host animal for harbouring the internal parasites may be identified from poor growth at young age, lose in body weight becoming pale and anemic, and passing of greenish black liquid dung with offensive odour. The affected animal lick objects of no interest and there is decrease in milk yield.

Besides these, there are some parasitic specific symptoms like exhibiting gripping pain and constipation in round worm infection in calves; rise in temperature and dysentery in coccidial infection; cough, pneumonia with thick mucous discharge in lung worm infection, high rise of temperature in blood parasite infection; passing of red or black urine in blood protozoan infection (babesiasis).

***Control measures of internal parasites:*** The farm animals can be safeguard against internal parasitic infections by the followings:

- ***Good management*** by providing healthy environment in keeping animal shed clean and dry, removal of dung properly and timely, avoiding the contamination of feed and water from dung, providing mineral mixture to increase resistance against worm infection
- ***Strategic deworming*** of farm animals at regular interval with anthelmintic parasitic specific medicines. The following medicines for deworming have been commended:

***Round worm and lung worms*** – *Nilverm* (50 ml/100kg body wt), *Thiabendole* (3.5 g/45 kg body wt), *Thiophanate* (25 mlof 14 % solution/50 kg body wt), *Panacur* (5ml of 6% solution/15 kg body wt);

***Round worms*** - *Curaminth* (15 mg/ kg body wt), *Vermex* (15-30 ml/30 kg body wt), *Piperazine adipate* (10 g/45 kg body wt), *Helatac* (0.75 g/kg body wt);

***Liver flukes*** – *Zanil* (30 ml/100kg body wt), *Distodin* (10 mg/40 kg body wt); *Avlothane* (5 g/45 kg body wt);

***Abomasum worms*** – *Phenovis* (10g/45 kg body wt).

- ***Integrated pest management (IPM)*** principle is used for better control of parasitic diseases. The young calves are treated with anthelmintics during epidemics and the cows are treated during terminal stages of their pregnancy. The incorporation of certain anthelmintics in the cattle feed pellets has been developed by Animal Biotechnology Lab of NDDB. The technique of deworming agent against roundworm infestation in ruminants has been patented and commercialized by Indian Immunological Ltd, Hyderabad, under the trade name "VETFEN -600" which has become very popular by using large numbers of farmers all over the country. Likewise, tissue culture vaccines have been developed against tropical theileriosis for its effective control. The vaccines developed to control these parasitic diseases are not so effective due to the complex life cycle of the causing parasites. Therefore, control of parasitic diseases is done mainly by the control of vectors (ticks, flies and mosquitoes) by spraying of insecticides or to have insect proof houses.

Some precautions are required in deworming the farm animals. Two treatments during June and August to prevent flukes in Northern India, deworming at 6-8 weeks interval in warm humid weather, deworming at 30-45 days in marshy areas, monthly deworming of calves till 6 months of age, snail control treatment in addition to deworming against liver flukes.

- ***Vaccination*** against lung worms to immunize the calves is also recommended. About 1000-2000 irradiated larvae of lung worms in 2 doses at monthly interval should be given orally.
- ***Biological control*** of helminthes can be done which focused on the free-living stages of the parasites on the ground and pasture. A number of fungal species and strains have been found effective to have antinematodal properties against larvae and eggs of Haemonchus species and other roundworms by the Biotechnological Lab of NDDB, Anand. The fungi have been to reduce the parasitic load of the treated pasture compared to control.

## 27.9 CONTROL OF CONTAGIOUS DISEASES

The contagious diseases are those which spread through contact. The contagious diseases are all infectious but all are not contagious. These contagious diseases are caused by bacteria and virus, and affect the health and productivity of the animals causing a great loss to animal owners and animal industry.

***Sources of infection:*** The farm animals get infection in various ways like contact with affected animals and their discharge, contact with animal died with disease, and contact with affected materials (utensils and equipment used for affected animals), and taking the contaminated feeds and water, dirty and narrow shed and overcrowding, dirty and unhealthy milkers.

Mode of infection (Dessemination): The infection enters in the animal body through digestive and respiratory systems, through reproductive and urinary systems, through udder and teats, through eyes, skin and dirty air, and through flies and mosquitoes. The poor management practices also invite infections and hence diseases.

**Symptoms of diseases:** The symptoms of a disease are disease specific. However, the followings are some of the common symptoms of disease -

- Rise in temperature
- Off feed and water by the animal
- Dullness and isolation from other animals
- Excessive salivation and shooting diarrhea.
- Disease specific symptoms viz. excessive salivation and appearance of ulcers or blisters in FMD, shooting diarrhea is rinderpest, swelling on neck and difficult breathing in H.S., swelling of the upper part of one of the limbs in black quarter.

***Prophylactic Measures of contagious diseases:*** It is desirable and economical to implement animal disease prevention and control programmes instead of treating the sick animals which is not only expensive but the sick animals and their products are risky to human beings. Therefore, the following two important steps should be undertaken to prevent and control of animal diseases:

(*i*) Disease diagnosis and generation of epidermiological data

(*ii*) Disease control programme formulation and implementation.

**I. *Disease Diagnosis***: The study of natural history of the diseases in a given population including nomenclature of the disease, determination of frequency, pattern and development of programme for prevention and control of disease is called the *Epidermiology*. The generation of epidermiological data regarding disease incidence, economic importance, prevalence and its zoonotic significance is necessary in prioritizing the disease to be prevented and controlled in view of the financial resources and infrastructure available.

The data regarding the disease can be collected from government epidermiological units, veterinary hospitals, abattoirs, and farm records, etc.

(*i*) The traditional methods of diagnosis of disease are based on clinical symptoms, morbidity and mortality rates, gross and histopathology, serological tests and isolation of etiological agents.

(*ii*) Now, there has been development of new techniques of tissue culture, monoclonal antibodies and molecular approach (recombinant DNA technology and DNA probes, and polymerase chain reaction) for quick and disease specific diagnosis. The PCR followed by RFLP is used in the diagnosis of genetic disorders caused by change in single base mutation generating the carriers of diseases and capable to pass on in the population.

(*iii*) It has become possible to detect and demonstrate the presence of toxins, toxic metals, pesticides in animals and animal products by gas-chromatography, liquid chromatography, and atomic absorption spectrometry.

**II. *Disease prevention and control measures (contagious diseases)***: The following preventive and control measures are useful in prevention of animals from various diseases:

(*i*) *Care and management of sick animals* should be done properly as outlined here. It is essential that proper care and hygienic management, proper feeding, exercise and treatment of sick animals should be provided to keep the farm animal healthy, disease free and productive.

*(a) Cleanliness:* The animals are infected from contaminated feed and water, utensils and equipments used for affected animals, contact with infected materials and areas, and contact with the affected animals. Healthy environment is essentially required to keep animals healthy and disease free. Therefore, it is essential to keep free from contamination the feeds and water, utensils and equipments used in animal feeding/milking, animal body and animal attendants. There should be proper disposal of faeces and dung, discharges from affected animals and wastes to avoid contamination.

(*b*) *Housing of animals: There should be* well ventilated and airy sheds with protection from adverse climatic conditions of excessive cold and heat. The floor should be well designed and constructed with cement-concrete for proper cleaning and disinfection. There should be provided the bedding with good material like paddy straw according to the season. The sick animals should be segregated and housed in different shed according to their age and the type of disease.

*(c) Feeding*: The diet should be of well appetite. The sick animals should be provided palatable easily digested, nutritious and laxative feeds. Feeding of high proteinous/nitrogenous diet should not be given. It is better not to provide any feed instead of giving unpalatable food (left over feed, etc) and the feed provided should be fresh, clean without any contamination. It is better practice to provide light and laxative feeds like wheat bran, starch gel, barley water, oat meal and green grasses. Provision of fresh water is must to sick animals with a provision of separate water trough.

*(d) Exercise:* Care should be taken for proper movement of joints by slight exercise of gentle walking and light grooming so as to eliminate the waste products of metabolism through skin.

*(ii) Screening of carrier animals* (like cittrulinemia and Bovine Leucocyte Adhesion Deficiency syndrome which is a serious genetic disorder of European cattle) before importation from other countries,

*(iii) Quarantining* the imported animals at the stations established at the port of the entry. Smilarly, any animal should not be added in the herd without quarantining at least for a fortnight.

*(vi) Segregation (isolation)* of affected animals with any disease and avoid the contact of healthy animals with diseased animals.

*(v) Separate bull station for prevention of sexually transmitted diseases:* There are some sexually transmitted diseases viz. Brucellosis, Vibriosis, Trichomoniasis, Infectious Bovine Rhinotrachetis, Blue tounge. The breeding bulls must be kept separately than female herd at A.I. station constructed away from cow sheds and bulls should be regularly screened for potential sexually transmitted diseases and remove those found positive. The egg yolk has Ig antibodies in higher concentration and hence the egg yolk from immunized hens against some of the sexually transmitted diseases of animals is useful semen diluents for inactivating the infective agents in semen.

*(vi) Disposal of urine, dung and discharge* of the affected animals properly to avoid the further spread of disease,

*(vii) Separate feeding and water trough* and bedding materials should be used for affected animals,

*(viii) Disinfection of sheds, housing area and equipments* should be done properly and timely,

*(ix) Disposal of died animals* with any contagious disease should be done by burying deep and spraying with lime without pealing-off the skin and in no case died animal should be thrown out at any place,

*(x)* ***Vaccinations-*** The effective vaccines are available for the control and prevention of many bacterial and viral diseases -

*(a)* The *eradication of rinderpest* from India by vaccination is an example to eradicate a disease.

(*b*) The *control of FMD* is difficult because of the highly contagious nature of causative virus of several types having the ability to mutate and present in almost all the secretions of the animals including milk and semen, and the FMD vaccines are expensive and require cold chain to remain effective. The recombinant vaccines developed recently may be helpful in making the control programme effective and economic by eliminating the need of cold chain for vaccine storage.

(*c*) *Neonatal calf diarrhea* caused by various strains of rotavirus and pathogenic E.coli in young calves causes heavy losses in cattle and buffaloes. The vaccination with traditionally available vaccines to control the diarrhea in neonatal calves is not effective because it occurs in neonatal calves, through passive immunization of calves by feeding colostrums from cows immunized with inactivated-adjuvanted rotavirus and E. coli has been found effective. The control and prevention of this ailment requires hygienic practices, and feeding of colostrums to weaned calves.

(*d*) *Rabies* is a fatal disease of warm blooded animals. The prevention and control of this disease can be attempted by regular vaccination of canine pets and post bite vaccinations with tissue culture vaccines, and destruction of stray dogs and wild canines.

(*e*) *Eradication of endemic bacterial diseases* of dairy animals like H.S., B.Q., anthrax, etc., is difficult due to their endemic nature and soil borne nature of the infectious agents. However, the mass vaccination of animals in the affected areas to control these diseases should be attempted prior to rainy season.

***Treatment of diseased animals:*** Veterinarian of the area should be called for giving disease specific treatment to the affected animals.

## 27.10 MASTITIS IN DAIRY ANIMALS

The mastitis is the most important disease affecting adversely the economy of the dairy farming and causing national loss of crores of rupees every year. It is caused by bacteria, virus and fungus, and more frequent in high milk producing animals. The infection generally enters through the teat canal as well through injury to the teats/udder.

The incidence of mastitis not only reduces the milk production but also involved huge lose through treatment. The mastitic animals also spread the disease to other animals. Sometimes, there is permanent loss of teat (teat block) and make the animal less productive/unproductive.

***Symptoms of mastitis:*** The symptoms of mastitis are the swelling of one or more quarters, presence of flakes or clots in the first stripping of milk in initial stages, and turbid thin milk and pus in advanced stage. The strip cup test is important to diagnose mastitis. The test is performed with strip cup which contain a flat black plate of metal/plastic on which the milk stream is milked to form a pool and careful examination is made for the presence of discolouration or clot/flake. Some chemical tests are also used to diagnose the mastitis viz. alkaline test, California mastitis test, Bromo-thymol Blue test, SLST test (Sodium Lawryl-sulfate and teepol), etc.

***Prevention and Control of mastitis:*** The mastitis can be prevented and controlled by the followings -

(*i*) The animal, milker and shed should be hygienic.

(*ii*) Regular and correct method of milking (full hand milking) is safe to avoid mastitis. Avoid slow or over milking and knuckle method of milking,

(*iii*) All precautions should be taken for clean milk production viz. clean and healthy animal, utensils, and milker; use of clean water for washing the animals and utensils (with preferable use of chlorine); proper, clean, dry and hygienic housing by proper and timely removal of dung, draining the washings to the gutter, white washing and disinfection of shed by some disinfectant liquid like phenyl (2 %0) throwing on the floor,

(*iv*) The teats and udder should be protected from injury by taking care that there is no pointing object in the milking shed, avoid knuckle milking, avoid uneven floor of shed and the nails of the milker should not be pointed,

(*v*) Segregation of animals suffering already from mastitis and take assistance of a veterinarian in case of injury, wounds on the teats and udder.

(*vi*) Treatment of the animals suffering with mastitis, preferably with the help of veterinarian.

## 27.11 ICAR RESEARCH INSTITUTES ON ANIMAL HEALTH

The ICAR has its following institutes for conducting research on various aspects of animal health -

**(1) Indian Veterinary Research Institute**, Izatnagar (Bareilly), U.P. 1889

Regional stations: Mukteswar (1893); Palampur (1959); Calcutta (1970);

Bangalore (1971); Srinagar (1973 till1992); High Security Animal

Disease Laboratory (HSADL), Bhopal (1998).

*Deemed University (1983)*

*Historical Background and Growth:* The IVRI was initially started with its name as Imperial Bacteriological Laboratory on December 9, 1889 at Pune. It was shifted in 1893 to Mukteswar in Kumaon hills, U.P. which is now in Uttrakhand with the objective of segregation and quarantine of highly contagious organisms. In 1913, the Izatnagar campus was started and the former IBL was re-designated as Imperial Veterinary Institute. The Biological Products Section was first started followed by other sections. The Animal Nutrition was established in 1936 followed by Poultry Research Division in 1938 while the Animal Genetics Section started functioning in 1944 for animal production, breeding and management. This institute was renamed as Indian Veterinary Research Institute in 1947 and conferred the status of the Deemed University in 1983 by the UGC, GOI. The M.V.Sc. degrees are awarded in 22 disciplines and Ph.D. degree in 20 disciplines.

The institute has a Centre for wildlife conservation Management and Disease Surveillance since 1984, K.V.K., and Communication and Computer Centre since

1986. In the same year of 1986 were established National Biotechnology Centre (Now Division of Veterinary Biotechnology and Immunology), Agricultural Research Information System (ARIS) cell and the Centre for Animal Disease Research and Diagnosis (CADRAD).

The institute has made significant contribution to develop technologies for *some* ***feed technologies*** (Urea molasses liquid feed, urea molasses impregnated straw, mineral mixture, UMMB (pashu chocolate), Compressed complete feed block); ***herbal preparations*** (post milking teat dip to prevent bovine sub-clinical mastitis), skin ointment, Herbo-mineral acaricide against tick infestation, treatment of mange, treatment of diarrhea, haemonchosis in sheep. The institute has developed a number of V***accines*** (Anti-rinder pest, rinderpest vaccine, African horse sickness vaccine, canine distemper vaccine, fowlpox vaccine, goat plague vaccine, FMD vaccine, goat pox vaccine, sheep pox vaccine, swine fever vaccine, rabbies vaccine, Ranikhet disease vaccine, BQ vaccine, Brucella strain 19 vaccine, HS vaccine, Anthrax vaccine, enteroxaemia vaccine, contagious caprine pleuropneumonia vaccine); some ***surgical techniques/ devices;*** and the ***services*** are provided to different stakeholders.

**(2) Project Directorate on Animal Disease Monitoring and Surveillance,** Bangalore. 2000.

The AICRP on Animal Disease Monitoring and Surveillance was started in 1987 with the establishment of 4 regional research units at Hyderabad, Bangalore, Pune and Ludhiana with central coordinating unit at Bangalore which was upgraded to an institute status in 2000 with 10 collaborating units, further addition of 5 more collaborating units in 10 $^{th}$ plan, and one more centre at Guwahati was included.

**(3) Project Directorate on FMD, Mukteswar,** Nainital (U.K.). 2000

The AICRP for FMD virus typing was started in 1968 at Mukteswar and 3 centres at Hissar, Hyderabad and Calcutta. The project was upgraded to a Project Directorate in July 2000 with its HQ at Mukteswar, 8 regional centres and 15 Network Units through the country. The mission of this directorate is active epidemiological surveillance through regularly monitoring antigenicity and genomic make up of FMD virus strains responsible for outbreaks of this disease and to eradicate the disease. The kits for FMD diagnosis and its surveillance/ serotyping have been developed among a number of other achievements.

**(4) Veterinary Type Culture Collection,** NRCE Campus, Hissar (Haryana) 2004

This was established at NRC on Camel in 2004 with the aim to act as a national repository of animal microbes including dairy and rumen microbes. The mandate with main activity included isolation, identification, characterization, accessioning and reposition, and utilization of microbes of animal origin. A total of 1352 accessioned microbes of animal origin belong to various species of animals (cattle, buffalo, sheep, goat, camel (single and double hump), equine (horse, donkey), pig, rabbit, deer, leopard, monkey, human, tick, fish, poultry and emu etc. These include rumen microbes (fungi and bacteria) and dairy microbes.

**(5) National Research centre on Meat,** Hyderabad (A.P.) 1999

The NRC on meat was started at Hyderabad in 1999 first in the campus of P.D. on Poultry and subsequently at Central Research Institute for Dryland Agriculture. It was finally shifted to its own campus in 2007 at Chengicherla (Hyderabad). The centre has developed technology for sex determination, for incorporating different natural ingredients in the value added meat products, and packing technology for improving the shelf-life of meat.

***(6) CCS Institute of Veterinary Health:*** This institute has been established at Bagpat (U.P.), under the administrative control of GOI.

**(7). *Projects on animal health*:** To reduce morbidity and mortality, a network of 26717 polyclinics/hospitals/dispensaries and 28195 ***Veterinary Aid Centres*** supported by 250 disease diagnostic labs are functioning in the states. There are 26 Veterinary Vaccine production units (19 are in public sector and 7 are in private sector).

The IVRI is assisting in monitoring the ***quality of vaccine and biological product***. One central and 4 regional ***disease diagnostic labs*** have been set up. The centre for animal disease research and diagnosis (CADRAD) of IVRI is functioning as central lab. The Disease Investigation Lab., Pune; Institue of Animal Health and Veterinary Biologicals, Calcutta, Bangalore and Animal Health Institute, Jallandhar are functioning as referral labs for western, north-east and eastern, southern and northern region, respectively.

The GOI has sponsored a scheme "***Livestock Health and Disease Control***" to implement disease control activities with the following objectives-

- Assistance to States for Control of Animal Diseases (ASCAD),
- National project on Rinderpest eradication (NPRE), since May, 1992
- Professional Efficiency Development (PED)
- FMD Control Programme (FMD-CP)

The country has been declared free from Rinderpest disease. The FMD-CP is being implemented in 54 specified districts in the country to control FMD.

❑❑❑

www.ingramcontent.com/pod-product-compliance
Ingram Content Group UK Ltd.
Pitfield, Milton Keynes, MK11 3LW, UK
UKHW020226290926
15470UKWH00001B/398

9 789351 309253